AF566957

Advances in Organic Chemistry

Methods and Results

VOLUME 8

Advances in Organic Chemistry

Methods and Results

VOLUME 8

Edited by

EDWARD C. TAYLOR

Princeton University, Princeton, New Jersey

1972

Wiley-Interscience Publishers

a division of John Wiley & Sons, Inc. New York • London • Sydney • Toronto

Library of Congress Catalog Card Number: 59–13036

ISBN 0–471–84644–9

Printed in the United States of America.

10 9 8 7 6 5 4 3 2 1

Preface

When this series was inaugurated eleven years ago, we felt it necessary to offer some special justification for the addition of still another treatise to the popular sphere of organic chemistry, which exercises such a perennial fascination for the scientific publisher. The stated goal was to provide a critical appraisal and evaluation of new discoveries in organic chemistry ripe for further development and of novel extensions to well-established methods. We aimed at bridging the time gap that inevitably intervenes between the evolution of new ideas, techniques, and methodologies and their appreciation and general use by organic chemists. It was recognized that, ideally, the authors of such articles should be either the originators of the methods described or practitioners expert in their use.

I hope that this volume, the eighth in this series, successfully fulfills these exacting objectives. The Birch reduction is finally reviewed, in collaboration with Subba Rao, by the master himself; this is the definitive treatment of this now classical synthetic method. The chapter by Story and Busch on new synthetic routes to macrocyclic compounds presents, for the first time, full details on their versatile ketone peroxide fragmentation procedure, which promises to be a very significant contribution (both scientifically and economically) to this field. Synthetic organic chemists should welcome the definitive, exhaustive, and critical discussion of silylation by Klebe, who has personally contributed so much to the development of this important technique. Comprehensive reviews by McOmie and by Boissonnas on the design and use of protecting groups in organic synthesis were published in Volume 3 of the series. In this volume, a complementary exposition of the latent functionality concept of protecting groups is given by Lednicer; not only has there been no previous review of this intriguing and widely applicable concept, but also the concept itself is formalized for the first time. Wiesner gives us a masterful presentation of a great achievement in the art of structure elucidation, the unraveling of the molecular intricacies of the alkaloid ryanodine. The volume concludes with the chapter by Jones and Bartle on the use of N.M.R. in structure identification of polycyclic hydrocarbons; as a comprehensive review of applications of a now broadly used physical technique to this specific field, it represents a unique and useful contribution.

I hope that these essays on the development and applications of synthetic and physical methodologies, complete with experimental details, will be of broad interest and utility to organic chemists. Opinions, criticisms, and suggestions for incorporation into future volumes are welcomed.

Dr. Johann F. Klebe died suddenly on September 22, 1971. His tragic death represents a profound loss to his family and friends, to General Electric, and to science. His chapter on silylation is his final contribution to organic chemistry.

EDWARD C. TAYLOR

Princeton, New Jersey
December 1971

Publisher's Note. The composition used for this book is a departure from that in the earlier volumes in the series. The change was made in the interest of continuing the series in the face of steadily rising costs. We hope the reader will accept this new format. There will be no change in the high quality of the articles already established for the series.

Contents

Advances in Organic Chemistry

Methods and Results

VOLUME 8

REDUCTIONS BY METAL-AMMONIA SOLUTIONS AND RELATED REAGENTS

By A. J. BIRCH and G. SUBBA RAO, *Research School of Chemistry, Australian National University, Canberra, Australia*

CONTENTS

I. Introduction

The aim of this chapter is to summarize those aspects of the organic reactions of solutions of metals in liquid ammonia, amines, or appropriate dipolar solvents that render them uniquely or outstandingly useful from a synthetic point of view. Reactions, where they are not greatly or usefully different from those of other dissolving metal reagents, are mentioned only for the understanding of theoretical points. Understanding of theory in this field is more important than in most, even from the angle of relatively routine experimental employment; experimental conditions frequently have a drastic effect on the reaction products, and a close control over products requires a good understanding of the reactions involved. It is now becoming increasingly possible to predict appropriate experimental conditions for achieving a given objective, to control degrees of reduction, and to isolate intermediates. It is not unusual to find reported in the literature the failure of a reaction, when this was carried out under obviously inappropriate conditions for the desired objective.

This chapter cannot cover all of the papers published in this field; we believe, however, that it covers the majority of those which contribute to a fundamental understanding of the reactions and their practical utility, and provides essential points of entry to the general literature. Organometallic compounds are not covered.

II. Historical

The alkali and alkaline-earth metals dissolve in liquid ammonia and in hexamethylphosphoramide to give solutions which behave as if they contained metal cations and "solvated electrons." Discussions as to the type of interaction implied by the last term are still inconclusive (1), but the solutions, particularly in ammonia, provide the organic chemist with a source of highly reactive electrons which can be used to carry out a number of difficult reduction reactions.

Reactions of the solutions with aromatic substances first drew attention to their unique character: in many cases saltlike substances were formed containing mesomeric carbanions; see, for example, Ref. 2. Several notable events in the development of their use were the demonstration of the reduction of naphthalene to 1,4-dihydronaphthalene (2), the reduction with calcium hexammoniate of benzene to 1,4-dihydrobenzene, and under different conditions, of benzene or substituted benzenes to cyclohexenes (3), of the production of 1,4-dihydrobenzene from benzene by the joint action of sodium and ethanol in liquid ammonia (4), and of 2,5-dihydroanisole from anisole (5). Synthetic use of these findings led to one of the two present major fields of utility: the partial reduction of aromatic systems. The other major utility lies in the frequent stereospecificity of the reductions first observed in the reduction of acetylenes (6), noted as a possibility in other cases (7), and generalized (8) with modifications (9,10). Theoretical ideas on distinctions between the products of reactions controlled by rates and by equilibria which arose from the reduction work (11) and which were later elaborated (12,13) had much wider applications.

On the experimental side, highlights include the completion of the total synthesis from estrone of 19-nortestosterone, the first potent synthetic hormone other than an estrogen (14). This type of reduction, with experimental modifications, is employed industrially on a large scale (15,16). Another noteworthy example is a stereospecific synthesis of steroids (17) with the efficient production of a number of centers of correct configurations. Ether fission reactions have enabled the structures and configurations of the bisbenzylisoquinoline alkaloids to be determined (18) and there have been other applications in structure work. In recent years such reductions have largely rendered unnecessary the catalytic hydrogenation of an aromatic compound as a step in synthesis.

III. The Solutions

A. *THE METALS; PURITY OF THE AMMONIA*

As already noted, alkali and alkaline-earth metals dissolve at the boiling point of ammonia (–33°C), but the solutions most frequently employed in practice are of lithium, which is very soluble and can form a concentrated solution of composition $Li(NH_3)_4$ with almost no vapor pressure even at room temperature, or of sodium, or occasionally of calcium. Evaporation of solvent ammonia usually regenerates the metal, although "ammoniates," for example, $Ca(NH_3)_6$ (19), may be formed as intermediates. Lithium, which is very soluble in ammonia, dissolves in some primary amines, for example, ethylamine (20) or ethylenediamine (21), but other metals are little soluble in such solvents with the occasional exception of the sodium-potassium alloy. There is some evidence that the electrons in amine solutions may be more chemically reactive than in ammonia, possibly because of a smaller degree of solvation stabilization.

Calcium solutions are usually employed when relatively low basicity is required, for example, in reduction of α-acetoxyketones (22). More extensive investigations of the uses of other metals might be profitable; for example, the production of insoluble salts may protect intermediates in some cases, a possible contributing factor to the utility of calcium. Europium and ytterbium have been used in ammonia at 60°C to reduce benzene to cyclohexa-1,3- and 1,4-diene, and cesium (2 atoms) adds to ethylene (23). Magnesium and alcohols in ammonia have been used to reduce benzoic esters or amides and some polycyclic hydrocarbons (24).

The solutions often form two immiscible phases (1), a tendency enhanced by the presence of salts or of organic solvents, globules of a concentrated "metallic" phase appearing. Phase separation is more marked with the alkaline earths than with the alkali metals. However, organic solvents can usually be admixed, particularly with lithium solutions (16,25). In fairly dilute solutions reactions of the metals with alcohols to evolve hydrogen gas are quite slow, after an initial fairly rapid reaction; the base formed in the reaction acts as an inhibitor (26). The rate of gas evolution increases with the acidity of an added proton source: ammonium salts react very rapidly, water and methanol quite rapidly, and *t*-butanol very slowly. This reaction, like that with pure ammonia itself which is normally very slow, is greatly subject to catalytic acceleration, particularly by traces of transition metals (5,16). Pure ammonia is therefore greatly superior as a solvent and this superiority manifests itself particularly for slow and difficult reductions in the presence of alcohols, such as that of some substituted benzenes to dihydrobenzenes (16). The deliberate addition of ferric chloride has been used to limit the extent of reductions; anthracene and phenanthrene, for example, react very readily in the 9,10 positions, and this reduction step is not notably inhibited by the presence of iron, although further reduction is prevented (27). Clean partial reduction can therefore be achieved.

If pure ammonia is used, the choice of the metal does not seem to be crucial (16) and in most cases where reduction proceeds with ease sodium can be efficiently used. With more difficult cases lithium still seems to have an advantage, probably because of a mixture of factors, including the effect of the cation. Ion-pair formation may be involved, since the addition of lithium bromide to sodium reductions was found to be favorable (28). The lower basicity of lithium salts, including the amide, results in fewer secondary processes; the greater solubility of the metal in mixed solvent systems helps in cases of difficult solubility, and lithium is less sensitive to catalysis in the competing reaction of the metal with ammonia or an alcohol. This last factor is possibly the main reason why lithium seemed to be outstanding before it became customary to use pure redistilled ammonia.

B. OTHER AMINES

The use of solutions of metals in other amines is limited to primary aliphatic amines of low molecular weight, such as methylamine or ethylamine (20) or anhydrous ethylenediamine (21), and to metallic lithium. The chemical nature of the blue solutions of alkali or alkaline earth metals in hexamethylphosphoramide is not clear; these solutions are rather unstable but can also be used for many reduction processes (29). However, the use of mixtures of solvents (30) could be further explored. For example, lithium dissolves readily in small volumes of ammonia; the resulting solution can then be added to ethylamine to give a solution or dispersion which is very reactive. In ethylamine alone the metal dissolves slowly; continuous shaking or stirring during the reaction is required.

Lithium can be used with mixed amines, including secondary plus primary amines (31). A solution of sodium, lithium, or barium in trimesitylborane has recently been used to reduce the conjugated octalone-2 to *cis*- and *trans*-decalone-2, the proportion of *cis* being much higher than with ammonia as solvent (32).

C. SOLUBILITY OF ORGANIC COMPOUNDS; MIXED SOLVENTS

Although many organic compounds, particularly those containing polar or hydrogen-bonding groups, are soluble to a useful extent in ammonia, solubility usually decreases in the presence of inorganic ions. Admixed solvents such as polar ethers (tetrahydrofuran, ethylene glycol dimethyl ether, and dioxane) or alcohols [ethanol, *t*-butanol, and *t*-amyl alcohol (5,33)] up to about 30% by volume and occasionally more can be helpful to solubility, if their presence is compatible with the desired process. Such admixture is usually not necessary for readily soluble compounds unless alcohols must be present for chemical reasons. Many substances which are rather insoluble can frequently be reduced in a sufficiently dispersed form; such dispersion can usually be achieved by adding them in solution in another compatible solvent which dissolves in ammonia. Reduction products of aromatic compounds are frequently more soluble in ammonia than the starting materials, so that insoluble compounds often dissolve as the reaction proceeds.

Many organic compounds are more soluble in amines such as methylamine, ethylamine, and particularly anhydrous ethylenediamine than in ammonia. Hexamethylphosphoramide is also a good solvent for many organic substances.

IV. The Reduction Process

Two basically different final consequences become apparent in reductions. A molecule can give rise to initial addition products with the metal, which then undergo further reactions without cleavage, or it can undergo cleavage into an anion and a radical or into two anions. In both cases further transformations occur.

A. *ADDITION REACTIONS*

The possible processes involving an unsaturated molecule A═B can be summarized as shown in Scheme 1.

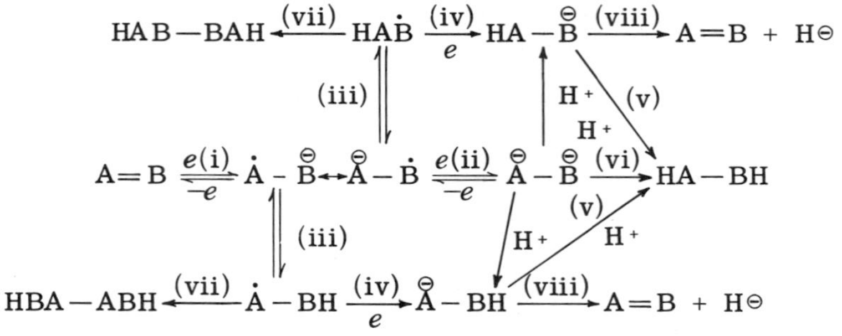

$$\mathrm{HAB{-}BAH} \xleftarrow{\text{(vii)}} \mathrm{HA\dot{B}} \xrightarrow[e]{\text{(iv)}} \mathrm{HA{-}\overset{\ominus}{\ddot{B}}} \xrightarrow{\text{(viii)}} \mathrm{A{=}B} + \mathrm{H}^{\ominus}$$

$$\text{(iii)} \qquad \mathrm{H^+} \qquad \mathrm{H^+}\ \text{(v)}$$

$$\mathrm{A{=}B} \underset{-e}{\overset{e\text{(i)}}{\rightleftharpoons}} \mathrm{\dot{A}{-}\overset{\ominus}{B} \leftrightarrow \overset{\ominus}{A}{-}\dot{B}} \underset{-e}{\overset{e\text{(ii)}}{\rightleftharpoons}} \mathrm{\overset{\ominus}{\ddot{A}}{-}\overset{\ominus}{\ddot{B}}} \xrightarrow{\text{(vi)}} \mathrm{HA{-}BH}$$

$$\text{(iii)} \qquad \mathrm{H^+} \qquad \text{(v)}\ \mathrm{H^+}$$

$$\mathrm{HBA{-}ABH} \xleftarrow{\text{(vii)}} \mathrm{\dot{A}{-}BH} \xrightarrow[e]{\text{(iv)}} \mathrm{\overset{\ominus}{\ddot{A}}{-}BH} \xrightarrow{\text{(viii)}} \mathrm{A{=}B} + \mathrm{H}^{\ominus}$$

(Vertical and diagonal arrows: HA$\dot{\mathrm{B}}$ ⇌ (iii) $\dot{\mathrm{A}}$–$\overset{\ominus}{\mathrm{B}}$ ↔ $\overset{\ominus}{\mathrm{A}}$–$\dot{\mathrm{B}}$ ⇌ (iii) $\dot{\mathrm{A}}$–BH; $\overset{\ominus}{\mathrm{A}}$–$\overset{\ominus}{\mathrm{B}}$ → H⁺ → HA–$\overset{\ominus}{\mathrm{B}}$; HA–$\overset{\ominus}{\mathrm{B}}$ → H⁺ (v) → HA–BH; $\overset{\ominus}{\mathrm{A}}$–$\overset{\ominus}{\mathrm{B}}$ → H⁺ → $\overset{\ominus}{\mathrm{A}}$–BH; $\overset{\ominus}{\mathrm{A}}$–BH → H⁺ (v) → HA–BH)

Scheme 1

B. *FISSION REACTIONS*

The possible processes involving A—B are shown in Scheme 2.

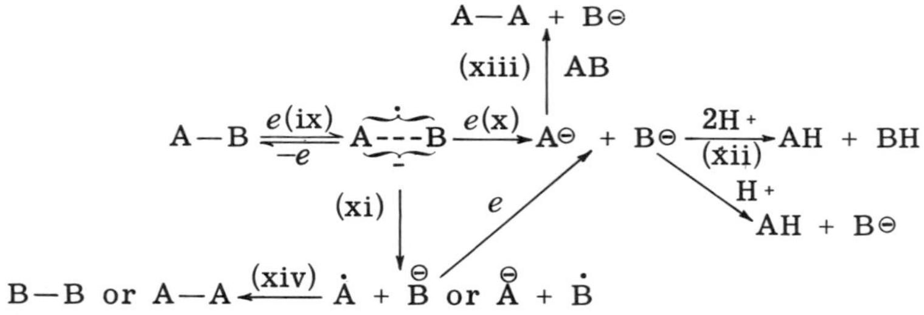

$$\mathrm{A{-}A} + \mathrm{B}^{\ominus}$$

$$\uparrow \text{(xiii)}\ \mathrm{AB}$$

$$\mathrm{A{-}B} \underset{-e}{\overset{e\text{(ix)}}{\rightleftharpoons}} \mathrm{[A{-}{-}{-}B]^{\dot{}}_{-}} \xrightarrow{e\text{(x)}} \mathrm{A}^{\ominus} + \mathrm{B}^{\ominus} \underset{\text{(xii)}}{\xrightarrow{2\mathrm{H}^+}} \mathrm{AH} + \mathrm{BH}$$

$$\mathrm{B}^{\ominus} \xrightarrow{\mathrm{H}^+} \mathrm{AH} + \mathrm{B}^{\ominus}$$

$$\text{(xi)} \downarrow \qquad e \nearrow$$

$$\mathrm{B{-}B\ or\ A{-}A} \xleftarrow{\text{(xiv)}} \mathrm{\dot{A}} + \mathrm{\overset{\ominus}{B}}\ \text{or}\ \mathrm{\overset{\ominus}{A}} + \mathrm{\dot{B}}$$

Scheme 2

1. *Stage i*

The electron addition stage is usually a reversible equilibrium, and the electron can often be removed completely in various ways from the adduct (34). The position of the equilibrium should be affected by the solvation of both ions involved, a probable reason for the favorable effect of ammonia as a dipolar molecule in encouraging the production of ions. The rate of the overall reduction can be affected by the rate of addition of the electron, or by the equilibrium position which determines the concentration of anion radical. The rate should be first order in metal and substrate if their concentrations are involved in this way. This has been claimed (28) but disputed (35) in the case of reduction of benzene with sodium and ethanol in ammonia. However, the original conclusions may well be correct so far as this stage of the process is concerned.

The position of the equilibrium is related to the structure of the substrate in predictable ways. The greater the extent of the unsaturation, the higher the electron affinity of the system, and the greater the ground-state strain, the more the equilibrium will tend toward addition. Conversely, the greater the stabilization of the ground state, as in aromatic compounds, the further the equilibrium will be towards dissociation. Since rate of reduction is dependent, as one factor at least, on the equilibrium concentration of anion radical, it can be qualitatively related to these factors. As examples of compounds of greater electron affinity, both naphthalene and pyridine are more readily reduced than benzene; as an example of stabilization of the substrate, benzene is less readily reduced than cyclohexa-1,3-diene. The effects of substituents are largely predictable. Isolated double bonds are difficult to reduce at all, and the rate of reduction depends on the degree of alkyl substitution in the order $CH{=}CH_2 > C{=}CH_2 > RCH{=}CHR' > RCH{=}CR'R'' > RR''C{=}CR'R''$ (36). Only the first two types are reducible in ammonia, but the others can be slowly reduced in amines (37), possibly because of the greater activity of the electron in these solvents. The order of reduction rates may be determined by the inductive effects of the attached groups and by their effects on solvation and the situation of the adjacent metal cation; all of these influences would retard reduction of more highly alkylated double bonds.

2. *Stage ii*

A second electron is added only slowly because of the negative charge already present. Solvation can be critical in formation of the dianion if the electron addition is reversible. This factor is clearly illustrated by the distinction between polar ethers and ammonia in the addition of sodium to naphthalene (38). In the ethers, one atom of sodium is added reversibly to give a green salt of an anion radical, and in ammonia two atoms give the red salt of a dianion, also reversibly. Whether two electrons can be added depends on the structure of the molecule; the addition is favored by extended unsaturation and by appropriate positioning of atoms such as nitrogen or oxygen, of higher electron affinities than carbon, so that they can stabilize the negative charge. Polycyclic aromatic compounds, in contrast to monobenzenoid ones, form dianions readily, and differences in the nature of products formed by reduction using two processes: (*1*) sodium and an alcohol; and (*2*) sodium alone followed by ammonium chloride or other proton source have been interpreted (13,39) as being due in the first case to protonation of an intermediate anion radical and in the second case to protonation of a dianion.

3. *Stage iii*

This reaction is often irreversible. Whether it is depends on the acidity of the proton source, that is, on the basicity of its conjugate base. The effect of alkoxide ion in slowing the reduction of benzene (26) is clear evidence that the protonation step is reversible under these conditions. Owing to electronic stabilization an anion radical is far less basic than a similar monoanion, and, experimentally, protonation of anion radicals derived from aromatic compounds is very slow and incomplete unless "acids" more acidic than ammonia (p$K_a \sim 34$), such as alcohols (pK_a = 16-18), are present. This explains the necessity for the presence of alcohols in the standard reduction of monobenzenoid compounds in ammonia; in their absence the low concentration of anion radical is not further transformed. With amine solvents the very slow reduction of monobenzenoid compounds may be due to the higher reaction temperatures (0-100°C), with slow protonation becoming possible; this view is supported by the occurrence of similar reductions in ammonia itself at higher temperatures than the boiling point under pressure (40). An alternative possibility in such cases is the production of small amounts of the more basic dianion, particularly in amines, where the electrons are probably more reactive. In amines the use of an added alcohol greatly speeds the reductions (41), and the combined reagent has more powerful reducing properties than the ammonia-lithium-alcohol system and can accomplish the efficient reduction of a tetrasubstituted ring in a hydrophenanthrene (42), a process which is very slow and difficult with lithium and *t*-butanol in ammonia (43). So far such solutions have been virtually unused for synthetic purposes.

4. *Stage iv*

Reduction of a radical would be expected to be very rapid in a medium which is an electrical conductor containing excess of reducing agent. Dimerization of such radicals is therefore likely to be a minor process.

5. *Stage v*

Protonation of a monoanion is likely to be rapid and complete for carbon systems because of the very high basicity of such anions. The mesomeric cyclohexadienyl anion, for reasons noted below, is experimentally found to be comparable in basicity to an amide anion, and clearly should undergo equilibrium protonation by solvent ammonia, and should be rapidly protonated by more acidic sources, with the equilibrium well towards the product.

6. *Stage vi*

The dianion should be very highly basic. Experimental evidence shows, for example, that the naphthalene dianion is rapidly protonated at the boiling point of ammonia by the solvent, although more slowly at −80°C (2). Protonation presumably occurs in two stages with the formation of an intermediate mesomeric monoanion, analogous to but not necessarily identical with that considered under Section IV.B.5, depending on the position taken up in the mesomeric system.

7. *Stage vii*

Anion radicals are unlikely to dimerize because of their charge; the protonated species can do so, but are unlikely to do this in competition with reduction.

8. Stage viii

As discussed below, there is a strong possibility with some monobenzenoid systems that expulsion of a hydride ion may occur to some extent, with regeneration of the aromatic system. The result is catalysis of hydrogen gas evolution.

9. Stages ix, x, and xi

The effects of substitution of the system in, for example, the fission of allyl alcohols, are in accord with expectation on the basis of the production of intermediate mesomeric anions as the rate-determining step. The products also usually correspond to the formation of two anions. It is not clear, however, whether this implies that the rate-determining step involves two electrons. The effect of substituents on a transition state involving the addition of one electron could well be that experimentally observed (see below), and any radical formed by fission in this way would clearly be rapidly converted into an anion in a subsequent process. An investigation of a number of cleavage reactions leads to the conclusion that probably a two-electron addition is in fact involved with allylic systems, and that in some cases at least dimeric products do not result from radicals, but by attack of an anion on the original molecule (44). The question still, however, seems open. The reaction normally occurs readily only when one of the resulting anions is the conjugate base of a fairly acidic acid: acetates, for example, cleave much more readily than alcohols.

10. Stages xii, xiii, and xiv

These represent possible alternative explanations for the dimeric products sometimes encountered. There is clear evidence for Stage xiii in some cases (44). Since one or both A^- and B^- are usually mesomeric if organic, the protonated products AH and BH (Stage xii) may be mixtures of isomers.

V. Reductions of Unsaturated Systems

Some characteristics in which the reagents differ from other systems in ways which render them useful are as follows:

1. Experimentally the stages of electron addition and of protonation can often be separated. This has consequences both in theoretical interpretations and in the arrest of reductions at intermediate stages.

2. The acidity of an added proton source can be varied widely and conveniently from ammonia or the amine solvent to alcohols and strong "acids" such as ammonium salts. Consequently the nature of intermediates can be altered in a number of instances, in the presence of reducing metal, leading to different final products.

3. Temperatures can be controlled and the lifetimes of unstable intermediates can be extended in some cases until work-up.

4. Control of the basicity of the reagent (buffering) can induce or inhibit conjugation of unsaturation in intermediates and alter the degree of reduction of final products.

5. Steric control can be exercised frequently, in the sense that predominantly a predictable isomer is obtained.

6. Selectivity in reduction of unsaturation, or in fission reactions, particularly of ethers, is frequently high and may differ from that observed with other reagents.

7. The reagents are the most powerful chemical reducing agents known, and correctly used can result in the specific partial reductions of monobenzenoid systems.

In order to achieve a desired result the experimental conditions must be very carefully chosen using two types of procedure:

1. Reaction with a metal under conditions where the anion radicals or anions resulting are not protonated until work-up.

2. Reaction with a metal in the presence of a proton source where the resulting anion radicals or anions are (*a*) irreversibly, or (*b*) reversibly, protonated according to the acidity of the proton source and the basicity of the anions.

Quite different products may result, for a variety of reasons, from these different procedures, and it is necessary to consider some fundamental questions before proceeding to detailed considerations of cases.

A. METAL ADDITION FOLLOWED BY PROTONATION

A mesomeric anion formed by electron addition is inhibited from further electron addition by its negative charge. Whether only one, or two, electrons can be added depends on the structure of the compound as discussed below. Monobenzenoid compounds, for example, are in equilibrium with anion radicals; many polycyclic compounds form dianions readily. There is no evidence that an intermediate carrying more than two electrons in the same system is ever involved. The earlier reductions of polycyclic aromatic compounds by Hückel and Bretschneider (2) were carried out by this method.

Because of survival of the intermediate, implying its inability to abstract a proton from solvents such as ammonia, the reduction products are those resulting from the addition of only one or two electrons. The protonation of some anion radicals on work-up leads to the formation of aromatic starting material and a dihydro product, apparently by disproportionation.

Examples are quoted below, but two illustrations can be given. The reduction of α,β-unsaturated ketones with lithium in ammonia leads to saturated ketones (45)

$$C{=}C{-}C{=}O + 2e \rightarrow \ominus C{-}C{=}C{-}O\ominus \rightarrow HC{-}C{=}C{-}O\ominus \rightarrow HCCHC{=}O$$

In this case the mesomeric carbanion probably is basic enough to abstract a proton from ammonia; the enolate anion is not ketonized until work-up and is therefore not reducible. Another example of a protective negative charge in directing the course of reaction is the reduction of *o*- and *p*-hydroxyacetophenones (46). These yield the pinacols, rather than the carbinols; presumably the formation of a dianion is inhibited by the *ortho* or *para* charges on the ring and the protonated anion radical dimerizes. In contrast, *m*-hydroxyacetophenone gives the carbinol.

The practical result of this general type of process is that, provided the electron-addition equilibrium lies towards a product, and provided the resulting anions are not protonated under the conditions of the reaction but only on work-up, it is possible to obtain final partially hydrogenated products even when such products are themselves reducible. Among compounds appropriate to the method are aromatic acids, amides, and polycyclic compounds.

Other effects of anion formation including inhibitions of cleavage reactions are noted later.

Theoretically, as discussed below, the products obtainable from protonation of an anion radical intermediate *in situ* or from a dianion by subsequent protonation could differ, as has been pointed out by Hückel (1).

In order to avoid reduction of protonated products by excess metal during work-up it is often advisable to destroy the latter. This frequently may be accomplished by addition of ammonium chloride or water; hydrogen gas is produced at a rapid rate, but this is without effect on the reduction product. However, the addition of benzoic acid, which is very rapidly reduced, is useful provided the product is not acidic. The addition of traces of ferric salts, or of sodium nitrite, also results in removal of metal.

B. *METAL ADDITION IN THE PRESENCE OF A PROTON SOURCE*

1. *Effects on Primary Reduction Stages*

The presence and acidity of a proton source other than ammonia or an amine can have profound effects on whether reductions occur at all and on the structures of the products.

As has been explained above (Section IV.C), the reduction of monobenzenoid compounds, other than carboxylic acids and amides, does not occur in the absence of a proton source such as ethanol. The observation of Wooster (4) that the joint action of sodium and ethanol in ammonia or benzene gives

1,4-dihydrobenzene was interpreted (28,47) as being due to the formation of the anion radical in low concentration, and the ability of alcohol, in contrast to ammonia, to protonate this with formation of the radical. This radical is then further converted by electron addition and protonation into the observed dihydrobenzene. The process can be summarized as follows:

$$\mathrm{Ar} \underset{-e}{\overset{+e}{\rightleftharpoons}} \mathrm{Ar}^{\ominus}\cdot \xrightarrow{\mathrm{ROH}} \mathrm{Ar}\cdot\mathrm{H} \xrightarrow{e} \mathrm{Ar}^{\ominus}\mathrm{H} \xrightarrow{\mathrm{ROH}} \mathrm{ArH_2}$$

The nature of the products with substituted benzenes is discussed below.

The possibility arises that different products may result from this procedure compared with the previous one, since initial protonation of an anion radical probably occurs in all cases, although this assumption has not been proved with polycyclic aromatic systems where a dianion could be produced. However, reduction of naphthalene through the dianion followed by protonation, and reduction through the simultaneous action of metal and alcohol, give the same initial product—1,4-dihydronaphthalene. Protonation of the preformed naphthalene anion radical salt gives a mixture of naphthalene and 1,4-dihydronaphthalene, probably due to electron transfer from the anion radical to the initial monoprotonated radical.

Other cases, however, indicate that different products may result from the two procedures; for example, *N*-methylindole with lithium in ammonia over a considerable period of time gives the 2,3-dihydro derivative, while rapid reduction with lithium and methanol in ammonia gives the 5,8-dihydro derivative (39). Such possible differences could be worth consideration with polycyclic aromatic compounds since the positions of initial protonation need not be the same in an anion radical as in a dianion.

Reduction of the tryptamine derivatives (**1**) (R = H or Me) is illuminating (48). In the presence of methanol the quaternary salt is stable to lithium but the benzene ring is reduced to the 4,7-dihydro derivative. In the absence of methanol, **1** (R = H) loses Me from the quaternary center, no ring reduction being observed (stabilization by formation of the anion from the acidic pyrrole ring). This is a usual type of cleavage with methyl ammonium salts (44). With **1** (R = Me), however, the indoline (**2**) is formed in 83% yield. The anion radical (or dianion) produced from the nucleus presumably displaces the NMe_3 group, and reacts on the pyrrole rather than the benzene ring despite the production of a 3-membered rather than a 5-membered ring. This is further evidence, if this is needed, that the charges are distributed over the whole ring system and can react in different positions according to circumstances.

Solutions of lithium in amines behave like the metal-ammonia-alcohol combination rather than the metal-ammonia combination since the reagents are capable of reducing monobenzenoid compounds, for example, benzene to cyclohexene (20). In view of the low acidity of the solvents this is at first sight puzzling if protonation of an anion radical is involved. We have noted, however, that the electrons in amines seem to be chemically more reactive than in ammonia and the reductions are also necessarily carried out at higher temperatures for prolonged periods of time. Benzene is reducible to cyclohexene with sodium in ammonia at elevated temperature and pressure (40) and also by calcium hexammoniate at room temperature (3). All such reductions are very slow, taking hours, compared with the few minutes required for the process with either ammonia or amine solvent in the presence of ethanol. We can assume that slow protonation of the anion radical does occur under these circumstances, or else that a small but ultimately significant proportion of the dianion is formed which is sufficiently basic to be capable of proton abstraction from the solvent. The reasons for the differing final products are noted below.

A highly acidic proton source, such as an ammonium salt, is usually undesirable for *in situ* reaction, because the very rapid competing evolution of hydrogen gas removes the reducing agent.

2. *Effects of Buffering on Secondary Reductions*

The "buffering" effects of added proton sources can be related to the removal of anionic products, either permitting further reduction to occur in some cases where the organic anion itself is stable (see above), or inhibiting further reduction where compounds containing unconjugated, but base-conjugable double bonds are produced. The first case is well illustrated by benzyl and allyl alcohols. It was early observed that allyl alcohol itself gives an equimolar proportion of propene and sodium allyloxide with sodium in ammonia (49), but in the presence of ethanol all of the allyl alcohol is reduced. This is explicable by equilibration of the allyloxide and ethoxide anions with their free alcohols, permitting removal of all of the allyl alcohol ultimately. Lithium in ethylamine reduces benzyl alcohol slowly to tetrahydrobenzyl alcohol (present as the anion) whereas lithium and ethanol in ammonia rapidly produces toluene; hydroxyl can be eliminated as an anion but loss of O^{2-} does not occur.

Another example of further reduction permitted by protonation is the reduction of benzoxazoles (50). In the absence of a proton source the anion (**3**) is stable until work-up and hydrolysis occurs to give **4** (R = H); in the presence of ammonium chloride protonation of **3** permits further reduction to **4** (R = Me).

Deliberate manipulation of "buffering" to achieve synthetic ends can be illustrated by some examples.

Usually amides containing NH give mainly salts with metal-ammonia solutions, together with hydrogen. The lactam group of strychnine was found to be reducible with sodium and ethanol in ammonia to a cyclic "aldehyde-ammonia" (51); if other amides behave similarly the process could lead to a general synthesis of aldehydes. However, acyclic amides were found to give the alcohols, presumably because of the base-catalyzed fission of **5** → **6**, the aldehyde then being reduced by metal still present. Stabilization in the

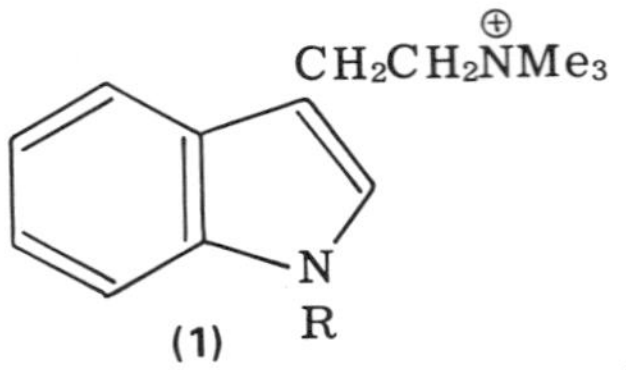

$CH_2CH_2\overset{\oplus}{N}Me_3$

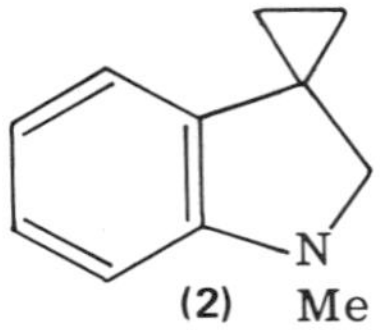

N R (1)

N Me (2)

N=CH⊖ O⊖ NHR OH

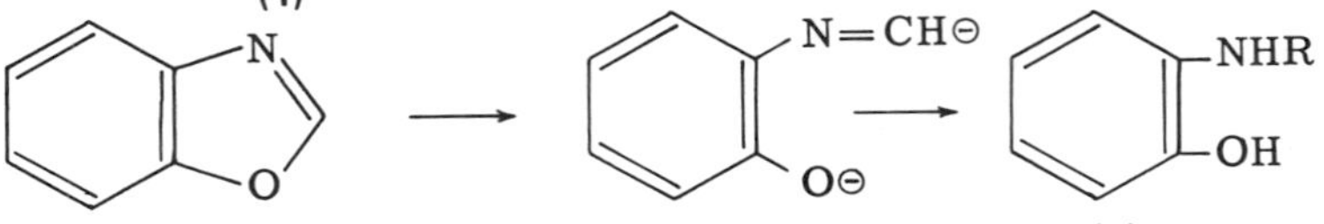

(3) (4)

strychnine case must be due to the ring. To stabilize **5** until work-up, the solution was buffered with ammonium acetate, and in this way good yields of aldehyde could finally be obtained. The same objective can be attained by using an amidine (**7**), instead of an amide; the low acidity of NH in the resulting *gem* diamine compared with OH in the aldehyde ammonia permits the presence of EtO^- and hence the use of ethanol as the proton source. In this way reduction of **7** to **8** could be achieved, the final product after work-up then being **9** (52). Benzamide is reducible at the amide group with lithium and ethanol in ammonia, but gives 1,4-dihydrobenzamide with lithium and *t*-butanol in ammonia (53), the latter process presumably representing a reduction of the salt of the amide.

Reduction of **10** with two equivalents of metal and alcohol in ammonia gives a mixture of **11** and **12** in competing processes. By initially preparing the salt of **10** and using metal alone, the product is **11** only (54). A terminal acetylene can be protected against reduction by formation of its salt; a disubstituted acetylene in the same molecule can then be preferentially reduced (55). In the presence of alcohol, terminal acetylenes are also reducible. Another example has been noted already with **3**. Anions can thus be deliberately produced or deliberately protonated according to the end in view.

Whether the protonation reaction is reversible or not under the conditions also has highly important experimental consequences. For reasons discussed later the initial kinetically controlled product of protonation of a mesomeric anion is frequently the thermodynamically least stable of possible products. As such it usually contains unconjugated unsaturated systems. Since isolated double bonds are difficult to reduce because of the very unfavorable electron-addition equilibrium, products with such bonds are frequently stable even in the presence of reducing agent. However, if equilibration favors the more thermodynamically stable conjugated products, these are then rapidly reduced (56). Such equilibrations are brought about by very basic ions through proton abstraction. The effects of the structures of olefins on their "acidities" has been discussed in this connection (56,57), and a good deal of subsequent work has been done (58,59). Of outstanding importance in the present connection is the fact that the doubly allylic CH_2 of systems like $C{=}CCH_2C{=}C$ is comparable in acidity to ammonia or an amine, but is much less acidic than an alcohol, even *t*-butanol.

$$\mathrm{RCONHR'} \rightarrow \mathrm{RCH(O{-}H)}\!-\!\mathrm{NHR'} \rightarrow \mathrm{RCH{=}O} \rightarrow \mathrm{RCH_2OH}$$

(5) (6)

(7) $\xrightarrow{}$ (8); $H_3O^{\oplus}$ → CHO; $H_3O^{\oplus}$ → CHO (9)

(10) PhCH=CHCMe$_2$OH
(11) PhCH$_2$CH$_2$CMe$_2$OH
(12) PhCH$_2$CH=CMe$_2$

(13) $\underset{\text{EtOH}}{\overset{\text{KNH}_2}{\rightleftharpoons}}$ (15) $\underset{\text{KNH}_2}{\overset{\text{NH}_3}{\rightleftharpoons}}$ (14)

The equilibration of 2,5-dihydroanisole (**13**) with 2,3-dihydroanisole (**14**) has been investigated (57), and is important also in illuminating the reduction process. The use of a proportion of potassamide greater than 1 mole gives chiefly the salt corresponding to the anion (**15**) and this on kinetic protonation gives **13**. The conjugated diene (**14**) similarly gives the same salt (**15**); the process can therefore convert **14** into **13**. If, however, a smaller proportion of amide is used, the anion (**15**) merely acts as a turntable to induce equilibration between **13** (20%) and **14** (80%). Even in the presence of a large proportion of amide, sufficient of **14** is produced to permit its reduction in the presence of metal and ultimately all of the material passes through it to further products. Experimentally the addition of **13** to sodium, or better potassium, in ammonia results in reduction in an autocatalytic process.

Any reaction which generates amide anion, as the result of protonation from ammonia or an amine, is therefore liable to lead to reduction past the initial stage. This was initially demonstrated with ammonia solutions (56) but is clearly true of the amine solutions and explains the reduction of monobenzenoid compounds to cyclohexenes, via cyclohexa-1,3-dienes, rather than to cyclohexa-1,4-dienes. Support is found from the fact that addition of ethanol to the amine solutions results in cyclohexa-1,4-dienes and also from the similarity of the further reduction products obtained by the two-step (metal-ammonia-alcohol, then metal-ammonia) and the one-step (metal-amine) sequences (see below).

The practical point to be emphasized is that in such cases, if further reduction is to be avoided, the conditions must not permit production of amide anions in the presence of the primary unconjugated protonation products, which can occur, for example, if the metal is present in higher molecular proportion than the alcohol. The use of potassium rather than lithium favors this type of process because of the greater reactivity of the former metal with ammonia and the greater basicity of potassamide compared with lithamide. Other secondary processes noted later may also occur in the presence of amide anions.

The discussion has been concerned with wholly carbon systems, but with appropriately placed hetero-atoms in the molecule, alkoxide ions may also be important isomerizing agents, for example, with dihydrobenzoic acids. It does not necessarily follow, however, for reasons discussed in connection with the nature of the products, that more acidic alcohols, for example, methanol, are always better than less acidic ones, for example, *t*-butanol, in reducing benzenoid systems; the reverse is frequently true (60). Alkoxide anions seem to have little isomerizing effect on carbon systems and specificity of protonation is the important factor. Sodium ethoxide isomerizes cyclohexa-1,4-dienes into 1,3-dienes in ammonia, but only after prolonged reaction at 16°C (61).

C. ELECTROLYTIC REDUCTIONS

Electrolysis in ammonia or amines is a potentially attractive process, particularly from an industrial point of view. Its possibility was demonstrated by the reduction in liquid ammonia of *m*-tolyl methyl ether to its dihydro derivative (62) and later it was systematically investigated in methylamine (63). The experimental method permitted preferential formation of either dihydro or tetrahydro derivatives in good yields, according to whether or not a divider was placed in the cell. Lithium chloride was used as electrolyte. The divider separates methylamine hydrochloride (anode) and lithium methylamide (cathode) and the latter then catalyzes conjugation and further reduction to tetrahydro derivatives. Alkyl and arylacetylenes give *trans* olefins (64). A practical use of the method is the reduction of estrone methyl ether to 1,4-dihydro-3,17β-estradiol-3-methyl ether in good yield with 44% current efficiency (65).

A more convenient medium for electrolytic reduction of aromatic rings is a mixture of ethanol and hexamethylphosphoramide (66). Reduction of tetrahydronaphthalene gives the hexahydro or the decahydro derivative as the principal product, according to conditions. Anhydrous ethylenediamine containing lithium chloride can also be used (67).

VI. Relation of Structures to Reducibility in Unsaturated Systems

It is clear from the preceding that the reducibility of an unsaturated system depends on whether it adds an electron and frequently on whether the resulting anion radical can then be irreversibly transformed. Also, despite uncertainties about the kinetic results of the metal-alcohol process, it is very probable that the rate is controlled to a large extent by the equilibrium position of the electron addition.

The first reported reduction of an unconjugated double bond by means of sodium and ethanol in ammonia was of the $CH{=}CH_2$ group in allylamines (68). Another example of terminal double-bond reduction is the conversion of *p*-allylphenol to *p*-propylphenol (69). Reductions of $C{=}CH_2$ also occur, but other olefin types, including 1,2-dialkyl derivatives, were thought not to be reducible by this reagent. The lithium-ammonia-methanol and lithium-methylamine systems were examined for their effects on substituted double bonds (36), and the conclusions support the assumption that the greater the substitution by electron-donating substituents and the greater the inhibition of solvation, the slower the rate. Conversely, the rate is increased by ground-state strain in the olefin: norbornene > cyclohexene; *cis* slightly greater than *trans*-olefins; cyclopentenes > cyclohexenes.

Lithium-amine solutions (particularly ethylamine and ethylenediamine) can cause slow reductions of even tetrasubstituted isolated double bonds, but the greater the substitution the slower the process (70). For example, lithium in methylamine converts benzene into cyclohexene and then more slowly into cyclohexane; naphthalene gives first tetralin, then a mixture of 9(10)- and 1(9)-octalin and finally *trans*-decalin. The basic conditions sometimes obscure the results by causing bond migrations, for example, of terminal olefins into internal olefins. It was shown early (56) that potassamide in ammonia converts oct-1-ene into oct-2-ene, and more highly substituted bonds can be caused to migrate by more drastic conditions, such as heating with lithium ethylenediamide in ethylenediamine (71). Less easily reducible bonds can result from such migrations to more substituted positions.

There are indications that reductions of isolated double bonds can be assisted by stabilization of the anion radical by means other than direct conjugation. The unconjugated 7,8 double bond in equilin (present as the phenolate anion) is reducible, apparently directly, with lithium in ammonia (72) (see later), and the unusual ease of formation of saturated cyclohexylamine derivatives from amines by lithium in methylamine (73,74) appears to indicate that the amino group assists the process. The products of reduction of **16**, for example, are **17** and **18**, indicating that the amino-substituted ring is the more difficult of the two to reduce, but that once it has been reduced the process continues to the tetrahydro stage. Moreover 3- and 4-dimethylaminocyclohexene are much more readily reducible than is cyclohexene. The bridged-ring ether **19**, which cannot undergo hydrogenolysis for steric reasons, undergoes reduction of the double bond with great ease using lithium and ethanol in ammonia (75). Donor atoms or groups in appropriate positions may interact nonclassically with the radical and assist in the electron addition.

Among "isolated" double bonds that are reduced is that in norbornadiene **(20)** giving norbornene **(22)** containing no norbornane. Stabilization as a

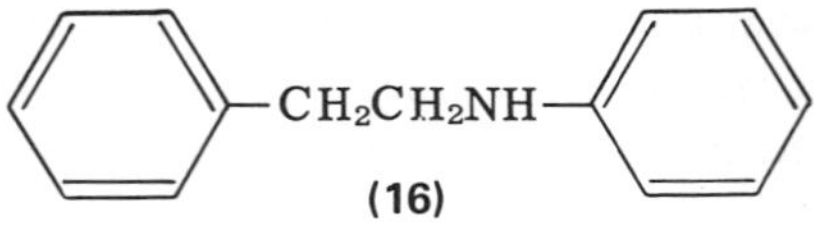

$-CH_2CH_2NH-$

(16)

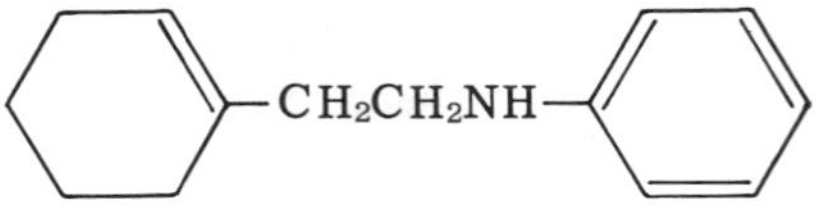

$-CH_2CH_2NH-$

(17)

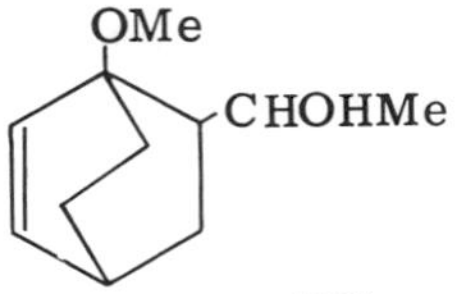

OMe

CHOHMe

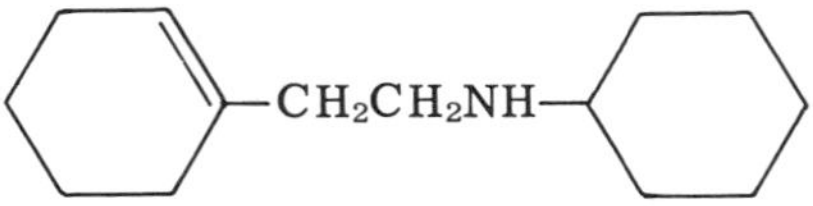

(19)

$-CH_2CH_2NH-$

(18)

⊖

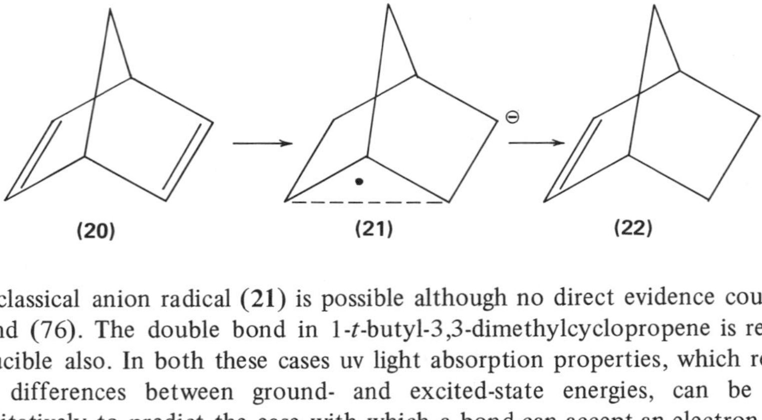

(20) **(21)** **(22)**

nonclassical anion radical **(21)** is possible although no direct evidence could be found (76). The double bond in 1-*t*-butyl-3,3-dimethylcyclopropene is readily reducible also. In both these cases uv light absorption properties, which reflect the differences between ground- and excited-state energies, can be used qualitatively to predict the ease with which a bond can accept an electron in its lowest lying antibonding orbital (76).

The situation warrants further study, but in the meantime some caution should be exercised in ruling out the possibility of direct reductions of 1,2 disubstituted or more highly substituted olefins.

Acetylenes as electron-deficient systems are usually readily reducible, except as the salt of a terminal group, and reduction of open-chain disubstituted acetylenes gives the *trans* isomer only (except in medium-sized rings, discussed later).

Allenes are readily reducible, the less-substituted double bond again being predominantly hydrogenated; mixtures of *trans* and *cis* isomers result (77). Conjugated dienes can be readily reduced, the products representing both α,β

and α,δ reduction (78-80). The relation between structure and the nature of the product has not been very closely investigated, although there is a tendency for less-substituted bonds to be reduced to a larger extent. The particular case of cyclohexadienes is discussed in more detail below.

Aromatic polycyclic hydrocarbons can be reduced readily, as might be expected from the ability of the extended unsaturation to stabilize charges. With aromatic systems, such as pyridine or quinoline where a seat can be found for a negative charge on an atom of higher electron affinity than carbon, reduction is correspondingly easier (81). However, with aromatic systems such as furan or *N*-methylpyrrole, where the heteroatom unshared electrons are involved in the aromatic stabilization, the addition of further electrons is more difficult, and reduction occurs slowly or not at all. In some heterocyclic cases the situation may be complicated by salt formation, for example, with an acidic NH as in pyrroles, indoles, or imidazoles. In such cases reduction is inhibited unless the solution is appropriately buffered, for example, with methanol in the case of indole (82). Thiophene can be reduced (83) but fission of carbon-sulfur bonds also occurs. Reduction of thianaphthene gives *o*-ethylphenol, and dibenzothiophene can give either 6,9-dihydrobenzothiophene (84) or a cyclohexenylthiophenol (47) according to the exact reduction procedure.

Other interesting effects of substituents on rates of reduction can clearly be seen in aromatic systems. The larger the number of saturated substituents on a ring, the slower is the rate; large groups such as *t*-butyl are highly inhibitory (29), although *p*-di-*t*-butylbenzene can be reduced to the cyclohexene by lithium in ethylenediamine (85). A tetrasubstituted aromatic B ring of steroids proved particularly difficult, but slow reduction could be achieved by lithium and *t*-butanol in ammonia (43) or better still in a similarly substituted ring by using methylamine (42). The inhibition of reduction could be due to destabilization of the anion radical both by the inductive effect of the substituents and by inhibition of solvation. Methoxybenzene is reducible about three times faster than benzene (43), a somewhat surprising conclusion if the mesomeric effect of the group in stabilizing the substrate is of importance. A predominant effect is probably the inductive effect of the methoxy group on formation of the anion radical. Reductions of anisole and some of its methyl derivatives with potassium and ethanol (*no* ammonia) showed, on the assumption of a constant competing reaction in each case to give hydrogen gas, that the rates of reduction based on yields of dihydro products were anisole $>$ *p*-methylanisole $>$ *m*-methylanisole $\gg$ *o*-methylanisole $>$ toluene (86). The positional effects are not readily explicable and more work seems necessary.

An effect of ring size is shown by the faster rate of reduction of hydrindane compared with tetralin (43).

Although it has not been extensively investigated, oxygen appears to be a catalyst in some ring reductions and may in traces assist electron-transfer processes (87).

A few examples of products other than those from direct reductions have been observed. Nitrogen-containing products are obtained by reaction of naphthalene with sodium and a secondary amine, for example, 2-piperidino-1,2,3,4-tetrahydronaphthalene (70% yield) (88). This is apparently due to addition of the amine salt to 1,2-dihydronaphthalene. Similar products do not appear to have been noted with ammonia, but some primary amines including ethylamine can react similarly.

The nature of the products obtained from the reduction of unsaturated systems is best defined by initial consideration of substituted benzenes.

VII. Reductions of Substituted Benzenes

Substituents on a benzene ring affect both the rate of reduction as noted previously, and the orientation of the added hydrogens. To understand the nature of the reduction products, let us look at the details of the reduction of methoxybenzene (**23**) to 1-methoxycyclohexa-1,4-diene (**29**).

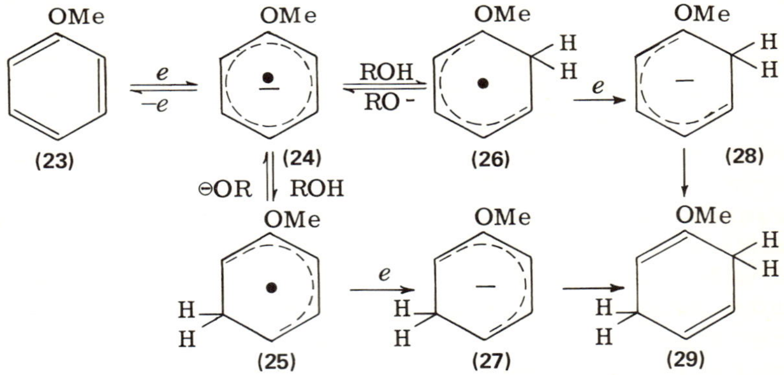

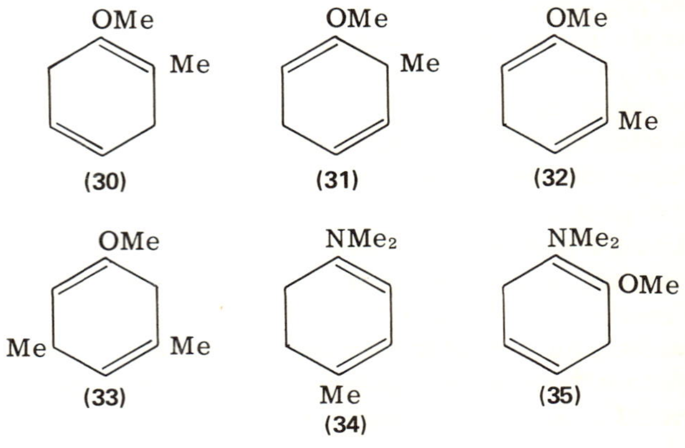

OMe OMe OMe OMe
e ROH H H
−e RO- H e H
(23) (24) (26) (28)
⊖OR ROH
OMe OMe OMe
H
e H
H H H
H H H
(25) (27) (29)

OMe OMe OMe
Me Me
Me
(30) (31) (32)

OMe NMe2 NMe2
OMe
Me Me
(33) Me (35)
(34)

The anion radical (**24**) could theoretically add a proton *ortho, meta,* or *para* to the substituent. Both the experimental evidence and theoretical considerations rule out the *para* position. Initially it was suggested that protonation of mesomeric anions in general (but not necessarily other types of reaction) should occur in a kinetically controlled process at this point of maximum charge density (11). Probably a better definition would have been at the position where the anionic charge can be localized most readily. Experimental evidence indicated initial protonation of **24** in the *meta* position to give **25** (89). This suggestion was based on the fact that 2-methylanisole gives two products: **30** and **31**, and 3-methylanisole only one: **32**. The nature of the products in other cases suggests that protonation does not occur readily on an alkylated position, but nevertheless 3,5-dimethylanisole gives **33** (90), leading to the conclusion that in this case initial *ortho* protonation probably occurs.

Theoretical calculations do not agree in distinguishing *ortho* (**26**,**28**) from *meta* (**25**,**27**) (91-94), although a recent treatment based on localization again favors *meta.* It now seems likely that both *ortho* and *meta* protonation occur at comparable rates and that the relative extents in different cases vary with structure. The theoretical calculations in any case ignore the fact that there is a counter ion and a dipolar solvating solvent, both of which factors carry steric implications. No experimental method at present exists capable of determining to what extent the first proton is added in one or other position, in other words, the proportion of **25** and **27** to **26** and **28** involved.

The problem of addition of the second proton to **27** or **28** is solved experimentally in the case of **27** and in principle in the case of **28** by the kinetic protonation of **15** = **27** which leads to **13** = **29**. Predominant reaction occurs in the middle of the system where the charge probably localizes most readily. Kinetic addition of a proton to give a conjugated diene such as **14** would lead by further reduction *in situ* to methoxycyclohexene and cyclohexene as discussed below.

Added protons avoid substituents in the order OMe, NMe_2 > alkyl > H, giving predominantly 2,5-dihydro derivatives.

In the case of aromatic amines, reported isolations of conjugated dienamines, for example, of **34** from *N,N*-dimethyl-4-methylaniline (95-97), were unexpected since such dienamines are themselves readily reducible to the monoenamines (98). Also the products of acid hydrolysis are the same ketones as those obtained from the corresponding anisole derivatives. The major difference in the two cases is the frequent occurrence of a larger proportion of saturated ketone, that is, of more reduction to tetrahydro derivative (99), with the aromatic amines. The initial reduction products are in fact almost certainly the unconjugated dienamines, and by careful work-up they can be isolated in some cases, particularly with *ortho*-substituted amines where they are reasonably stable, for example, **35**. Conjugation occurs with great ease at temperatures above 30°C and almost invariably on distillation.

The reduction of primary or secondary aromatic amines in ammonia leads both to enamines and to cyclohexylamines, probably in part due to tautomerizations and subsequent reductions of imines (73,74,100).

Two extreme cases which warrant further discussion are those of phenolate anions (inhibition), and carboxylic acids and other related substrates bearing electron-stabilizing groups (acceleration). The simple view, largely borne out in practice as previously noted, is that phenolate (and enolate) anions should inhibit electron addition. However, phenols do reduce under appropriate conditions, possibly in the form of undissociated salts. The first observed case was that of 2-naphthol, which with sodium and *t*-amyl alcohol in ammonia at –33°C gives about 75% of β-tetralone and 25% of reduced phenolic material (5). By contrast, reduction with sodium at –70°C in the absence of alcohol gives a 3:2 ratio of 5,8-dihydro-2-naphthol and 5,6,7,8-tetrahydro-2-naphthol (72). The latter could be produced from the former by reduction with potassium, but not sodium, at –33°C, without intermediate conjugation. Equilenin 17-ethylene ketal (**36**) with sodium, lithium, or potassium at –70°C gives 55% yield of the equilin ketal (**37**) in contrast to calcium which reduces ring A. At –33°C with potassium the 8α-estrone derivative (**38**) is formed with direct reduction of the 7,8 double bond. The importance of metal, temperature, and proton source is obvious, and apart from the reduction of the isolated double bond, which is discussed elsewhere, the nature of the product is determined by the position of initial protonation of the anion or anion radical in the naphthalene system.

Monobenzenoid phenols are not reducible in ammonia under the usual conditions, but the use of lithium in very high concentration (4*M*) has led to reduction of phenols, including that of oestradiol to **39** (101). In amine solvents phenol is reducible to cyclohexanone (41). Inhibition of reduction is therefore relative and not absolute.

Benzoic acids are rapidly reducible, despite the formation of the carboxylate anion. The high rate (28) and the nature of the products both attest to the fact that the attached group behaves as an anion-stabilizing one. The products are normally 1,4-dihydro derivatives (53,57,102,103), although further reduced products are rather readily formed by facile conjugation reactions. Investigation of the reduction of *p*-isopropylbenzoic acid clearly shows the sensitivity of the reaction to exact experimental conditions, but good yields of the 1,4-dihydro derivative are obtained under conditions where conjugation is avoided (104). Groups such as methoxy, in *para* positions in particular but also to some extent *ortho*, are readily eliminated: *p*-methoxybenzoic acids, for example, give reduced benzoic acids, and 3,4,5-trimethoxybenzoic acid gives 1,4-dihydro-3,5-dimethoxybenzoic acid (102). Reduction of benzamide in the presence of *t*-butanol gives the 1,4-dihydro derivative (53).

Other electron-stabilizing groups in the monobenzenoid series (e.g., acyl, nitro) appear to be reduced before the ring, although *p*-acyl can also labilize a *p*-methoxy group to hydrogenolysis to some extent (105). Although it has not

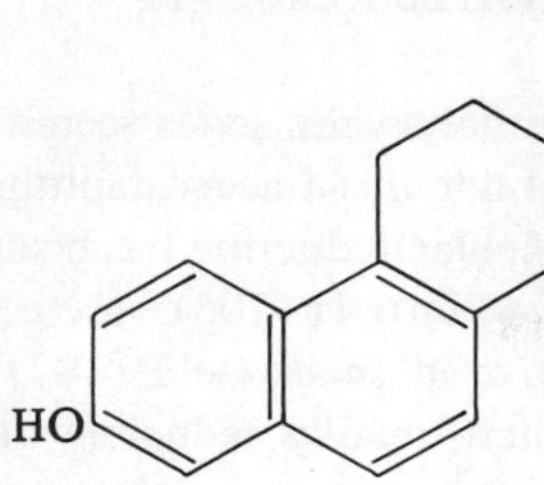

(36) 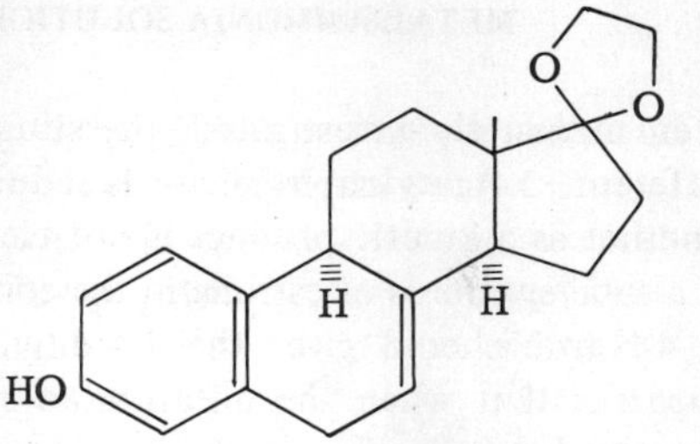 (37)

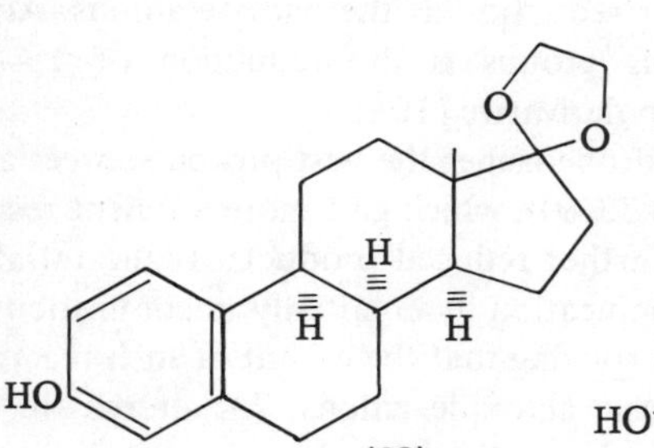

(38) 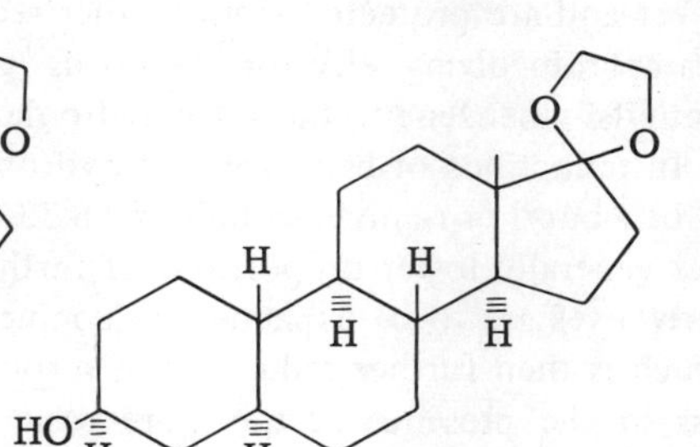 (39)

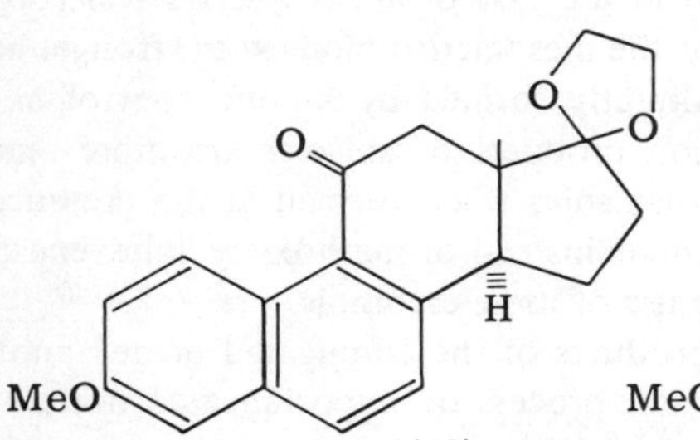

(40) 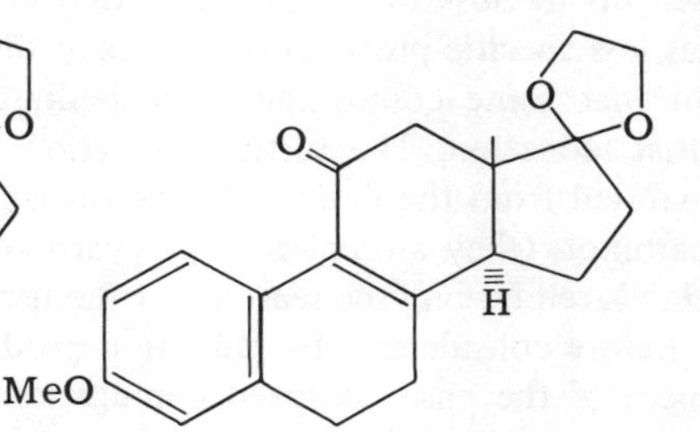 (41)

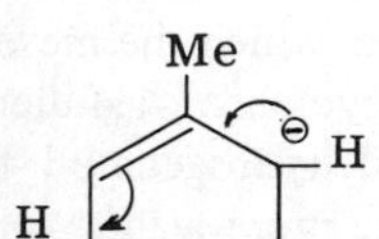

(42) 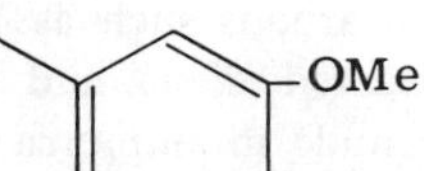 (43) 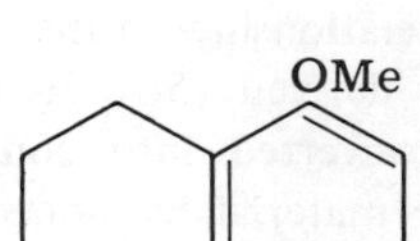 (44)

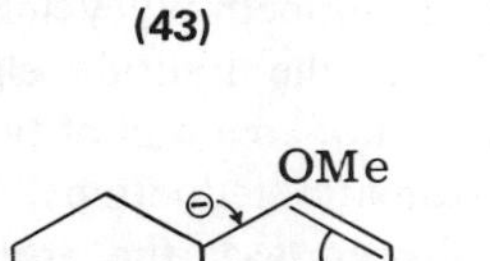

(45)

been intensively investigated, the situation in the polycyclic series seems to be different. 1-Acetylnaphthalene is reducible to 3,4-dihydro-1-acetylnaphthalene, whether as a kinetic product is not clear, and a similar reduction has been used in a total synthesis of estrone in the conversion of **40** into **41** (106).

4-Nitrobiphenyl gives the 1,4-dihydro derivative in small yield (107). It is apparent that when the unsaturated system is more readily reducible than an isolated benzene ring, such anion-stabilizing groups may exert the expected effect and are protected from further reduction as the enolate anions. Another example involving electron-stabilizing groups is the reduction of 1,4-bistrimethylsilylbenzene to the 1,4-dihydro derivative (108).

In reductions of benzenes to dihydrobenzenes the best proton sources appear to be *t*-butyl or *t*-amyl alcohols (5,16,33,60), which give more efficient reactions and generally lower proportions of further reduced products. If the tetrahydro derivatives are to be explained by conjugation in an initially unconjugated diene which is then further reduced, it is surprising that the extent of such reactions is less in the presence of the more basic alkoxide anions. An alternative, more acceptable explanation, based on the observation that the more acidic methanol gives up to 30% of further reduction in the case of anisole derivatives (60), is that less specific protonation occurs of the mesomeric anions with stronger acids and that some conjugated diene is directly formed by kinetic control in the initial reduction. The further reduction products of anisoles are more readily separated from the desired 2,5-dihydroanisoles when formed in the presence of *t*-carbinols (they are chiefly hydrocarbons instead of methoxycyclohexenes; see below), reinforcing the reasons for the use of these carbinols.

Before considering the reduction products of the conjugated dienes, another aspect of the base-catalyzed conjugation process of importance to note is the observed loss of hydride ion from anions of type **42** with regeneration of the aromatic starting material. The tendency to lose hydride seems to be in an inverse relationship to the "acidity" of the 1,4-diene from which the mesomeric anion is derived (56). Hydrocarbons such as 1-methylcyclohexa-1,4-diene are partly converted into conjugated diene and largely dehydrogenated to the aromatic material by potassamide in ammonia (56). The more acidic 1-methoxycyclohexa-1,4-dienes are dehydrogenated only after leaving for several hours and the even more acidic 1,3-dimethoxycyclohexa-1,4-dienes give very stable salts. As noted previously, if the hydride elimination is sufficiently rapid compared with the protonation of an anion of type **42** (which is almost certainly an intermediate in the aromatic reductions), the overall result will be the catalytic evolution of hydrogen gas, the aromatic material being recovered unchanged. In other words, overall reduction may be aborted and the extent of reversal could differ with the structure of the anion. The anion **27**, for example, is quite stable, probably more so than **28**, which could therefore undergo reversal to a greater extent. In cases where different anions result from protonation in two different *ortho* or *meta* positions in unsymmetrical systems, the final reduction product could therefore be decided not by the relative rates of initial protonation in the alternative positions but by the rates of reversal. In simple cases the efficiency of reduction is high and the effect is unlikely to be important, but it might also be an explanation of the difference in rates of reduction of **43** and **44**. The former is reducible in good yield to the 1,4-dihydro derivative with only a small excess of lithium and ethanol in ammonia, while the latter requires about 80 equivalents of metal with ethanol to produce a 40-50% yield of the 3,9-dihydro derivative, although the metal rapidly disappears and hydrogen is evolved. The intermediate anion **45** may be unstable, expelling a hydride ion to regenerate **44**, and leading to inefficient reduction.

Until the importance of some of the effects noted above can be properly assessed, arguments based on product structures must be treated with caution in discussing intermediates.

VIII. Reductions of Cyclohexa-1,3-dienes and Other Secondary Processes

Some further reduction products obtained from derivatives of cyclohexa-1,3-diene are relevant to the by-products of the aromatic reductions above and also to the products of the metal-amine reductions. Of particular interest are the 1-methoxycyclohexa-1,3-dienes (e.g., **13**) as shown. Such conjugated dienes on reduction with metals in ammonia give a mixture of the 1-methoxycyclohexene and the cyclohexene, the ratio varying with the conditions (89,110). In the absence of an alcohol or in the presence of *t*-butanol the hydrocarbon is the dominant product; in the presence of methanol or ethanol the methoxycyclohexene predominates. The difference in the results may be due either to rapid protonation of an anion radical with the more acidic alcohol, leading to a shorter life and lower opportunity to expel methoxide anion, or possibly to the production of a dianion under conditions of slow protonation, with greater tendency to lose the methoxide anion. Another possibility, the formation of an intermediate allyl ether (28), seems to be ruled out as the sole explanation by the nature of the hydrocarbon product from 1-methoxy-4-methylcyclohexa-1,3-diene (**46**). The former reaction sequence should lead to this product via the hydrocarbon (**47**), and in fact the mixture of 1- and 3-methylcyclohexene obtained is closely similar to that from **47** (110). The allylic intermediate (**48**) in the second sequence could give no 1-methylcyclohexene (observed) and should give some 4-methylcyclohexene (not observed). However, reductions of dialkylaminocyclohexa-1,3-dienes do give some of the 1-aminocyclohex-2-ene type as well as the predominant 1-aminocyclohex-1-ene type (98). The amines do not undergo hydrogenolysis, in accord with expectation, because of the more difficult elimination of $R_2N^{\ominus}$ compared with $MeO^{\ominus}$.

The use of lithium in different amines (methylamine, ethylamine, and *n*-propylamine) (31) at –7°C with toluene, isopropylbenzene, and *t*-butylbenzene leads to tetrahydro derivatives with some interesting solvent effects on the

OMe
e
-OMe
e
Me Me Me Me
(46) (47)
OMe
Me Me Me Me Me
(48)

products. The ratio of the 1-alkylcyclohexene to the others formed increases, and the rate of reduction decreases, with increasing molecular weight of the amine. Addition of secondary amines such as morpholine was found also to increase considerably the proportion of the 1-alkylcyclohexene; a small proportion of primary amine is required for reduction to take place at all. Up to 93% of dimethylamine (with methylamine) could be used to give 82% of 1-isopropylcyclohexene compared with 46% in pure methylamine. Isomerization of the monoolefins does not occur. Assuming (see below) that cyclohexa-1,3-dienes are intermediates, the results are probably to be explained either by the formation of different dienes, depending on the solvent, or of differing modes of reduction of the dienes according to the nature of the proton source (in this case the solvent). Further investigations of conditions on the nature of products of reduction of conjugated dienes seem desirable.

A further secondary factor which may have to be taken into consideration in connection with the nature of further reduction products is the action of strong bases such as amide anion on the conjugated cyclohexa-1,3-dienes, notably hydrocarbons. Although the hydrocarbon 1,4-dienes with potassamide in ammonia lose hydride ion to regenerate aromatic material and some 1,3-diene, the 1,3-dienes undergo a further disproportionation reaction: potassamide in ammonia converts cyclohexa-1,3-diene to benzene and cyclohexene (56). Similar reactions have been noted with potassium *t*-butoxide in dimethylsulfoxide with both cyclohexa-1,3-diene and 1,4-dihydronaphthalene (59,111). The disproportionation is probably due to a direct reaction of the carbanion on the diene, rather than to reduction by hydride ion. In cases where strongly basic anions are formed as the result of a reduction process, particularly of aromatic systems, both dehydrogenation and disproportionation reactions could take place. Although no evaluation of these factors has been made, it is possible that some of the divergent results reported, particularly with polycyclic hydrocarbons, may be due to the operation of both dehydrogenation and disproportionation on primary products. The dienes themselves, and particularly the anions, are also very sensitive to atmospheric oxidation to aromatic material. Reoxidation by air of 1,4-dihydro derivatives to aromatic rings takes place readily and in cases where more than one ring has been reduced, rearomatization of one of them could occur preferentially. Examination by thin-layer chromatography or gas chromatography, or extended exposure to air during attempts at crystallization, may lead to such oxidations and thus to spurious products.

IX. Reductions of Polycyclic Aromatic Compounds

The reductions of polycyclic compounds are sometimes quite complex, the products depending on the exact conditions and the amount of reducing agent used. Owing to the relative insolubility of the compounds, particularly hydrocarbons, in ammonia and the often increased solubility of the reduction products, the ratio of metal to compound in *solution* is difficult or impossible to control. Even naphthalene which on complete reduction with metal and alcohol in ammonia produces 1,4,5,8-tetrahydronaphthalene (103,112) cannot be stopped at the 1,4-dihydro stage with any degree of efficiency, despite the slower intrinsic rate of reduction of the dihydro compound. The combination of lithium and *t*-butanol in amines might overcome solubility difficulties in some instances.

A method of avoiding reduction of anthracene, phenanthrene, and similar hydrocarbons beyond a dihydro stage is to add ferric chloride (27). The iron produced removes the reducing agent, by catalysis, in successful competition with later reduction stages, but not with the initial very rapid reaction.

Polycyclic compounds give products consistent with the original assumption that initial protonation occurs at the positions of highest charge density in the anions. The complexity of the products is increased not only by the factor above, but also by what seems to be the initial production of kinetically controlled conjugated isomers in some cases, and by the greater ease of base-catalyzed migration of double bonds, since hydrocarbon "acids" involved may be both benzylic and allylic. The nature of the products obtained depends upon the exact experimental conditions including the factors: (*1*) whether the substance is soluble and whether co-solvents are employed; (*2*) whether the process involves addition of metal (dianion) followed by protonation or simultaneous reaction (protonation of anion radical?); (*3*) whether ammonia is at the boiling point or lower temperatures, affecting solubility, possible protonation of dianions at the boiling point, and the extent of secondary migration processes; (*4*) the nature of the alcohol (selectivity in protonation) and the basicity of the resulting alkoxide anion (rates of bond migrations); (*5*) the period of time involved and whether the mixture is left after metal is all consumed (dehydrogenation and disproportionation); (*6*) whether air is rigorously excluded during both reduction and work-up; and (*7*) whether the substance is fully in solution or so finely divided that it dissolves rapidly. Few reductions have been investigated with all of these points in mind.

Since aromatic rings act as charge-stabilizing substituents on other aromatic rings, some of the results are of considerable theoretical as well as experimental interest. Let us consider the reductions of some biphenyls. Biphenyl itself would on this basis be expected to give 1,4-dihydrobiphenyl. A dianion prepared at −70 to −75°C indeed gives 1,4-dihydrobiphenyl (113). Under varying experimental conditions (114,115) varying results are obtained; the products include 1,4-dihydro and 3,4-dihydro derivatives and under further reducing conditions tetra and hexahydro derivatives. Under apparently the same conditions materially different results can be obtained (2,112), indicating a great sensitivity to exact experimental conditions. The factors chiefly involved in this case are probably the rapidity with which the reduction is performed and the method of work-up in excluding or permitting conjugation processes.

Some biphenyls substituted with hydroxyl and methoxy groups have been studied (98). Hydroxyl has the expected effect of inhibiting reduction of the attached ring, since it forms an alkoxide ion; 4-hydroxybiphenyl gives a mixture of the 2′,5′-dihydro, the 2′,3′,4′,5′-tetrahydro, and the 1′,2′,3′,5′,6′-hexahydro derivatives. When present in a 9,10-dihydrophenanthrene system (116) or a fluorene ring system (117), results are much more clear-cut; the product (**49**) is obtained in good yield. The substituent acts in the normal manner for an electron-donating one. 4-Hydroxy-4′-methoxybiphenyl gives principally the 2′,5′-derivative (**50**) without oxygen removal.

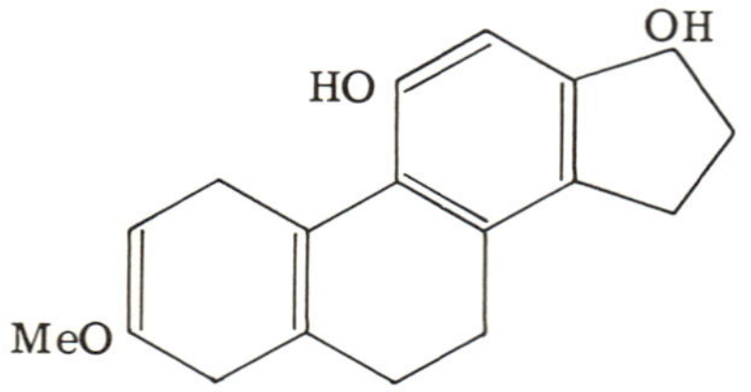

OH HO MeO

(**49**)

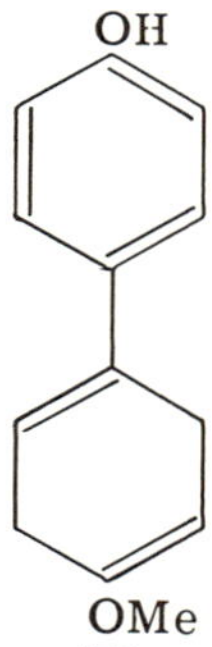

OH OMe

(**50**)

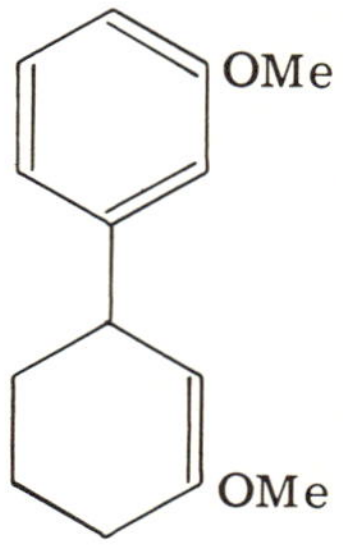

OMe OMe

(**51**)

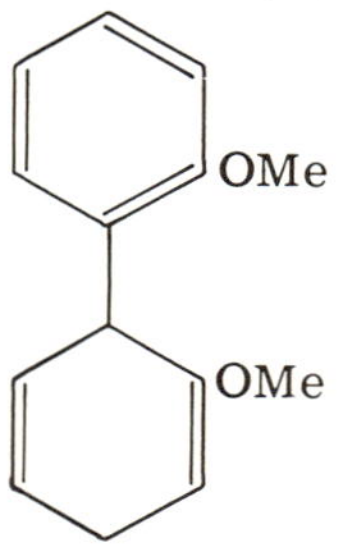

OMe OMe

(**52**)

4-Methoxybiphenyl loses the methoxy group completely, giving a mixture of products similar to that obtained from biphenyl. 2-Methoxybiphenyl gives a mixture resulting from about an equal proportion of reduction in either ring with about 30% consisting of 1,4-dihydro-2-methoxybiphenyl. 3-Methoxybiphenyl gives products corresponding to preponderant reduction in the substituted ring, but the major one (45%) is a tetrahydro derivative, hydrolyzed to 3-phenylcyclohexanone. 3,3′-Dimethoxy- and 2,2′-dimethoxybiphenyl show a more clear-cut but differing picture. The main product from the former is **51** and from the latter **52**. The reason for production of the tetrahydro (**51**) and the dihydro derivatives (**52**) under apparently closely similar conditions is not clear,

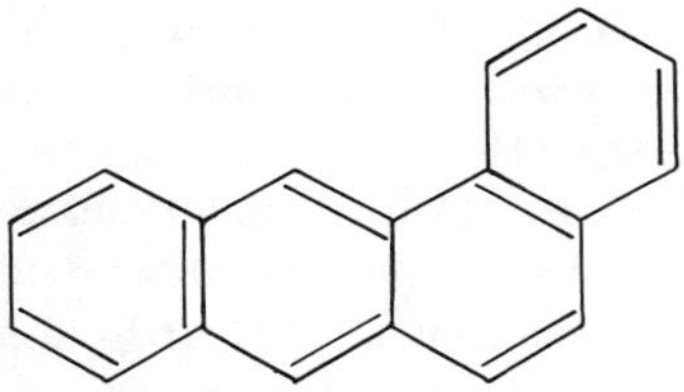

(53)

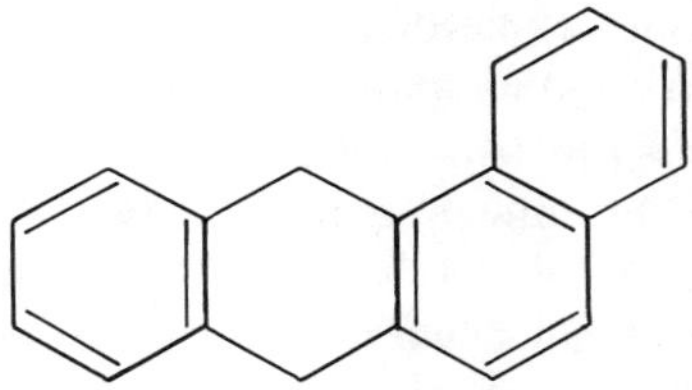

(54)

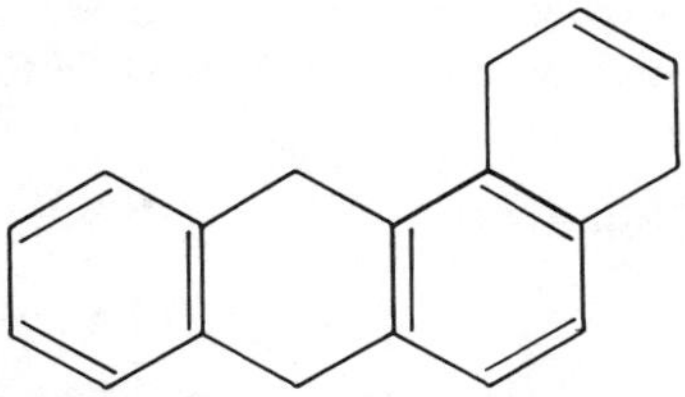

(55)

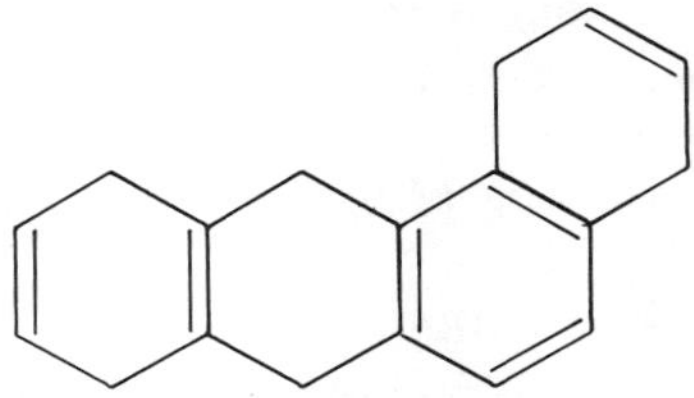

(56)

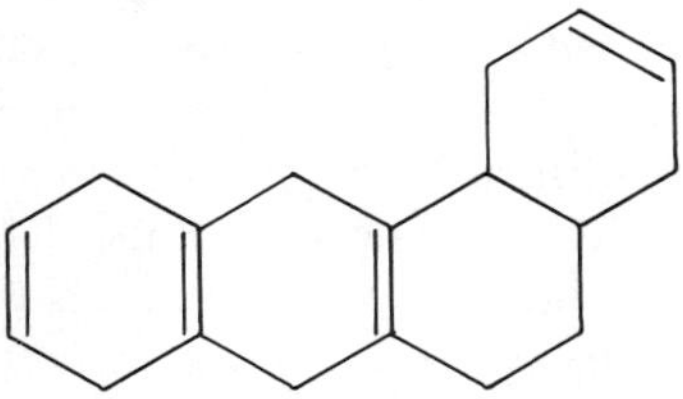

(57)

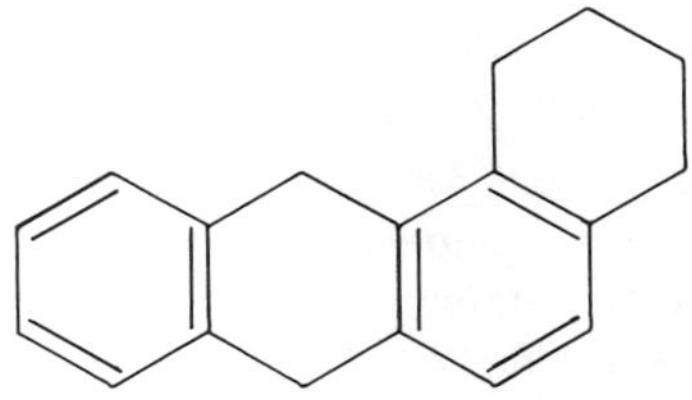

(58)

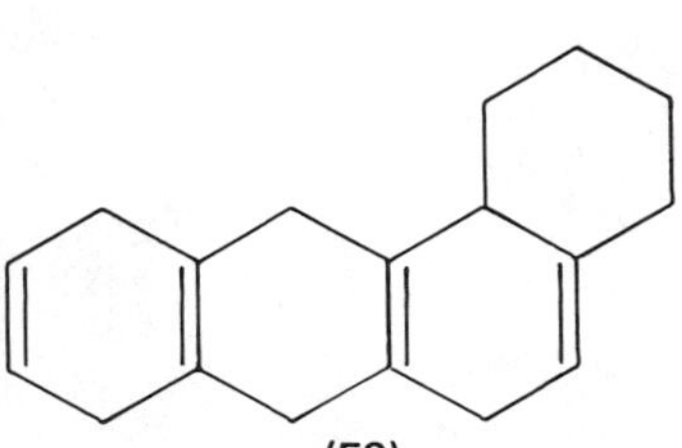

(59)

O

Me

(60)

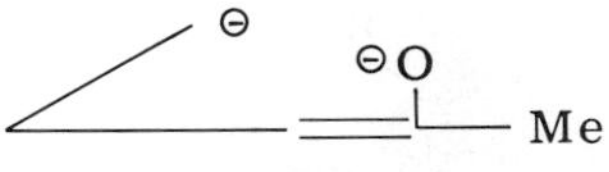

⊖

⊖O

Me

(61)

but **51** appears to be a kinetically controlled product. 4,4′-Dimethoxybiphenyl gives a mixture including hydrocarbon products. It is notable that 2-methoxybenzoic acid produces the 1,4-dihydro but 3-methoxybenzoic acid the tetrahydro derivative. The analogy between the charge-stabilizing benzene ring and the carboxyl group has been noted previously.

From these results it seems clear that a substituent on one ring can also affect the reduction, including the protonation stage, of the other ring.

N-Arylpyrroles are reducible in the benzene ring to give 2,5-dihydro derivatives; in other words, the system does not behave like a biphenyl (118).

A particularly interesting series of reduction products was observed with benz[*a*]anthracene **(53)** (42). With lithium in ammonia in the presence of ferric chloride the anthracene portion is reduced giving **54**. This on reaction with lithium and ethanol in ammonia gives first **55**, with the less substituted ring of the naphthalene system attacked, and then **56**. The last is obtainable in 80% yield by direct reduction of **53** with sufficient lithium and *t*-butanol in ammonia. Reduction of **56** with lithium and *t*-butanol in methylamine produces **57** containing two more than the expected number of hydrogens. The reduction of **58** with lithium and *t*-butanol in methylamine gives a mixed product, the principal constituent of which is **59**. Reduction of the tetrasubstituted aromatic ring of **56** and **58** in methylamine was unexpected and was not observed in ammonia. The method clearly merits serious consideration in other difficult cases.

X. Reductive Fissions

A. CARBON-CARBON BONDS

These can be broken under circumstances where the anions formed are particularly stable or the starting material is particularly strained. Examples are the reduction of tetraphenylethylene to diphenylmethane, the tetraphenylethane intermediate being cleaved (119). The reduction of cyclopropylmethylketone **(60)** gives pentan-2-ol and -2-one (120). Cyclopropanes are themselves stable and can be generated from dihalocarbene adducts of olefins by means of lithium in ammonia. The reduction of **60** must therefore depend on the stability of the enolate anion in **61**. An interesting carbon-carbon cleavage occurs in the cyclophane **(62)** with lithium in ammonia, probably because of the strain in the system (121).

Some other cleavages, such as the conversion of acetophenone pinacol to the ketyl radical, are probably not direct reductions, but depend on the base present.

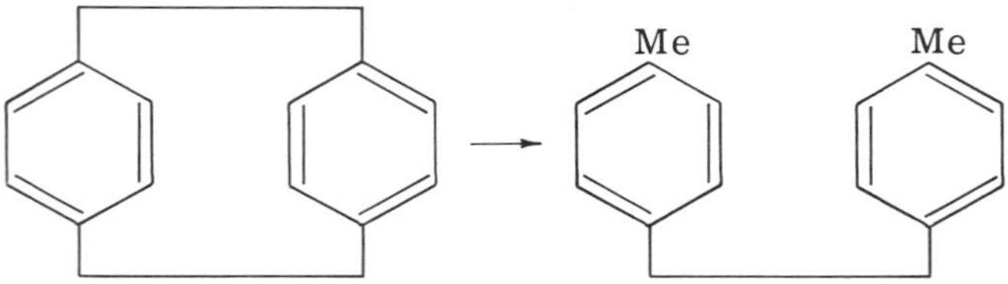

Me Me

(62)

B. CARBON-OXYGEN, CARBON-SULFUR, AND CARBON-HALOGEN BONDS

Saturated alcohols form salts by reaction with the metals at a rate increasing with their acidities; saturated ethers are unaffected. Allylic or benzylic alcohols, ethers, or esters are frequently cleaved, as are many phenol ethers. The basic theory has already been noted.

Allyl alcohols, but not allyloxide anions, may be cleaved with varying degrees of ease depending on substitution. It was pointed out earlier (56) that ready hydrogenolysis of allylic alcohols could only be achieved if the intermediate mesomeric anion contains a CH_2 or CHAr group at one or the other end (not necessarily at the point where the hydroxyl group was situated), and that this could be correlated with the "acidity" of the conjugate hydrocarbon acid (11). For example, geraniol **(63)** (R = OH) and linalool **(63a)** generate what is structurally the same hydrocarbon **(63)** (R = H), through the mesomeric anion shown, whereas carveol **(64)** (R = OH) is unchanged. Reduction of aucubin **(65)** occurs in two stages (122); the hydroxymethyl group is reduced first, followed by the cyclopentenol. By conversion of the hydroxyl group into a derivative which gives a more stable anion, for example, an acetate, reduction is greatly facilitated and in all cases hydrogenolysis occurs (123). Competition with ester reduction occurs in some cases, but may be minimized by the use of lithium. The original double bond in such cases retains its stereochemistry (see below). Reduction of (+)-carveol methyl ether **(64)** (R = OCH_3) gives the (±)-hydrocarbon **(64)** (R = H), thus confirming the mesomeric nature of the anion (124).

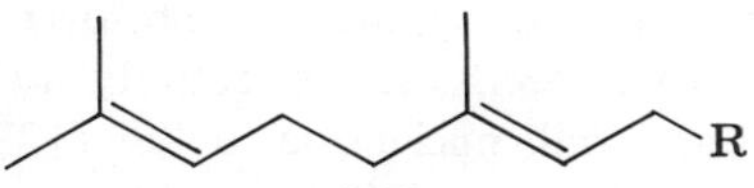

R

(63)

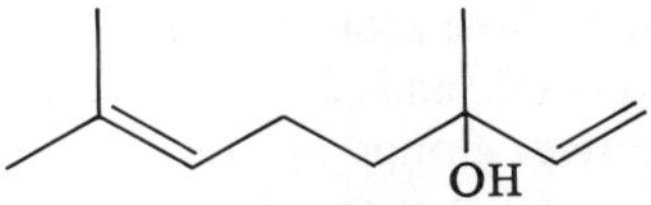

OH

(63a)

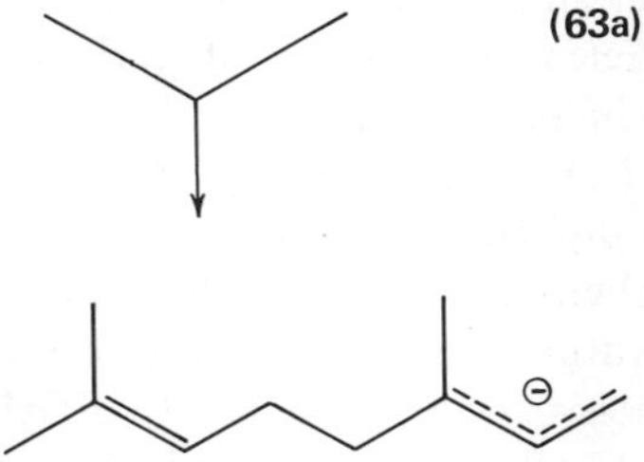

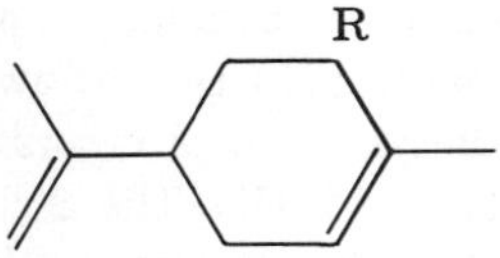

R

(64)

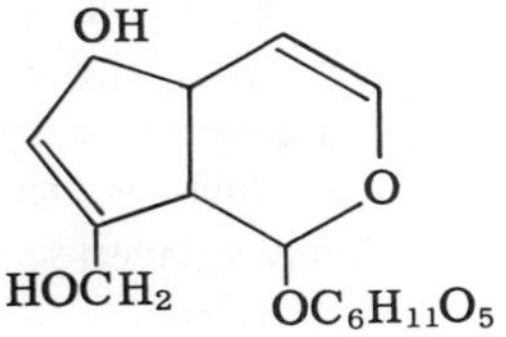

⊖

OH

HOCH₂ → $HOCH_2$ $OC_6H_{11}O_5$

(65)

Benzyl alcohol is reducible, initially to toluene, but *p*-methoxybenzyl alcohols undergo minimal hydrogenolysis under appropriate conditions and the ring is reduced instead (125). *Ortho*-OMe has no effect in inhibiting hydrogenolysis, illustrating the frequent difference between *o*- and *p*-OMe in this type of process. 4-Methylphenylmethylcarbinol can be made to undergo nuclear reduction in the presence of *t*-butanol by preliminary formation of the alkoxide, hydrogenolysis being inhibited by the charge on the oxygen (126). The final product is the 2,5-dihydro derivative of the carbinol. For the same reason, lithium in ethylamine on benzyl alcohol yields the tetrahydrobenzyl alcohol, not the hydrocarbon, conjugation and further reduction of the diene occurring under the basic conditions (127).

Saturated ethers are stable and are frequently used as co-solvents in reductions. Allylic and benzylic ethers are normally readily cleaved.

Vinyl ethers such as the 1-methoxycyclohexadienes discussed above are frequently cleaved if terminal to a polyene system, but the process clearly depends on initial addition of electrons to the unsaturated system. If cleavage is desired, no alcohol, or a nonacidic one such as *t*-butanol, should be present.

Epoxides are cleaved; for example, propylene oxide gives solely isopropanol (125). The reagent of choice appears to be lithium in ethylamine (128) which leads in the case of steroidal epoxides to the stereospecific production of axial alcohols. Sterically hindered epoxides are attacked which are unaffected by lithium aluminium hydride.

Sulfides are normally reducible, even in the saturated series; the cleavage of benzylthiol ethers has been extensively used in the protection of thiols and removal of the protecting benzyl group in peptide synthesis (see below).

O-Benzyl and *S*-benzyl cleavages have been used effectively to remove protecting groups in the carbohydrate and nucleoside fields (129). It is noteworthy that an adenine ring survived the treatment. Benzylidene groups are also cleaved. Selective cleavage of carbon-sulfur rather than carbon-oxygen bonds can be achieved (130).

A number of aryl ethers undergo fission. The pioneer work of Sartoretto and Sowa (131) on diphenyl ethers led to some important applications, notably the cleavage of bisbenzylisoquinoline alkaloids by Tomita (18) (see below). The effects of substitution were studied (131) using unsymmetrical ethers ROR′, with examination of the products RH, ROH, R′H, and R′OH. The aryl group most capable of stabilizing an anionic charge on the ring appears as RH (89). *Ortho*-OMe (stabilizing) again contrasts with *p*-OMe (destabilizing). Cleavage is inhibited by salt formation of phenolic groups as expected. It is possible to predict on empirical grounds where fission will occur in a given structure. What little evidence is available suggests that the results in such cases are not materially altered by the presence or absence of a proton source.

In the absence of proton sources such as alcohols, alkyl ethers of phenols, particularly methyl ethers, are slowly converted into phenols, particularly by potassium in ammonia (132, 133), by lithium in ethylamine (127), and by calicum hexammoniate (134). This slow reaction contrasts with the rapid nuclear reduction in the presence of alcohols. Some caution has to be exercised in interpretations of results, since strong bases such as potassamide in ammonia can form phenols from higher alkyl ethers by elimination rather than by reduction processes.

1,2-Dimethoxybenzenes undergo notably facile cleavage with potassium in ammonia to 1-hydroxy-2-methoxybenzenes, and 1-isopropoxy-2-methoxybenzene gives exclusively 1-hydroxy-2-isopropoxybenzene, the inverse cleavage to that produced by Lewis acids. Methylenedioxybenzenes (and catechol acetals) undergo a facile reduction of oxygen from the ring, for example, **66** (R = H) to **67** (R = H) (133). The effects of substitution on the ring are again those to be expected of anion formation, for example, **66** (R = Me) gives predominantly **67** (R = Me) in contrast to **66** (R = CO_2H) which forms the *meta*-substituted acid.

A difficult reaction for which there are few general processes is the removal of oxygen from a phenolic ring: this occurs by reaction of arylphosphates with lithium in ammonia (135). Enol diethylphosphates similarly give the corresponding olefin: the enol diethylphosphate from dihydrotestosterone is thus converted into androst-2-en-17β-ol (136). An interesting synthetic use is the conversion of **68** into **69** by lithium-ammonia reduction to the correct enolate anion, reaction with $(EtO)_2POCl$ and further lithium reduction (137).

Cleavage of carbon-halogen bonds frequently occurs readily, but in order to obtain a high yield of the reduction product, uncontaminated by elimination and dimerization products, the solutions should contain alcohol, and the metal be added in excess (138). Vinyl halides are reduced stereospecifically (139).

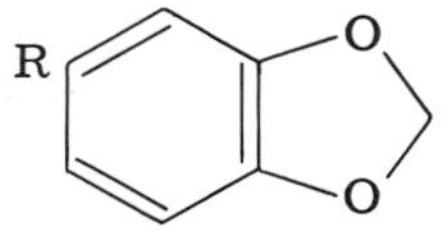

R O O
(66)

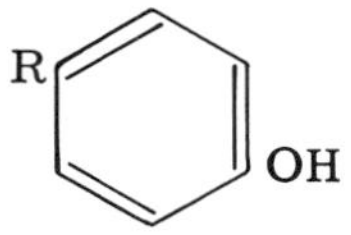

R OH
(67)

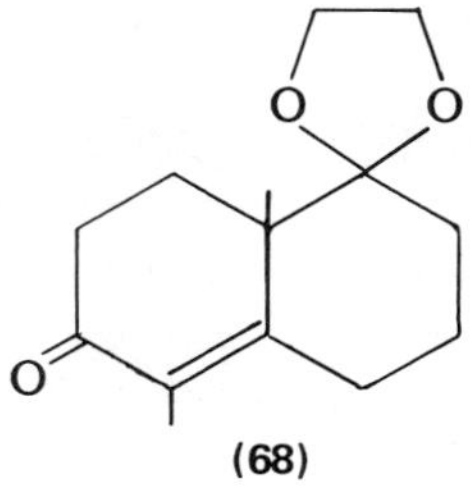

O O O
(68)

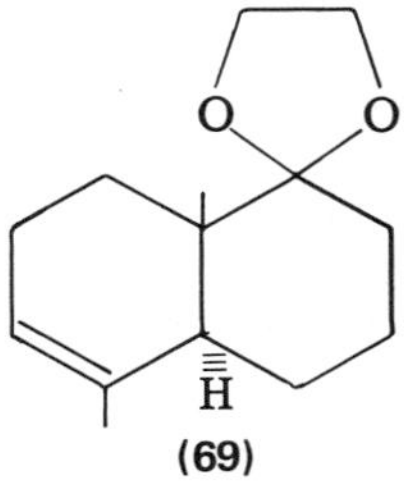

O O H
(69)

C. CARBON-NITROGEN BONDS

Benzyl and allylamines do not experience cleavage, but the corresponding quaternary salts, including also arylammonium salts, undergo Emde-type fissions and the process is frequently experimentally superior to the standard method in avoiding competing base-catalyzed reactions (140). The presence of traces of water have been shown to affect the course of such reactions (141), probably by buffering, ensuring clean fission at an aryl-nitrogen bond in the case of arylammonium salts. In its absence a proportion of aromatic amine results. Similar reductions have been carried out electrolytically in ammonia (142).

XI. Reductive Cyclizations

Several types of cyclization depend on the reductive formation of anions or radicals. Reductions of esters in ammonia follow the usual course in the presence of alcohols, except for highly hindered esters when acids result (143). In the absence of alcohols, dimeric products often result. A practical application is the conversion of diesters into cyclic acyloins in good yield, for example, **70** → **71** (96%) (144). Other examples, probably involving anions, are the conversions of **72** into **73** (145), of **74** into **75** containing the bridged-ring system of gibberellic acid (146), and of **76** into **77** (147). A somewhat different type of example is the conversion of **78** into **79** (148). A key step in the total synthesis of colchicine is the formation of a seven-membered ring by an acyloin reduction involving a lactone and an ester (149).

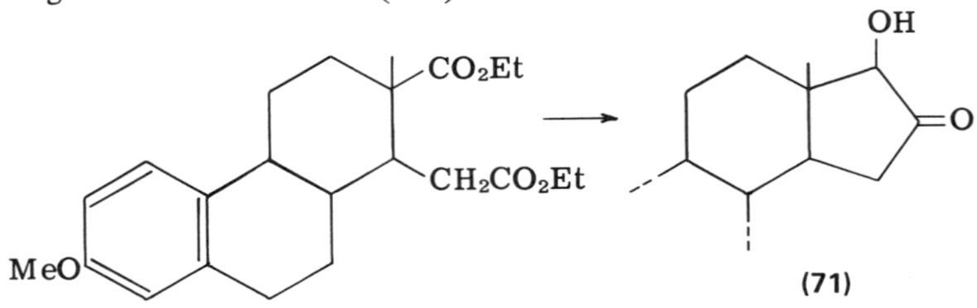

MeO, CO_2Et, CH_2CO_2Et → OH, O

(70) **(71)**

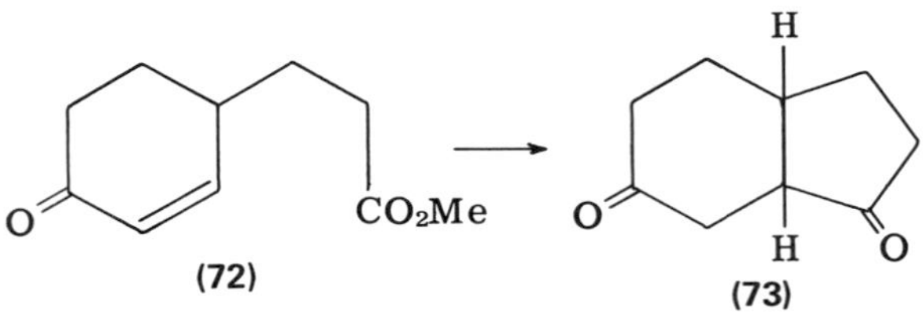

O, CO_2Me → H, O, H, O

(72) **(73)**

MeO, O, $CH_2C{\equiv}CH$ → MeO, OH

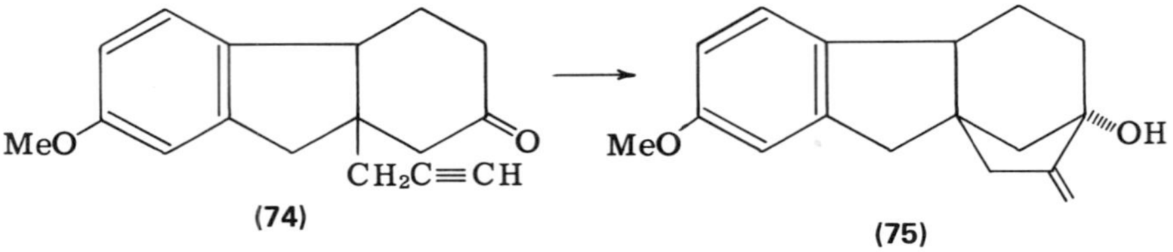

(74) **(75)**

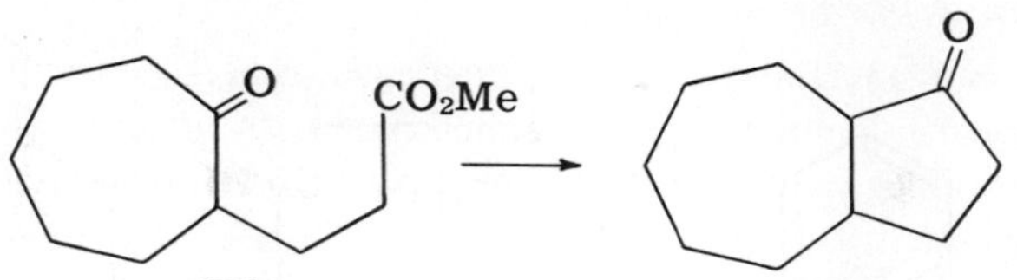

CO$_2$Me

(76) → (77)

CH$_2$OTs

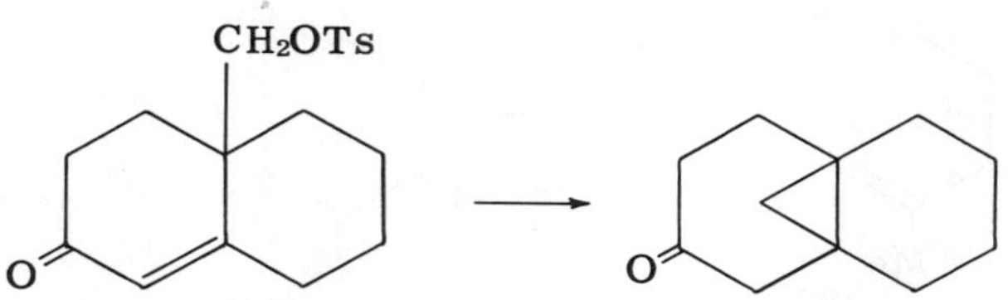

(78) → (79)

XII. Stereochemistry

As a consequence of the mechanistic considerations discussed above, a large measure of stereospecificity is to be expected. The intermediates as anions or anion radicals should possess the ability to invert at the charged centers, and should therefore be capable of adopting the sterically most stable configuration and conformation under the reaction conditions. The products obtained from them may or may not correspond to these conformations. The conformations of the charged intermediates are usually closely similar to those predicated for the protonated products and therefore a simple rule would be that this type of reduction would lead to the thermodynamically most stable *product.* Such a generalization was made (8) but turns out to be an oversimplification. Quite early we had considered this possibility and had attempted to use the above principle elaborated in steroid synthesis (7), only to find that in the cases we examined a mixture of *cis* and *trans* isomers was obtained in a ratio clearly different from that expected on the basis of thermodynamic stability considerations (see also 150). Moreover, reduction of **80** gives **81**, although the *trans* isomer is demonstrably the more stable. With steric mobility existing between the conformations and configurations of the anion, the result depends on the relative rates of protonation of the different forms as well as on the predominant form. The preceding generalization is still a first approximation which covers a large proportion of cases; results are now largely predictable using more refined considerations (12,13).

A. *SATURATED KETONES*

Reduction of ketones either by metal addition followed by protonation, or by simultaneous protonation and reduction, gives yields of the stable stereoisomer (94-99%) higher than the equilibrium ratio of the isomers in the case of 2-, 3-, or 4-alkylcyclohexanones (151). Norcamphor, in contrast, gives the

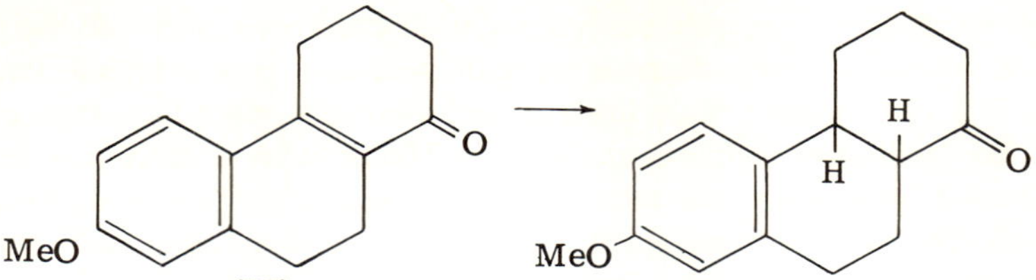

(80) (81)

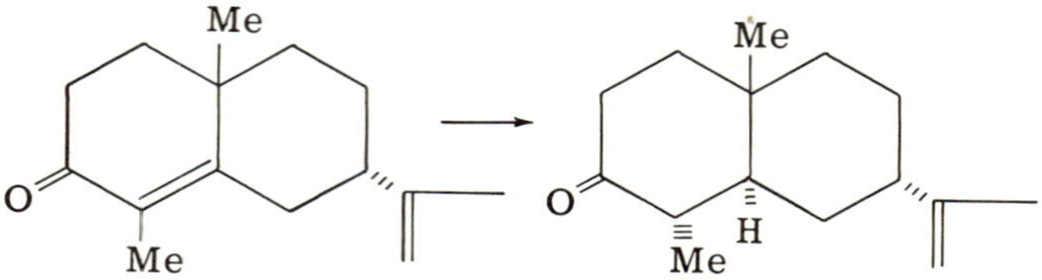

(82) (83)

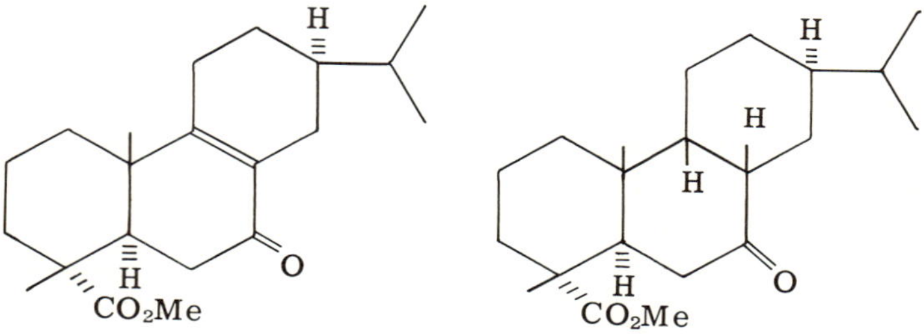

(84) (85)

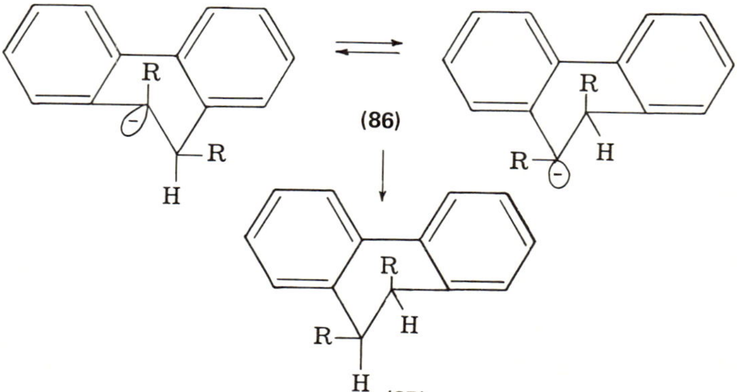

(86)

(87)

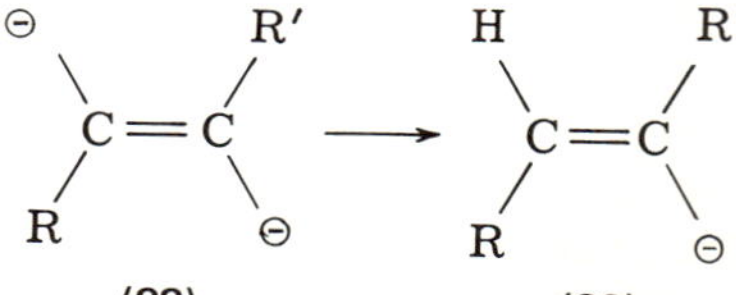

(88) (89)

thermodynamically unstable alcohol under all conditions, although the proportion of possible isomers depends on both metal and proton source: Potassium gives less *exo* (stable) than lithium, and *t*-butanol more *exo* than ethanol. Camphor itself gives mainly borneol (*exo*). The addition of metal first, followed by alcohol, gives more *exo* then does simultaneous reaction. Pinacols are also formed. The suggested mechanism is the formation of an anion radical which is protonated from the less-hindered side only when hindrance is marked (*exo* attack for norcamphor, but axial attack for simple cyclohexanones). If a dianion is formed this would be expected to be more reactive and therefore unspecific in its direction of protonation and the product would then correspond more to the equilibrium forms of the anion and hence roughly to that of the products. The ratio of anion radical to dianion under different conditions could therefore determine the result.

B. α, β-UNSATURATED KETONES

The reactivity of the anion involves a stereo-electronic factor, maximizing orbital overlap with the addendum, and in some cases possibly a steric approach factor of the protonating reagent. In the reduction of an α,β-unsaturated ketone, the more basic β position is protonated first, leaving a more stable enolate anion. This step has been discussed (152) and several crucial examples examined, including the reduction of **82**. Stork is of the opinion that the proton should add axially at the ring junction because of the overlap requirements of the orbitals, and that even if the *cis* isomer were the more stable, the *trans* isomer should result. It was demonstrated that this is true of **82** which gives **83** although in this example the latter is the less stable isomer formed from the less stable carbanion.

More recently the assumption of tetrahedral symmetry of the β carbon has been questioned (9) and the results of reduction of 1(9)-octalone-2 explained by assumption of a trigonal β carbon. The results of reductions of 3,4-dialkylcyclohex-2-enones (10b) appear to show that the β carbon in the transition state can vary between the limits of trigonal (leading to the *cis* isomer) and tetrahedral (leading to the *trans* isomer). The 3,4-dimethyl derivative gives 16% *cis;* the 3,4-diethyl, 56.5% *cis;* and the 3,4-diphenyl, 98% *cis.* The last case differs materially in that the 3-phenyl group delocalizes the charge, and the resulting geometry probably forces the other phenyl group into an axial orientation.

A further interesting example (10a) is the conversion of **84** into the *cis* C-D compound (**85**) which appears to arise from protonation of the most stable carbanion, rather than by maintenance of orbital overlap. The difference compared with steroids is probably due to the lack of the D ring of the abietanes which permits them to assume other conformations. The authors (10a) discuss further the factors involved in determining configurations.

Reduction with lithium in deuterioammonia incorporates deuterium stereospecifically into α,β-unsaturated ketones of the steroid series (153).

C. *PHENANTHRENES AND ANTHRACENES*

9,10-Dialkylphenanthrenes undergo stereospecific *cis* reduction at the 9,10 positions (154), and the 9,10-dialkylanthracenes give solely the *trans*-9,10-dihydro isomers (155). In both series the favored isomer contains one substituent axial and the other equatorial. The result is probably due to protonation (quasiaxial) with equilibration of the monoanion and further protonation (quasiaxial): for example, **86** → **87**.

D. *ACETYLENES*

The reduction of open-chain 1,2-disubstituted acetylenes gives the *trans* isomer in virtually quantitative yield. This is a far higher proportion than the equilibrium ratio of the double bond produced (6). The explanation seems to be the formation of a dianion (**88**) as an intermediate, with the charges *trans* disposed because of mutual repulsion. Protonation may occur in two stages, but the monoanion (**89**) must be sterically stable (compare the vinyl halides above).

The reduction of cyclic acetylenes is of considerable interest (156). With medium rings, for example, cyclodecyne, and sodium in ammonia, reaction is initially slow and the ratio of *trans:cis* olefin is 1:1. Reduction rapidly accelerates as sodamide accumulates, due to isomerization into cyclodeca-1,2-diene and its rapid reduction, and the ratio of *trans:cis* changes to 1:47. With lithium, due to the lower basicity of $LiNH_2$, the isomerization reaction is much slower and the process gives a *trans:cis* ratio of 10:1. In the presence of ethanol the *trans:cis* ratio is 9:1. The large-ring acetylene cyclotetradecyne gives the *trans* isomer only, but reduction is very slow compared with open-chain analogs. Medium-ring acetylenes are therefore difficult to reduce and apparently give some *cis* isomer directly, particularly in the presence of ethanol, but most of the *cis* isomer appears to be formed from the 1,2-ene produced by isomerization. A *trans* dianion is clearly formed with difficulty in such systems; the *cis* olefin could arise in the ethanol case through an anion radical rather than a dianion. The transoid dianions are not only strained but solvation of the charges would be inhibited. Di-*t*-butylacetylene is also resistant to reduction by sodium in ammonia (157), presumably for solvation reasons.

E. *RETENTION OF CONFIGURATION OF ALLYLIC ANIONS*

In the reduction of allylic alcohols and derivatives thereof, mesomeric anions are generated which are protonated at both positions, although predominantly at the less alkylated (or more arylated one) in line with our earlier views (56) on the importance of charge density or charge localization. Two products are therefore usually obtained, one with a double bond in the original position. The steric configuration of this is found to be unchanged (123,158), that is, retention occurs in this position of the mesomeric anion. The new double bond is found in both *cis* and *trans* configurations, the latter predominating (123). In the reductions of open-chain dienes and of allenes, the mesomeric allylic anions which are intermediates are mixtures, but predominantly *trans* (e.g., Ref. 77).

F. STERIC HINDRANCE

A useful feature of metal-ammonia reductions, although they are clearly subject to steric effects, is the small degree of hindrance by substituents of reduction of appropriately activated double bonds; thus highly arylated styrenes which are difficult to hydrogenate catalytically are readily reducible.

XIII. Examples of Use of the Reagents

A. DETERMINATIONS OF STRUCTURE AND STEREOCHEMISTRY

Simplifications of structures to provide evidence about basic skeletons or stereochemistry can sometimes be achieved, particularly as the result of cleavage reactions.

1. Aucubin (122)

A problem in the structure of aucubin (**65**) was the instability of the aglycone, and the difficulty in consequence of obtaining structural information about it. Reduction of aucubin, or better still of its hexaacetyl derivative, with lithium and alcohol in ammonia gave a good yield of **90** which was hydrolyzable to a product, **91**, the structure of which could be ascertained. The structure was independently derived by other methods (159) which, however, were more complex. A clue to the skeleton of catalposide was provided by reduction to the same deoxyaucubin (160).

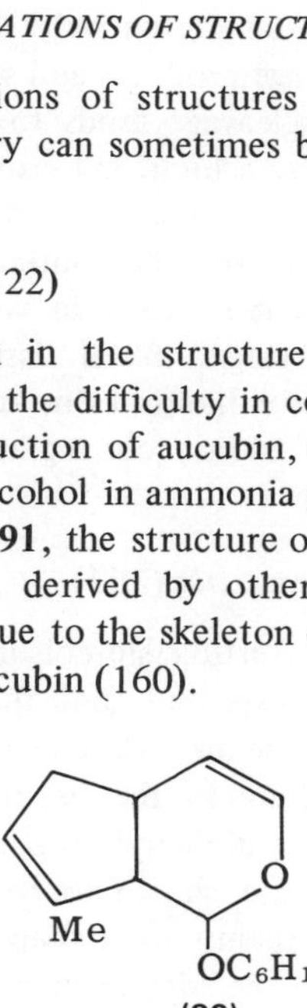

Me O $OC_6H_{11}O_5$

(90)

CHO CHO Me

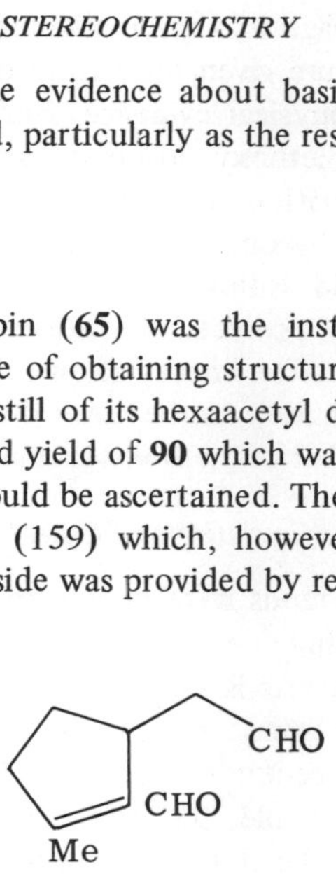

(91)

2. Mycelianamide (161)

The ether side chain of mycelianamide (**92**) was not only shown to be a geranyl group by cleavage to give "methylgeraniolene," but the steric configuration of the double bond adjacent to the ether group is demonstrated to be *trans* since this stereochemistry is retained in the hydrocarbon product (123,162). The side chains of bergamottin and of umbelliprinin were similarly shown to be *trans*-geranyl ethers (162). Acid cleavage of the side chain of mycelianamide had previously given misleading information because of secondary cyclizations.

3. Bisbenzylisoquinoline Alkaloids (18)

These consist of two similar benzylisoquinoline units linked by ether groups of the diphenyl ether type. By using the results of Sartoretto and Sowa (131) on

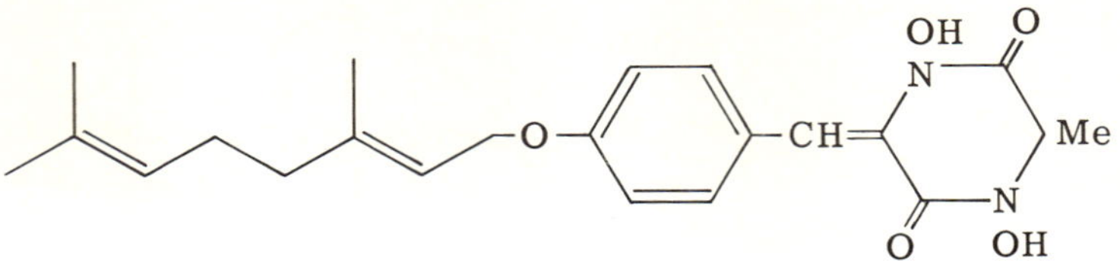

(92)

the fissions of substituted diphenyl ethers it is usually possible to predict the course of cleavage in a given alkaloid structure, or alternatively to decide on a probable structure given the nature of the cleavage products and supplementary chemical and physical evidence. For example, cleavage tends to occur readily adjacent to a methoxy group and is inhibited by a hydroxyl group because of phenoxide formation. The absolute stereochemistry of the fission products can be determined by observing their optical activity, and the configuration of the dimeric alkaloid follows. An example is berbamine **(93)** in which the free phenolic group protects the adjacent ether bridge, initial fission giving **94** (R = H) (163). After methylation to remove the inhibition due to the phenolic group, **94** (R = Me) can be cleaved as shown, fission occurring adjacent to the maximum number of methoxy groups.

4. Absolute Configurations of Catechin and Epicatechin (164)

These compounds were known to be interconvertible stereoisomers [e.g., the pair (+)-catechin-(+)-epicatechin] and it was suspected that the asymmetric center which inverts is that of the cyclic ether. The methylated (+)-catechin and (–)-epicatechin were reduced in order to eliminate the asymmetry of the benzylic ether center by cleavage. The resulting diphenyl-propanol-type compounds, for example, **95** from (+)-catechin, were shown to be enantiomeric (164). That is, the 3-hydroxyl groups in (+)-catechin and (+)-epicatechin have the *same* absolute configuration, which can be determined by standard methods. The result therefore led to the absolute configurations **96** and **97**, respectively. In this instance a possible alternative cleavage using lithium aluminum hydride and aluminium chloride failed because of a carbonium ion rearrangement (165). Similar benzyl-ether cleavages have been used in the determination of lignan structure and stereochemistry (166). One benzyl ether-type linkage in neogmelinol **(98)** failed to reduce in this way, giving rise to misleading conclusions as to structure (167). The method therefore has to be applied with some caution.

5. Galcatin (168)

The lignan galcatin **(99)** could have had the methylenedioxy and dimethoxy groups in the alternative positions, so far as the initial chemical evidence was concerned. Cleavage of the methylenedioxy group with sodium in ammonia gave

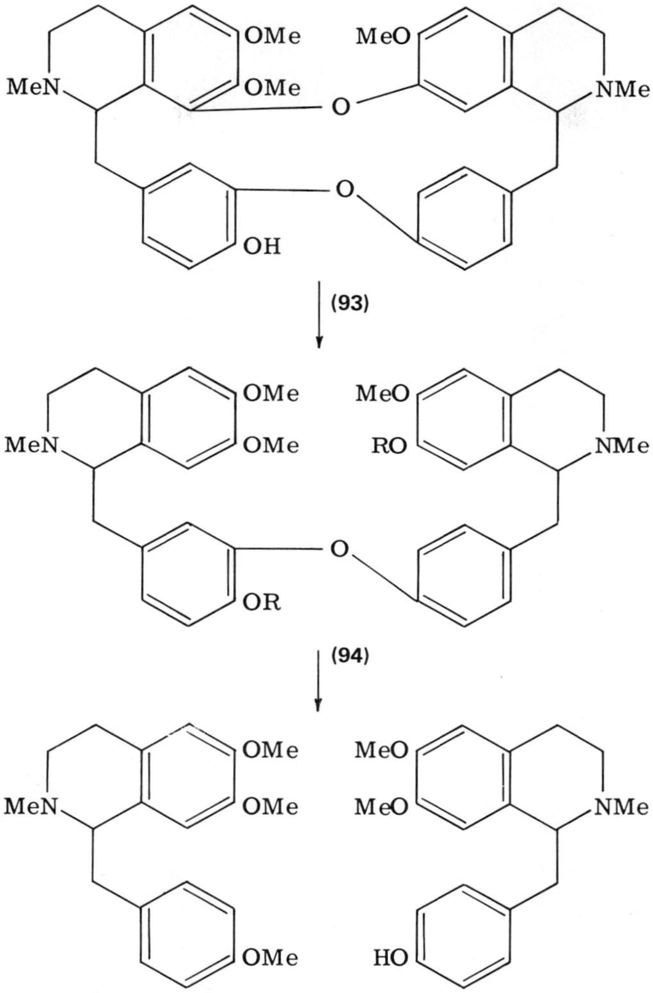

OMe MeO
MeN NMe
OMe
O
O
OH

(93)

OMe MeO
MeN NMe
OMe RO
O
OR

(94)

OMe MeO
MeN NMe
OMe MeO
OMe HO

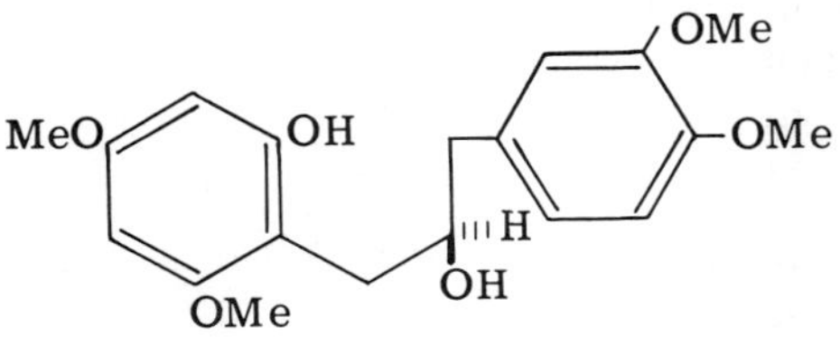

OMe
MeO OH
OMe
H
OH
OMe

(95)

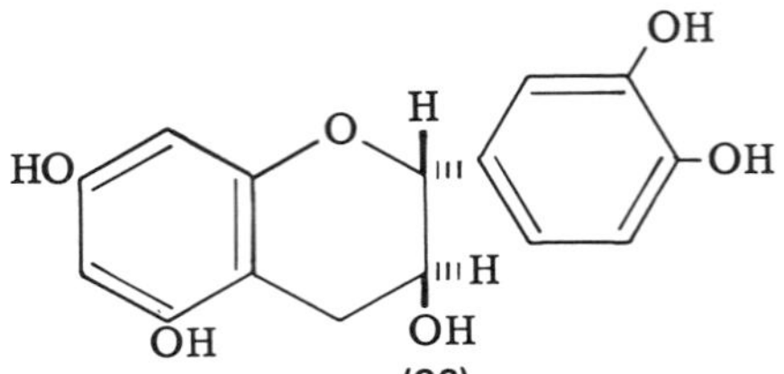

(96)

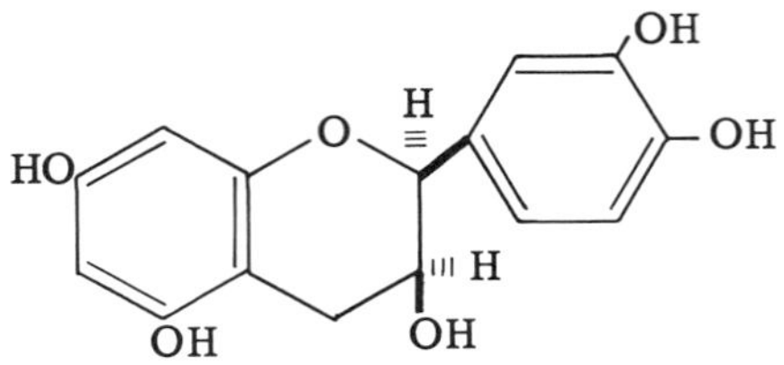

(97)

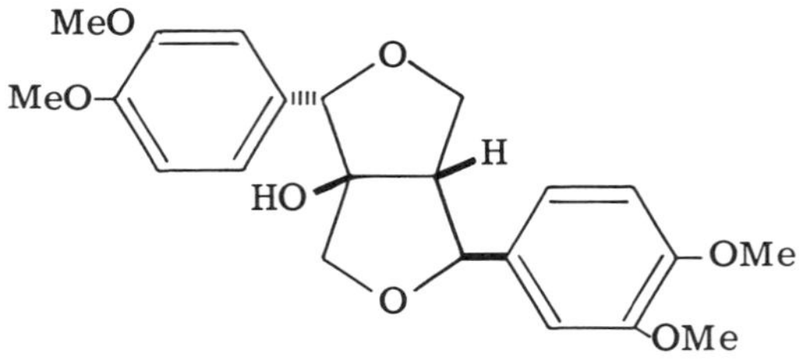

(98)

the monophenol, exposing the ring to oxidative degradation. 3,4-Dimethoxybenzoic acid, not formed from galcatin itself, was obtained by oxidation, thus confirming **99**.

6. *Natural Dimethylchromenes and Related Compounds* (169)

The cleavage of dimethylchromenes and some related ring systems produces *o*-isopentenylphenols which are probably their biogenetic precursors. The relationship between scandenin **(100)** and lonchocarpic acid **(101)** was confirmed by their reduction to the same compound **(102)**.

7. *Position of* ^{14}C *Label in the Clavine Series* (170)

The methiodide of elymoclavine **(103)** undergoes clean reductive fission to give **104**, and this reaction has been used in degradation to examine the positions of label in material biosynthesized from 2[^{14}C]-mevalonate.

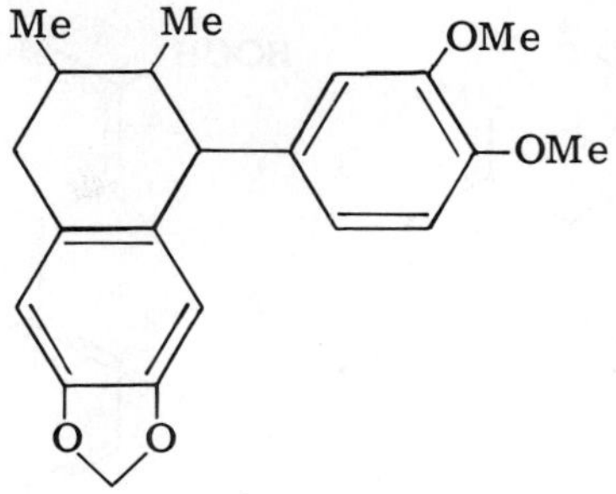

(99)

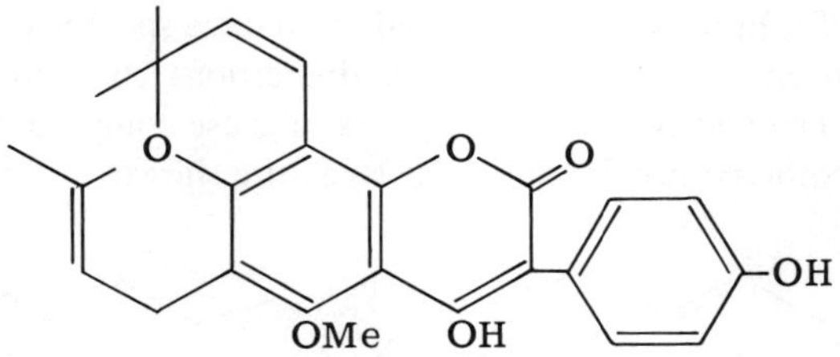

(100)

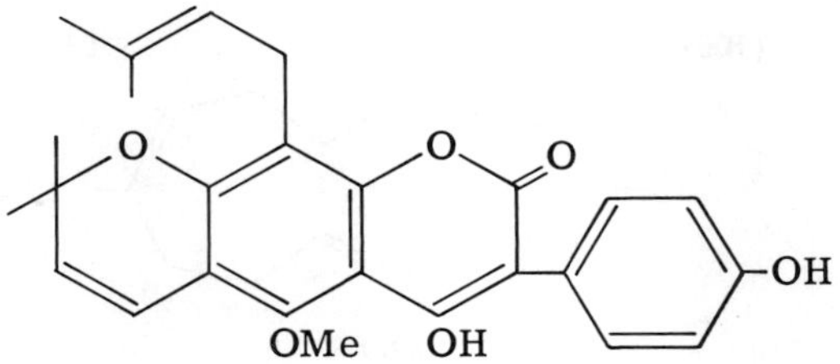

(101)

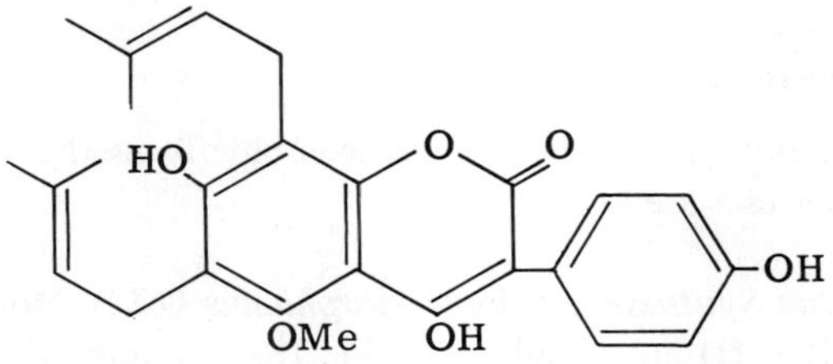

(102)

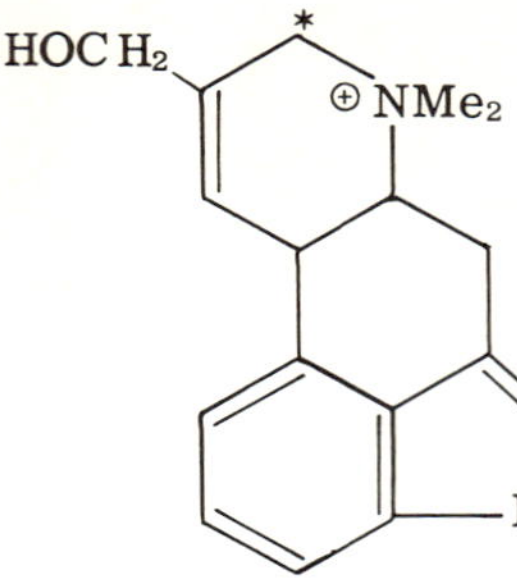

(103) 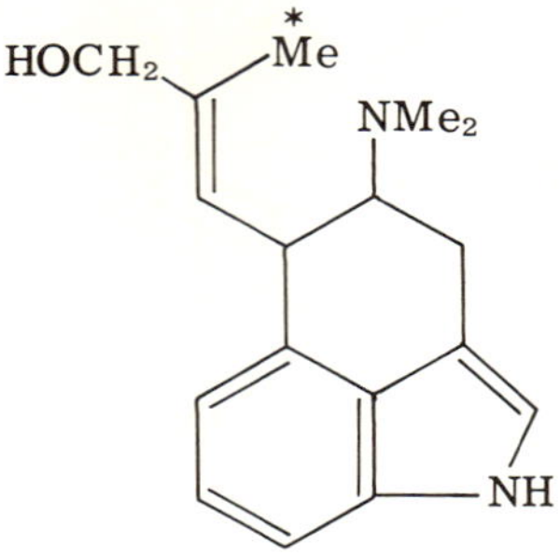 (104)

8. Cembrene (171)

The elucidation of the structure of this unusual diterpene (**105**) was assisted by the ability of lithium in liquid ammonia to reduce specifically the conjugated diene system to give a mixture of two hydrocarbons **106** and **107**, the former predominating. The degradations and spectra of these compounds in conjunction with those of cembrene itself support the structure shown.

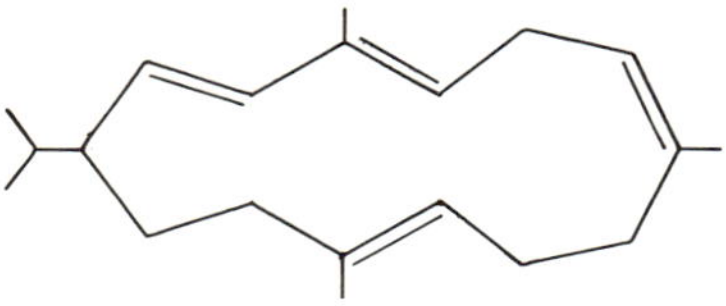

(105) 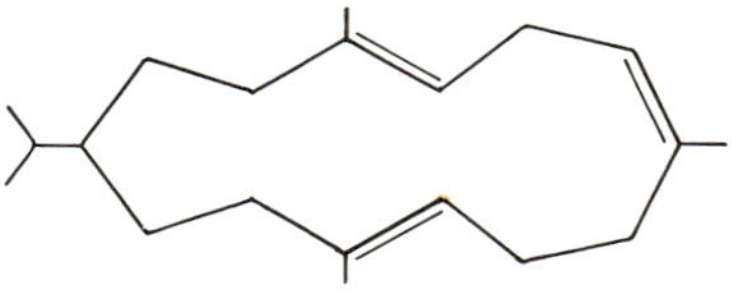 (106)

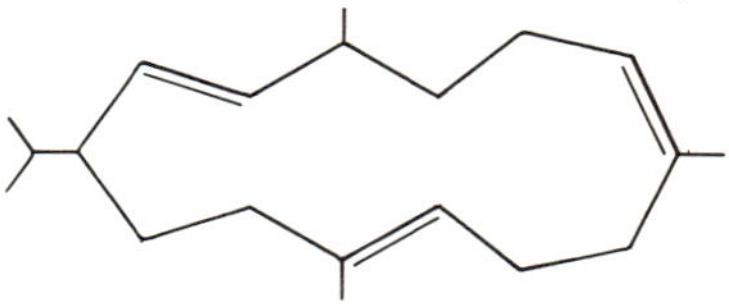

(107)

B. APPLICATIONS IN SYNTHESIS

1. Cleavage Reactions

Cleavage reactions in synthesis have been chiefly used to remove protecting groups, but other uses are known.

(a) The Partial Syntheses of Some Morphinans (172). Morphinan derivatives of type **108** (R = H) are readily available from natural alkaloids, and it was desired to remove the hydroxyl group from the ring. This was accomplished by conversion into a diphenyl ether-type group, in the knowledge that the *o*-OMe

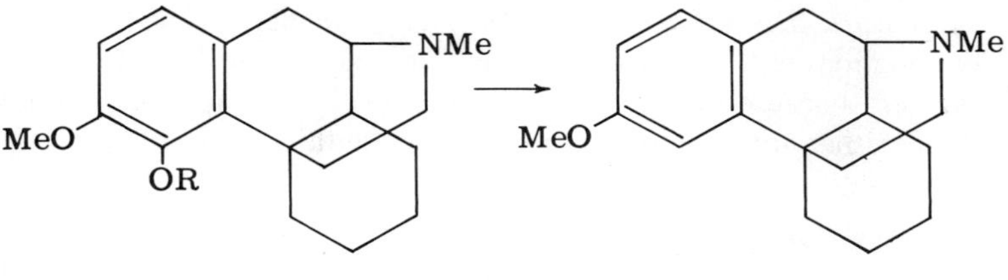

(108)

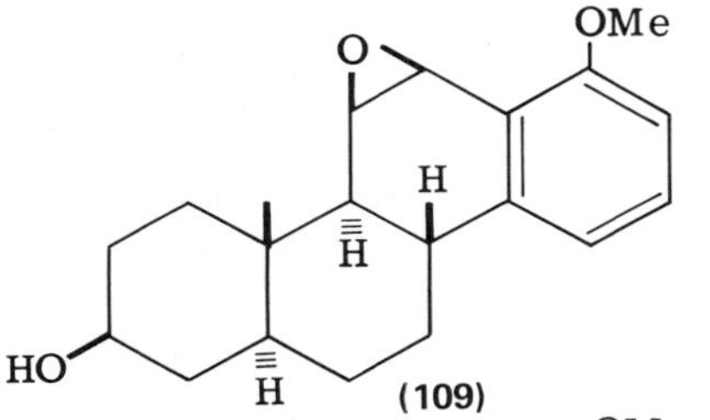

(109)

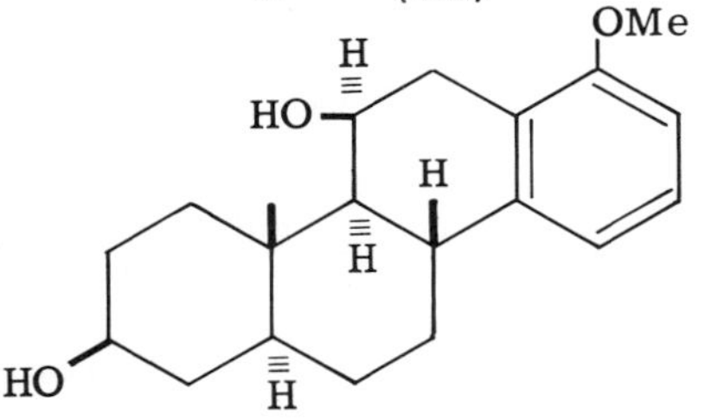

(110)

group would promote cleavage in the correct position (172). The best method consists in reaction with 2,4-dinitrofluorobenzene, hydrogenation of the ether to **108** (R = 2,4-diaminophenyl) and sodium-ammonia cleavage of this to give the desoxy derivative shown. The 2,4-diamino groups are additional factors in ensuring cleavage of the correct carbon-oxygen bond (131,172). The process could probably be a much more general one, although the *o*-OMe group clearly facilitates the process.

(b) A Total Synthesis of 11-Hydroxysteroids (173). The tetracyclic compound (**109**), obtained by epoxidation of the appropriate olefin, was reduced with lithium and ethanol in ammonia to cleave the benzylic carbon-oxygen bond, giving **110.** This was subsequently submitted to metal-ammonia reduction of the aromatic ring.

(c) A Specific Demethylation (174). The 3,4-dimethoxyphenylethylamine (**111**) on reaction with potassium in ammonia was specifically demethylated at the 3 position because of the effect of the side-chain substituent (133). Other methods are either not specific or tend to show the reverse specificity.

(d) Synthesis of Iboga Alkaloids (175). The ammonium ion (**112**) when treated with sodium in liquid ammonia gave the known dihydro cleavamines (**113**), stereoisomeric about the center carrying the ethyl group. The sequence led finally to a total synthesis of the iboga alkaloid system. The process is reminiscent of a previous model in the quebrachamine series (176).

(e) Peptide Synthesis. The sodium-liquid ammonia cleavage of protecting groups (e.g., tosyl, benzyl, carbobenzyloxy) from peptide units, pioneered by du Vigneaud (177), has had extensive use. Recent discussions of the applicability of the method have appeared in connection with the synthesis of a nonadecapeptide amide portion of ACTH (178) and of insulin (179). To avoid peptide cleavage the ammonia must be pure and dry. It is possible that the amide groups are stabilized by salt formation and that in the presence of traces of water they are cleaved reductively rather than by hydrolysis. A synthesis of coenzyme-A involves benzyl cleavage and also disulfide cleavage (180).

(f) Synthesis of Bulnesol (181). An indirect carbon-oxygen cleavage is illustrated by the following. A blocking group for hydroxyl was required which, unlike benzyl, would be stable to the strongly acidic conditions involved in the synthetic sequence. The *p*-chlorophenyl group was used in the precursor of **114** and this was then reduced with LAH to **115** (Y = OH) with some loss of chlorine, converted into the tosylate, and by LAH reduction further into **115** (Y = H). This on Li-EtOH-NH_3 reduction and hydrolysis with acid gave the diol, **116**, in 80% yield, which was further transformed into the sesquiterpene alcohol bulnesol.

(g) Prevention of Hydrogenolysis of Allylic and Benzylic Groups. This is frequently of some importance. An indication has already been given that this can be done by making the alkoxide anion of an allylic or benzylic alcohol, but this process may be incompatible with the presence of a more acidic alcohol required, for example, for the reduction of a benzene ring. It does appear in some cases, particularly with the use of *t*-butanol, that the cleavage of benzyl alcohols can be inhibited by initial salt formation, and that the aromatic ring can be reduced. In one case at least (**117**), the use of the tetrahydropyranyl ether resulted in retention of the oxygen ring reduction (**118**) (107). Another possible approach is to use the primary benzylamine, which is not generally subject to hydrogenolysis, and later to remove the nitrogen by diazotization. Benzylamine thus gives **119** converted by nitrous acid chiefly into **120**, but with some rearranged product (**121**) (182).

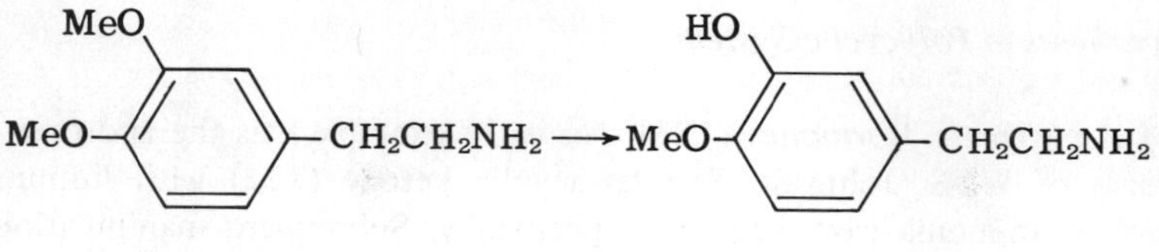

(111)

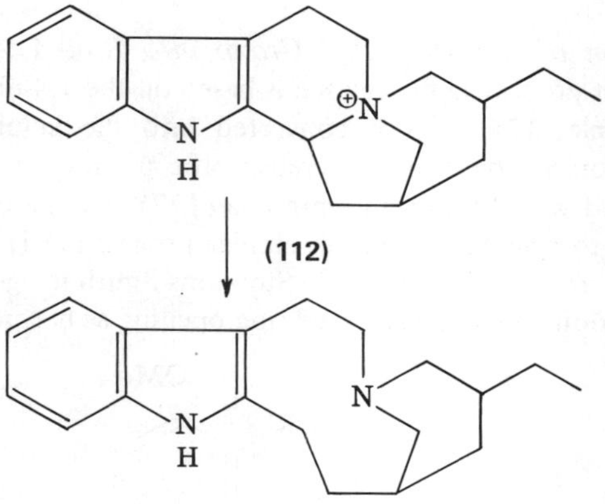

(112)

(113)

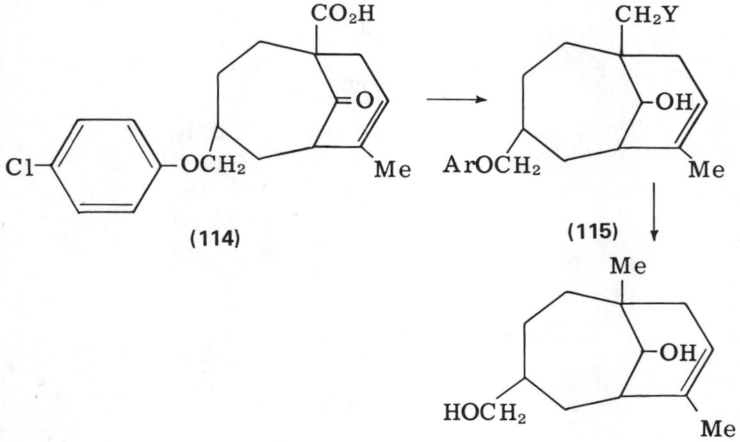

(114)

(115)

(116)

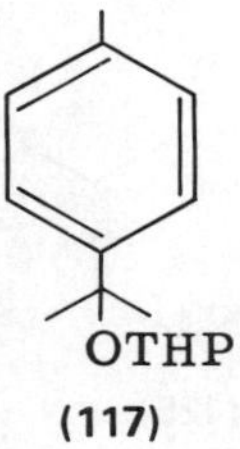

(117)

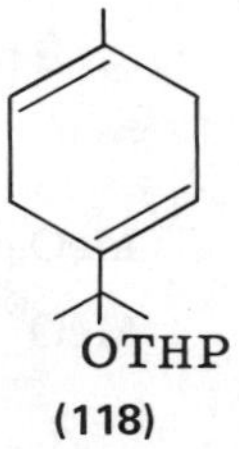

(118)

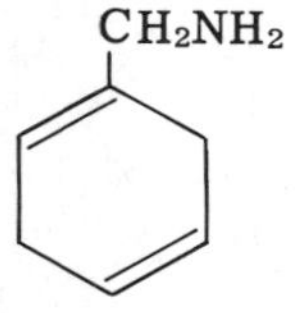

(119)

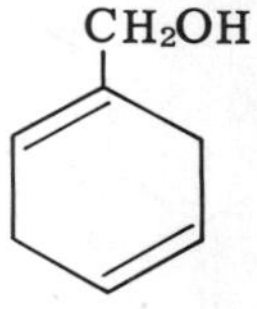

(120)

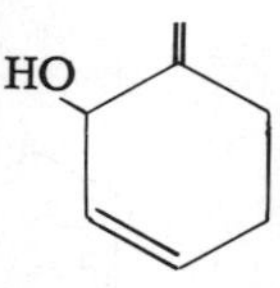

(121)

2. *Syntheses in Polycyclic Systems*

(a) Androgenic Hormones (183). An early example was the androgen total synthesis of W. S. Johnson. The tetracyclic ketone (**122**) with lithium and ethanol in ammonia gave **123** stereospecifically. Subsequent manipulations led to androgenic hormones.

(b) Introduction of the 19-Methyl Group into Ring-A Aromatic Steroids (138). The simplest process so far devised is based on the 1,4-dihydro-3-methoxy steroid, for example, **124**. This is converted into the ketal (**125**), which is reacted with dibromocarbene in a stereospecific β addition to give **126**. The bromine is removed with lithium in ammonia (**127**) and the cyclopropane then reacted with acid to give **128**. A rather similar process (184) is to convert **124** into **129** and to react this with the Simmons-Smith reagent forming **130**, followed by oxidation and acid-catalyzed ring opening as before.

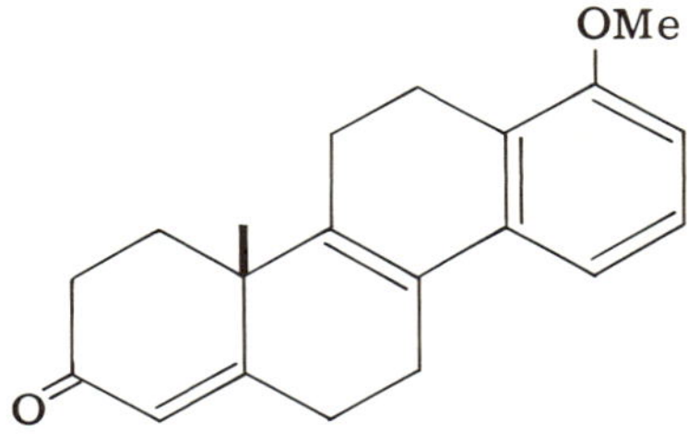

OMe
O

(122)

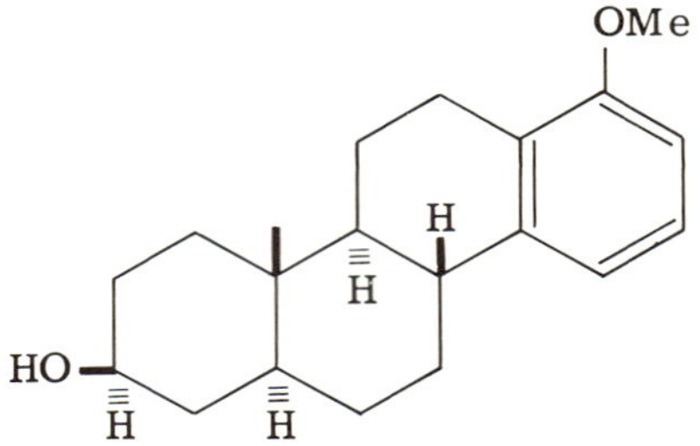

OMe
H
H
HO
H
H

(123)

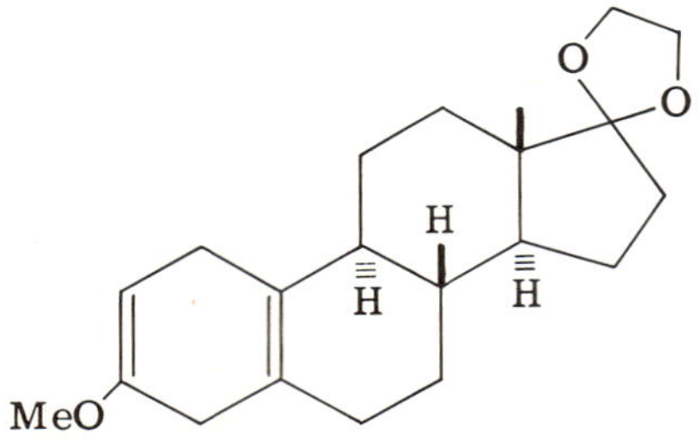

O
O
H
H
H
MeO

(124)

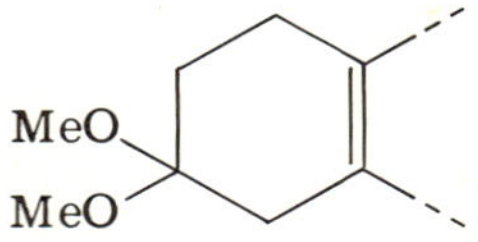

MeO
MeO

(125)

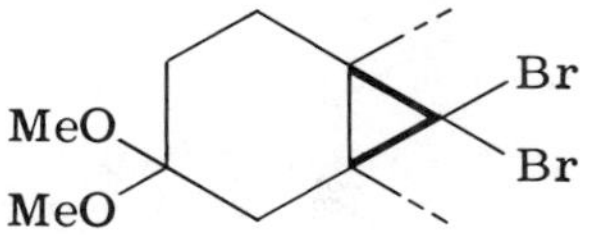

(126)

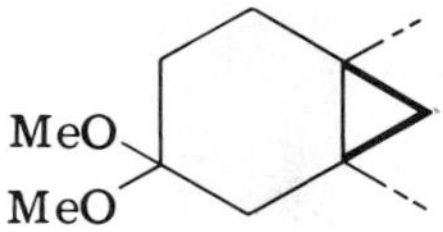

(127)

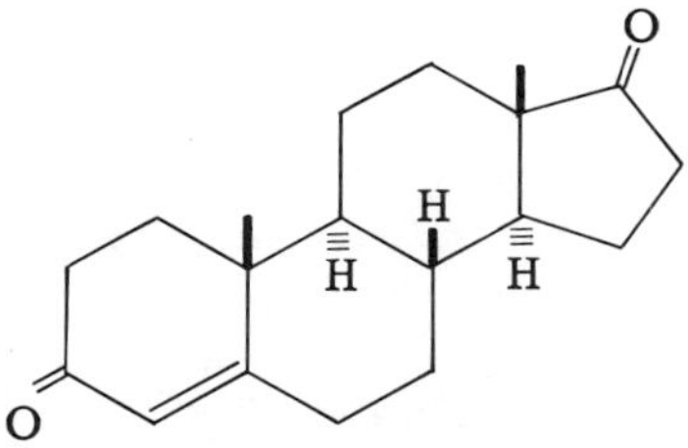

(128)

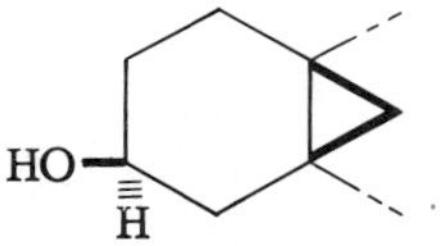

(129)

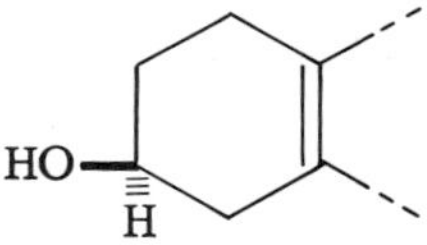

(130)

(c) The Total Synthesis of 19-Nortestosterone (14). A considerable number of syntheses depend on the reduction of anisole rings to dihydro derivatives, and subsequent hydrolysis to cyclohexenones. The anisole ring therefore acts as a stable cyclohexenone precursor, which can be unmasked at a suitable stage. The first example was the conversion of estrone into 19-nortestosterone.

The reductions of anisole derivatives were initially examined in 1942 as models for this process, for example, **131** into **132** (5), but the synthesis was only completed in 1948. 19-Nortestosterone was then the first totally synthetic androgenic hormone, and the process has been the prototype for later industrially important 19-norsteroids. Owing to the low solubilities of the estrone or estradiol methyl ether, they were not initially reducible by sodium and ethanol in ammonia, and consequently the more soluble β-hydroxyethyl ether **(133)** (R = CH_2CH_2OH) was used. This technique has not been used since, but could still offer solubility advantages in some cases; also the intermediate enol-ethers are convertible quantitatively with great ease into ethylene ketals **(134)**. The reduction of the ethers **(133)** (R = $HOCH_2CH_2$— or $HOCH_2CHOHCH_2$—) and hydrolysis gave first compound **(135)**, also the parent of a commercially important series of hormones, and then 19-nortestosterone **(136)**. With the use of lithium the methyl ethyl of estrone became reducible (15), and later it was shown that with pure ammonia and the use of *t*-butanol as the proton source, ring reduction could be efficiently carried out with sodium (16).

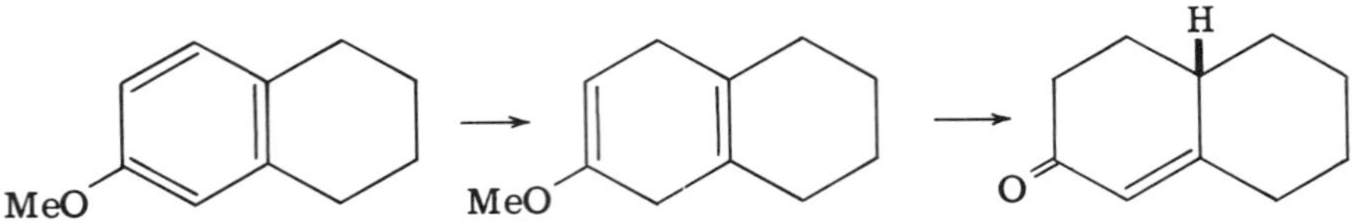

MeO MeO O H

(131) (132)

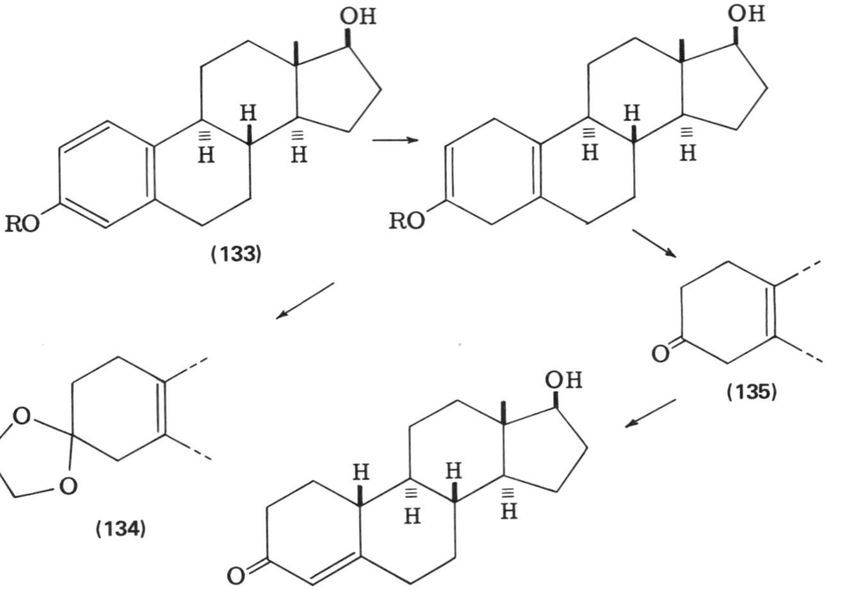

OH OH

H H

H H H H

RO RO

(133)

O O

OH

(135)

O

H H

(134) H H

O

(136)

(d) Synthesis of (±)-Fichtelite (185). The acetylene **138** is synthesized by fairly standard methods, starting from **137**, and on reduction with sodium in ammonia, removal occurs of the protecting benzyl group (necessary during the synthesis of **138**) and of the acetylene to give the pure *trans*-olefin **(139)**, in one experimental step. Further reactions led to the stereospecific synthesis of (±)-fichtelite.

(e) Stereospecific Synthesis of Lycopodine (186). This synthesis uses an anisole ring in order to carry out a ring-closure reaction, and then converts it into a saturated propionic acid chain.

The precursor **(140)** is cyclized to **141**, the amide reduced (LAH) to the amine, and the anisole ring reduced first to the unconjugated dihydro derivative, which is then conjugated to **142**. Protection of the amino group and specific ozonolysis of the enol-ether bond gives **143** converted into the ketone **(144)**. Several further steps give (±)-lycopodine.

$HOCH_2CH_2C{\equiv}CH$

(137)

CH_2
$\|$
$PhCH_2OCH_2CH_2C{\equiv}CCH_2CH_2CCHMe_2$

(138)

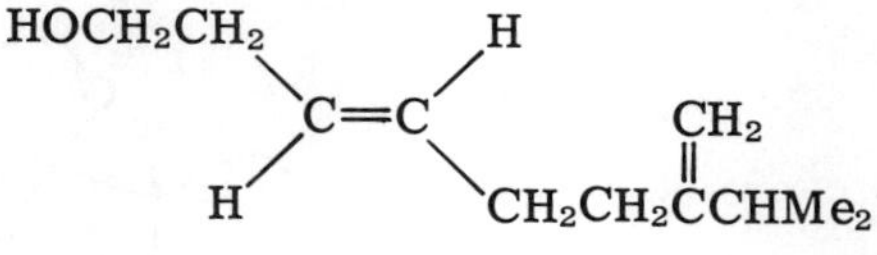

$HOCH_2CH_2$ H
C=C CH_2
H $CH_2CH_2CCHMe_2$

(139)

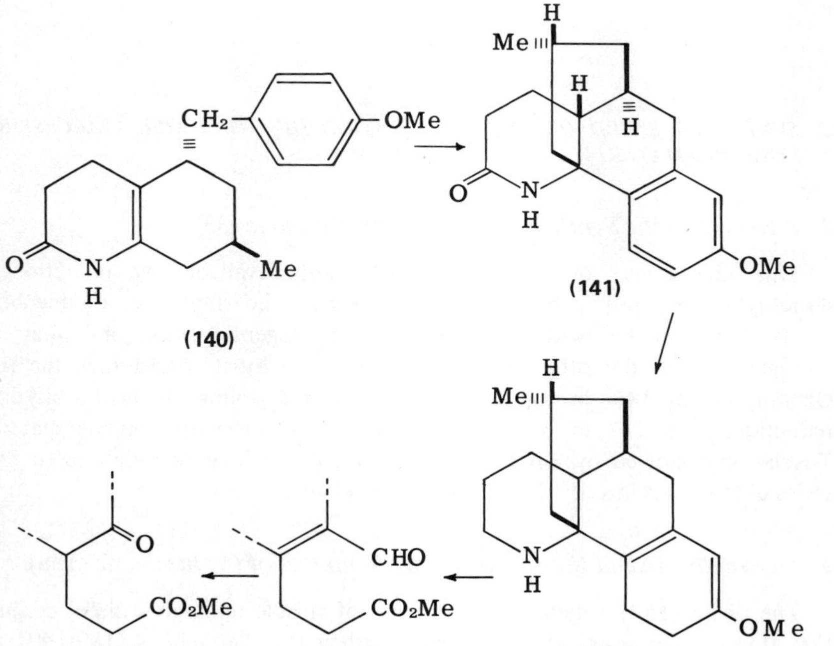

CH_2 OMe
O N Me
H

(140)

H
Me H
H
O N
H
OMe

(141)

H
Me
N
H
OMe

(142)

O
CO_2Me

(144)

CHO
CO_2Me

(143)

(f) Estrone and 18-Methylsteroids. The stereospecific reduction of the styrenoid double bond of **145** (R = Me or Et) using sodium and aniline in ammonia is an important industrial step in hormone synthesis (187). The aniline probably minimizes reduction of the aromatic ring, but is capable of protonating the anions derived from the stereo styrene.

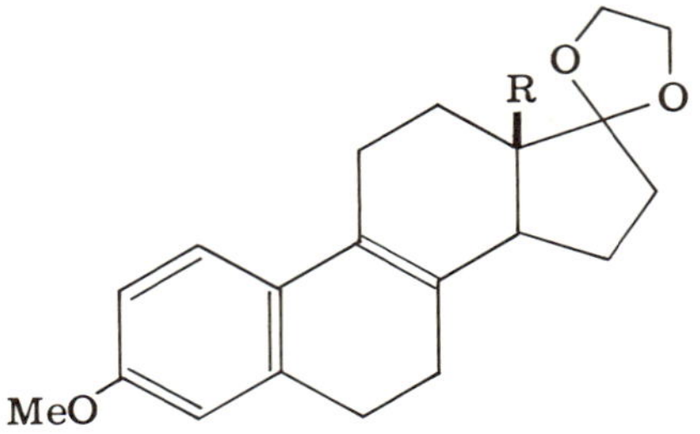

(145)

C. *SYNTHESES BASED ON THE PREPARATION OF OTHERWISE INACCESSIBLE STARTING MATERIALS*

1. *A Stereospecific Synthesis of the Juvenile Hormone* (33)

This makes use of **146**, which is readily available by reduction of 4-methylanisole, and is based on the facts that the vinyl-ether double bond reacts preferentially with the electrophilic reagent ozone, and that the configuration of the other double bond is fixed by its position in the ring. Ozonolysis of **146** in methanol-dimethylsulfide followed by borohydride reduction gives **147** in 52% yield based on the aromatic starting material. Tosylation followed by LAH reduction gives **148** which was elaborated by a series of reactions into the juvenile hormone (**149**).

2. *Stereoselective and Nonstereoselective Syntheses of (±)-Juvabione* (188)

The diene (**150**) obtained by reduction of anisole undergoes direct conjugation through charge-transfer complexes with many dienophiles (189,190) and consequently behaves as **151**. The Diels-Alder reaction shown leads to a mixture of the adducts **152** and **153** which can be separated by distillation. Acid-catalyzed ring opening leads from the correct isomer to (±)-**154** which can then be further elaborated to give (±)-juvabione (**155**). The mixture of **152** and **153** can be used in a very simple series of reactions to give a mixture of (±)-juvabione stereoisomers.

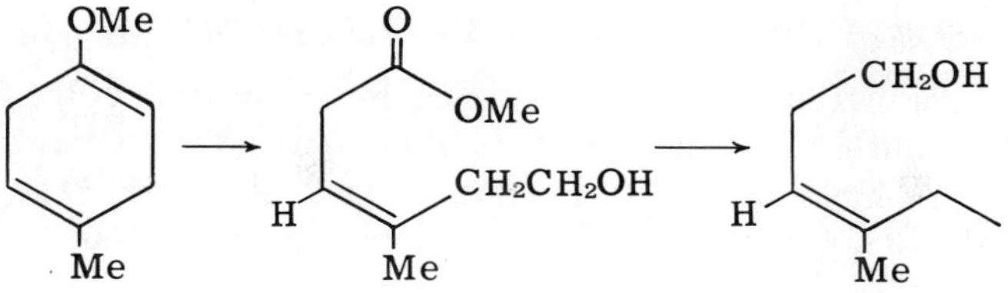

(146) (147) (148)

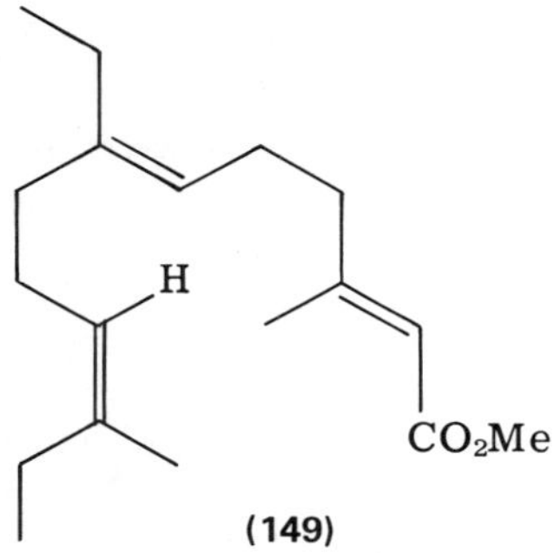

(149)

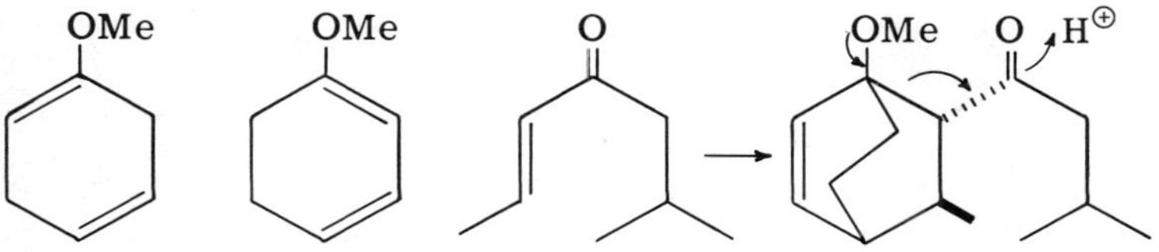

(150) (151) (152)

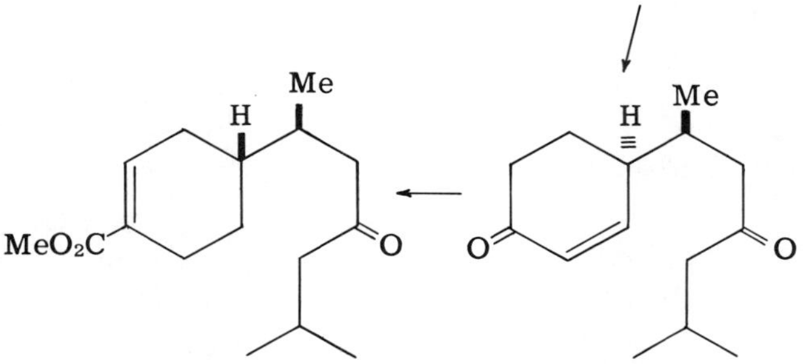

(155) (154)

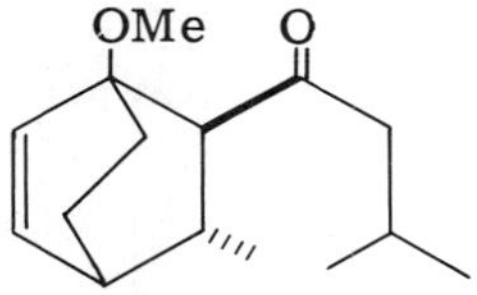

(153)

3. *The Synthesis of Tricarbonylironcyclohexa-2,4-dienone* (191)

The reaction of **150** with $Fe_3(CO)_{12}$ gives rise to a mixture of the complexes **156** and **157** which on treatment with tritylfluoroborate produces a mixture of the stable carbonium salt (**158**) and the unstable salt (**159**). The latter on reaction with water produces the cyclohexadienone complex (**160**).

4. *A Synthesis of (±)-Epiibogamine* (192)

Reduction of 3,4,5-trimethoxybenzoic acid gives **161** (R = CO_2H) (53,102) reducible (LAH) to **161** (R = CH_2OH), converted into **162**. The action of ethylmagnesium bromide followed by acid gives **163** which undergoes Wolff-Kishner reduction to **164**. This was used as the source of the nonindolic portion of epiibogamine.

5. *Bridged-Ring Annulenes* (193)

The synthesis of the very interesting bridged-ring 10-annulenes, for example, **166**, is made possible by the ready availability of the 1,4,5,8-tetrahydronaphthalenes (**165**) and analogous compounds, and the fact that the more highly substituted double bond reacts preferentially with dihalocarbenes. Removal of the bromine is also accomplished by metal-ammonia reduction.

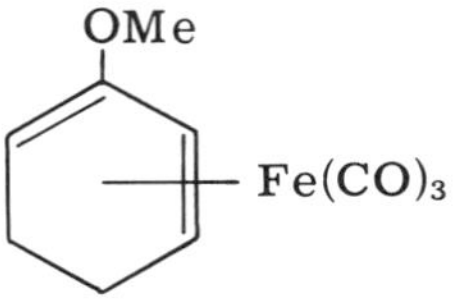

(156) 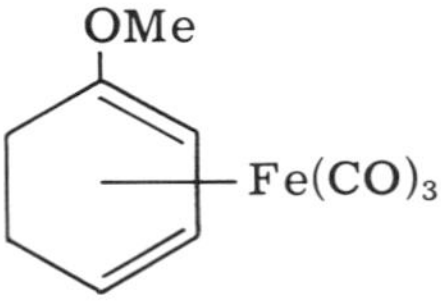 (157) 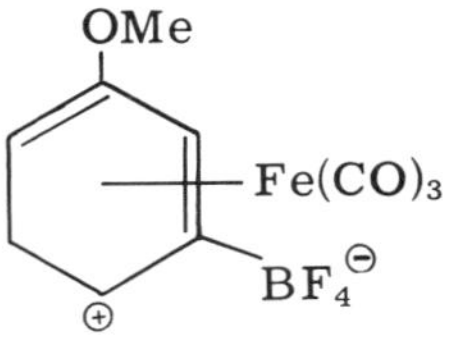 (158)

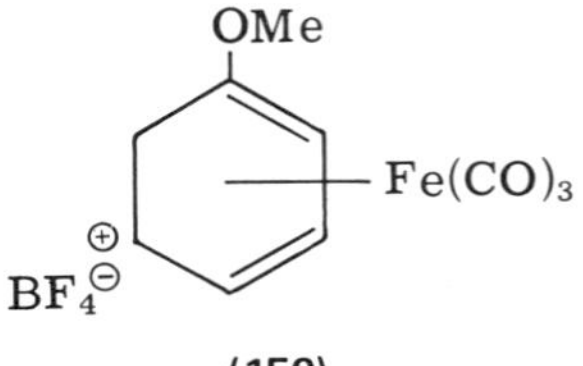

(159) 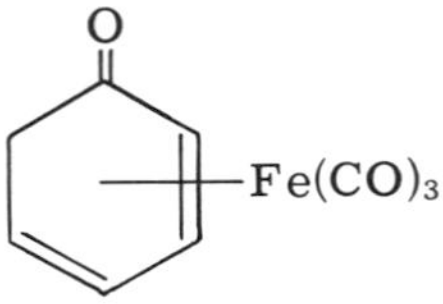 (160)

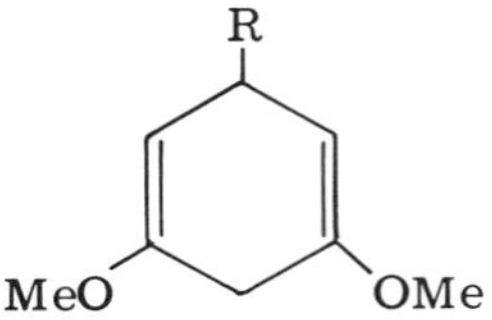

(161) 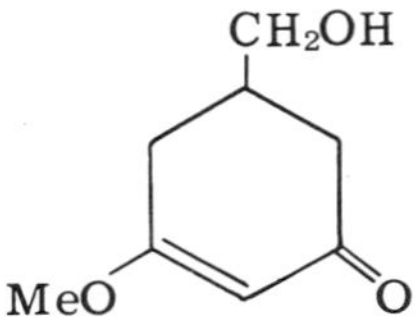 (162)

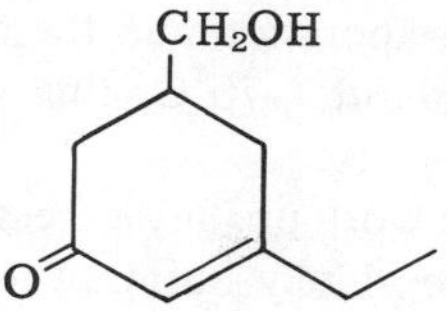

CH_2OH CH_2OH

O

(163) 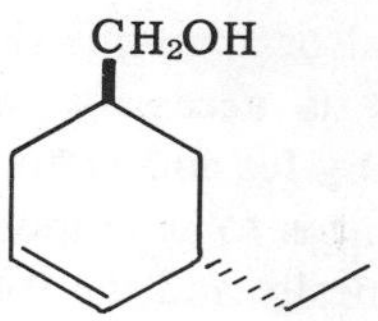 (164)

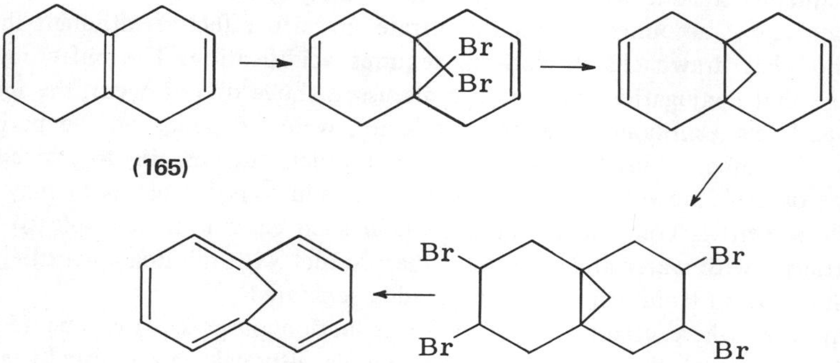

Br
Br

(165)

Br Br

Br Br

(166)

D. EXPERIMENTAL ASPECTS

1. General

Some typical experimental procedures are given below, but some general points may be noted first. The purity of the ammonia seems of little importance where reduction is easy, as it is with α,β-unsaturated ketones. However, the presence of traces of transition metals, particularly iron, catalyzes vigorously the competing evolution of hydrogen gas in more difficult cases, for example, the reduction of benzene rings, and it is usually advisable to distill the ammonia from sodium, condensing it with a solid carbon dioxide-acetone or alcohol condenser in the apparatus to be used. The ammonia is then also dry (but extremely hygroscopic). In most cases traces of water are unimportant unless salts of very weak acids have to be conserved. Since many of the secondary processes are highly time-dependent, the best results are usually obtained from rapid reduction and rapid and mild work-up.

During the reduction the solution is preferably stirred with a stirrer which is capable of reaching most of the solution. Sodium alkoxides are not very soluble, and in their presence pastes may result; furthermore, phase splitting frequently occurs and a concentrated solution of metal in the form of light globules may float on the surface. The ammonia can be allowed to boil under reflux using a solid carbon dioxide condenser. In some cases no condenser is necessary; the high latent heat of ammonia causes slow evaporation and the solvent is readily replaced. If it is necessary to cool to about −70°C, this can be done by immersion in Dry Ice and acetone.

If the solvent is to be evaporated before work-up, any excess metal should be removed chemically and the solution appropriately buffered unless it is certain that the product will be stable to strongly basic conditions. If the desired product is not acidic, removal of the metal is often conveniently achieved with benzoic acid; the reduction product and excess benzoic acid may be removed in the aqueous alkaline work-up. Addition of catalytic amounts of ferric nitrate or of equivalent amounts of sodium nitrate are also effective although the last method has drawbacks if work-up requires acidification. The buffering is to ensure that conjugations of the type discussed above do not occur; the agent to be used (e.g., ammonium chloride, ethanol, water) depends on the particular case. A number of low molecular weight products such as dihydrobenzene and dihydroanisole are volatile in boiling ammonia and considerable losses may occur if the solvent is completely evaporated; in such cases it is desirable to dilute cautiously with water and to extract the product with minimal quantities of an organic solvent from which it can be readily separated.

In cases where ethanol is added for reduction purposes it can be added to 10% or more of the volume of the ammonia although this is usually a great molecular excess. The alcohol helps to keep substances in solution, particularly the alkoxide, whose separation, sometimes sudden, is occasionally a nuisance. As much as 40% by volume has been used with rather insoluble substrates. Frothing can be momentarily controlled by adding some light petroleum; it is also advisable to add the metal slowly to the stirred mixture.

Lithium and potassium dissolve rapidly in ammonia, especially if the former is added as cut sections of wire or ribbon. Sodium is slower to dissolve and it is best used in thin slices. In the case of amines such as ethylamine, lithium dissolves quite slowly, and the reaction mixture usually has to be stirred well or shaken continuously in a closed flask. If the lithium is dissolved in a little ammonia first, good solutions in ethylamine can readily be obtained. So far the results obtained with such mixed solutions do not seem to differ significantly from those with pure amines. Ethylene diamine is difficult to obtain completely anhydrous and this is best accomplished by saturation with potassium hydroxide and by final distillation from either lithium or sodium-potassium alloy.

In many cases metal-ammonia reductions are best worked-up without complete evaporation of the ammonia by cautious addition of water. In any case, with ammonia present in an aqueous work-up, extraction with organic solvents should not be done by shaking in a closed funnel until it is quite certain that the solution is not supersaturated. Furthermore, with sodium particularly, there is some danger from residual sodium pieces, which can be disguised with a white coating of rather insoluble alkoxide or hydroxide. Residual metal may react vigorously at a later stage when the ammonia has been considerably diluted and result in an explosion or fire hazard. If solid metal is recognized, it can be removed mechanically or by a moderate addition of water. Complete evaporation of the ammonia increases this hazard, which can also be avoided by adding ethanol, but the presence of large proportions of this may be awkward for further work-up.

Anhydrous potassium carbonate is an efficient alkaline drying agent for organic extracts which do not contain major amounts of ethanol.

Frosting on the reaction vessel may be removed by washing with ethanol, which facilitates observation of the interior.

In cases where initial salt formation is desired, for example, with alcohols, it should be noted that the standard method of preparing metal amide in ammonia cannot be used since the iron catalyst inhibits reductions. Available metal hydrides or alkoxides may be used, or amide generated by the reduction of an appropriate unsaturated compound, for example, butadiene or naphthalene, chosen so that the by-products are readily separable.

2. *Reaction Procedures*

There are two fundamentally different reaction procedures described in the literature:

1. The substance to be reduced is added to a solution of the metal in ammonia, and after some time a proton source such as an alcohol or ammonium chloride is added. The temperature is usually either at the boiling point (–33°C) or maintained about –70 to –80°C by an acetone-Dry Ice bath.

2. The metal is added to a mixture of the substrate, proton source, co-solvent, and ammonia. A usually less desirable variant is to add the substrate, proton source, and co-solvent to the metal solution. This may be necessary if the metal is required to be in excess, but carries some danger of over-reductions in some cases.

Which method is chosen depends on the type of substance and the desired result. Indications of this may be obtained from the foregoing review, but usually Method 1 is preferable for very readily reducible compounds, where a controlled intermediate degree of reduction is desired, and Method 2 is preferable for all difficult reductions and for reactions where highly basic conditions are undesirable. Careful thought must be given in all cases to the exact details of the process to be employed; there are no exact standard conditions to suit all cases. However, fairly general conditions for Method 1 involve the use at about –70°C of lithium or sodium (about 2.1 equivalents for a polycyclic hydrocarbon, for example), reaction being terminated by the addition of solid ammonium chloride, when the blue color of the metal is replaced by deep red or brown. For Method 2 good conditions frequently involve lithium and *t*-butyl or *t*-amyl alcohol at 33°C; the proportion of metal employed is dependent on the desired degree of the reduction and is very efficiently used except in very difficult cases. The proportion of alcohol to metal should be at least equivalent but is not otherwise crucial.

The apparatus consists of a three-necked flask (A), fitted with a mechanical stirrer and an acetone-Dry Ice condenser (cold finger condenser). The third neck is connected to a round-bottomed flask (B) through a PVC tube. Flask B is charged with commercial ammonia and sodium added (about 2% or until a deep blue color is maintained). Acetone is placed in the cold finger condenser (about one-fourth full), and a slow stream of ammonia gas is passed through the apparatus for a short time (30 sec) to displace the air. Powdered Dry Ice is added to the condenser with care (rapid addition causes bumping of acetone) and ammonia is distilled into Flask A from Flask B by warming the latter on a steam bath. When sufficient solvent has been condensed the inlet is replaced by a stopper through which the dry metal, cut into small portions, can be introduced. The mixture of the substrate in the co-solvent and the proton donor is added, followed by the metal, or vice versa. The reaction mixture is stirred under reflux. The reaction may be terminated when the blue color has disappeared, or it may be necessary to destroy excess of the metal by addition of finely powdered sodium benzoate, ammonium chloride, or methanol as appropriate. After completion of the reduction the ammonia is allowed to evaporate slowly, water is added, and the products isolated by customary methods. Alternatively water may be added very cautiously, particularly initially, to the ammonia solution. Several volumes of water are usually required in order to give a solution which can be safely handled and extracted.

All operations should be carried out in a hood, including product extractions, with eye, face, and hand protection.

3. *Some Detailed Examples*

(a) 3-(4′-Methoxyphenyl) Propionic Acid (194). Sodium (10 g, 0.43 g mole) was added in small pieces to a solution of 4′-methoxycinnamic acid (35.6 g, 0.2 g mole) in liquid ammonia (300 ml) under stirring and as rapidly as the metal dissolved. Ammonium chloride (22 g) was added and the ammonia was left to evaporate overnight. Water (200 ml) was added and the mixture extracted with ether, but no neutral product was observed. The aqueous extract was acidified with 2 *N* hydrochloric acid, the precipitated acid extracted with ether (3 × 50 ml) and worked up by washing with brine and drying ($MgSO_4$) giving a cystalline product (35.0 g, 96%) on evaporation. Recrystallization from water gave the acid as needles, mp 104° C.

(b) Reduction of Estrone Methyl Ether (16). A solution of estrone methyl ether (6.5 g) in tetrahydrofuran (80 ml) was added to redistilled ammonia (200 ml) and *t*-butanol (80 ml) followed by lithium (2.31 g). After stirring for 1 hr, methanol (40 ml) was added carefully to discharge the blue color. Work-up in the usual way gave 3-methoxyestra-2,5-dien-17-β-ol (95%).

(c) Reduction of Anisole (5). Sodium (11.5 g, 0.5 g mole) was added in portions to a stirred solution of anisole (21.6 g, 0.2 g mole) and ethanol (46.1 g, 1.0 g mole) in pure ammonia (200 ml). Ten minutes after completion of the addition, excess sodium was destroyed by adding sodium benzoate to removal of the blue color. Water (200 ml) was added after evaporation of ammonia to half the volume and the mixture was extracted with pentane (100 ml). The dried extract (K_2CO_3) was evaporated and distilled to give 1-methoxycyclohexa-1,4-diene (17.4 g).

(d) Reduction of 9-Methylanthracene (155). To a stirred solution of lithium (872 mg, 12.5 g-atoms) dissolved in liquid ammonia (150 ml) was added 9-methylanthracene (961 mg) in 75 ml of THF. The solution was maintained at reflux for 1 hr, then the deep red color was discharged by addition of alcohol (5 ml), followed by water (10 ml). Evaporation of the ammonia followed by partition of the residue between ether and water led to recovery of an oily solid; by thin-layer chromatography it appeared to consist of a single product (R_f 0.35) free of 9-methylanthracene (R_f 0.18). Recrystallization from methanol provided white needles (680 mg) melting at 59-60°C.

(e) Reduction with Lithium and Ethylamine (20). Lithium amine reductions are usually carried out under nitrogen or helium in a three-necked flask fitted with a mechanical stirrer and Dry Ice condenser at the reflux temperature of ethylamine or methylamine. Alternatively a ground-glass shaken flask with the stopper sealed in place can be used with ethylamine, although pressure build-up can occur. Other amines are treated as normal for liquids of their boiling point. If the substrate is a liquid, it is directly added to the blue solution of the amine containing lithium and stirred vigorously with cooling. In the case of solids, the material and lithium are placed in the flask and the amine added with stirring. The reaction is usually complete after 8-10 hr. At the end of this period, the excess lithium metal is removed mechanically and the solution worked up by careful addition of ethanol or ammonium chloride to discharge the blue color. The solvent is then left to evaporate and water added. The product is extracted with ether, washed with saturated brine and water, dried, the solvent removed, and the product isolated by distillation.

(f) Phenol to Cyclohexanol (127). Phenol (0.2 mole) was reduced with lithium (1.6 g atom) and ethylamine (250 ml). After 6 hr stirring, the unreacted lithium was not removed but instead ethanol (20 g) was added slowly over a 2-hr period. The nonphenolic product (bp 160-164° C; 14 g; n_a 1.4618) was identified as a mixture consisting of cyclohexanol (82%) and mixed ketones (18%). Unreacted phenol (16%) was also recovered.

REFERENCES

1. (a) *Solvated Electron,* American Chemical Society Publication, 1965; (b) U. Schindewolf, *Angew. Chem. Int. Ed.,* **7,** 190 (1968); (c) J. J. Lagowski, *The Chemistry of Nonaqueous Solvents,* Vol. 2, Academic Press, New York, 1967, pp. 265-317; (d) G. Lepoutre and M. J. Sienko, eds., *Metal Ammonia Solutions,* Benjamin, New York, 1964; (e) W. Hückel, *Fortsch. Chem. Forsch.,* **6,** 197 (1966).
2. W. Hückel and H. Bretschneider, *Justus Liebigs Ann. Chem.,* **540,** 157 (1939).
3. B. A. Kazanskii and N. F. Glushnev, *J. Gen. Chem. USSR,* **8,** 642 (1938); *Chem. Abstr.,* **33,** 1280 (1939).
4. C. B. Wooster, U.S. Pat. 2, 182, 242, (1939), *Chem. Abstr.,* **34,** 1993 (1940).
5. A. J. Birch, *J. Chem. Soc.,* **1944,** 430.
6. (a) K. N. Campbell and L. T. Eby, *J. Amer. Chem. Soc.,* **63,** 216, 2683 (1941); (b) A. L. Henne and K. W. Greenlee, *J. Amer. Chem. Soc.,* **65,** 2020 (1943).
7. A. J. Birch and H. Smith, *J. Chem. Soc.,* **1951,** 1882.
8. D. H. R. Barton and C. H. Robinson, *J. Chem. Soc.,* **1954,** 3045.
9. M. J. T. Robinson, *Tetrahedron,* **21,** 2475 (1965).
10. (a) W. Herz and J. J. Schmid, *J. Org. Chem.,* **34,** 3473 (1969); (b) S. K. Malhotra, D. F. Moakley, and F. Johnson, *Tetrahedron Lett.,* **1967,** 1089.
11. A. J. Birch, *"The Labile Molecule," Discussions Faraday Soc.,* (2), 246 (1947).
12. A. G. Catchpole, E. D. Hughes, and C. K. Ingold, *J. Chem. Soc.,* **1948,** 246.
13. A. J. Birch, *J. Chem. Soc.,* **1950,** 1551.
14. A. J. Birch, *J. Chem. Soc.,* **1950,** 367.
15. A. L. Wilds and N. A. Nelson, *J. Amer. Chem. Soc.,* **75,** 5360 (1956).
16. H. L. Dryden, G. M. Webber, R. R. Burtner, and J. Cella, *J. Org. Chem.,* **26,** 3237 (1961).
17. W. S. Johnson, J. Ackermann, B. K. Bhattacharya, R. A. Clement, B. Bannister, B. M. Bloom, E. R. Rogier, and J. Szmuszkovicz, *J. Amer. Chem. Soc.,* **78,** 6289 (1956).
18. M. Curcumelli-Rodastamo and M. Kulka, in *Alkaloids,* Vol. 9 (R. H. F. Manske, ed.), Academic Press, New York, 1967, Chap. 4.
19. C. A. Kraus, *J. Amer. Chem. Soc.,* **29,** 1557 (1907).
20. R. A. Benkeser, R. E. Robinson, D. M. Sauve, and O. H. Thomas, *J. Amer. Chem. Soc.,* **77,** 3230 (1955).
21. L. Riegel, R. A. Friedel, and I. Wender, *J. Org. Chem.,* **22,** 891 (1957).
22. J. H. Chapman, J. Elks, G. H. Philips. and L. J. Wymann, *J. Chem. Soc.,* **1956,** 4344.
23. L. Hackspill and R. Rohmer, *C. R. Acad. Sci., Paris,* **217,** 152 (1943).
24. (a) P. Markov and C. Ivanoff, *Tetrahedron Lett.,* **1962,** 1139; (b) P. Markov and C. Ivanoff, *C. R. Acad. Sci., Paris, Ser. C,* **264** (19), 1605 (1967).
25. W. F. Short, Boots Laboratories, unpublished work.
26. E. J. Kelly, H. V. Secor, C. W. Keenan, and J. F. Eastham, *J. Amer. Chem. Soc.,* **84,** 3611 (1962).
27. R. G. Harvey, *J. Org. Chem.,* **33,** 2570 (1968).
28. A. P. Krapcho and A. A. Bothner-By, *J. Amer. Chem. Soc.,* **81,** 3658 (1959).
29. H. Normant, *Angew. Chem. Int. Ed.,* **6,** 1046 (1967).
30. A. J. Birch and S. M. Mukherji, *Nature,* **163,** 766 (1949).
31. R. A. Benkeser, R. K. Agnihotri, M. L. Burrous, E. M. Kaiser, J. M. Mallan, and P. W. Ryan, *J. Org. Chem.,* **29,** 1313 (1964)
32. S. D. Darling, O. N. Devgan, and R. E. Cosgrove, *J. Amer. Chem. Soc.,* **92,** 696 (1970).
33. E. J. Corey, J. A. Katzenellenbogen, N. W. Gilman, S. A. Roman, and B. W. Erickson, *J. Amer. Chem. Soc.,* **90,** 5618 (1968).
34. A. J. Birch, *Quart. Rev.* (London), **4,** 69 (1950).
35. O. J. Jacobs and J. F. Eastham, *J. Amer. Chem. Soc.,* **87,** 5799 (1965).

36. A. P. Krapcho and M. E. Nadel, *J. Amer. Chem. Soc.*, **86**, 1096 (1964).
37. R. A. Benkeser and E. M. Kaiser, *J. Org. Chem.*, **29**, 955 (1964).
38. N. D. Scott, J. F. Walker, and V. L. Hansley, *J. Amer. Chem. Soc.*, **58**, 2442 (1936).
39. W. A. Remers, G. J. Gibs, C. Pidacks, and M. J. Weiss, *J. Amer. Chem. Soc.*, **89**, 5513 (1967).
40. Brit. Pat. 1,025,426 (1966); *Chem. Abstr.*, **64**, 19446c (1966). Netherlands Pat. 6,507,378 (1965); *Chem. Abstr.*, **64**, 19447a (1966).
41. R. A. Benkeser, M. L. Burrous, J. L. Hazdra, and E. M. Kaiser, *J. Org. Chem.*, **28**, 1094 (1963).
42. R. G. Harvey and K. Urberg, *J. Org. Chem.*, **33**, 2206 (1968).
43. E. J. Bailey, A. Gale, G. H. Philips, P. T. Siddons, and G. Smith, *Chem. Comm.*, **1967**, 1253.
44. E. Grovenstein, S. Chandra, C. E. Collum, and W. E. Davis, *J. Amer. Chem. Soc.*, **88**, 1275 (1966).
45. (a) F. Sondheimer, R. Yashin, G. Rosenkranz, and Carl Djerassi, *J. Amer. Chem. Soc.*, **75**, 1282 (1953); (b) A. J. Birch, H. Smith, and R. E. Thornton, *J. Chem. Soc.*, **1957**, 1339.
46. A. J. Birch and G. Subba Rao, *Aust. J. Chem.*, **22**, 761 (1969).
47. (a) A. J. Birch, *J. Roy. Inst. Chem.*, **1957**, 100; (b) A. J. Birch and D. Nasipuri, *Tetrahedron* **6**, 148 (1959); (c) A. J. Birch, D. Nasipuri and H. Smith, *Experientia,* **15**, 126, (1959).
48. W. A. Remers and M. J. Weiss, *Tetrahedron Lett.*, **1968**, 81.
49. E. Chablay, *Ann. Chim.*, **8**, 192 (1917).
50. C. M. Knowles and G. W. Watt, *J. Org. Chem.*, **7**, 60 (1942).
51. G. R. Clemo and T. J. King, *J. Chem. Soc.*, **1948**, 1661.
52. A. J. Birch, J. Cymerman-Craig, and M. Slaytor, *Aust. J. Chem.*, **8**, 512 (1955).
53. M. E. Kuehne and B. F. Lambert, *J. Amer. Chem. Soc.*, **81**, 4278 (1959).
54. A. J. Birch, *J. Chem. Soc.*, **1945**, 809.
55. N. A. Dobson and R. A. Raphael, *J. Chem. Soc.*, **1955**, 3558.
56. A. J. Birch, *J. Chem. Soc.*, **1947**, 1642.
57. A. J. Birch, *J. Chem. Soc.*, **1950**, 1551.
58. A. I. Shatenshtein, *Tetrahedron,* **18**, 95 (1962), and references cited therein.
59. J. E. Hofmann, P. A. Argabright, and A. Schriesheim, *Tetrahedron Lett.*, **1964**, 1005.
60. M. Smith, Ph.D. Thesis, University of Manchester, England, 1961.
61. M. Amagasa, T. Yamaguchi, and M. Tanaka, Brit. Pat. 1,033,760 (1966); *Chem. Abstr.*, **65**, 7075h (1966).
62. A. J. Birch, *Nature,* **158**, 60 (1946)
63. R. A. Benkeser, E. M. Kaiser and R. F. Lambert, *J. Amer. Chem. Soc.*, **86**, 5272 (1964).
64. R. A. Benkeser and C. A. Tincher, *J. Org. Chem.*, **33**, 2727 (1968).
65. A. F. Fentiman and R. H. Poirier, *Chem. Ind.* (London), **1966**, 813.
66. H. W. Sternberg, R. E. Markby, I. Wender, and D. M. Mohilner, *J. Amer. Chem. Soc.*, **89**, 186 (1967).
67. H. W. Sternberg, R. E. Markby, I. Wender, and D. M. Mohilner, *J. Electrochem. Soc.*, **113**, 1060 (1966).
68. T. J. King, *J. Chem. Soc.*, **1951**, 898.
69. M. J. Baldwin and R. K. Brown, *Can. J. Chem.*, **43**, 2621 (1965).
70. R. A. Benkeser, G. Schroll, and D. M. Sauve, *J. Amer. Chem. Soc.*, **77**, 3378 (1955).
71. B. S. Tyagi, B. B. Ghatge, and S. C. Bhattacharya, *Tetrahedron,* **19**, 1189 (1963).
72. D. J. Marshall and R. Deghenghi, *Can. J. Chem.*, **47**, 3127 (1969).
73. R. A. Benkeser, R. F. Lambert, P. W. Ryan, and D. G. Stoffey, *J. Amer. Chem. Soc.*, **81**, 228 (1959).

74. R. A. Benkeser, J. J. Hazdra, R. F. Lambert, and P. W. Ryan, *J. Org. Chem.*, **24**, 854 (1959).
75. A. J. Birch, P. L. Macdonald, and G. Subba Rao, unpublished work.
76. B. R. Ortiz de Montellano, B. A. Loving, T. C. Shields, and P. D. Gardner, *J. Amer. Chem. Soc.*, **89**, 3365 (1967).
77. G. Nagendra, R. K. Srivastava, and D. Devaprabhakara, *J. Org. Chem.*, **35**, 347 (1970).
78. J. Castells, G. A. Fletcher, E. R. H. Jones, G. D. Meakins, and R. Swindells, *J. Chem. Soc.*, **1960**, 2785.
79. N. L. Bauld, *J. Amer. Chem. Soc.*, **84**, 4347 (1962).
80. R. Y. Levina, V. R. Skvarchenko, V. N. Kostin Treschova, and A. S. Okunevich, *Chem. Abstr.*, **49**, 829 (1955).
81. W. Hückel and L. Hagedorn, *Chem. Ber.*, **90**, 752 (1957).
82. S. O'Brien and D. C. C. Smith, *J. Chem. Soc.*, **1960**, 4609.
83. S. F. Birch and D. T. McAllan, *Nature*, **165**, 899 (1950).
84. W. Hückel, S. Gupte, and M. Wartini, *Chem. Ber.*, **99**, 1388 (1966).
85. R. D. Stolow and J. A. Ward, *J. Org. Chem.*, **31**, 965 (1966).
86. A. J. Birch, A. Fogiel, and G. J. Harvey, *Aust. J. Chem.*, **7**, 261 (1954).
87. J. F. Eastham and D. R. Larkin, *J. Amer. Chem. Soc.*, **81**, 3652 (1959).
88. E. J Eisenbraun, R. C. Bansal, D. V. Hertzler, W. P. Duncan, P. W. K. Flanagan, and M. C. Hamming, *J. Org. Chem.*, **35**, 1265 (1970).
89. A. J. Birch and H. Smith, *Quart. Rev.*, **12**, 17 (1958).
90. A. J. Birch, *J. Chem. Soc.*, **1946**, 593.
91. H. Zimmerman, *Tetrahedron*, **16**, 169 (1961).
92. J. K. Brown, D. R. Burnham, and N. A. J. Rogers, *Tetrahedron Lett.*, **1966**, 2621.
93. A. Streitwieser, Jr., *Molecular Orbital Theory for Organic Chemists*, Wiley, New York, 1961.
94. D. R. Burnham, *Tetrahedron*, **25**, 897 (1969).
95. G. Stork and W. N. White, *J. Amer. Chem. Soc.*, **78**, 4604 (1956).
96. B. B. Millward, *J. Chem. Soc.*, **1960**, 26.
97. A. J. Birch, E. G. Hutchinson, and G. Subba Rao, *J. Chem. Soc., D*, **1970**, 657; *J. Chem. Soc., C*, **1971**, 637.
98. A. J. Birch, E. G. Hutchinson, and G. Subba Rao, unpublished work.
99. A. J. Birch, unpublished work.
100. A. J. Birch, *Nature*, **160**, 754 (1947).
101. J. Fried, N. A. Abraham, and T. S. Santhanakrishnan, *J. Amer. Chem. Soc.*, **89**, 1044 (1967).
102. A. J. Birch, P. Hextall, and S. Sternhell, *Aust. J. Chem.*, **7**, 256 (1954).
103. A. J. Birch, A. R. Murray, and H. Smith, *J. Chem. Soc.*, **1951**, 1945.
104. F. Camps, J. Coll, and J. Pascual, *J. Org. Chem.*, **32**, 2563 (1967)
105. A. J. Birch, J. A. K. Quartey, and H. Smith, *J. Chem. Soc.*, **1952**, 1768.
106. (a) S. Mejer and S. Respondek, *Bull. Acad. Pol. Sci. Ser. Sci. Chim.*, **14**, 611 (1966); *Chem. Abstr.*, **66**, 105088a (1967); (b) A. J. Birch and G. Subba Rao, *Aust. J. Chem.*, **23**, 547 (1970).
107. A. J. Birch and G. Subba Rao, unpublished work.
108. H. Alt, E. R. Franke, and H. Bock, *Angew. Chem. Int. Ed.*, **8**, 525 (1969).
109. (a) H. Bock, H. Alt, and H. Siedl, *J. Amer. Chem. Soc.*, **91**, 355 (1969); (b) H. Alt, H. Bock, F. Gerson, and J. Heinzer, *Angew. Chem. Int. Ed.*, **6**, 941 (1967).
110. A. J. Birch and G. Subba Rao, *Aust. J. Chem.*, **23**, 1641 (1970).
111. L. G. Wideman, *J. Org. Chem.*, **35**, 1698 (1970).
112. W. Hückel and R. Schwen, *Chem. Ber.*, **89**, 150 (1956).
113. D. F. De Tar and R. A. J. Long, *J. Amer. Chem. Soc.*, **80**, 4742 (1958).
114. Iv. Egorov, E. P. Kaplan, Z. I. Letina, V. A. Shliapochnikov, and A. D. Petrov, *J. Gen. Chem. USSR*, **28**, 3284 (1958).

115. P. J. Grisdale, T. H. Regan, J. C. Doty, J. Figueras, and J. L. R. Williams, *J. Org. Chem.*, **33**, 1116 (1968).
116. A. J. Birch and G. Subba Rao, *Tetrahedron Lett.*, **1967**, 857.
117. J. Fried and N. A. Abraham, *Tetrahedron Lett.*, **1965**, 3505.
118. A. J. Birch, R. W. Rickards, and K. S. J. Stapleford, *Aust. J. Chem.*, **22**, 1321 (1969).
119. C. B. Wooster and J. F. Ryan, *J. Amer. Chem. Soc.*, **56**, 1133 (1934).
120. W. G. Dauben and E. J. Deviny, *J. Org. Chem.*, **31**, 3794 (1966).
121. C. B. Wooster, *J. Amer. Chem. Soc.*, **56**, 2436 (1934).
122. (a) J. Grimshaw and H. R. Juneja, *Chem. Ind.* (London), **1960**, 656; (b) A. J. Birch, J. Grimshaw, and H. R. Juneja, *J. Chem. Soc.*, **1964**, 5194.
123. J. M. Brown, Ph.D. Thesis, University of Manchester, England, 1963.
124. A. S. Hallsworth, H. B. Henbest, and T. I. Wrigley, *J. Chem. Soc.*, **1957**, 1969.
125. A. J. Birch, *J. Proc. Roy. Soc. N.S.W.*, **83**, 245 (1950).
126. A. J. Birch and G. Subba Rao, *Aust. J. Chem.*, **22**, 2037 (1969).
127. R. A. Benkeser, C. Arnold, Jr., R. F. Lambert, and O. H. Thomas, *J. Amer. Chem. Soc.*, **77**, 6042 (1955).
128. A. S. Hallsworth and H. B. Henbest, *J. Chem. Soc.*, **1957**, 4604.
129. E. J. Reist, V. J. Bartuska, and L. Goodman, *J. Org. Chem.*, **29**, 3725 (1964).
130. U. G. Nayak and R. K. Brown, *Can. J. Chem.*, **44**, 591 (1966).
131. P. A. Sartoretto and F. J. Sowa, *J. Amer. Chem. Soc.*, **59**, 603 (1937).
132. K. Freudenberg, W. Lautsch, and G. Piazolo, *Chem. Ber.*, **74**, 1879 (1941).
133. A. J. Birch, *J. Chem. Soc.*, **1947**, 102.
134. A. J. Birch and S. Dunstan, unpublished work.
135. G. W. Kenner and N. R. Williams, *J. Chem. Soc.*, **1955**, 522.
136. M. Fetizon, M. Jurion, and N. T. Anh, *J. Chem. Soc.*, **1969D**, 112.
137. R. E. Ireland and G. Pfister, *Tetrahedron Lett.*, **1969**, 2145.
138. A. J. Birch, J. M. Brown, and G. Subba Rao, *J. Chem. Soc.*, **1964**, 3309.
139. M. C. Hoff, K. W. Greenlee, and C. E. Boord, *J. Amer. Chem. Soc.*, **73**, 3329 (1951).
140. D. B. Clayson, *J. Chem. Soc.*, **1949**, 2016.
141. J. T. Wrobel, A. S. Bien, and K. M. Pazdro, *Chem. Ind.* (London), **1966**, 1759.
142. J. T. Wrobel, K. M. Pazdro, and A. S. Bien, *Chem. Ind.* (London), **1966**, 1760.
143. E. Wenkert and B. G. Jackson, *J. Amer. Chem. Soc.*, **80**, 217 (1958).
144. W. J. Adams, D. K. Patel, V. Petrow and I. A. Stuart-Webb, *J. Chem. Soc.*, **1956**, 297.
145. R. G. Carlson and R. G. Blecke, *Chem. Commun.*, **1969**, 93.
146. G. Stork, S. Malhotra, H. Thompson, and M. Uchibayashi, *J. Amer. Chem. Soc.*, **87**, 1148 (1965).
147. C. D. Gutsche and I. T. C. Tao, *J. Org. Chem.*, **28**, 883 (1963).
148. G. Stork and T. Tsuji, *J. Amer. Chem. Soc.*, **83**, 2783 (1961).
149. E. E. Van Tamelen, T. A. Spencer, Jr., D. S. Allen, Jr., and R. L. Orvis, *Tetrahedron*, **14**, 8 (1961).
150. P. A. Robins and J. Walker, *J. Chem. Soc.*, **1956**, 3249.
151. J. W. Huffman and J. T. Charles, *J. Amer. Chem. Soc.*, **90**, 6486 (1968).
152. G. Stork and S. D. Darling, *J. Amer. Chem. Soc.*, **86**, 1761 (1964).
153. D. H. Williams, J. M. Wilson, H. Budzikiewicz, and C. Djerassi, *J. Amer. Chem. Soc.*, **85**, 2091 (1963).
154. P. W. Rabideau and R. G. Harvey, *J. Org. Chem.*, **35**, 25 (1970).
155. R. G. Harvey, L. Arzadon, J. Grant, and K. Urberg, *J. Amer. Chem. Soc.*, **91**, 4535 (1969).
156. J. Sicher, M. Svoboda, and J. Zavada, *Collect. Czech, Chem. Commun.*, **30**, 413, 421 (1965).
157. G. F. Hennion and T. F. Banigan, *J. Amer. Chem. Soc.*, **68**, 1202 (1949).
158. R. B. Bates, D. M. Gale, B. J. Gruner, and P. Nicholas, *Chem. Ind.* (London), **1961**, 1907.

159. S. Fujise, H. Obara, and H. Uda, *Chem. Ind.* (London), **1960,** 289.
160. W. H. Lunn, D. W. Edward, and J. T. Edward, *Can. J. Chem.,* **40,** 104 (1962).
161. (a) A. J. Birch, R. A. Massy-Westropp, and R. W. Rickards, *Chem. Ind.* (London), **1955,** 1599; (b) A. J. Birch, R. A. Massy-Westropp, and R. W. Rickards, *J. Chem. Soc.,* **1956,** 3717.
162. R. B. Bates, J. H. Schauble, and M. Soucek, *Tetrahedron Lett.,* **1963,** 1683.
163. M. Tomita, E. Fujita, and F. Murai, *J. Pharm. Soc. Japan,* **71,** 226, 1035 (1951).
164. A. J. Birch, J. W. Clark-Lewis, and A. V. Robertson, *J. Chem. Soc.,* **1957,** 3586.
165. B. R. Brown and G. A. Somerfield, *Proc. Chem. Soc.,* **1958,** 236.
166. A. J. Birch, B. Milligan, E. Smith, and R. N. Speake, *J. Chem. Soc.,* **1958,** 4471.
167. A. J. Birch, P. L. Macdonald, and A. Pelter, *J. Chem. Soc.,* **1967C,** 1968.
168. G. K. Hughes and E. Ritchie, *Aust. J. Chem.,* **7,** 104 (1954).
169. A. J. Birch, M. Maung, and A. Pelter, *Aust. J. Chem.,* **22,** 1923 (1969).
170. (a) S. C. Bhattacharji, A. J. Birch, A. Hoffman, H. Kobel, D. C. C. Smith, H. Smith and J. Winter, *J. Chem. Soc.,* **1962,** 421; (b) B. Naidoo, J. M. Cassady, G. E. Blair, and H. G. Floss, *Chem. Commun.,* **1970,** 471.
171. W. G. Dauben, W. E. Thiessen, and P. R. Resnick, *J. Org. Chem.,* **30,** 1693 (1965).
172. (a) W. H. Pirkle and M. Gates, *J. Org. Chem.,* **30,** 1769 (1965); (b) Y. Sawa, N. Tsuji, and S. Meada, *Tetrahedron,* **15,** 144, 154 (1961).
173. W. S. Johnson, A. D. Kemp, R. Pappo, J. Ackerman, and W. F. Johns, *J. Amer. Chem. Soc.,* **78,** 6312 (1956).
174. K. E. Hamlin and F. E. Fischer, *J. Amer. Chem. Soc.,* **75,** 5119 (1953).
175. J. P. Kutney, W. J. Cretney, P. Le Quesne, B. McKague, and E. Piers, *J. Amer. Chem. Soc.,* **92,** 1712 (1970).
176. E. Wenkert, S. Garratt, and K. G. Dave, *Can. J. Chem.,* **42,** 489 (1964).
177. V. du Vigneaud and O. K. Behrans, *J. Biol. Chem.,* **117,** 27 (1937).
178. J. Ramachandran, D. Chung, and C. H. Li, *J. Amer. Chem. Soc.,* **87,** 2696 (1965).
179. P. G. Kastoyanis, K. Fukuda, A. Tometso, K. Suzuki, and M. Tilak, *J. Amer. Chem. Soc.,* **86,** 930 (1964).
180. J. G. Moffatt and H. G. Khorana, *J. Amer. Chem. Soc.,* **83,** 663 (1961).
181. J. A. Marshall and J. J. Partridge, *Tetrahedron,* **25,** 2159 (1969).
182. K. S. J. Stapleford, Ph.D. Thesis, University of Manchester, England, 1967.
183. (a) W. S. Johnson, B. Bannister, R. Pappo, and J. E. Pike, *J. Amer. Chem. Soc.,* 78, 6354 (1956); (b) W. S. Johnson, W. A. Vredenburgh, and J. E. Pike, *J. Amer. Chem. Soc.,* **82,** 3409 (1960).
184. (a) R. Ginsig and A. D. Cross, *J. Amer. Chem. Soc.,* 87, 4629 (1965); (b) H. D. Brendt and R. Wiechert, *Angew. Chem. Int. Ed.,* **8,** 376 (1969).
185. W. S. Johnson, N. P. Jensen, J. Hooz, and E. J. Leopald, *J. Amer. Chem. Soc.,* **90,** 5872 (1968).
186. G. Stork, R. A. Kretchmer. and R. H. Schlessinger, *J. Amer. Chem. Soc.,* **90,** 1647 (1968).
187. (a) G. A. Hughes and H. Smith, Fr. Pat. 1,394,701 (1965); *Chem. Abstr.,* **63,** 4371a (1965); (b) G. H. Douglas, J. M. H. Graves, D. Hartley, G. A. Hughes, B. J. McLoughlin, J. B. Siddall, and H. Smith, *J. Chem. Soc.,* **1963,** 5072.
188. (a) A. J. Birch, P. L. MacDonald, and V. H. Powell, *Tetrahedron Lett.,* **1969,** 351; (b) A. J. Birch, P. L. MacDonald, and V. H. Powell, *J. Chem. Soc.,* **1970C,** 1469.
189. A. J. Birch and P. L. MacDonald, unpublished work.
190. N. A. J. Rogers, personal communication.
191. A. J. Birch, P. E. Cross, J. Lewis, D. A. White, and S. B. Wild, *J. Chem. Soc.,* **1968A,** 332.
192. W. Nagata, S. Hirai, T. Okumura, and K. Kawata, *J. Amer. Chem. Soc.,* **90,** 1650 (1968).

193. (a) E. Vogel and H. D. Roth, *Angew. Chem. Int. Ed.,* **3,** 228 (1964); (b) E. Vogel and W. A. Boll, *Angew. Chem. Int. Ed.,* **3,** 642 (1964).
194. G. Subba Rao, unpublished work.

Supplementary References

C. A. Kraus, *Chem. Rev.,* **8,** 251 (1931).
K. N. Campbell and B. K. Campbell, *Chem. Rev.,* **31,** 77 (1942).
A. J. Birch, *Quart. Rev.* (London), **4,** 69 (1950).
G. W. Watt, *Chem. Rev.,* **46,** 317 (1950).
A. J. Birch and H. Smith, *Quart. Rev.* (London), **12,** 17 (1958).
W. Hückel, *Fortsch. Chem. Forsch.,* **6,** 197 (1966).
H. Normant, *Angew. Chem. Int. Ed.,* **6,** 1046 (1967).
U. Schindewolf, *Angew. Chem. Int. Ed.,* **7,** 190 (1968).
R. G. Harvey, *Synthesis,* **1970,** 161.
H. Smith, *Organic Reactions in Liquid Ammonia,* Wiley-Interscience, New York, 1963.
C. Djerassi, ed., *Steroid Reactions,* Holden-Day, San Francisco, 1963, Chaps. 6 and 7.
G. Lepoutre and M. J. Sienko, eds., *Metal Ammonia Solutions,* Benjamin, New York, 1964.
Solvated Electron, American Chemical Society Publication, 1965.
H. O. House, *Modern Synthetic Reactions,* Benjamin, New York, 1965, Chap. 3.
J. J. Lagowski, ed., *The Chemistry of Nonaqueous Solvents,* Vol. 2, Academic Press, New York, 1967, pp. 265-317.
M. Smith, in *Reduction* (R. L. Augustine, ed.), Dekker, New York, 1967.

MODERN METHODS FOR THE SYNTHESIS OF MACROCYCLIC COMPOUNDS

By PAUL R. STORY and PETER BUSCH, *Department of Chemistry, The University of Georgia, Athens, Georgia*

CONTENTS

I. Introduction

Macrocyclic compounds contain as an essential component a large ring; large rings are generally considered to be those containing 12, 13, and more atoms, while medium-size rings fall in the range of 8 to 12 atoms (1). The new syntheses treated here make this categorization awkward and we will consider that any ring size of 9 atoms or greater will fall within the scope of our discussion of macrocyclic systems.

The chemistry of many-membered rings is relatively unexplored, primarily because of the considerable difficulties attendant to their synthesis. This brief chapter will review several recent developments which have served to reduce entry to macrocyclic, and especially carbocyclic, systems almost to the ordinary. These new methods are placed in perspective by a summary of the history of macrocyclic synthesis.

The new methods include our own synthesis of macrocyclic compounds by ketone peroxide fragmentation (2), the cyclic olefin metathesis reaction of Calderon (3,4) and Wasserman (5), Corey's allylic dibromide cyclization (6), and acetylene cyclopolymerization developed by Sondheimer (7). The most significant aspect of several of the new macrocyclic syntheses is that large rings can now be generated directly from readily available monomers, thus obviating the necessity, as with all previously existing methods, of using long-chain difunctional precursors obtained only with difficulty. The ketone peroxide fragmentation, olefin metathesis, and acetylene cyclopolymerization methods all offer this important advantage.

II. Historical

Macrocyclic chemistry dates from 1926 when Leopold Ruzicka (8) established the large ring structures for the musk ketones, muscone and civetone, prized as perfume conponents. Muscone had been isolated earlier by Walbaum (9) from natural musk and characterized as having the empirical formula $C_{16}H_{30}O$ or $C_{15}H_{28}O$. In 1912 civetone had been isolated from civet [Civet cat musk] by Sack and found to have a pleasant musk odor (10).

Ruzicka's establishment of the existence of large ring compounds in the face of von Baeyer's strain theory, which implied that such compounds would be too unstable to exist, and his subsequent studies of large ring compounds led to the Nobel Prize in 1939.

The first synthesis of macrocyclic compounds (ketones) was achieved by Ruzicka and his students, Stoll, Schinz, and Brugger, in rather low yield through pyrolysis of heavy metal salts of long-chain dicarboxylic acids (11), a method previously applied by Zelinsky (12) and by Willstätter (13) to the synthesis of smaller ring systems.

Subsequently, Karl Ziegler, applying Ruggli's observation (14) that intramolecular ring closure is favored under conditions of high dilution, used the Thorpe reaction (15) to synthesize muscone and other macrocyclic ketones in high yield (16). The yields of even-numbered ring ketones were especially high (eq. 1).

The high dilution technique was extended by Hunsdiecker in the synthesis of both muscone and civetone, utilizing ω-halo-β-keto esters in another carbanionic cyclization reaction (17).

$$(CH_2)_n\begin{cases}CH_2-C\equiv N\\ CH_2-C\equiv N\end{cases}\xrightarrow[\text{ether}]{LiEtNC_6H_5}(CH_2)_n\begin{cases}C-C\equiv N\\ \;\;\|\\ C-NH_2\\ CH_2\end{cases}\longrightarrow (CH_2)_n\begin{cases}CH_2\\ \;\;\;\;\;\;\;>C=O\\ CH_2\end{cases} \quad (1)$$

Again, at high dilution, Blomquist developed a synthesis of macrocyclic ketones via cyclization of diketenes generated, *in situ,* from acid chlorides (18) (eq. 2).

$$(CH_2)_n\begin{cases}CH_2-\overset{O}{\overset{\|}{C}}-Cl\\ CH_2-\underset{O}{\underset{\|}{C}}-Cl\end{cases}\xrightarrow[Et_2O]{Et_3N}\left[(CH_2)_n\begin{cases}CH=C=O\\ CH=C=O\end{cases}\right]\longrightarrow \begin{matrix}CH=C-O\\ | \quad\;\; | \quad\;\; |\\ (CH_2)_nCH-C{=}O\end{matrix}$$

$$\xrightarrow{KOH} \begin{matrix}CH_2-C{=}O\\ / \qquad\quad |\\ (CH_2)_n-CH-COO^{\ominus}\end{matrix}\xrightarrow[\Delta]{-CO_2}(CH_2)_{n+2}\;C=O \quad (2)$$

All of the synthetic methods, to this point in time, suffered two serious disadvantages: (*1*) a long-chain difunctional precursor was required; and (*2*) the cyclizations had to be carried out at high dilution. The latter disadvantage was finally overcome in 1947 by Prelog (19) and by Stoll (20) using the acyloin condensation originally developed by Hansley (21). High dilution is not a requirement for the acyloin condensation since the reaction takes place on the surface of finely divided sodium. The yields of cyclic products are spectacularly high. The initial product of the condensation, an acyloin, is readily converted to a macrocyclic ketone by zinc-hydrochloric acid reduction (eq. 3).

$$(CH_2)_{14}\begin{cases}COOR\\ COOR\end{cases}\xrightarrow[\text{hot toluene}]{Na}(CH_2)_{14}\begin{matrix}C=O\\ |\\ CHOH\end{matrix}\xrightarrow[HCl]{Zn}(CH_2)_{15}\;C=O \quad (3)$$

Wasserman has used the acyloin condensation in an interesting application of this technique to prepare catenanes (interlocked large rings) (22). Catenane formation was effected by carrying out an acyloin condensation with diethyl tetratriacontanedioate in the presence of partially deuterated C-34 cyclic hydrocarbon. A very low yield of the topological isomer was obtained, as evidenced by subsequent ring opening to release the deuterated hydrocarbon.

III. Sondheimer Synthesis

Development of the acyloin condensation for macrocyclic synthesis was followed some 9 years later, in 1956, by the next significant finding in the synthesis of large-ring compounds. Sondheimer successfully applied the oxidative coupling of terminal diacetylenes (23) to the preparation of macrocyclic compounds (24). Eglinton and Galbraith, at about the same time, also generated large-ring compounds from the oxidative coupling of α-ω diynes at high dilution (25). The Sondheimer synthesis, although it is characterized by low yields in most examples, was the first macrocyclic synthesis which required neither conditions of high dilution nor expensive and difficult to obtain long-chain difunctional precursors. Illustrative is the first Sondheimer report which dealt with the oxidative coupling of octa-1,7-diyne in aqueous ethanol at 55°C with air using cuprous chloride and ammonium chloride as catalysts (24a,c) (eq. 4). The cyclic dimer was obtained in a total yield of 8% and could be hydrogenated to the known cyclohexadecane.

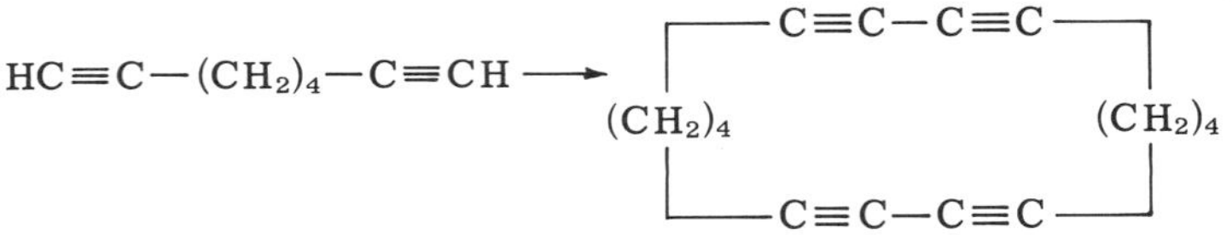

$$HC{\equiv}C{-}(CH_2)_4{-}C{\equiv}CH \longrightarrow \text{cyclo}[{-}C{\equiv}C{-}C{\equiv}C{-}(CH_2)_4{-}C{\equiv}C{-}C{\equiv}C{-}(CH_2)_4{-}] \qquad (4)$$

The reaction was subsequently repeated using cupric acetate in pyridine at 55°C to give the cyclic dimer in 8.6% yield along with a mixture of several cyclic polymers which were characterized up through the cyclic hexamer (24e).

Cyclopolymerization of 1,5-hexadiyne with cupric acetate in pyridine led to formation of the cyclic trimer, cyclooctadeca-1,3,7,9,13,15-hexayne, in approximately 6% yield. Treatment of this material with potassium *t*-butoxide in *t*-butyl alcohol at 90°C for 25 min generated the rather interesting fully conjugated hexaene-triyne system which showed considerable aromatic character consistent with its 18-π-electron planar structure (eq. 5). The Sondheimer synthesis is not only applicable to acetylenic hydrocarbons but to acetylenic alcohols, amines, nitro compounds, carboxylic acids, and esters as well.

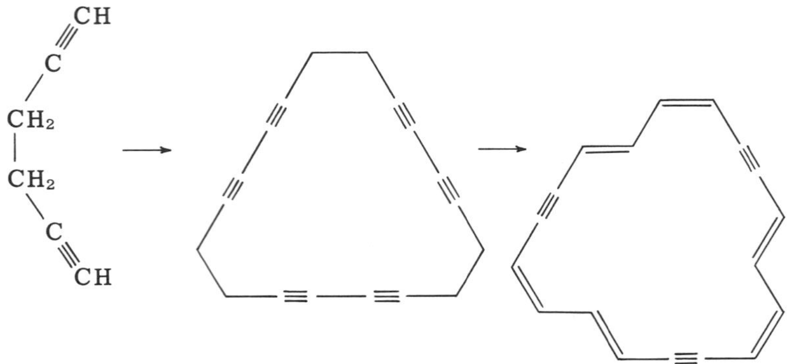

$$HC{\equiv}C{-}CH_2{-}CH_2{-}C{\equiv}CH \longrightarrow \longrightarrow \qquad (5)$$

IV. Special Syntheses

A number of syntheses for particular macrocyclic compounds have been developed; a few have some general applicability but have received only limited use, while others are clearly special cases. One example of a synthesis which would be expected to have some general applicability is provided by an application of the Wittig reaction to generate macrocyclic systems (26) (eq. 6). The reaction has also been employed for the preparation of macrocyclic ketones (27).

$$(CH_2)_n\begin{cases}CH_2Br\\CH_2Br\end{cases} \longrightarrow (CH_2)_n\begin{cases}CH{=}PPh_3\\CH{=}PPh_3\end{cases} + \begin{matrix}O{=}CH\\|\\(CH_2)_m\\|\\O{=}CH\end{matrix} \longrightarrow (CH_2)_n\begin{matrix}CH{=}CH\\ \\CH{=}CH\end{matrix}(CH_2)_m \quad (6)$$

Goldfarb and his co-workers, in an unusual synthesis, condensed thiophene with ω-chloroalkanoic acids and converted the reaction product, in several steps, to the cyclic ketone (28) (eq. 7).

Thiophene $\xrightarrow{HO_2C(CH_2)_8Cl}$ 2-$CO(CH_2)_8Cl$-thiophene $\xrightarrow[HCl]{Zn}$

2-$(CH_2)_9Cl$-thiophene $\longrightarrow$ 2-$(CH_2)_9Cl$-5-$COCH_2COOEt$-thiophene $\longrightarrow$

S-ring bridged by $(CH_2)_9$ — CH–COOEt — C=O $\xrightarrow[CH_3OH]{HCl}$ S-ring bridged by $(CH_2)_9$ — CH_2 — C=O $\quad (7)$

$\xrightarrow[H_2]{Ni}$ $(CH_2)_{14}$ C=O (cyclic)

Hubert and Dale have used the Gilman procedure (29) for the synthesis of large-ring compounds by condensing disodio-diynes with α,ω-dibromides (30).

Finally, the Dieckmann condensation has been applied to macrocyclic synthesis with success by Leonard and Schimelpfenig, who prepared several large-ring ketones and diketones in good yields (31).

V. Ring Enlargements, Ring Contractions

A number of important ring enlargement reactions have been devised within the last 10 years for the synthesis of macrocyclic compounds. These reactions, and for that matter, ring contraction reactions, assume an even greater importance with the increasing availability of certain medium- and large-ring compounds. Some of the most important of the new methods will be described here.

A. *BERCHTOLD-BRANNOCK ENAMINE-ACETYLENE CONDENSATION*

Berchtold (32) and Brannock (33) developed a synthetic method which consists of the condensation of a cyclic ketone, in the form of its enamine, with acetylene dicarboxylic ester to effect a ring enlargement by two carbon atoms (eq. 8).

N—
C
$(CH_2)_n$
C
+
COOR
C
C
COOR
N
COOR
$(CH_2)_n$
COOR
(8)
Δ
N
COOR
$(CH_2)_n$
COOR

B. *ACYLOIN-TRIMETHYLCHLOROSILANE RING ENLARGEMENT, NOZAKI*

Nozaki was able to effect a two-carbon ring enlargement through a modified intramolecular acyloin condensation reaction (34). In this way, 11-, 12-, and 13-membered rings were converted to 13-, 14-, and 15-membered rings, respectively. The acyloin condensation was carried out in the presence of trimethylchlorosilane, a procedure first described by Schräpler and Rühlmann (35). In this way, cyclododecanone, for example, was first converted by known procedures to dimethylcyclotridecane-1,2-dicarboxylate and then subjected to acyloin conditions in the presence of trimethylchlorosilane to yield, after work-up, 2-hydroxycyclopentadecanone. Alternatively, there is now a new method available for the conversion of α-diketones to ketones in good yield by treatment of the diketone with hydroiodic acid in refluxing acetic acid (36) (eq. 9).

$$(CH_2)_{11}\left[\begin{array}{l}CH{-}COOCH_3\\ |\\ CH{-}COOCH_3\end{array}\right. \xrightarrow[(CH_3)_3SiCl]{Na} (CH_2)_{11}\left[\begin{array}{l}CH{-}C{-}OSi(CH_3)_3\\ |\quad\ \ \|\\ CH{-}C{-}OSi(CH_3)_3\end{array}\right. \xrightarrow{\Delta}$$

$$(CH_2)_{11}\left[\begin{array}{l}CH{=}C{-}OSi(CH_3)_3\\ \qquad\ |\\ CH{=}C{-}OSi(CH_3)_3\end{array}\right. \xrightarrow{H^+} (CH_2)_{13}\left[\begin{array}{l}C{=}O\\ |\\ C{=}O\end{array}\right. \xrightarrow[2)\ KOH]{1)\ (CH_3CH_2O)_3P} (CH_2)_{13}\left[\begin{array}{l}CHOH\\ |\\ C{=}O\end{array}\right. \tag{9}$$

C. PHOTOCHEMICAL CLEAVAGE OF BICYCLIC KETONES, NOZAKI

Nozaki has also accomplished a three-carbon ring enlargement by photochemical cleavage of an appropriate bicyclic ketone (37). Cyclododecanone was converted into 2-carbethoxycyclododecanone (38), which yielded 1-carbethoxybicyclo[9.4.1]hexadecan-16-one on condensation with 1,4-dibromobutane in the presence of sodium hydride. Hydrolysis of the ketone, followed by decarboxylation, gave bicyclo[9.4.1]hexadecan-16-one in 65 to 72% yield. Photolysis of this ketone in the presence of oxygen gave cyclopentadecanone in 20 to 30% yield (eq. 10).

$$(CH_2)_{11}\ C{=}O \longrightarrow (CH_2)_{10}\left[\begin{array}{l}CH{-}COOEt\\ |\\ C{=}O\end{array}\right. \longrightarrow$$

$$\text{[bicyclo[9.4.1]hexadecan-16-one, 1-COOEt]} \xrightarrow[3)\ h\nu]{1)\ H^+,\ 2)\ \Delta} \left[(CH_2)_{14}\ C{=}C{=}O\right] \longrightarrow (CH_2)_{14}\ C{=}O \tag{10}$$

D. CYCLODODECANONE AS A BUILDING BLOCK

The now inexpensive cyclododecanone (produced from 1,5,9-cyclododecatriene which is readily available commercially from butadiene cyclotrimerization) has been used in several other syntheses, in addition to those already mentioned, to produce the more expensive macrocyclic musk compounds (39). In general terms, the cyclododecanone is used to construct a bicyclic intermediate containing the proper number of ring members; this intermediate is fragmented to the monocyclic ring-enlarged product.

E. OHLOFF-ESCHENMOSER SYNTHESIS

Ohloff (40) converted cyclododecanone via a Stobbe condensation, and two additional steps, to bicyclo[10.3.0]pentadec-14-en-1-one (41). Hydrogenation, followed by dehydration of the intermediate saturated alcohol, led to bicyclo[10.3.0]-1(12)-pentadecene. Treatment of this olefin with ozone or, alternatively, generation of the hydroperoxide by photooxidation followed by its reaction with boron trifluoride etherate, yields 5-oxocyclopentadecanone. Partial reduction and dehydration gave an isomeric mixture of cyclopentadecenones. The final product, cyclopentadecanone, is easily obtained, of course, by hydrogenation of the mixture.

A somewhat shorter route to cyclopentadecanone from bicyclo[10.3.0]pentadec-14-en-1-one was subsequently devised by Eschenmoser (42) who treated the tosylhydrazone of the epoxy ketone with either acid or base, thereby yielding 4-cyclopentadecyn-1-one. The acetylenic ketone can be hydrogenated to either the olefinic ketone, *cis*-4-cyclopentadecen-1-one, or to cyclopentadecanone. Neither Ohloff nor Eschenmoser has yet published any detailed experimental data (eq. 11).

F. RING CONTRACTION, GARBISCH

Up until recently the use of a ring contraction reaction in the synthesis of large rings had not been very interesting because of the obvious difficulty in obtaining the necessary precursor molecule. However, now, because of the ready commercial availability of cyclododecanone, and because our new peroxide method will provide very inexpensive macrocyclic systems such as cyclopentadecane and cyclodocosane, the modified Favorskii rearrangement developed by Garbisch becomes much more attractive. Garbisch reports, for example, a 77% overall yield in the preparation of cycloundecanone from cyclododecanone (43).

VI. 1,2-Cycloaddition Reactions

Several examples of a new class of syntheses best described as involving a 1,2-cycloaddition reaction have recently appeared.

A. BRYCE-SMITH SYNTHESIS

A procedure of limited applicability has been reported by Bryce-Smith (44) who irradiated a mixture of cyclooctene and chloranil to give a mixture of the oxetane and the appropriate pentacyclic addition product. The latter, on heating in tetralin solution, rearranged to the C-22 tetrachlorodiketone (eq. 12).

B. ADDITION TO ACETYLENES

Heimbach reports the synthesis of cycloeicosan-1,10-dione through the nickel-catalyzed addition of two molecules of butadiene to cyclododecyne followed by selective hydrogenation and oxidative cleavage of the resulting olefin (45) (eq. 13).

$h\nu$

Δ (12)

$(CH_2)_{10}$ + Ni → $(CH_2)_{10}$

(13)

C. *THE OLEFIN METATHESIS REACTION*

The most important, potentially most useful, new synthesis which falls into the category of a 1,2-cycloaddition is that provided by the olefin metathesis reaction developed by Calderon (46,47) and investigated also by Wasserman (48). Wasserman, using the modified Ziegler catalyst developed by Natta (49) and co-workers, reported the conversion of cyclooctene to cyclohexadecadiene, cyclotetracosatriene, cyclotetracontatetraene, and so on, through the macrocyclic polyene containing 120 carbon atoms. Cyclooctene was chosen for the initial work on the expectation that the ring strain of the C-8 system would drive the reaction to the relatively strainless C-16 and higher cyclic polyolefins (eq. 14). No reports have yet appeared concerning optimization of yield for any particular ring size or extension of the reaction to other cycloolefinic starting materials.

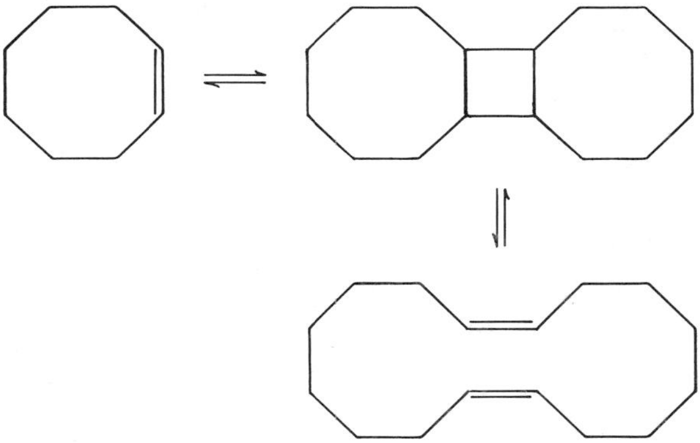

(14)

D. *1,4-CYCLOPOLYMERIZATION*

Conceptually, but not categorically, related to the 1,2-cycloaddition reaction is the cyclopolymerization of butadiene, a process which has placed both cycloocta-1,5-diene and 1,5,9-cyclododecatriene on a readily available commercial basis and thereby contributed materially to the study of macrocyclic systems (50). Many large-ring systems have been generated from the C-8 and C-12 intermediates so derived, using one of the several ring expansion and ring contraction reactions now available.

VII. The Corey Synthesis

Using what is, in effect, a high dilution technique, Corey has been able to carry out the cyclization of allylic dibromides by the slow addition of the allylic dibromide to a dimethylformamide solution of nickel carbonyl (6). The dibromides required for this synthesis were obtained from the corresponding acetylenic diols which, in turn, were prepared by alkylation of the appropriate dibromide with the sodio derivative of the tetrahydropyranyl ether of propargyl alcohol, using either liquid ammonia-ether or tetrahydrofuran-dimethylsulfoxide as solvent.

In the case of the lower members of this series of allylic dibromides, in other words, where n = 2 or 4, the predominate course of the reaction was found to be formation of six-membered rings through 1,6 or 3,8 coupling; the geometry of the starting dibromide appeared to be essentially irrelevant to the product composition. For n = 2, using nickel carbonyl as the catalyst, the products were 4-vinylcyclohexene (42%) and *cis,cis*-1,5-cyclooctadiene (58%). Using triphenylphosphine nickel tricarbonyl, the same products were obtained but with somewhat greater yields of the 1,5-cyclooctadiene. However, in the case of dibromide where n = 4, using either catalyst, only the *cis*- and *trans*-1,2-divinylcyclohexane was obtained. For the homologous allylic dibromides of greater chain length, n = 6, 8, or 12, the yield of the macrocyclic compound becomes extremely interesting. The yield of cyclotetradecadiene for n = 8 was found to be 70 to 74% (eq. 15).

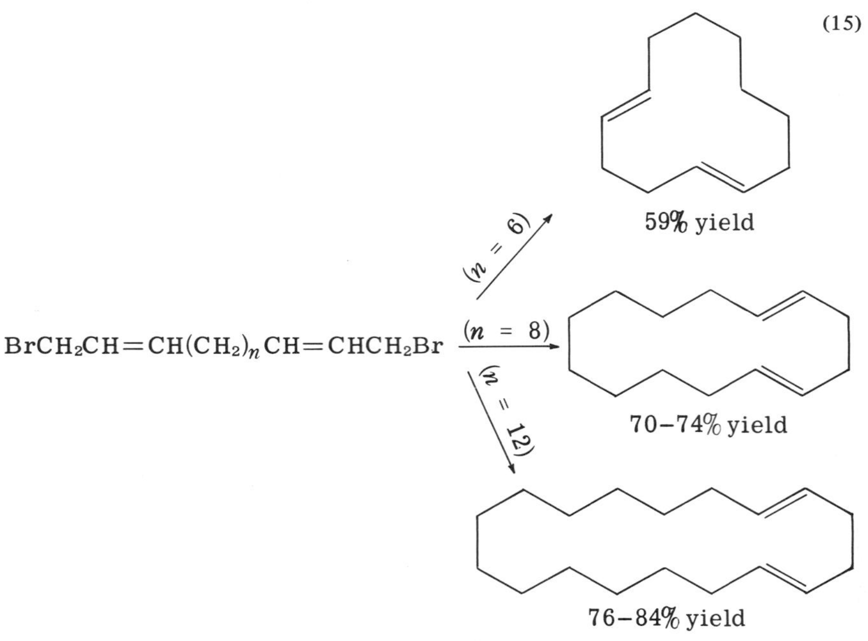

Corey has already applied this synthetic method to the total synthesis of humulene and has indicated its potential applicability to the synthesis of several other natural products (6). In addition, he is studying the application of the method to the synthesis of very large ring systems.

VIII. Macrocyclic Polyethers

In 1967 C. J. Pederson at duPont disclosed several methods for the preparation of cyclic polyethers (51). Few examples of this type of macrocyclic compound were known at that point, and those due primarily to the work of A. Lüttringhaus (a literature survey is provided in Ref. 51). The unusual solvent properties exhibited by Pederson's macrocyclic polyethers have caused them to be considered more than just an idle curiosity. Pederson has prepared a total of 33 cyclic polyethers, derived from vicinal diols, and containing 9 to 60 atoms, including 3 to 20 oxygen atoms in the ring. Some of these have been prepared in reasonably good yields even without the use of a high dilution technique. Those ethers containing 5 to 10 oxygens were found to have very interesting solvent properties by virtue of their ability to form stable complexes with many different metallic cations. Pederson described four synthetic methods; a typical example is provided by eq. (16).

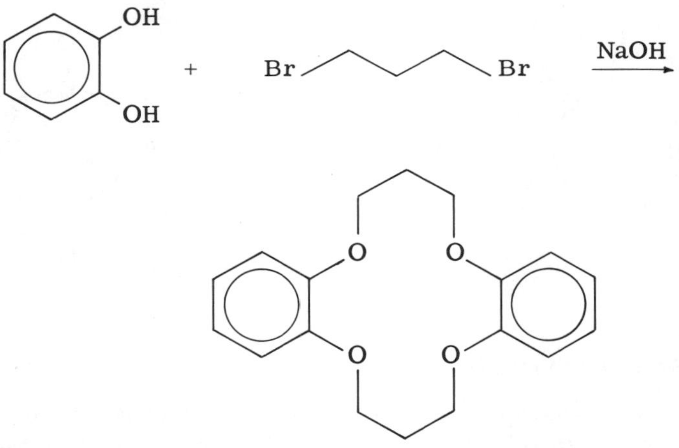

OH OH + Br Br NaOH → O O O O (16)

"dibenzo-14-crown"

IX. The Story Synthesis

None of the synthetic methods yet discussed has the useful generality necessary to permit vigorous research activity in macrocyclic chemistry. No one method is applicable to the synthesis of every ring size and beyond that, the synthesis of multisubstituted macrocyclic compounds is almost an impossibility.

The peroxide fragmentation synthesis developed in our laboratories in 1968 (2) gives every promise of being a method capable of providing virtually any macrocyclic compound. We have now used the method to synthesize every carbocyclic system from C-8 through C-33 plus C-42 and one macrocyclic ether system. In addition, many substituted macrocyclic compounds have been synthesized. The method is best illustrated by the thermal decomposition of tricyclohexylidene peroxide (trimeric peroxide of cyclohexanone) to yield cyclopentadecane, 16-hexadecanolide (dihydroambrettolide), and a small amount of the isomeric dilactones (eq. 17).

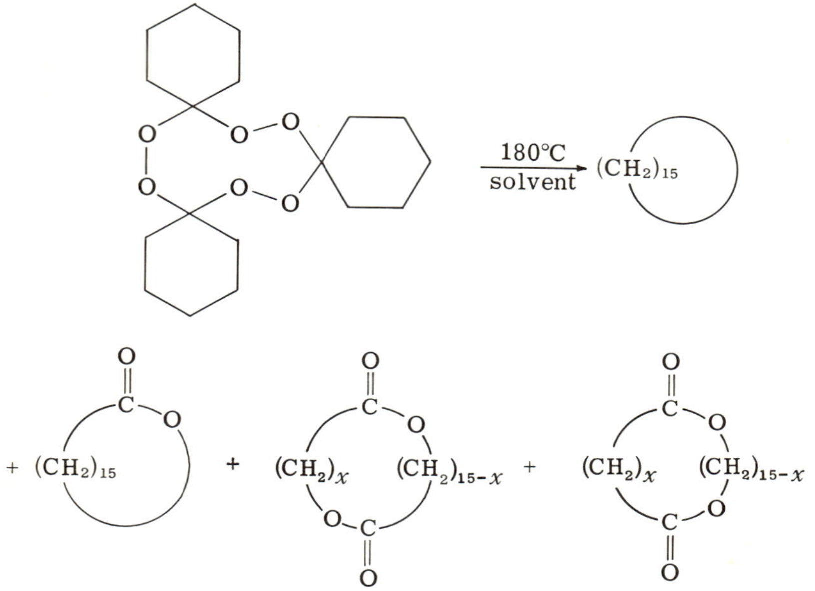

$$\text{tricyclohexylidene peroxide} \xrightarrow[\text{solvent}]{180°C} (CH_2)_{15} + (CH_2)_{15}\text{–C(=O)–O (lactone ring)} + (CH_2)_x,\ (CH_2)_{15-x} \text{ dilactones} \tag{17}$$

A. PEROXIDE SYNTHESIS

As may be anticipated, decomposition, either thermal or photochemical, of the requisite peroxide, once it is obtained, is the relatively trivial part of the synthetic method. The success or failure of any given synthetic attempt will invariably hinge upon one's ability to obtain the appropriate peroxide. In many cases, there will exist more than one possible peroxide precursor to a particular macrocyclic compound. Consequently, we have devoted much of our research effort to the synthesis and development of synthetic methods for peroxides.

At this point the most important single group of ketone peroxides are those derived from cyclohexanone, both because of their versatility and their ready availability. Considerable information exists in the literature concerning the synthesis of the cyclohexanone peroxides but much of it is ambiguous at best. Furthermore, the known syntheses tend to be somewhat dangerous and do not give sufficiently high yields.

We have investigated the synthesis of the cyclohexanone peroxides in considerable detail and find that much of the ambiguity of the earlier methods disappears with our observation that, in every synthesis now known, the treatment of cyclohexanone with hydrogen peroxide and acid *first* yields tricyclohexylidene as the kinetic product and that the trimer is subsequently converted into the dimeric peroxide, dicyclohexylidene peroxide (52) (eq. 18).

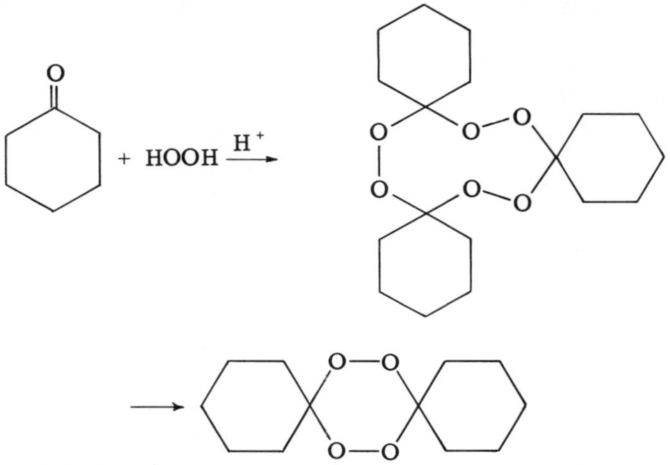

+ HOOH $\xrightarrow{H^+}$

(18)

Both the Kharasch-Sosnovsky synthesis (53) of dicyclohexylidene peroxide and the more recent Ledaal method (54) have been found to produce the trimeric peroxide initially (52). In fact, the tricyclohexylidene peroxide can be isolated in reasonably good yield from either synthetic method if the reaction is interrupted sufficiently early. Pertinent to these observations is our finding that 1,1′-dihydroperoxydicyclohexyl peroxide is spontaneously converted to trimeric peroxide in high yield on its dissolution in methanol (52) (eq. 19).

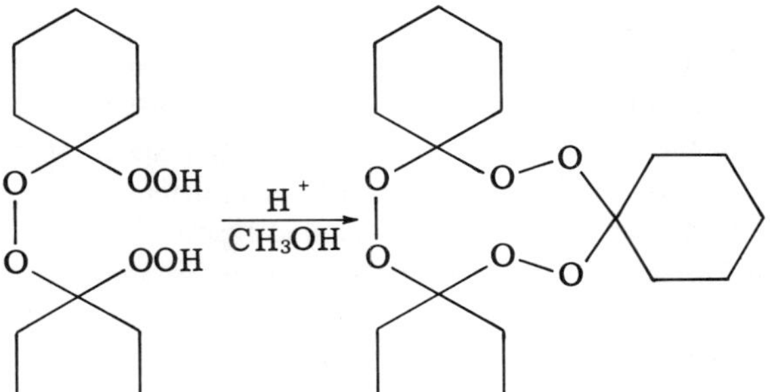

$\xrightarrow[CH_3OH]{H^+}$ (19)

Rarely does a peroxide synthesis yield only pure dimeric peroxide or pure trimeric peroxide. Unless the reaction is very carefully controlled the resulting product will very likely consist of a mixture of the dimeric and trimeric peroxides. Such a mixture is very difficult to assess since ordinary analytical methods will not distinguish between dimeric and trimeric peroxides. Infrared analysis is insufficiently sensitive and since parent ions are difficult to observe, mass spectrometry is of little value (55,56).

We have found that such peroxidic mixtures are readily and reliably analyzed by thermal decomposition of the mixture followed by gas chromatographic and mass spectrometric analysis of the macrocyclic products resulting therefrom (57). Generally, the most reproducible and high-yield method for the decomposition of ketone peroxides to yield macrocyclic products is thermal decomposition in a refluxing solvent (see later) (57). Using this technique we have examined the decomposition of well over 100 different peroxides. Invariably, the macrocyclic hydrocarbons are formed in 10 to 40% yield and the corresponding lactones in 10 to 25% yield. In a few cases decomposition of the ketone peroxide is effected photochemically with similar results.

The reaction of cycloheptanone with hydrogen peroxide in acetonitrile, using perchloric acid as catalyst, is reported to yield dicycloheptylidene peroxide (mp 103°C) (54). Thermoanalysis of this product (thermal decomposition in refluxing decane, bp 174°C, for 1 hr, followed by gas chromatographic analysis and isolation of the products) revealed not only the presence of cyclododecane and 13-tridecanolide, which would result from decomposition of dicycloheptylidene peroxide, but also cyclooctadecane and 19-nonadecanolide, which can only arise from tricycloheptylidene peroxide (57) (eq. 20).

Similarly, cyclododecanone yields a peroxide, mp 201°C, which is described as dicyclododecylidene peroxide. Thermolysis of this product in refluxing decane for 1 hr gave cyclodocosane and 23-tricosanolide along with some tritriacontane and 34-tetratriacontanolide, indicating that the peroxide, in fact, consisted of a mixture of dimeric and trimeric peroxides (57) (eq. 21).

The dimer/trimer ratio is critically dependent upon the temperature of the ketone-hydrogen peroxide reaction. Lower reaction temperatures favor a lower dimer/trimer ratio, in agreement with our detailed findings in the case of cyclohexanone peroxides (52).

It is often difficult to obtain either pure dimeric or trimeric peroxide; for example, in an attempt to prepare tri-*t*-butylcyclopentadecane we found that 4-*t*-butylcyclohexanone reacted with hydrogen peroxide to give predominantly the dimeric peroxide, under initial reaction conditions. However, systematic changes in the peroxide synthesis accompanied by "thermoanalysis" enabled us to improve the trimer/dimer ratio to 80:20 (57) (eq. 22).

B. *SYNTHESIS OF MIXED TRICYCLOALKYLIDENE PEROXIDES*

While it has proved possible to prepare several symmetrical dimeric and trimeric peroxides and thereby obtain the corresponding macrocyclic compounds, the method could not provide a source for all macrocyclic systems without a synthesis of mixed peroxides (Structures **1** and **2**).

Building upon the earlier work of Criegee (58) and of Rieche (59), we have developed syntheses of both mixed trimeric and mixed dimeric peroxides, thereby enabling us to synthesize, at this time, every size carbocyclic system from C-8 through C-33 and as high as C-42, including many substituted

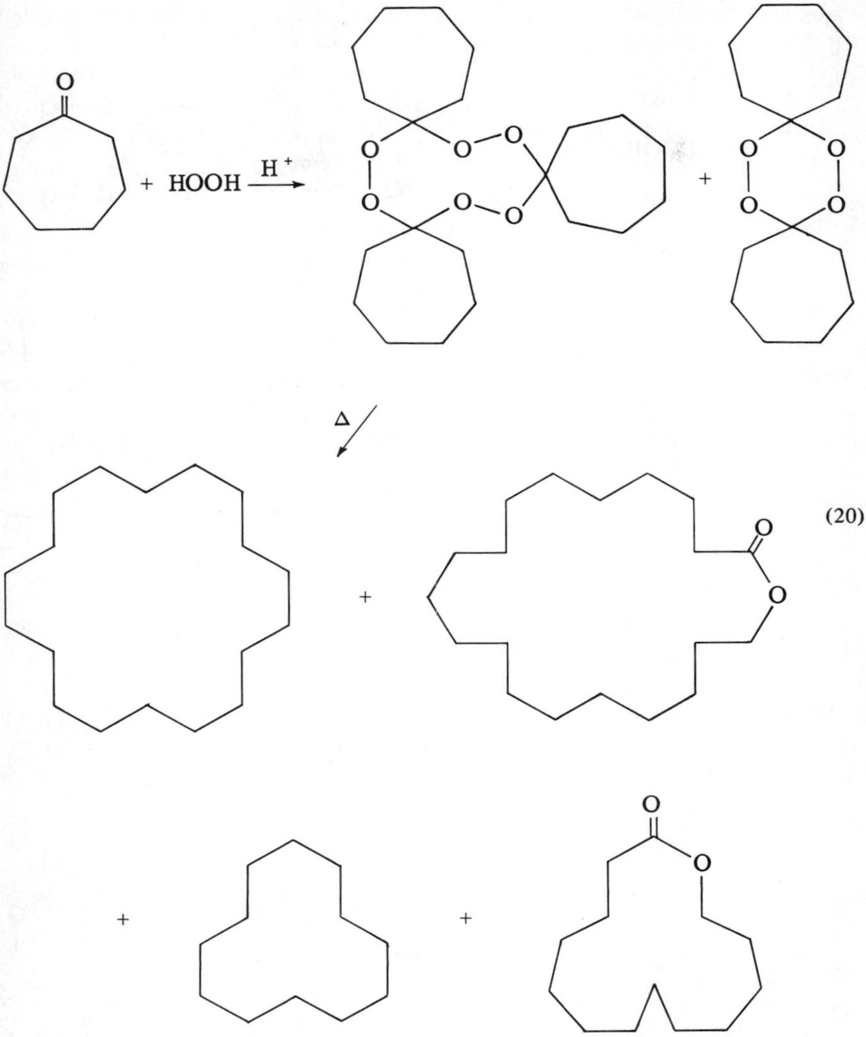

(20)

macrocyclic systems. The method has, therefore, provided a truly general synthesis of macrocyclic compounds.

The cornerstone for the synthesis of mixed peroxides is the 1,1′-dihydroperoxydicycloalkyl peroxides and, more particularly, 1,1′-dihydroperoxydicyclohexyl peroxide (eq. 23). This peroxide was first prepared by Criegee (60), but by a somewhat dangerous and low-yield method requiring 2.7 molar hydrogen peroxide in ether. We have developed a synthesis which is relatively safe and gives the peroxide in very high yield (see experimental section) (61).

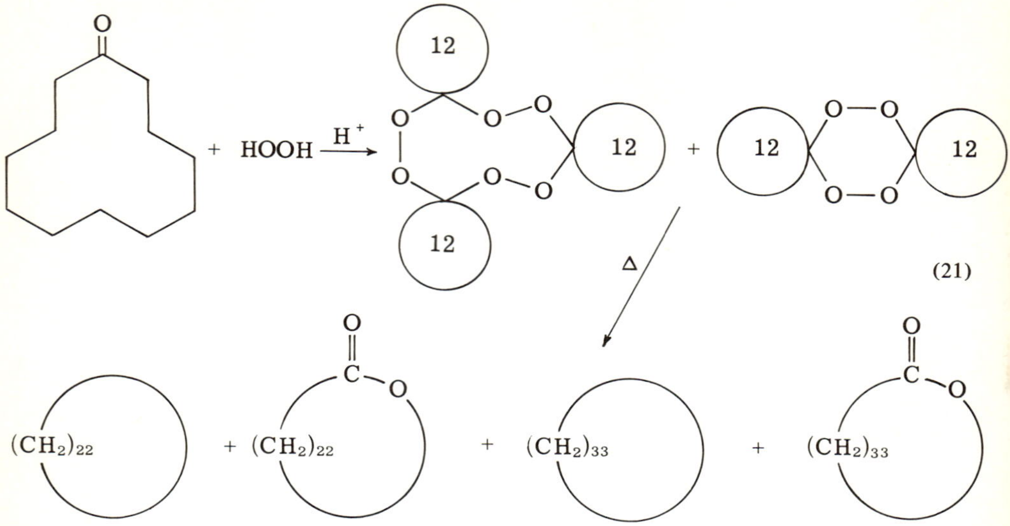

+ HOOH $\xrightarrow{H^+}$ + (21)

Δ

$(CH_2)_{22}$ + $(CH_2)_{22}$ + $(CH_2)_{33}$ + $(CH_2)_{33}$

$\xrightarrow[H^+]{HOOH}$

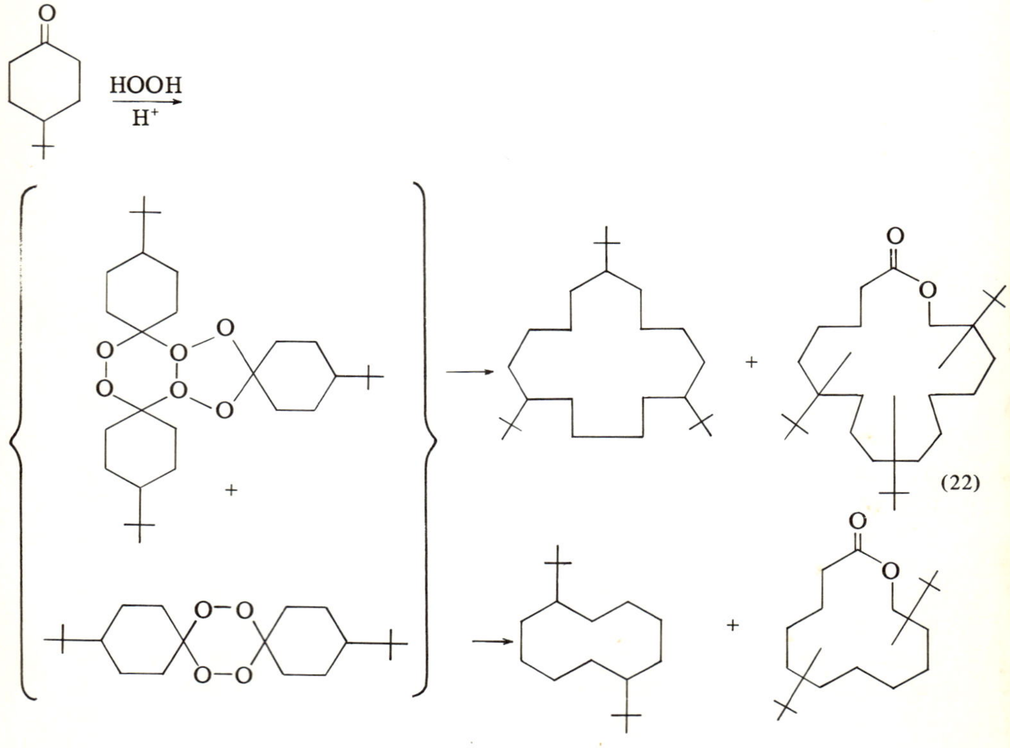

(22)

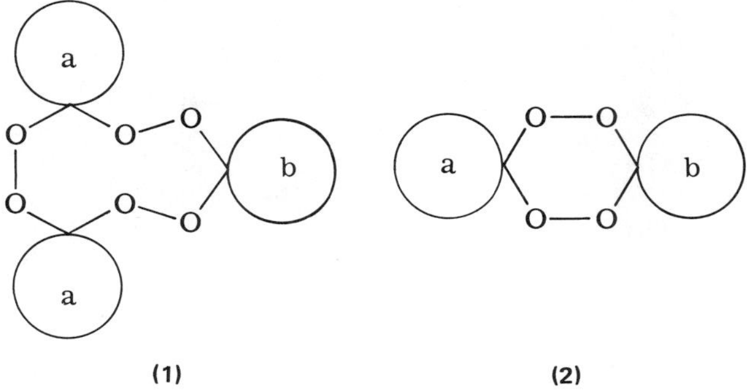

(1) (2)

Further, the method has been extended to the synthesis of the 1,1′-dihydroperoxide of several cyclic ketones. Condensation of the 1,1′-dihydroperoxydicycloalkyl peroxide with the cyclic ketones provides the mixed tricycloalkylidene peroxide. The method which we have reported (61) is a modification of a reaction developed originally by Criegee and extended more recently by Ol'dekop and Moiseichuck (62) in which a mixture of 1,1′-dihydroperoxydicyclohexyl peroxide and a liquid ketone is treated for a lengthy period with anhydrous cupric sulfate (eq. 24).

O

+ HOOH $\xrightarrow{H^+}$ O OOH | O OOH (23)

This method, however, is limited in application to liquid ketones which can be used in excess. Furthermore, the yields are not spectacular and reaction times are quite high. We subsequently developed an improved method for the production of mixed peroxides which permits the use of solid ketones and provides for much shorter reaction times (63). The improved method utilizes propionic acid as a solvent with perchloric acid catalyst at low temperatures. This technique has permitted the synthesis of a wide variety of mixed peroxides including Structures **3-7** (63).

The latter synthesis of mixed tricycloalkylidene peroxides is also applicable with slight modification to the synthesis of mixed dimeric peroxides (63) (eq. 25). To this point all such peroxides provide good yields of the appropriate macrocyclic compounds on either thermal or photochemical decomposition. It has now become possible, in fact, to synthesize many macrocyclic systems from several different peroxide precursors.

C. *PEROXIDE DECOMPOSITION*

A word of CAUTION: All of the peroxides considered in this discussion are to some degree shock sensitive if not handled properly. The proper precautions should be taken.

While any of these ketone peroxides may be decomposed to give good yields of macrocyclic compounds using either thermal or photochemical methods, with few exceptions, thermal decomposition is the preferred technique. Thermal decomposition is generally preferred because of higher yields, shorter reaction times, and because little if any solvent is required. Photochemical decompositions must usually be carried out at rather high dilution.

Initially, we reported the thermal decomposition of neat peroxides (2). Subsequently, we have found that thermal decomposition is effected more safely and gives higher yields of macrocyclic products if a solvent is used (64). Accordingly, we have carried out the decomposition of a large number of peroxides in many common solvents and in concentrations as high as 100 g of peroxide per liter of solvent and higher. Since the peroxide is usually

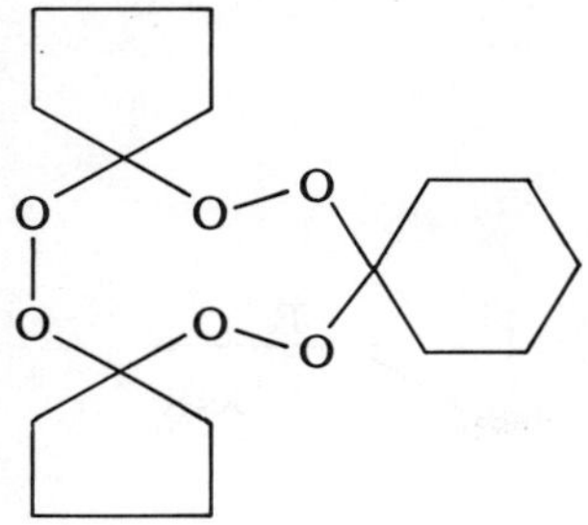

O O—O O O—O
(3)

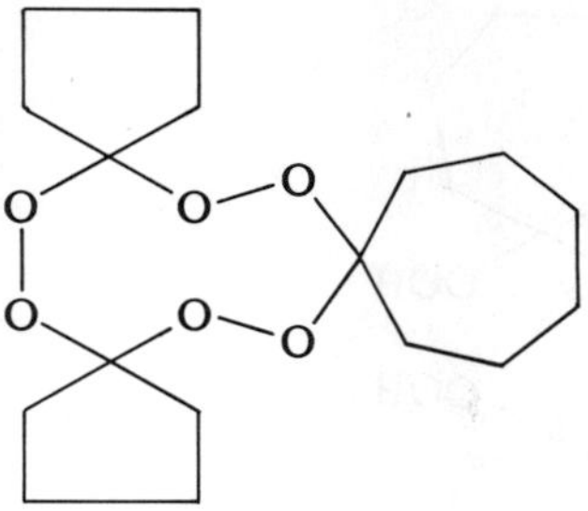

O O—O O O—O
(4)

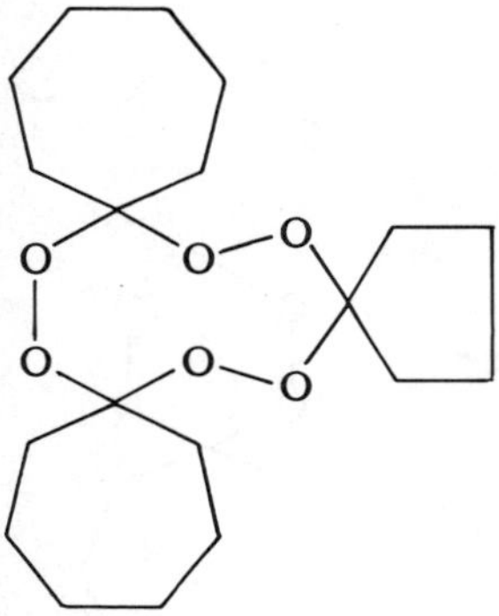

O O—O O O—O
(5)

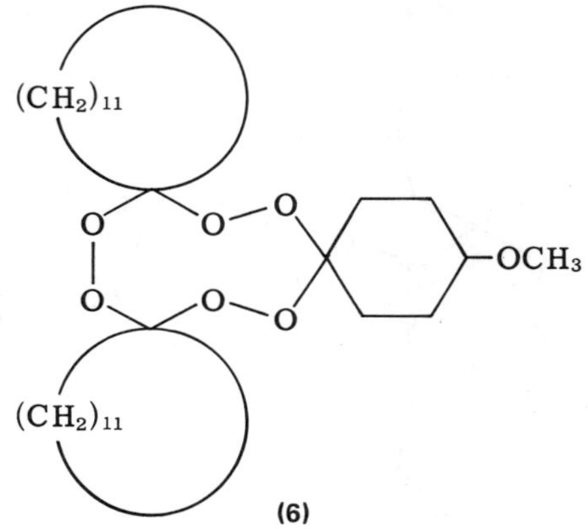

$(CH_2)_{11}$ O O—O —OCH_3 O O—O $(CH_2)_{11}$
(6)

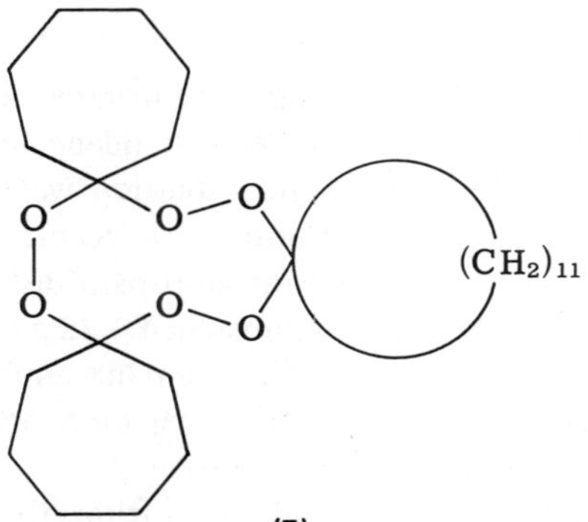

O O—O $(CH_2)_{11}$ O O—O
(7)

decomposed in the temperature range of 170-200°C, a solvent boiling in this range permits the decomposition to take place smoothly at atmospheric pressure. The proper amount of peroxide can be dissolved in the solvent before heating begins or, alternatively, peroxide can be added slowly to the refluxing solvent; the latter technique, of course, minimizes the explosive hazard. However, at this writing we have never experienced any difficulty with solution decomposition of peroxides. A favorite solvent for the decomposition of most peroxides is decan, boiling at 174°C (64).

OOH OOH + O H+ COOH (25) O O O—O O O—O O Δ H+ O O O O

Such pyrolyses are effected in the following manner. A solution of 180 ml of decane and 40 g of dicyclohexylidenecyclopentylidene peroxide was refluxed at 175°C for 3 hr after which the reaction solution was left standing at room temperature overnight. The decane solution was decanted from a small amount of heavy oil and the decane removed by short-path distillation. Distillation of the product using a spinning band column yielded 4.22 g of cyclotetradecane, bp 77-78°C/0.4 mm (17.2%), and 6.17 g of 15-pentadecanolide, bp 114°C/0.5 mm (22.2%). Both products were identified by ir, nmr, and mass spectrometric analysis. The reaction is shown (64) in eq. (26).

Table I summarizes the results obtained from the pyrolysis of several peroxides in decane as solvent unless otherwise noted. The peroxides and products are represented schematically in eq. (27).

We have also examined the decomposition of several special peroxides, for example, octalin diperoxide, which is readily obtained on ozonolysis of octalin. Thermal decomposition of the neat peroxide affords 1,10-cyclooctadecadione in 43% yield (isolated) (65) (eq. 28).

Using the techniques described here we have also generated heterocyclic systems (66) (eq. 29).

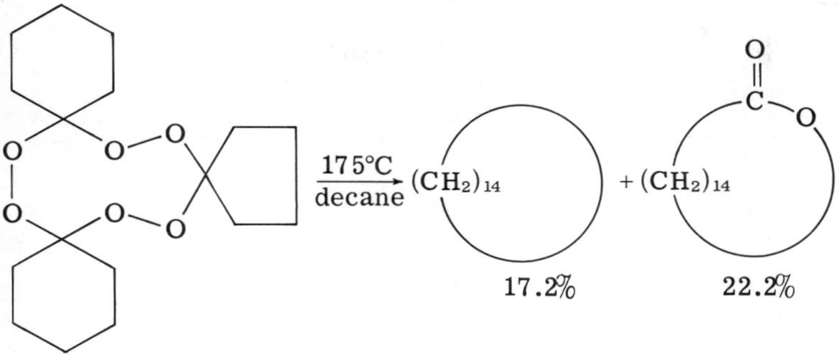

$$\xrightarrow[\text{decane}]{175^\circ C} (CH_2)_{14} \quad + (CH_2)_{14}\ (C{=}O,\ O) \qquad (26)$$

17.2% 22.2%

The effect and fate of substituent groups is of prime importance in planning a synthesis using this method. With this in mind we have examined the decomposition of the dimeric peroxide of 2-methylcyclohexanone. The peroxide was found to have the structure shown here by x-ray analysis. Thermal decomposition of the neat material gave 1,2-dimethylcyclodecane (possibly a *cis*/*trans* mixture) exclusively as the hydrocarbon product; no 1,6-dimethylcyclodecane was detectable (67) (eq. 30).

In summary, it has become quite evident that this synthetic method is generally applicable and now makes macrocyclic compounds readily available. We have efforts underway to extend the reaction to the synthesis of very large ring systems and to the highly substituted macrolide antibiotics. Pertinent experimental features follow.

X. Experimental

A. PREPARATION OF TRICYCLOHEXYLIDENE PEROXIDE BY A KHARASCH-SOSNOVSKY-TYPE REACTION (52)

To an Erlenmeyer flask containing 19.6 g (0.20 mole) of cyclohexanone, 22.6 g (0.20 mole) of 30% hydrogen peroxide was slowly added with cooling, if necessary, to maintain room temperature. After stirring for 15 min, 100 ml of glacial acetic acid was added followed by 2.4 ml of 10% perchloric acid/acetic acid. This mixture was stirred at room temperature for 24 hr. After this time the solution was diluted with 400 ml of water to yield a heavy oil and some solid material. (*Note:* Best results are obtained if the temperature is held near 20°C throughout the reaction period.) The cloudy aqueous solution was decanted and diluted with an equal volume of water. The diluted decantate was cooled at 10°C for 24 hr after which time the solution had cleared considerably and a white solid was formed. All the resulting oils and solids were combined and recrystallized from methanol to yield 14.8 g (72%) of tricyclohexylidene peroxide, mp 91-92°C. Infrared (CCl_4, cm^{-1}): 2940(s), 2860(m), 1453(s), 1366(m), 1279(m), 1169(m), 1078(s), 951(s), 932(s), 858(w).

TABLE I

Thermal Decomposition of Ketone Peroxides

Peroxide	Melting point, °*C*	Amount, *mg*	Decane, *ml*	Reflux time, *hr*	Hydrocarbon	Yield, %	Lactone	Yield, %
5, 5, 5	170	600	10	2.67	12	30	C(=O)–O, 13	20
6, 5, 5	51-81	670	10	1.67	13	20	C(=O)–O, 14	15
7, 5, 5	81-86	500	10	2.0	14	20	C(=O)–O, 15	15
6, 6, 6	oil	100	2	4.25	15	15	C(=O)–O, 16	25
6, 6, 6	oil	420	10	1.0	15	15	C(=O)–O, 16	25
7, 7, 12	oil	500	10	2.0	23	25	C(=O)–O, 24	20
7, 7, 7	78-80	500	10	2.0	18	15	C(=O)–O, 19	15
6, 6, 6	91-92	700	nonane, 11 ml 156°C	24	15	20	C(=O)–O, 16	17

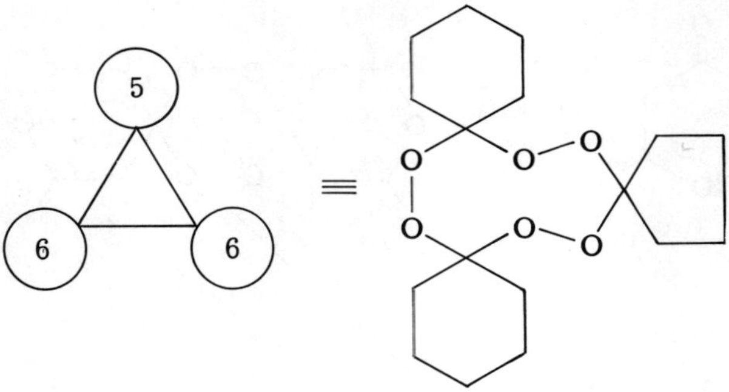

5

6 6

≡

O O—O O O—O

(27)

14 ≡ $(CH_2)_{14}$

O ‖ C O 15 ≡ $(CH_2)_{14}$

O_3

O O O O O O

Δ 180°C

O O

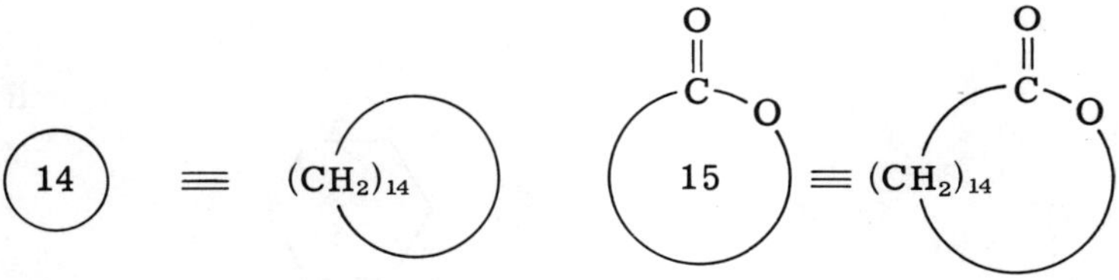

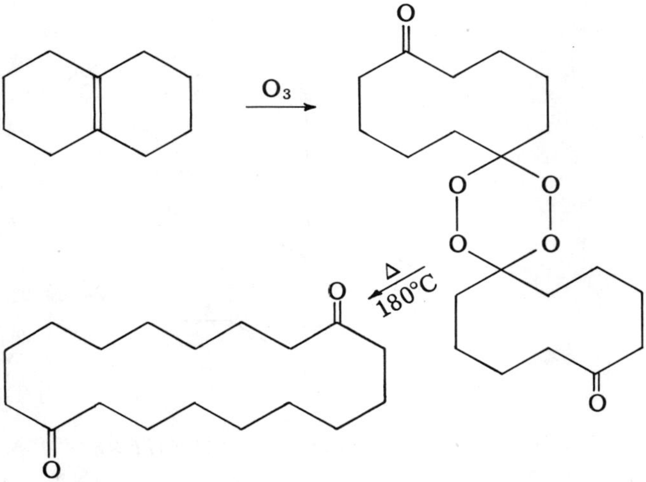

(28)

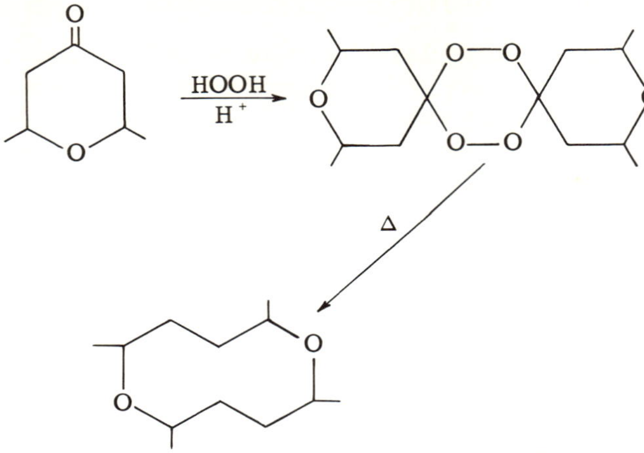

HOOH
H^+

Δ

(29)

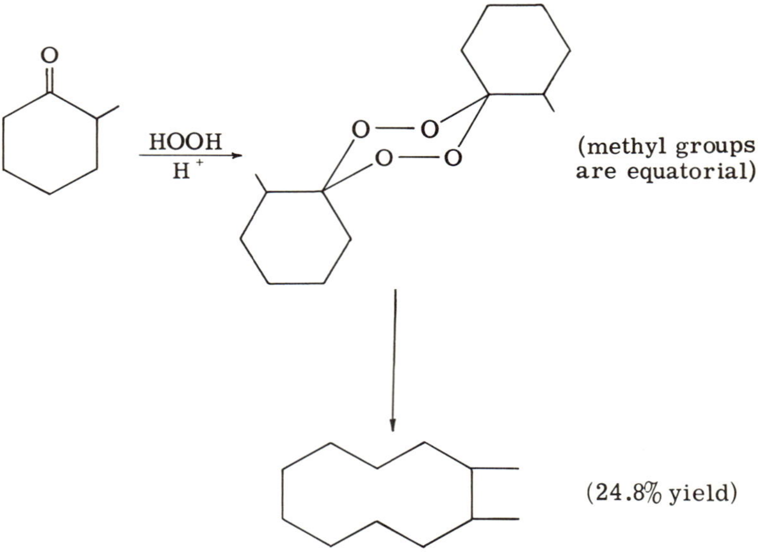

HOOH
H^+

(methyl groups are equatorial)

(30)

(24.8% yield)

B. TRICYCLOHEXYLIDENE PEROXIDE BY A LEDAAL-TYPE REACTION (52)

A mixture of 9.80 g (0.10 mole) of cyclohexanone, 11.3 g (0.10 mole) of 30% hydrogen peroxide, 20 drops of 70% perchloric acid, and 100 ml of acetonitrile was placed in a shallow dish in a fume hood and allowed to stand at room temperature (preferably under 24°C) for 5 days. After this time there remained only a white solid, 9.46 g (83%), which was shown to be "pure" trimer by ir comparison, mp 87-90°C.

C. RECRYSTALLIZATION OF TRICYCLOHEXYLIDENE PEROXIDE (52)

Recrystallization of tricyclohexylidene peroxide can be effected in methanol, 80% acetic acid, or in acetone. If the trimer is dissolved in the minimum amount of hot methanol and then cooled, an oil will usually separate. On standing overnight or sometimes longer, the oil will solidify. The oil is pure trimer by infrared. If the trimer is dissolved in a sufficient amount of methanol at room temperature and the solution is allowed to stand open to the air so as to permit evaporation of the methanol, the resulting very large crystals take two forms which can be separated by hand. One crystalline form is clear and melts at 87 to 87.5°C. The other is white and melts at 92 to 92.5°C. The two forms are identical by ir analysis.

D. PREPARATION OF 1,1-DIHYDROPEROXYDICYCLOHEXYL PEROXIDE (61)

A mixture of cyclohexanone (2.94 g, 30 mmoles), 90% hydrogen peroxide (2 ml, 60 mmoles), acetonitrile (30 ml), and 70% perchloric acid (6 drops) was stirred at room temperature in an open crystallization disk for 5 hr; the mixture became a crystalline mass. The solid was washed with 40 ml of ice water and with 600 ml of water and was recrystallized from hexane to yield 3.86 g (94%) of pure product, mp 83-84°C.

E. PREPARATION OF DICYCLOHEXYLIDENE-CYCLOHEPTYLIDENE PEROXIDE (61)

A mixture of 13.22 g, 51 mmoles, of 1,1-dihydroperoxydicyclohexyl peroxide, 30 ml of cycloheptanone, and 7 g of anhydrous cupric sulfate was stirred at 0-4°C for 24 hr. A large excess of water was added, and the solid product was filtered and recrystallized from methanol to yield 7.56 g (42%) of the mixed trimer, mp 90.4-92.4°C. Thermal decomposition of the peroxide in refluxing decane gave cyclohexadecane in 25% yield and 17-heptadecanolide in 20% yield.

F. PREPARATION OF DICYCLOHEXYLIDENE-CYCLOPENTYLIDENE PEROXIDE (63)

A solution of 5 mmoles of cyclopentanone in 10 ml of propionic acid was placed in a 50-ml Erlenmeyer flask containing a spinbar for magnetic stirring and cooled to just below −20°C by adding Dry Ice to the solution. To the cooled solution was added with stirring 0.5 ml of perchloric acid solution (10% in acetic acid) and 5 mmoles of 1,1-dihydroperoxydicyclohexyl peroxide. The temperature was maintained at about −20°C for 1 hr after addition was complete. The temperature was then raised to 0°C and kept there for 4 hr. After this time, 35 ml of water was added, resulting in a separation of crystals. The crystals were washed and dried and identified as 1.02 g of dicyclohexylidene-cyclopentylidene peroxide, mp 65-67°C.

ACKNOWLEDGMENTS

The authors are grateful to the National Air Pollution Control Administration, to the Donors of the Petroleum Research Fund of the American Chemical Society, and to the National Science Foundation for support of the work described herein from our laboratories.

REFERENCES

1. *Rodd's Chemistry of Carbon Compounds*, 2nd ed., Vol. 2, "Alicyclic Compounds, Part B," 1968, p. 380.
2. P. R. Story, D. D. Denson, C. E. Bishop, J.-C. Farine, and B. C. Clark, Jr., *J. Amer. Chem. Soc.*, **90,** 817 (1968).
3. N. Calderon, E. A. Ofstead, and W. A. Judy, *J. Polym. Sci., Part A-1,* **5,** 2209 (1967).
4. N. Calderon, E. A. Ofstead, J. P. Ward, W. A. Judy, and K. W. Scott, *J. Amer. Chem. Soc.*, **90,** 4133 (1968).
5. E. Wasserman, D. A. Ben Efraim, and R. Wolovsky, *J. Amer. Chem. Soc.*, **90,** 3286 (1968).
6. E. J. Corey and E. K. W. Wat, *J. Amer. Chem. Soc.*, **89,** 2757 (1967); E. J. Corey and E. Hamanaka, *ibid.*, **89,** 2758 (1967).
7. F. Sondheimer, et al., *Chem. Soc. (London), Spec. Publ.*, **21,** 75-107 (1967).
8. L. Ruzicka, *Helv. Chim. Acta*, **9,** 715, 1008 (1926).
9. H. Walbaum, *J. Prakt. Chem.*, **[2], 73,** 488 (1906).
10. E. Sack, *Chem. Ztg.*, **39,** 538 (1915); Ger. Pats. 279 and 313 (Jan. 3, 1912).
11. L. Ruzicka, M. Stoll, and H. Schinz, *Helv. Chim. Acta,* **9,** 249 (1926); **11,** 496, 670 (1928); L. Ruzicka and W. Brugger, *ibid.*, **9,** 339, 389 (1926).
12. N. D. Zelinsky, *Ber. Deut. Chem. Ges.*, **40B,** 3277 (1907).
13. R. Willstätter, see Ref. 12.
14. P. Ruggli, *Ann.*, **392,** 92 (1912).
15. E. C. Taylor and A. McKillop, *The Chemistry of Cyclic Enaminonitriles and o-Aminonitriles,* Wiley Interscience, New York, 1970.
16. K. Ziegler, H. Eberle, and H. Ohlinger, *Justus Liebigs Ann. Chem.*, **504,** 94 (1933); K. Ziegler and R. Aurnhammer, *ibid.*, **513,** 43 (1934).
17. H. Hunsdiecker, *Ber.*, **75B,** 1190, 1197 (1942).
18. A. T. Blomquist and R. D. Spencer, *J. Amer. Chem. Soc.*, **69,** 472 (1947); A. T. Blomquist, R. W. Holley, and R. D. Spencer, *ibid.*, **70,** 30 (1948); A. T. Blomquist, *ibid.*, **70,** 34 (1948).
19. V. Prelog, L. Frenkiel, M. Kobelt, and P. Barman, *Helv. Chim. Acta,* **30,** 1741 (1947).
20. M. Stoll and J. Hulstkamp, *Helv. Chim. Acta,* **30,** 1815 (1947); M. Stoll and A. Rouvé, *ibid.*, **30,** 1822, 1837 (1947); M. Stoll and M. Commarmont, *ibid.*, **31,** 1077 (1948).
21. V. L. Hansley, *J. Amer. Chem. Soc.*, **57,** 2303 (1935); U.S. Pat. 2,228,628 (Jan. 14, 1941).
22. E. Wasserman, *J. Amer. Chem. Soc.*, **82,** 4433 (1960).
23. C. Glaser, *Chem. Ber.*, **2,** 422 (1869); *Justus Liebigs Ann. Chem.*, **154,** 159 (1870).
24. (a) F. Sondheimer and Y. Amiel, *J. Amer. Chem. Soc.*, **78,** 4178 (1956); (b) Y. Amiel, F. Sondheimer, and R. Wolovsky, *Proc. Chem. Soc.*, **1957**, 22; (c) F. Sondheimer and Y. Amiel, *J. Amer. Chem. Soc.*, **79,** 5817 (1957); (d) F. Sondheimer, Y. Amiel, and R. Wolovsky, *ibid.*, **79,** 4247 (1957); (e) *idem,* **81,** 4600 (1959).
25. G. Eglinton and A. R. Galbraith, *Chem. Ind.* (London), **1956,** 737; *J. Chem. Soc.*, **1959,** 889.
26. G. Wittig, H. Eggers, and P. Duffner, *Justus Liebigs Ann. Chem.*, **619,** 19 (1958); G. Wittig and U. Schöllkopf, *Chem. Ber.*, **87,** 1318 (1954).
27. H. O. House and H. Babad, *J. Org. Chem.*, **28,** 90 (1962).

28. Ya. L. Goldfarb, S. Z. Taits, and L. I. Belenku, *Bull. Acad. Sci., USSR*, **1957**, 1287; S. Z. Taits and Ya. L. Goldfarb, *ibid.*, **1960**, 1574; L. I. Belenku, S. Z. Taits, and Ya. L. Goldfarb, *ibid.*, **1961**, 1589.
29. H. Gilman and N. J. Beaber, *J. Amer. Chem. Soc.*, **45**, 839 (1923).
30. A. J. Hubert and J. Dale, *Chem. Ind.* (London), **1961**, 1224.
31. N. J. Leonard and C. W. Schimelpfenig, *J. Org. Chem.*, **23**, 1708 (1958).
32. G. A. Berchtold and G. F. Uhlig, *J. Org. Chem.*, **28**, 1459 (1963).
33. K. C. Brannock, R. D. Burpitt, and V. W. Goodlett, *J. Org. Chem.*, **28**, 1480 (1963).
34. T. Mori, T. Nakahara, and H. Nozaki, *Can. J. Chem.*, **47**, 3266 (1969).
35. U. Schräpler and K. Rühlmann, *Chem. Ber.*, **97**, 1383 (1964).
36. W. Reusch and R. LeMahieu, *J. Amer. Chem. Soc.*, **86**, 3068 (1964).
37. H. Nozaki, T. Mori, and R. Noyori, *Tetrahedron Lett.*, **1967**, 779.
38. S. J. Rhoads, J. C. Gilbert, A. W. Decora, T. R. Garland, R. J. Spangler, and M. J. Urbigkit, *Tetrahedron Lett.*, **1963**, 1625.
39. For an excellent review of macrocyclic musks see: G. Ohloff, *Fortschr. Chem. Forsch.*, **12**, 185 (1969).
40. G. Ohloff and K. H. Schulte-Elte, *Helv. Chim. Acta*, **50**, 705 (1967).
41. K. Biemann, G. Buchi, and B. H. Walker, *J. Amer. Chem. Soc.*, **79**, 558 (1957).
42. A. D. Felix, A. Eschenmoser, M. Winter, F. Gautschi, K. H. Schulte-Elte, E. Sundt, G. Ohloff, J. Kalvoda, H. Kaufmann, P. Wieland, and G. Anner, *Helv. Chim. Acta*, **50**, 2101 (1967).
43. E. W. Garbisch and J. Wohllebe, *J. Org. Chem.*, **33**, 2157 (1968).
44. D. Bryce-Smith and A. Gilbert, *Tetrahedron Lett.*, **1964**, 3471.
45. D. Heimbach and W. Brenner, *Angew. Chem.*, **78**, 983 (1966).
46. N. Calderon, E. A. Ofstead, and W. A. Judy, *J. Polym. Sci., Part A-1* **5**, 2209 (1967).
47. N. Calderon, E. A. Ofstead, J. P. Ward, W. A. Judy, and K. W. Scott, *J. Amer. Chem. Soc.*, **90**, 4133 (1968).
48. E. Wasserman, D. A. Ben Efraim, and R. Wolovsky, *J. Amer. Chem. Soc.*, **90**, 3286 (1968).
49. G. Natta, G. DallAsta, I. W. Bass, G. Carella, *Makromol. Chem.*, **91**, 87 (1966).
50. R. E. Foster and R. S. Schreiber, *J. Amer. Chem. Soc.*, **70**, 2303 (1948); K. Ziegler and H. Wilms, *Justus Liebigs Ann. Chem.*, **567**, 1 (1950); J. C. Hillyer and J. V. Smith, *Ind. Eng. Chem.*, **45**, 1133 (1953); U.S. Pat. 2,631,173 (1953); or in the presence of a nickel carbonyl catalyst, see: H. W. B. Read, *J. Chem. Soc.*, **1954**, 1931; Brit. Pat. 701,106 (1953); U.S. Pat. 2,686,208 (1954).
51. C. J. Pederson, *J. Amer. Chem. Soc.*, **89**, 7017 (1967).
52. P. R. Story, B. Lee, C. E. Bishop, D. D. Denson, and P. Busch, *J. Org. Chem.*, **35**, 3059 (1970).
53. M. S. Kharasch and G. Sosnovsky, *J. Org. Chem.*, **23**, 1322 (1958).
54. T. Ledaal, *Acta Chem. Scand.*, **21**, 1656 (1967).
55. M. Bertrand, S. Fliszar, and Y. Rousseau, *J. Org. Chem.*, **33**, 1931 (1968).
56. T. Ledaal, *Tetrahedron Lett.*, **1969**, 3661.
57. P. Busch and P. R. Story, unpublished results.
58. R. Criegee and G. Lohaus, *Chem. Ber.*, **86**, 1 (1953).
59. A. Rieche and R. Meister, *Chem. Ber.*, **65**, 1274 (1932).
60. R. Criegee, W. Schnorrenberg, and J. Becke, *Justus Liebigs Ann. Chem.*, **565**, 7 (1949).
61. P. Busch and P. R. Story, *Synthesis*, **1970**, 181.
62. Y. A. Ol'dekop and K. L. Moiseichuk, *Zh. Org. Khim.*, **1**, 1934 (1965); Engl. ed., p. 1971.
63. P. Busch, B. Lee, and P. R. Story, unpublished results.
64. P. R. Story, P. Busch, D. D. Denson, and C. E. Wright, unpublished results.
65. D. D. Denson and P. R. Story, unpublished results.
66. P. R. Story and R. H. Turner, unpublished results.
67. P. R. Story, J. A. Alford, and D. D. Denson, unpublished results.

SILYLATION IN ORGANIC SYNTHESIS

By JOHANN F. KLEBE,* *General Electric Research and Development Center, Schenectady, New York*

CONTENTS

*Deceased September 24, 1971

I. Introduction

The explosive growth of organosilicon chemistry during the last two decades has created in its wake a steadily growing appreciation of a new tool in synthetic organic chemistry. "Silylation" of organic compounds–substitution with triorganosilyl groups–has already found wide application as an aid for chromatographic separation procedures. The value of silylation in organic synthesis has become apparent in numerous investigations throughout the realm of organic chemistry during recent years.

This appreciation is based on a combination of factors: reactive protons can be replaced by silyl groups–generally trialklysilyl groups–by use of several simple procedures; they can readily be removed from the organic compound by hydrolysis if the synthetic reaction has not involved the displacement of the silyl substituent. The newly introduced bond between silicon and the molecule in question may show an increased and specific reactivity toward certain reagents. The silyl substituent may exert an activating influence on other parts of the molecule. And finally, the silyl group may protect sensitive parts of the molecule from unwanted attack or simply serve as an inert substitute for a proton which itself would interfere with the attempted reaction.

These organic synthetic applications of organosilicon chemistry are reviewed in this chapter as comprehensively as possible in view of the fact that the pertinent publications are widely scattered through the literature and important contributions may thus have been overlooked. The preparation of silylated starting materials is only alluded to now and then, since this topic has been comprehensively covered (e.g., Ref. 1) in connection with gas chromatographic investigations. The great variety of synthetic reactions utilizing, and the frequent dualfold purpose of, silylation–for example, protection and activation–which makes this tool so interesting and valuable also posed a problem in organizing the material for the purpose of this chapter.

Two reviews on silylation as an aid in organic synthesis by Birkofer and Ritter (2) and by Rühlmann (3) appeared in 1965. These investigators and their groups have made many contributions to this field; in particular Birkofer and his collaborators recognized and early exploited the synthetic benefits of silylation. The present survey shows that often only a few examples of any given reaction involving silylated compounds are known and further significant expansion of this field can be expected.

II. Synthesis by Displacement of Silyl Groups

The majority of synthetically useful displacement reactions at silicon involve compounds in which silicon is attached to a more electronegative element. The reasons for the increased reactivity of the silicon-heteroatom bond as compared with analogous bonds to carbon are well understood (4). The most important factor appears to be the ability of silicon to expand its octet shell by utilizing *3d* orbitals, thereby increasing the maximal number of ligands to five or six. The transition states accessible to silicon in displacement reactions thus are of lower energy than those of analogous carbon compounds. Detailed studies of displacement reactions at silicon by Sommer and co-workers (5) using optically active silicon compounds have elucidated the mechanism of many of these reactions.

Section II describes reactions in which an attacking group displaces a silyl group without further rearrangement of the molecule. The material was organized in this way for the sake of convenience, and it is not intended to preclude the possibility that some of the reactions described in later chapters may also represent displacements at silicon in a mechanistic sense.

A. CLEAVAGE OF THE SILICON-HALOGEN BOND

Halosilanes are the starting materials for practically all other functional silanes in which silicon is bound to nitrogen, oxygen, sulfur, hydrogen, and carbon, to name the most common combinations, the last one being "functional" only under exceptional conditions. The use of halosilanes for silylations, usually carried out in the presence of acid acceptors, is a standard laboratory practice which has been described in ample detail (e.g., Ref. 1). Silylation in this sense involves displacement of an active hydrogen by a silyl group (sometimes the active hydrogen is first replaced by an alkali metal which in turn is displaced by silyl); the resulting ***organosilicon compounds*** can generally be hydrolyzed to starting materials or else they can serve as substrates for various synthetic modifications which are the subject to the present review.

A number of reactions of halosilanes are known, however, in which groups or atoms other than protons are displaced by silyl groups; these reactions are discussed in the following section.

1. *Formation of Alkylhalides and Hydroxyalkylhalides*

The reaction of halosilanes with aliphatic ethers in the presence of catalysts such as pyridine hydrochloride or aluminum chloride at elevated temperatures affords alkyl halides and siloxanes (6,7) (eq. 1). Cyclic ethers such as epoxides, tetrahydrofuran, and tetrahydropyran are cleaved more readily; the products may be dihaloalkanes or siloxyalkylhalides which can be hydrolyzed to the corresponding hydroxyalkylhalides (6-9) (eqs. 2 and 3). Reaction of halosilanes with tetrahydrofuran in the presence of equivalent amounts of $MgBr_2$ or MgI_2 affords 1-bromo- (or iodo-) 4-siloxybutane; halogen exchange between halosilane and the magnesium salt occurs prior to ring opening (10) (eqs. 4 and 5).

$$Et_2O + Me_2SiBr_2 \xrightarrow{\text{pyr}\cdot\text{HCl}} 2\,EtBr + (Me_2SiO)_n \tag{1}$$

$$\text{(tetrahydrofuran)} + Me_2SiCl_2 \xrightarrow[\text{reflux, 14 hr}]{ZnCl_2} Cl(CH_2)_4Cl \quad (68\%) \tag{2}$$

$$PhOCH_2\underset{\diagdown O \diagup}{CH-CH_2} + Me_3SiCl \xrightarrow[\text{1.5 hr}]{110°C}$$

$$PhOCH_2\underset{\substack{|\\ OSiMe_3}}{CH}CH_2Cl \quad (60\%) \quad + \quad PhOCH_2\underset{\substack{|\\ Cl}}{CH}CH_2OSiMe_3 \tag{3}$$

$$Me_3SiCl + MgI_2 \xrightarrow[\text{60°C, 4 hr}]{THF} Me_3SiO(CH_2)_4I \quad (63\%) \tag{4}$$

$$SiCl_4 + MgBr_2 \xrightarrow[\text{60°C, 22 hr}]{THF} [Br(CH_2)_4O]_2SiCl_2 \quad (47\%) \tag{5}$$

2. *Formation of* gem- *and* vic-*Disilanes*

Chlorosilanes have recently been found to react with carbonyl compounds and magnesium metal in hexamethylphosphoramide solution; good yields of geminal disilanes were obtained (11) (eq. 6). Under different reaction conditions, oxygen-containing compounds were obtained (**1**, **2**). These compounds may be intermediates in the reduction to *gem*-disilanes.

Olefins such as 1,1-diphenylethylene and stilbene afforded vicinal disilanes under the same conditions (12) (eq. 7).

$$\text{Ph(R)C=O} + 2\,\text{Mg} + 4\,\text{Me}_3\text{SiCl} \xrightarrow[100-110^\circ\text{C}]{\text{HMPA}} \text{Ph(R)C(SiMe}_3)_2 + 2\,\text{MgCl}_2 + (\text{Me}_3\text{Si})_2\text{O} \quad (6)$$

R = Ph 65%; CH_3 75%; H 50%

$$\text{Ph}_2\text{C(OSiMe}_3)(\text{SiMe}_3) \quad \textbf{(1)} \qquad \text{Ph(H)C(OSiMe}_3)(\text{SiMe}_3) \quad \textbf{(2)}$$

$$\text{Ph}_2\text{C=CH}_2 + \text{Mg} + 2\,\text{Me}_3\text{SiCl} \xrightarrow[100-110^\circ\text{C}]{\text{HMPA}} \text{Ph}_2\text{C(SiMe}_3)\text{-CH}_2(\text{SiMe}_3) + \text{MgCl}_2 \quad (7)$$

(80%)

B. CLEAVAGE OF THE SILICON-NITROGEN BOND

Aminosilanes represent, next to the halosilanes, the most reactive class of organosilanes in which silicon is bound to a more electronegative element. Mineral acids, carboxylic acids, phenols, alcohols, and mercaptans are all capable of displacing the amino moiety at silicon.

For synthetic purposes, aminosilanes have mainly been used in combination with reactive halides such as benzyl halides, acyl halides (or azides), and halides of Lewis-acid type to produce compounds in which halogen is replaced by the amine moiety. The attractive features in all of these reactions are the facility of the displacements which generally proceed at a rapid rate at low temperatures, and the fact that the low-boiling halosilane formed as the second reaction product is generally readily separated from the desired amino compound.

The alternative approach to the same amino end products generally is the reaction of the respective halide with the corresponding amine with hydrogen in place of the silyl group. Apart from the fact that the aminosilane has in some cases been found to be more reactive than the parent amine, the former has often been the starting material of choice because the halosilane by-products are unable to form salts or stable complexes with amines, again favoring easy separation of the products, generally by distillation.

A number of reviews describe synthetic routes to, and the properties of, numerous aminosilanes (13-15). The simplest method of preparation involves reaction of a commercially available chlorosilane with the amine in the presence of an acid acceptor such as excess of the amine used for the reaction or an added tertiary amine such as pyridine or triethyl amine. Similar methods are applicable for the synthesis of acyloxy-, alkoxy-, aryloxy-, mercaptosilanes, and other functional silanes. The synthesis of such compounds has been described in numerous instances and will not be detailed in this chapter.

1. Formation of Amides

The formation of amides from aminosilanes and acid chlorides was first exploited synthetically in the preparation of *N*-acyl heterocycles (16) (eq. 8). *N*-Trimethylsilyl imidazole reacts more readily with acid chlorides than imidazole itself. *N*-Acetyl-1,2,4-triazine was similarly prepared in 97% yield; benzoyl chloride, phosgene, and ethyl chloroformate gave yields of 95-97% of **3**, **4**, and **5** from the corresponding *N*-trimethylsilyl derivatives.

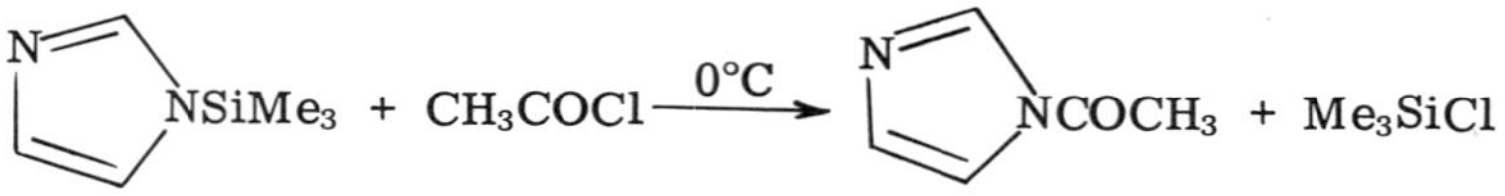

$$\text{(Im)NSiMe}_3 + CH_3COCl \xrightarrow{0^\circ C} \text{(Im)NCOCH}_3 + Me_3SiCl \quad (96\%) \tag{8}$$

(Im)NCOPh

(3)

(Im)NCON(Im)

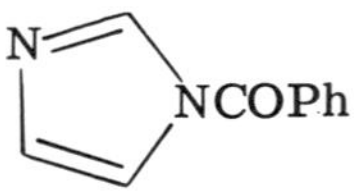

(4)

(Im)NCOOEt

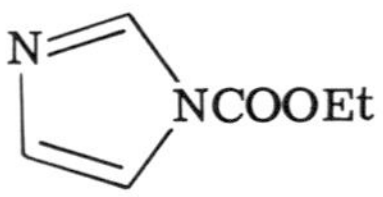

(5)

All reactions were carried out in anhydrous benzene at 0°C and were found to be complete soon after the acid chloride had been added. Removal of the volatiles furnished the products in a nearly pure state.

It was shown that isolation of the silylated heterocycle was not necessary for the preparation of the acylated product; the silylation with hexamethyldisilazane in refluxing toluene was followed by addition of the acyl chloride to the reaction solution, resulting in a high yield of product.

N,N'-Thiocarbonyldiimidazole has been prepared in essentially quantitative yield from trimethylsilylimidazole and thiophosgene (17) (eq. 9).

$$2\ \text{(imidazol-1-yl)}SiMe_3 + CSCl_2 \xrightarrow[0°C]{PhH} \text{(imidazol-1-yl)}CS\text{(imidazol-1-yl)} + 2\ Me_3SiCl \tag{9}$$

High yields of amides by reaction of silyl amines with acid chlorides have also been described by other workers (18). Instead of acid chlorides, azides may be used for acylations (19) (eq. 10). A catalytic amount of aluminum chloride enhances the rate of reaction; without the catalyst, higher temperatures are necessary, leading to decomposition of the azide to isocyanate and formation of a silylated urea by addition of the silylamine (eq. 11).

$$PhCON_3 + CH_3CH(HNSiMe_3)COOSiMe_3 \xrightarrow[40°C]{AlCl_3} CH_3CH(HNCOPh)COOSiMe_3 + Me_3SiN_3 \tag{10}$$

$$PhCON_3 + Me_3SiNHC_4H_9 \xrightarrow{-N_2} PhN(SiMe_3)CONHC_4H_9\ \ (85\%) \tag{11}$$

The reactions of silylamines with halides of elements of Groups III-VI in the Periodic Table resemble those of carboxylic acid halides in that the silyl group is displaced by the electrophilic moiety of the attacking halide. The numerous reactions known in this category are somewhat outside the scope of this chapter which is more concerned with syntheses in carbon chemistry. Moreover, a comprehensive review of such cleavage reactions appeared in 1968 (20). Only a few examples of this type of reaction are given in eqs. 12-20 (Refs. 21-28 and 16, respectively).

$$Me_3SiNMe_2 + (C_3H_7)_2BCl \longrightarrow (C_3H_7)_2BNMe_2 + Me_3SiCl \quad (12)$$

$$2\ Ph_2C{=}NSiMe_3 + 2\ C_6H_4O_2BCl \longrightarrow [C_6H_4O_2B(\mu\text{-}N{=}CPh_2)]_2 + 2\ Me_3SiCl \quad (13)$$

$$(Me_3Si)_2NH + AlCl_3 \longrightarrow Me_3SiNHAlCl_2 + Me_3SiCl \quad (14)$$

$$Me_3SiNHC_2H_5 + Me_3SnBr \longrightarrow Me_3SnNHC_2H_5 + Me_3SiBr \quad (15)$$

$$(CH_2NEt)_2SiMe_2 + PhPCl_2 \longrightarrow (CH_2NEt)_2PPh + Me_2SiCl_2 \quad (16)$$

$$2(CH_3)_3CN(Li)SiMe_3 + 2\ PCl_3 \longrightarrow [(CH_3)_3CNPCl]_2 + 2\ Me_3SiCl + 2\ LiCl \quad (17)$$

$$C_3H_3N_2SiMe_3 \xrightarrow{PCl_3} (C_3H_3N_2)_3P;\quad \xrightarrow{POCl_3} (C_3H_3N_2)_3PO;\quad \xrightarrow{SOCl_2} (C_3H_3N_2)_2SO \quad (18)$$

$$2\ Me_3SiNEt_2 + SO_2Cl_2 \longrightarrow (Et_2N)_2SO_2 + 2\ Me_3SiCl \qquad (19)$$

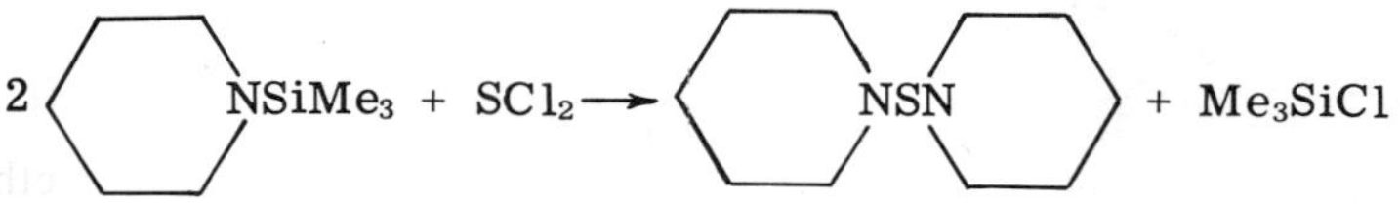

$$2\ C_5H_{10}NSiMe_3 + SCl_2 \longrightarrow C_5H_{10}NSNC_5H_{10} + Me_3SiCl \qquad (20)$$

2. *Formation of Amines, Amidines, and Guanidines*

Halides of considerably lower reactivity than exhibited by carboxylic acid halides or Lewis acids are also capable of cleaving the silicon-nitrogen bond. Benzyl chloride and silylamine undergo a displacement reaction at above 100°C and yield benzylamine and chlorosilane quantitatively (29) (eq. 21).

$$PhCH_2Cl + R_2NSiMe_3 \xrightarrow{100°C} PhCH_2NR_2 + Me_3SiCl \qquad (21)$$

Phenylisocyanide dichloride contains sufficiently activated halogen to react with silylamines at elevated temperatures with formation of chloroformamidines or guanidines (30) (eqs. 23 and 24). The reaction failed with trimethylsilyldiethylamine and trimethylsilyldipropylamine, presumably because of steric hindrance.

$$PhN{=}CCl_2 + Me_3SiN(C_3H_7)_2 \xrightarrow[16\ hr]{75°C} PhN{=}C(Cl)N(C_3H_7)_2 + Me_3SiCl \qquad (23)$$

(84%)

$$PhN{=}CCl_2 + 2\ Me_3SiNMe_2 \xrightarrow[24\ hr]{85°C} PhN{=}C(NMe_2)_2 + 2\ Me_3SiCl \qquad (24)$$

A similar displacement reaction is observed when silylamides are treated with imide chlorides without a solvent at 110-130°C; trimethylchlorosilane and acylated amidines are the reaction products (31) (eq. 25). Symmetrically substituted acylamidines are obtained by reaction of imide chlorides with sodium silanolate; the exothermic reaction is believed to involve metalated amide intermediates (eq. 26).

$$PhC(Cl){=}NEt + CH_3CON(SiMe_3)Ph \longrightarrow PhC(PhNCOCH_3){=}NEt + Me_3SiCl \qquad (25)$$

(83%)

$$\underset{\displaystyle\mathrm{Cl}}{\mathrm{PhC}}{=}\mathrm{NCH_3} \xrightarrow[-\mathrm{NaCl},\ (\mathrm{Me_3Si})_2\mathrm{O}]{2\ \mathrm{NaOSiMe_3}}$$

$$\left[\mathrm{PhC}^{\ominus}\begin{matrix}\cdots\mathrm{NCH_3}\\ \cdots\mathrm{O}\end{matrix}\right]\mathrm{Na}^{\oplus} \xrightarrow[-\mathrm{NaCl}]{\mathrm{PhC(Cl)}=\mathrm{NCH_3}} \mathrm{PhC(=NCH_3)}\text{–}\mathrm{CH_3NCOPh}\ (91\%) \qquad (26)$$

C. CLEAVAGE OF THE SILICON-SULFUR BOND

Silylmercapto compounds can be prepared by the same general methods as their oxygen analogs; silylation of thiols is readily achieved with chlorosilane/acid acceptor or with one of the silylamines or silylamides (1). A recent review contains a detailed description of organosulfur derivatives of silicon (32). The silicon-sulfur bond has received relatively little attention from a synthetic point of view, although it appears that its reactivity may approach that of the silicon-nitrogen bond and advantages may be found in displacing silyl groups from sulfur with reactive halides and the like, similar to the attractive features discussed in the group of silicon-nitrogen cleavage reactions.

1. Formation of Thiolesters and Thioethers

When equimolar amounts of acid chlorides and trimethylsilylesters of trimethylsilylmercapto carboxylic acids were allowed to react, the corresponding trimethylsilyl acyl mercapto carboxylates were formed in 62-77% yield; no evidence for cleavage of the silylester bond was found. The analogous methyl esters were formed in the same way in 80-85% yield (33) (eq. 27). The free acyl thiocarboxylic acids were obtained by hydrolysis of the trimethylsilyl esters in wet ether.

$$\mathrm{RCOCl} + \mathrm{Me_3SiSCHR'COOR''} \longrightarrow \mathrm{RCOSCHR'COOR''} + \mathrm{Me_3SiCl}$$

$$\mathrm{R{=}CH_3,\ Ph;\quad R'{=}H,\ CH_3;\quad R''{=}CH_3,\ Me_3Si} \qquad (27)$$

Thioethers can be obtained when alkylthiotrimethylsilanes are heated with alkyl halides, preferably bromides or iodides (34) (eqs. 28 and 29). Iodobenzene was found unreactive. The experiments were conducted in such a way that the silyl halide was distilled off as it was formed over an extended period of time. Evidence was found for the intermediacy of sulfonium salts. The crystalline colorless product which was formed at room temperature (eq. 30) decomposed on heating to methylbutylthioether and trimethyliodosilane.

$$n\text{-}C_4H_9SSiMe_3 + n\text{-}C_7H_{15}I \xrightarrow{15\ hr} \underset{(68\%)}{C_4H_9SC_7H_{15}} + Me_3SiI \tag{28}$$

$$\text{(cyclic } [Me_2SiS]_3\text{)} + 3\ Br(CH_2)_4Br \xrightarrow{48\ hr} 3\ \text{(tetrahydrothiophene, } \overline{CH_2CH_2CH_2CH_2S}\text{)} + 3\ Me_2SiBr_2 \tag{29}$$

$$n\text{-}BuSSiMe_3 + CH_3I \longrightarrow \left[Me_3Si\underset{\underset{CH_3}{|}}{S}Bu \right]^{\oplus} I^{\ominus} \tag{30}$$

Triphenylthiosilanol was reacted with metallic sodium or potassium and converted to silyl alkyl thioethers by reaction with alkylbromide or iodide in THF at room temperature in 50-80% yield (35) (eq. 31).

$$Ph_3SiSNa + C_3H_7Br \longrightarrow \underset{(77\%)}{Ph_3SiSC_3H_7} + NaBr \tag{31}$$

2. *Formation of Thiocarbamoyl Chlorides*

Trimethylsilyldithiocarbamates are readily accessible by addition of silylamines to carbon disulfide (Section IV.A). These thermally unstable adducts can be used for the synthesis of thiocarbamoyl chlorides which are not accessible in equally good yields by other methods.

Addition of thionyl chloride to the dithiocarbamate at room temperature initially leads to a solid adduct which on heating decomposes to thiocarbamoyl chloride, trimethylchlorosilane, sulfur dioxide, and sulfur (36) (eq. 32). Yields of 80-88% were obtained with five silylamines.

$$Et_2NSiMe_3 \xrightarrow{CS_2} Et_2N\overset{\overset{S}{\|}}{C}SSiMe_3 \xrightarrow[-Me_3SiCl,\ -SO_2,\ -S]{SOCl_2} Et_2N\overset{\overset{S}{\|}}{C}Cl \tag{32}$$

The oxygen analogs were obtained in 90-94% yield by reaction of trimethylsilylcarbamates with phosphorus pentachloride (eq. 33).

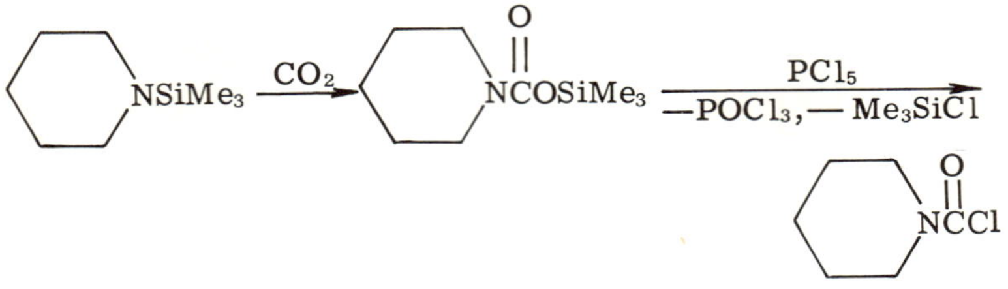

$$\text{C}_5\text{H}_{10}\text{NSiMe}_3 \xrightarrow{CO_2} \text{C}_5\text{H}_{10}\text{NC(=O)OSiMe}_3 \xrightarrow[-POCl_3, -Me_3SiCl]{PCl_5} \text{C}_5\text{H}_{10}\text{NC(=O)Cl} \quad (33)$$

(93%)

D. CLEAVAGE OF THE SILICON-OXYGEN BOND

1. Formation of Esters, Amides, and Ethers

Alkoxysilanes react with carboxylic acids and a number of acid derivatives to form esters (Ref. 3 and references cited therein) (eqs. 34 and 35). Most of the work dealing with these reactions has been carried out with the attention focused on the silicon moiety rather than on the resulting ester, and the synthetic value for the latter is generally limited.

$$Me_3SiOR \begin{array}{l} + R'COOH \\ + R'COOCOR' \\ + R'COCl \\ + R'COOSiMe_3 \end{array} \longrightarrow R'COOR \begin{array}{l} + Me_3SiOH \\ + Me_3SiOCOR \\ + Me_3SiCl \\ + Me_3SiOSiMe_3 \end{array} \quad (34)$$

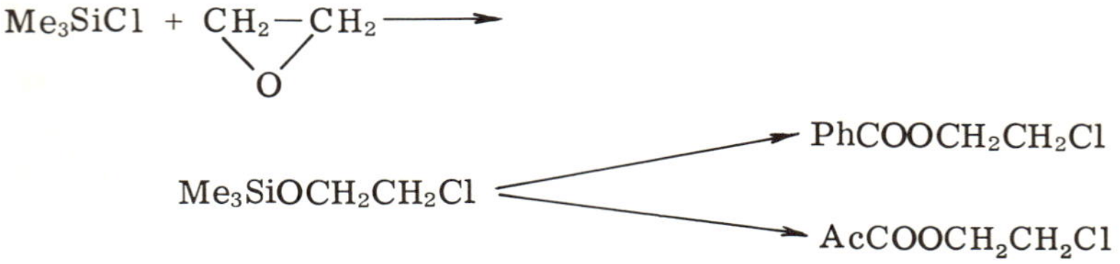

$$Me_3SiCl + \underset{\diagdown O \diagup}{CH_2{-}CH_2} \longrightarrow$$

$$Me_3SiOCH_2CH_2Cl \begin{array}{l} \nearrow PhCOOCH_2CH_2Cl \\ \searrow AcCOOCH_2CH_2Cl \end{array} \quad (35)$$

(79%)

In special cases, ester formation via silylated intermediates may have advantages over other methods; for example, silylation of a carboxylic acid with bis(trimethylsilyl)acetamide (BSA) and subsequent reaction with an alcohol containing anhydrous HCl yielded 80-100% of the corresponding ester at room temperature in less than 30 min total reaction time (37) (eqs. 36 and 37).

HOOC–C₆H₈(=O)–COOH $\xrightarrow{\text{BSA}}$ Me$_3$SiOOC–C$_6$H$_7$(OSiMe$_3$)–COOSiMe$_3$ $\xrightarrow{\text{MeOH, HCl}}$ MeOOC–C$_6$H$_8$(=O)–COOMe (36)

C$_6$H$_5$CH(CH$_3$)COOH $\xrightarrow{\text{BSA}}$ C$_6$H$_5$CH(CH$_3$)COOSiMe$_3$ $\xrightarrow{i\text{–PrOH, HCl}}$ C$_6$H$_5$CH(CH$_3$)COOCHMe$_2$ (37)

Earlier work with acyloxysilanes and alcohols in the absence of acids required higher temperatures for the preparation of esters because the condensation of silanol to disiloxane and water is faster than the acylation of alcohols at low temperature, and the water caused hydrolysis of part of the acyloxysilane (3) (eq. 38).

$$
\begin{aligned}
&\mathrm{RCOOSiMe_3 + R'OH \longrightarrow RCOOR' + Me_3SiOH}\\
&\mathrm{2\ Me_3SiOH \longrightarrow Me_3SiOSiMe_3 + H_2O}\\
&\mathrm{RCOOSiMe_3 + H_2O \longrightarrow RCOOH + Me_3SiOH}
\end{aligned}
\tag{38}
$$

In addition, the formation of free acids and alkoxysilane is a competing pathway which is favored at low temperatures (eq. 39).

$$RCOOSiMe_3 + R'OH \longrightarrow RCOOH + Me_3SiOR' \qquad (39)$$

Similarly, aminolysis of acyloxysilanes may lead to amides and silanol as well as to free acids and aminosilanes. High temperatures and large groups on silicon favor amide formation, and high yields of amides can be realized under such conditions (38) (eq. 40). Generally, however, mixtures of amides and amine salts are obtained (eq. 41).

$$CH_3COOSi(C_3H_7)_3 + \text{piperidine (NH)} \xrightarrow{150°C} \text{piperidine-NCOCH}_3 + (C_3H_7)_3SiOH \qquad (40)$$

$$2\ Me_3SiOCOR + 2\ HNR'_2 \longrightarrow RCONR'_2 + RCOOH \cdot HNR'_2 + (Me_3Si)_2O \qquad (41)$$

Related in a formal way to the ester formation from alkoxy silanes and carboxylic acid derivatives is the recently reported synthesis of an alkyl aryl ether from aryloxysilane and an alkyl halide. The reactions were carried out by refluxing the components in acetone in the presence of potassium carbonate (39) (eq. 42).

$$Me_3Si\text{-}C_6H_4\text{-}OSiMe_3 + Br(CH_2)_4Br \xrightarrow[18\ hr,\ -Me_3SiBr]{K_2CO_3} Me_3Si\text{-}C_6H_4\text{-}O(CH_2)_4Br \quad (67\%) \qquad (42)$$

Loss of the *p*-trimethylsilyl group was observed when *p*-trimethylsilylphenol instead of its silyl ether was used under the same conditions. Comparison of the preparation of **6** and **7** from methylbromoacetate and the free phenols and their silylethers, respectively, in the presence of K_2CO_3 showed that comparable yields were obtained, the silyl ethers requiring shorter reaction times.

$$CH_3S\text{-}C_6H_4\text{-}OCH_2COOCH_3 \quad \textbf{(6)}$$

$$CH_3S\text{-}C_6H_3(CH_3)\text{-}OCH_2COOCH_3 \quad \textbf{(7)}$$

Displacement of the trimethylsilyl group and formation of lactones or polyesters are observed when silyl esters of ω-bromo fatty acids are heated with silver cyanate or perchlorate (40) (eq. 43). Small amounts of polyester are found as by-products in these reactions; polyesters are the exclusive products when lactone formation is impossible; for example, in the reaction of trimethylsilyl-α-bromoacetate with silver salts.

$$Br(CH_2)_3COOSiMe_3 + AgOCN \xrightarrow[\text{toluene}]{130-140°C} \underset{(74\%)}{\text{γ-butyrolactone } (CH_2-CH_2-CH_2-C(=O)-O, \text{ ring})} + AgBr + Me_3SiNCO \quad (43)$$

The reaction of 4-chloro-1-trimethylsiloxybutane with silver cyanate at 140-160°C afforded tetrahydrofuran in an analogous displacement reaction (eq. 44).

$$Cl(CH_2)_4OSiMe_3 + AgOCN \longrightarrow \text{tetrahydrofuran} + AgCl + Me_3SiNCO \quad (44)$$

2. *Friedel-Crafts Acylations*

Silyl esters can be used in much the same way as acid chlorides and acid anhydrides for acylation reactions catalyzed by Lewis acids. Silicon tetrachloride serves as the starting material for the tetraacyloxysilane commonly used for these reactions; all four acyl groups can be utilized for the acylation (eq. 45).

$$Si(OCOR)_4 + 4\ ArH \xrightarrow{AlCl_3} 4\ ArCOR + SiO_2 + 2\ H_2O \quad (45)$$

These reactions have been extensively investigated by Russian workers (41-43). Acyloxysilanes are generally less reactive than acid chlorides and anhydrides; this has been taken advantage of in the preparation of acetylferrocene. The common acylating agents generally yield diacetylferrocene (44).

E. CLEAVAGE OF THE SILICON-HYDROGEN BOND

1. *Formation of Dihydropyridine Derivatives*

The large field of transition-metal-catalyzed addition reactions of silanes to double bonds is not covered in this chapter since the products of such reactions have generally been prepared for their own sake rather than as transitory aids for organic synthesis, which these adducts are usually less well suited for due to their nonfunctional character.

However, of greater interest here is the palladium- or rhodium-catalyzed addition of silanes to pyridines, providing entry into the class of dihydropyridines. When trimethylsilane is added by pyridine containing palladium or rhodium, adducts are formed. Depending on the reaction conditions, up to 60% of 1-trimethylsilyl-1,4-dihydropyridine can be obtained (45,46) (eq. 46).

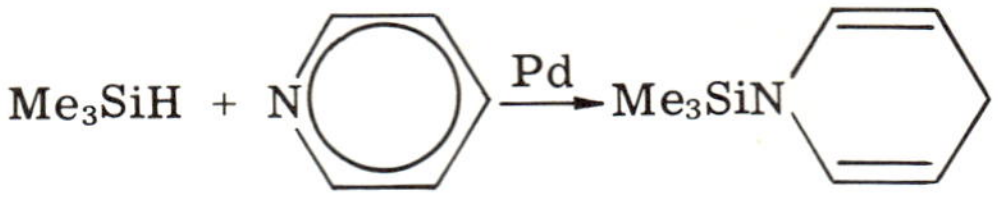

$$Me_3SiH + \text{N(pyridine)} \xrightarrow{Pd} Me_3SiN\text{(1,4-dihydropyridine)} \tag{46}$$

The balance of the products is made up by positional isomers, tetrahydropyridine derivatives, ring-silylated, and coupled products such as **8**, **9**, **10**, and **11**. Analogous products were obtained by silane addition to 3- and 4-methylpyridine; 2-methylpyridine did not react under comparable conditions, presumably due to steric hindrance. The silyl groups can be displaced by protons under carefully controlled hydrolytic conditions.

N–$SiMe_3$ **(8)** N–$SiMe_3$ **(9)** $SiMe_3$ N–$SiMe_3$ **(10)** $SiMe_3$ N … N $SiMe_3$ **(11)**

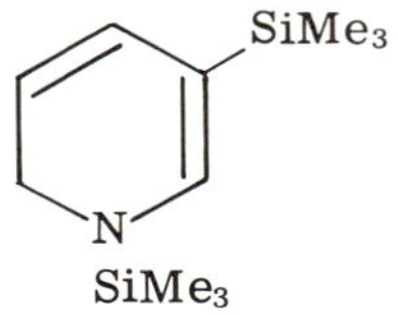

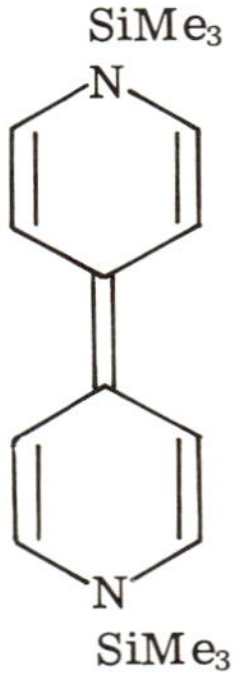

Some reactions of 1-trimethylsilyl-1,4-dihydropyridine have recently been reported. Reaction with ethyl azidoformate results in attack at the 2 position, and treatment with *p*-tosylisocyanate affords a nicotinamide derivative (47) (eqs. 47 and 48).

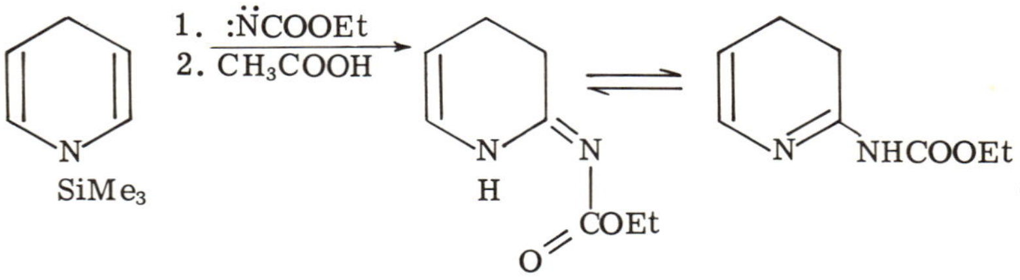

$$\text{N–}SiMe_3 \xrightarrow[\text{2. } CH_3COOH]{\text{1. :}\ddot{N}COOEt} \text{N(H)=N–C(=O)OEt} \rightleftharpoons \text{N} \text{ NHCOOEt} \tag{47}$$

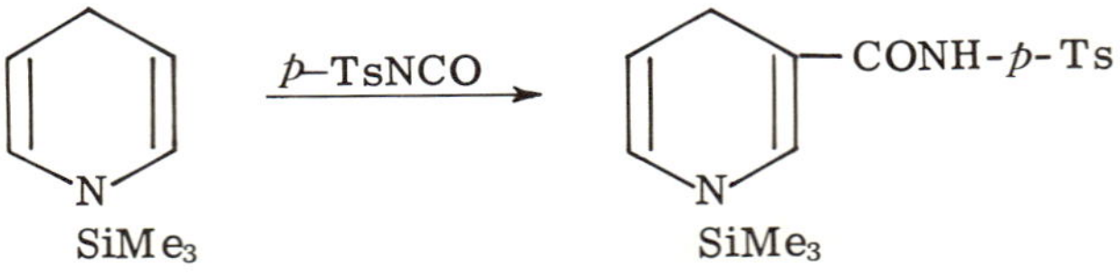

$$\text{N–}SiMe_3 \xrightarrow{p\text{–TsNCO}} \text{N–}SiMe_3 \text{ CONH-}p\text{-Ts} \tag{48}$$

N,N'-Bis(trimethylsilyl)tetrahydrodipyridyls were also obtained by reaction of pyridine or methyl pyridines with bis(trimethylsilyl)mercury (48) (eq. 49).

$$(Me_3Si)_2Hg + 2\ \text{(2-X-4-Y-pyridine)} \xrightarrow{-Hg} Me_3SiN\text{(X)(Y)}\!-\!\text{(Y)(X)}NSiMe_3 \qquad (49)$$

$$X, Y = H, CH_3$$

Oxidation with permanganate in acetone afforded the corresponding dipyridyls (eq. 50).

$$Me_3SiN\text{<}\!\!=\!\!\text{>}\!=\!\text{<}\!\!=\!\!\text{>}NSiMe_3 \xrightarrow[\text{aq. } CH_3COCH_3]{KMnO_4} \text{4,4'-bipyridyl} + (Me_3Si)_2O \qquad (50)$$

2. Metal-Catalyzed Silylation of Amine and Carboxyl Functions

In contrast to the unusual addition reaction observed with pyridine, the interaction of silanes with primary and secondary amines in the presence of palladium or platinum results in formation of *N*-silyl substituted compounds (49) (eq. 51.)

$$PhNH_2 + Me_2PhSiH \xrightarrow[30°C,\ 2\ hr]{Pd} PhNHSiMe_2Ph \xrightarrow[160°C,\ 24\ hr]{Me_2PhSiH,\ Pd} PhN(SiMe_2Ph)_2 \qquad (51)$$

Under the same conditions, carboxyl groups are converted to their silyl esters, and carbobenzoxy groups are displaced by trialkylsilyl (50) (eq. 52).

$$\underset{\displaystyle NHCO_2CH_2Ph}{RCHCOOH} + Et_3SiH \xrightarrow{PdCl_2} \underset{\displaystyle HNSiEt_3}{RCHCOOSiEt_3} + H_2 + CO_2 + PhCH_3 \qquad (52)$$

The fact that the decarbobenzoxylation with triethylsilane can be carried out in *N*-carbobenzoxy-glycine-trimethylsilylester shows that the cleavage is effected by the silane and not by the hydrogen evolved in the esterification in reaction (52).

A large number of aminosilanes have been prepared from amines and silanes in the presence of alkali metal hydrides; the metalated amine component presumably is an intermediate in this reaction (51) (eq. 53).

$$H_2N(CH_2)_4NH_2 + 4\ MePh_2SiH \xrightarrow[80-200^\circ C,\ 2\ hr]{NaH} (MePh_2Si)_2N(CH_2)_4N(SiMePh_2)_2 + 4\ H_2 \quad (88\%) \tag{53}$$

3. Reductions with Silanes

Silanes have been used in numerous instances for the reduction of carbonyl and cyano groups, generally in the presence of zinc chloride or other catalysts. These reactions are similar in many respects to those of other metal hydrides; however, they have been shown to take a unique direction in some cases. The method has been reviewed recently (52,53). A number of examples is shown here to illustrate these reactions (54,55) (eqs. 54 and 55).

$$RCOCl + R'_3SiH \rightarrow RCHO + R'_3SiCl \tag{54}$$

$$RCONR''_2 + 2\ R'_3SiH \xrightarrow{ZnCl_2} RCH_2NR''_2 + R'_3SiOSiR'_3 \tag{55}$$

Monoalkyl-substituted amides react in the same manner (56); however, unsubstituted amides are converted to nitriles in 80-90% yield (57) (eq. 56).

$$RCONH_2 + 2\ R'_3SiH \rightarrow RCN + 2\ H_2 + R'_3SiOSiR'_3 \tag{56}$$

Under more stringent conditions, aliphatic nitriles are reduced by silanes in the presence of zinc chloride to silylamines whereas aromatic nitriles are converted to silylated imines which can be hydrolyzed to aldehydes (58) (eqs. 57 and 58).

$$RCH_2CN + R'_3SiH \xrightarrow{ZnCl_2} RCH_2CH_2NHSiR'_3 \tag{57}$$

$$ArCN + R'_3SiH \xrightarrow{ZnCl_2} ArCH{=}NSiR'_3 \xrightarrow{hydrol.} ArCHO \tag{58}$$

The γ- and δ-lactones are converted into mixtures of disiloxyalkanes and cyclic ethers (59) (eq. 59).

$$(CH_2)\left\langle\begin{matrix}C{=}O\\ |\\ O\end{matrix}\right. + R'_3SiH \xrightarrow{ZnCl_2} R'_3SiO(CH_2)_nCH_2OSiR'_3 + (CH_2)\left\langle\begin{matrix}CH_2\\ |\\ O\end{matrix}\right. + (R'_3Si)_2O \tag{59}$$

Silanes can add to ketones in 1,2- or 1,4-fashion under the catalytic influence of platinum compounds or uv light; the resulting alkoxysilanes can be hydrolyzed to alcohols or saturated ketones (60-64) (eqs. 60 and 61).

$$R_2C{=}O + R'_3SiH \xrightarrow[\text{or uv}]{H_2PtCl_6} R_2CHOSiR'_3 \tag{60}$$

$$RCOCH{=}CHR + R'_3SiH \longrightarrow \underset{\displaystyle OSiR'_3}{\underset{|}{RC}}{=}CHCH_2R \tag{61}$$

In the presence of nickel catalysts, enoxy silanes are the major products (53) (eq. 62).

$$EtCOEt + Et_3SiH \xrightarrow{Ni/NiS} \underset{(2\%)}{Et_2CHOSiEt_3} + \underset{(98\%)}{CH_3CH{=}\underset{\displaystyle OSiEt_3}{\underset{|}{C}}Et} \tag{62}$$

The reduction of aldehydes can lead to dimeric products in addition to the expected alkoxysilanes (65) (eq. 63).

$$PhCHO + Et_3SiH \xrightarrow{Ni/NiS} PhCH_2OSiEt_3 + \underset{\displaystyle Et_3SiO}{\underset{|}{PhCH}}{-}\underset{\displaystyle OSiEt_3}{\underset{|}{CHPh}} \tag{63}$$

Acetals and ketals are converted to ethers by silanes in the presence of zinc chloride (66) (eqs. 64 and 65).

$$R_2C(OR'')_2 + R'_3SiH \xrightarrow[100^\circ C]{ZnCl_2} R_2CHOR'' + R'_3SiOR'' \tag{64}$$

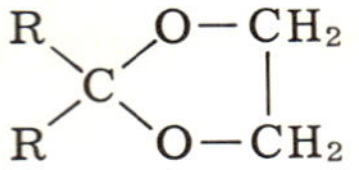

$$R_2C\langle O-CH_2 / O-CH_2 \rangle + R'_3SiH \xrightarrow{ZnCl_2} R_2CHOCH_2CH_2OSiR'_3 \quad (65)$$

α-Halogen-substituted ethers are reduced in high yield by silanes catalyzed by zinc chloride (eq. 66); in the presence of aluminum chloride, alkyl halides can be reduced to alkanes (67-69) (eq. 67).

$$CH_3OCH_2Cl + R_3SiH \xrightarrow{ZnCl_2} CH_3OCH_3 + R_3SiCl \quad (66)$$

$$RCl + R'_3SiH \xrightarrow{AlCl_3} RH + R'_3SiCl \quad (67)$$

Phosphine oxides and phosphates have been reduced to phosphines using poly(methylsiloxane), organosilanes, or silicochloroform (70-72) (eqs. 68 and 69).

$$Ph_3PO + 2\ HSiCl_3 \xrightarrow{80°\ C} Ph_3P + SiCl_4 + H_2 + (Cl_2SiO)_n \quad (68)$$

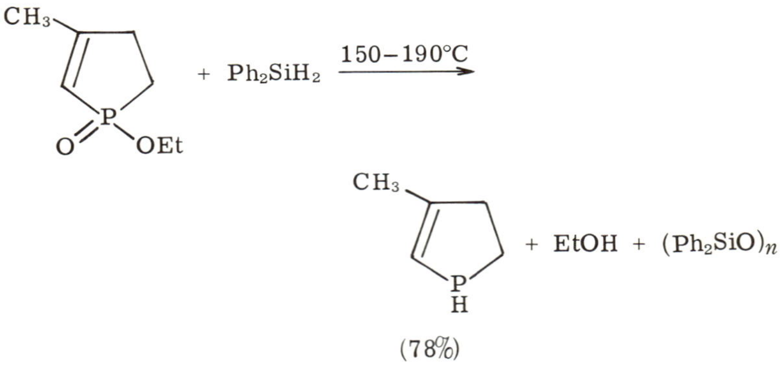

$$\text{3-methyl-1-ethoxy-3-phospholene 1-oxide } (CH_3, P(=O)OEt) + Ph_2SiH_2 \xrightarrow{150-190°C} \text{3-methyl-3-phospholene } (CH_3, PH) + EtOH + (Ph_2SiO)_n \quad (69)$$

(78%)

Recently, the reduction with trichlorosilane has been extended to sulfoxides. Aromatic sulfoxides are reduced to sulfides in high yield under mild conditions; sulfoxides with α-hydrogens yield mixtures of sulfides and mercaptals (73) (eqs. 70-72).

$$Ph_2SO + 2.5\ HSiCl_3 \xrightarrow[1\ hr]{25°C} PhSPh \quad (70)$$

(98%)

$$(PhCH_2)_2SO + 2\ HSiCl_3 \xrightarrow[3\ hr]{0°C} PhCH_2SCH_2Ph + PhCH(SCH_2Ph)_2 \quad (71)$$

(13%) (85%)

$$(PhCH_2)_2SO + 4\ HSiCl_3 \xrightarrow[3\ hr]{0°C} \underset{(46\%)}{PhCH_2SCH_2Ph} + \underset{(52\%)}{PhCH(SCH_2Ph)_2} \quad (72)$$

F. CLEAVAGE OF THE SILICON-CARBON BOND

1. Displacement of Silyl Groups from Aromatic Nuclei

Synthetic procedures involving cleavage of the silicon-carbon bond have so far been very limited in number. Silyl groups can be displaced from aromatic compounds under acidic conditions (4). "Protodesilylation" has been shown to resemble electrophilic aromatic substitution and was used to determine electronic effects at a particular ring position in aromatic compounds (74) (eq. 73).

$$R\text{-}C_6H_4\text{-}SiMe_3 \xrightarrow{H_3O^+} R\text{-}C_6H_5 + Me_3SiOH + H^+ \quad (73)$$

Related to this reaction is the "mercuridesilylation" which allows displacement of trimethylsilyl groups by mercuric acetate in acetic acid (75) (eq. 74).

$$R\text{-}C_6H_4\text{-}SiMe_3 \xrightarrow[AcOH]{Hg(OAc)_2} R\text{-}C_6H_4\text{-}HgOAc + Me_3SiOAc \quad (74)$$

The elimination of halosilane by reaction of halogen with silylated naphthalene was used for the preparation of trimethylsilyl bromide and iodide (76) (eq. 75). The reaction is carried out without isolation of intermediates.

The carbon-metal bond in aryl metal compounds is more reactive to electrophilic attack by aryl sulfonyl halides than the carbon-hydrogen bond. Use of this fact has been made in the preparation of substituted diaryl sulfones; the method affords only one isomer with halo or phenyl-substituted aryl silanes whereas normal Friedel-Crafts acylation with aryl sulfonyl halides leads to ortho and para mixtures (77) (eq. 76).

Electron-donating substituents such as methyl apparently lead to some acid cleavage of the carbon-silicon bond; the subsequent acylation of the aromatic nucleus results in formation of isomer mixtures.

$$\text{naphthalene} \xrightarrow[\text{2. } Me_3SiCl]{\text{1. Na}} \text{1,4-bis(trimethylsilyl)-1,4-dihydronaphthalene}$$

$$+ \text{1,2-bis(trimethylsilyl)-1,2-dihydronaphthalene} \xrightarrow{I_2} \left[\text{1,4-bis(trimethylsilyl)-2,3-diiodotetralin} \right. \qquad (75)$$

$$\left. + \text{1,2-bis(trimethylsilyl)-3,4-diiodotetralin} \right] \longrightarrow \text{naphthalene} + 2\ Me_3SiI\ (91\%)$$

$$Cl\text{-}C_6H_4\text{-}SiMe_3 + PhSO_2Cl \longrightarrow Cl\text{-}C_6H_4\text{-}SO_2\text{-}C_6H_5\ (55\%) + Me_3SiCl \qquad (76)$$

2. Formation of Acetylenic Ketones and Ethers; Cyclopropanes

Ethynylsilanes are accessible by a number of methods (e.g., Ref. 78). These compounds are stable against dilute acid and alkali; acid chlorides and anhydrides react in the presence of Lewis acids (79) (eq. 77).

$$Me_3SiC{\equiv}CSiMe_3 + CH_3COCl \xrightarrow[2-8°C]{AlCl_3} Me_3SiC{\equiv}CCOCH_3\ (90\%) + Me_3SiCl \qquad (77)$$

Monosilyl-substituted alkynes give the analogous 1-acylalkynes. The failure of the second trimethylsilyl group in bis(trimethylsilyl)acetylene to react with excess acetyl chloride is attributed to deactivation of the molecule by the acetyl group against further electrophilic attack. In contrast, α-chloroalkyl ethers are able to displace both silyl groups from the acetylenic residue (eq. 78).

$$Me_3SiC{\equiv}CSiMe_3 + 2\ ClCH_2OC_3H_7 \xrightarrow[2°C]{AlCl_3} C_3H_7OC{\equiv}COC_3H_7 + 2\ Me_3SiCl \quad (41\%) \tag{78}$$

The trimethylsilyl ester of chlorosulfonic acid reacts with silyl acetylenes to give the corresponding ethynyl sulfonates (80) (eq. 79). Hydrolysis affords the free sulfonic acids.

$$RC{\equiv}CSiMe_3 + ClSO_3SiMe_3 \rightarrow RC{\equiv}CSO_3SiMe_3 + Me_3SiCl \tag{79}$$

$R{=}Ph$ (55%)
CH_2Cl (55%)
$SiMe_3$ (80%)

γ-Bromoalkylsilanes undergo an intramolecular displacement of the silyl group in the presence of a Lewis acid (81) (eq. 80). The starting material is conveniently accessible by addition of trimethylsilane to allylbromide.

$$Me_3Si(CH_2)_3Br \xrightarrow{AlCl_3} \underset{\diagdown\ \diagup \atop CH_2}{CH_2{-}CH_2} + Me_3SiBr \tag{80}$$

III. Activation by Silyl Substitution

The heading of this section is general enough to cover the whole chapter. However, "activation" is meant here in the restricted sense that silyl substitution of certain molecules is able to increase the reactivity of adjacent sites of the molecule toward attack by a reagent. Often this is the result of polarization of an existing double or triple bond of the molecule due to d_π-p_π overlap with silicon, or stabilization of enol forms by silyl enolate formation and attack of a reagent at the enolic double bond, or else the loss of a proton due to the electron-withdrawing character of the silyl group on a heteroatom initiates a subsequent reaction. As in previous sections, a variety of mechanisms may govern the reactions discussed in this section.

A. *ACYLATION OF SILYLATED PRIMARY AMINES*

1. *Formation of Nitrogen Heterocycles*

The reaction of *N*-trialkylsilyl-substituted secondary amines with acid chlorides leads to displacement of the silyl group under mild conditions and amide formation, as discussed in Section II. Monosilylated primary amines, in the presence of tertiary amine as acid acceptor, display a different mode of reaction with acid chlorides; hydrogen chloride is eliminated and the resulting amide retains the silyl substituent. The fact that these reactions are more facile than the acylation of the more basic parent primary amines suggests that loss of the proton may initiate the reaction, followed by attack of carbonyl at the nitrogen.

The synthesis of trimethylene and tetramethylene urea from phosgene and the corresponding diamines was found to be unsatisfactory; however, after silylation, the ureas were obtained in 75 and 60% yield, respectively (82) (eq. 81). The nitrogen-hydrogen bond proves to be the most reactive bond under these conditions; the presence of triethylamine presumably prevents the cleavage of the nitrogen-silicon bond by the HCl liberated in the reaction.

$$(CH_2)_n(NHSiMe_3)_2 + COCl_2 \xrightarrow{Et_3N,\ 0^\circ C} (CH_2)_n(NSiMe_3)_2C{=}O \xrightarrow{\text{hydrol.}} (CH_2)_n(NH)_2C{=}O \qquad n = 3, 4 \tag{81}$$

Benzimidazolone-2 was synthesized in analogous fashion (eq. 82). Treatment of bis(trimethylsilyl)benzimidazolone with acetyl chloride at reflux afforded the mono- and the bis-acetylated product in successive steps (83) (eq. 83).

$$C_6H_4(NHSiMe_3)_2 \xrightarrow{COCl_2} C_6H_4(NSiMe_3)_2C{=}O\ (95\%) \longrightarrow C_6H_4(NH)_2C{=}O \tag{82}$$

$$\text{1,3-bis(trimethylsilyl)benzimidazol-2-one} \xrightarrow[-Me_3SiCl]{CH_3COCl} \text{1-acetyl-3-(trimethylsilyl)benzimidazol-2-one} \xrightarrow[-Me_3SiCl]{CH_3COCl} \text{1,3-diacetylbenzimidazol-2-one} \tag{83}$$

N,N'-Bis(trimethylsilyl)-*o*-phenylene diamine was found more reactive toward phosgene than the unsilylated diamine; thus the carbonylation of a silylated diamine was attempted which prior to silylation does not react with phosgene. Tetrakis(trimethylsilyl)uric acid was indeed formed from persilylated diamino uracil in 80% yield (82) (eq. 84). The uric acid derivative could also be obtained in 87% yield by carbonylation with CO_2 at 165°C, 17 atm.

$$\text{2,6-bis(trimethylsilyloxy)-4,5-bis(NHSiMe}_3\text{)pyrimidine} \xrightarrow[Et_3N,\ 0°C]{COCl_2} \text{2,6-bis(OSiMe}_3\text{)-7,9-bis(SiMe}_3\text{)purin-8-one (80\%)} \longrightarrow \text{uric acid} \tag{84}$$

The reaction of tetrakis(trimethylsilyl)-4,5-diaminouracil with oxalyl chloride provides an entry into the pteridine system. The reaction was carried out in the presence of a slight molar excess of triethylamine at 0°C (84) (eq. 85). The mild reaction conditions are in contrast to the reaction of oxalyl chloride with diaminouracil which requires melt temperatures or formylation.

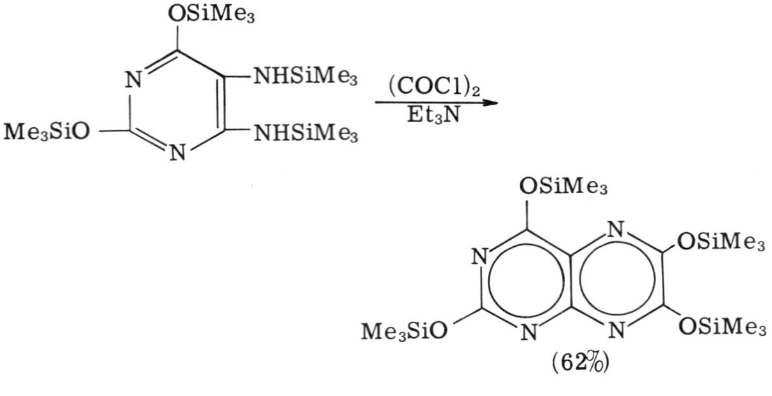

(85)

The formation of the heterocycle with oxalyl chloride is not confined to the uracil derivative (85) (eq. 86). Surprisingly, reaction of oxalyl chloride with the very similar 2,4,5-tris(trimethylsilylamino)-6-alkoxypyrimidine under identical conditions yielded a purine derivative with elimination of CO rather than the expected pteridine (84) (eq. 87).

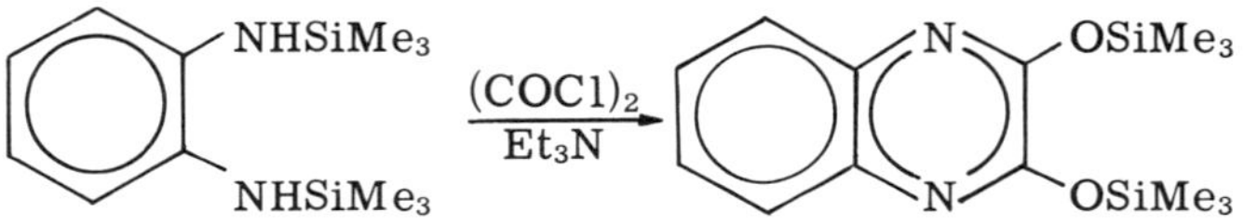

(86)

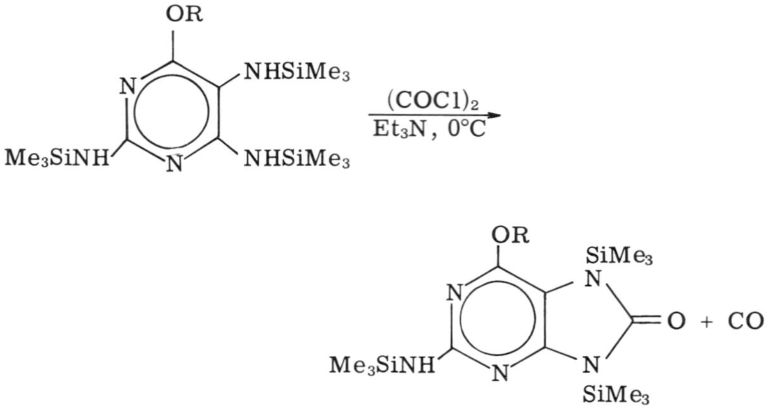

(87)

2. *Formation of Peptides*

Application of Anderson and Paul's imidazolide method for peptide synthesis to silylated di- and tripeptides allowed the preparation of a series of tri- and tetrapeptides in good yield (86). Imidazole is eliminated in these reactions; the silyl group thus has an activating function (eq. 88). The imidazolides were prepared from the carbobenzoxy amino acid or -peptide by reaction with

N,N'-carbonyldiimidazole at 50°C in THF; the silylated di- or tripeptide was added to the mixture and refluxed for 3 hr. Treatment with dilute acid afforded the carbobenzoxylated tri- or tetrapeptides in yields ranging from 60 to 93% for the 16 examples described.

$$\text{CboNHCH(R)COIm} + \text{HN(SiMe}_3\text{)CH(R')CONHCH(R'')COOSiMe}_3 \longrightarrow \text{CboNHCH(R)CON(SiMe}_3\text{)CH(R')CONHCH(R'')COOSiMe}_3 + \text{ImH} \tag{88}$$

$$\text{Cbo} = \text{PhCH}_2\text{OCO—}$$

$$\text{Im} = \text{imidazol-1-yl (N—)}$$

Whether or not in these reactions the activating role of the silyl group toward displacement of a proton is cleanly separated in all cases from a displacement of the silyl group itself is of no practical consequence; the fact that displacement of a silyl group is possible under similar conditions is shown (86) by eqs. (89) and (90).

$$\text{CboNHCH(CH}_3\text{)COIm} + \text{Me}_3\text{SiNHSiMe}_3 \xrightarrow{-\text{Me}_3\text{SiIm}} \text{CboNHCH(CH}_3\text{)CONHSiMe}_3 \tag{89}$$

$$\text{CboNHCH}_2\text{COIm} + \text{Me}_3\text{SiN(CH}_3\text{)SiMe}_3 \xrightarrow{-\text{Me}_3\text{SiIm}} \text{CboNHCH}_2\text{CON(CH}_3\text{)SiMe}_3 \tag{90}$$

Instead of imidazolides, the mixed anhydrides prepared from carbobenzoxylated amino acids and ethyl chloroformate can be used for peptide formation with silylated amino acids (86) (eq. 91).

$$\mathrm{CboNH\overset{R}{\overset{|}{C}}HCOOCOOEt + Me_3SiNH\overset{R'}{\overset{|}{C}}HCOOSiMe_3} \longrightarrow \mathrm{CboNH\overset{R}{\overset{|}{C}}HCONH\overset{R'}{\overset{|}{C}}HCOOSiMe_3 + CO_2 + Me_3SiOEt} \qquad (91)$$

The yield of *N*-Cbo-glycyl-glycyl-β-alanylglycine prepared by this method using *N*-silylglycine silyl ester was higher (94%) than using glycine ethyl ester (61%) because of the more facile silyl ester hydrolysis.

Since silyl esters can be used in the same fashion as free amino acids in the synthesis of mixed anhydrides with ethyl chloroformate, the method lends itself to the stepwise addition of amino acids to peptides without isolation of intermediates (86) (eq. 92). The reactions are carried out in THF or THF-DMSO mixtures; anhydride formation proceeds at low temperature with triethylamine as acid acceptor. The condensation with the silyl ester is carried out at or below room temperature with a short period at reflux to complete the reaction.

$$\mathrm{CboNH\overset{R}{\overset{|}{C}}HCONH\overset{R'}{\overset{|}{C}}HCOOSiMe_3} \xrightarrow[-\mathrm{Me_3SiCl}]{\mathrm{ClCOOEt}} \mathrm{CboNH\overset{R}{\overset{|}{C}}HCONH\overset{R'}{\overset{|}{C}}HCOOCOOEt} \xrightarrow[-\mathrm{CO_2},\ -\mathrm{Me_3SiOEt}]{\mathrm{Me_3SiNH\overset{R''}{\overset{|}{C}}HCOOSiMe_3}} \mathrm{CboNH\overset{R}{\overset{|}{C}}HCONH\overset{R'}{\overset{|}{C}}HCONH\overset{R''}{\overset{|}{C}}HCOOEt} \quad \text{and so on} \qquad (92)$$

The yields of six tri- or tetrapeptides without isolation of intermediates ranged from 68 to 90%; the connection of two tri- or tetrapeptide residues by this method afforded the carbobenzoxyhexa- or heptapeptide in 74-92% yield.

The mixed anhydride of an *N*-Cbo-amino acid and phosphorus oxychloride can also be used for formation of peptide bonds (87) (eq. 93). Eleven di- and tripeptides were synthesized without racemization in yields of 40-90% by this method. *N*-Silylsarcosine silyl ester did not react, indicating that the mixed anhydride reacts with the activated nitrogen-hydrogen rather than with the nitrogen-silicon bond.

$$\mathrm{CboNHCH(R)COOPOCl_2 + Me_3SiNHCH(R')COOSiMe_3 \xrightarrow[\text{10°C, Et}_3\text{N}]{\text{THF}} CboNHCH(R)CON(SiMe_3)CH(R')COOSiMe_3 + HPOCl_2} \quad (93)$$

B. ALKYLATION OF SILYLATED HETEROCYCLES

1. Methylation of α-Pyridone, Uric Acid, and Xanthine; N-*Vinyl Uracil*

N-Alkylation of purines and pyrimidines has been of interest for many years, primarily as a path to nucleosides. The discovery by Hilbert and Johnson in 1930 (88) that *O*-alkylated pyrimidines can be reacted with alkyl halides to form products alkylated at the adjacent nitrogen with loss of the *O*-alkyl group was the main route to these products until the recent discovery that *O*-silylated starting materials can be used for the same purpose.

Silylated α-pyridone yields *N*-methyl pyridone on short heating with methyl iodide (84) (eq. 94).

$$\text{2-(Me}_3\text{SiO)pyridine} + \mathrm{CH_3I} \longrightarrow \text{N-methyl-2-pyridone} + \mathrm{Me_3SiI} \quad (94)$$

An intermediate with quaternary nitrogen may be postulated which should lose the silyl substituent very readily (**12**).

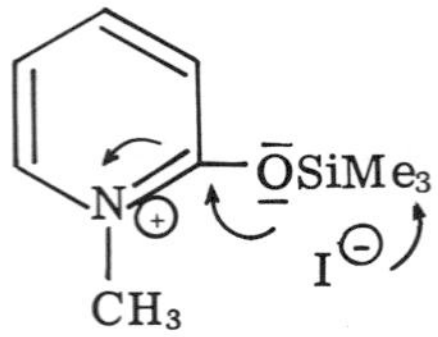

(12)

Bis(trimethylsiloxy)quinoxaline was converted to the bis-*N*-methylated product (eq. 95) and persilylated uric acid yielded the 1-methyl derivative which had been synthesized before by a multistep synthesis (eq. 96).

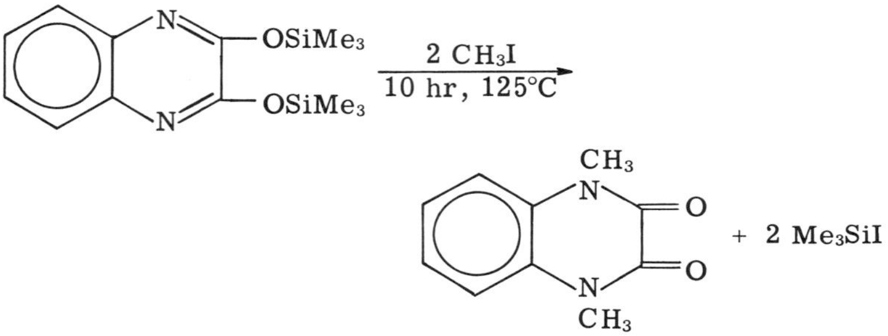

$$\xrightarrow[10\ \text{hr},\ 125^\circ\text{C}]{2\ CH_3I} \quad + 2\ Me_3SiI \qquad (95)$$

(71%)

$$\xrightarrow[2.\ \text{hydrolysis}]{1.\ CH_3I,\ AgClO_4,\ 10\ \text{min},\ 20^\circ\text{C}} \qquad (96)$$

(85%)

Similarly, tris(triethysilyl)xanthine was methylated quantitatively at room temperature at the 7 position; at 60°C the 3 position was also methylated and theobromine was obtained upon hydrolysis (84) (eq. 97).

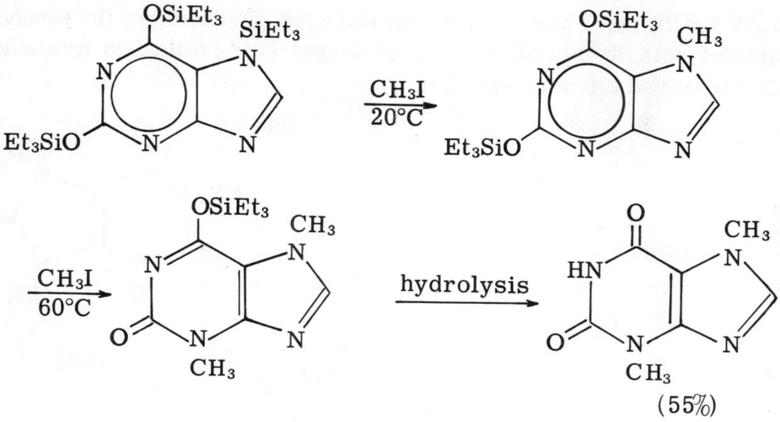

CH_3I, 20°C; CH_3I, 60°C; hydrolysis (55%) (97)

2,4-Bis(trimethylsilyl)thymine is alkylated in the 1 position by methyl iodide/silver perchlorate (89) (eq. 98). The formation of 1,3-dimethylthymine requires heating of the silyl derivative with dimethylsulfate.

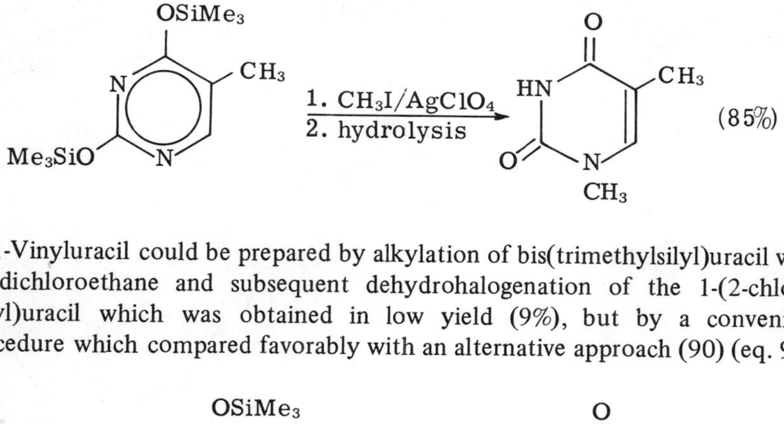

1. $CH_3I/AgClO_4$ 2. hydrolysis (85%) (98)

1-Vinyluracil could be prepared by alkylation of bis(trimethylsilyl)uracil with 1,2-dichloroethane and subsequent dehydrohalogenation of the 1-(2-chloroethyl)uracil which was obtained in low yield (9%), but by a convenient procedure which compared favorably with an alternative approach (90) (eq. 99).

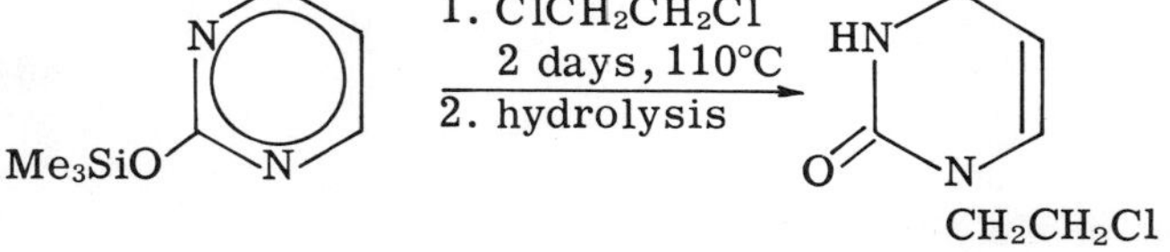

1. $ClCH_2CH_2Cl$, 2 days, 110°C 2. hydrolysis (99)

Dimethylsulfate and diethylsulfate have also been used for alkylations of silylated uric acid (84). The 1,7,9-trialkyl derivatives were obtained in high yield upon short heating of tetrakis(triethylsilyl)uric acid and hydrolysis or alcoholysis under neutral conditions. However, if the hot reaction mixture was poured into $2N$ NaOH, the 3 position was also alkylated. Treatment of the persilylated compound with diethylsulfate for a prolonged time resulted in formation of 1,2,7,9-tetraethyl uric acid (eq. 100).

Reactant: tetrakis(triethylsilyl) uric acid (OSiEt$_3$, Et$_3$SiO, N-SiEt$_3$, N-SiEt$_3$)

1. Me_2SO_4, 1 hr, 150°C; 2. H_2O → 1,7,9-trimethyl uric acid (CH_3N, CH_3, CH_3; N–H) (90%)

1. Me_2SO_4, 1 hr, 150°C; 2. 2N NaOH → 1,3,7,9-tetramethyl uric acid (CH_3N, CH_3, CH_3, CH_3) (90%)

1. Et_2SO_4 (Me_2SO_4), 10 hr, 130°C; 2. H_2O → (Me) EtN, EtO (Me), Et (Me), Et (Me) (80%)

(100)

2. Formation of N-*Glycosides*

The reaction of acylated glycosyl halides with silylated purines and pyrimidines was found to provide convenient access to a series of glycosides.

Tetrasilyl uric acid was alkylated in the 3 position; the reactions were carried out in toluene in the presence of an equivalent amount of silver perchlorate at 20-70°C; yields between 45 and 90% were obtained (84,91) (eq. 101).

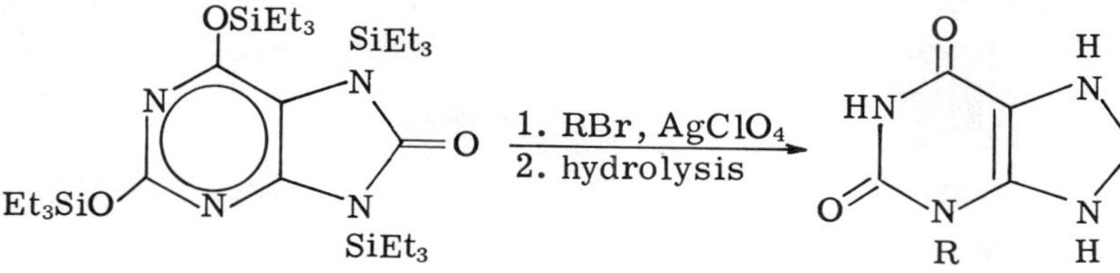

$$\xrightarrow[\text{2. hydrolysis}]{\text{1. RBr, AgClO}_4} \tag{101}$$

R = 1-(2, 3, 4-triacetyl) ribopyranosyl
1-(2, 3, 5-tribenzoyl) ribofuranosyl
1-(2, 3, 4, 6-tetraacetyl) glucosyl

The glycosyl bromide is presumed to form initially the very reactive glycosyl perchlorate which acts as the alkylating agent. Silyl perchlorate can be distilled off. 2-Trimethylsiloxypyridine was converted to the *N*-tetraacetyl glucoside under the same conditions in nitromethane.

Several papers deal with the glycoside formation of silylated pyrimidines (92-98). The formation of thymine nucleosides from silylated intermediates and glycosyl halides was studied under various reaction conditions: heating of the compounds in inert solvents; melt condensation at 90-110°C *in vacuo*; reaction at room temperature in the presence of silver perchlorate; and reaction in the presence of mercuric salts. The reaction in the presence of equimolar amounts of mercuric bromide and mercuric oxide gave the best yields (95) (eq. 102).

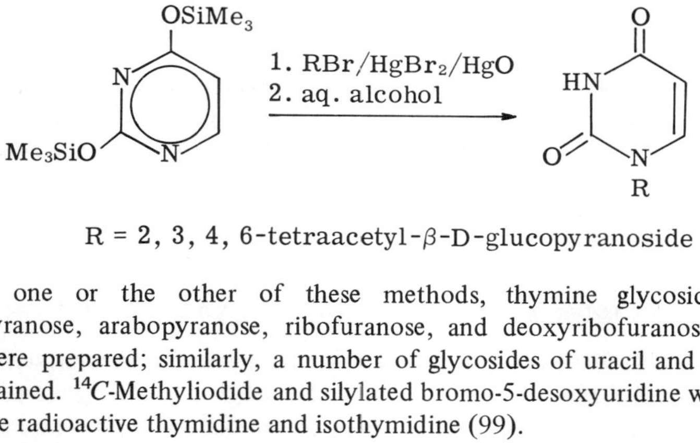

$$\xrightarrow[\text{2. aq. alcohol}]{\text{1. RBr/HgBr}_2\text{/HgO}} \tag{102}$$

R = 2, 3, 4, 6-tetraacetyl-β-D-glucopyranoside

Using one or the other of these methods, thymine glycosides with galactopyranose, arabopyranose, ribofuranose, and deoxyribofuranose substituents were prepared; similarly, a number of glycosides of uracil and cytosine were obtained. ^{14}C-Methyliodide and silylated bromo-5-desoxyuridine were used to prepare radioactive thymidine and isothymidine (99).

Glycoside formation of two trimethylsilyl-substituted quinoxalines (**13** and **14**) with α-acetobromoglucose in inert solvents or in the melt at temperatures of 90-115°C were unsuccessful; the reactions were not attempted in the presence of silver or mercuric salts (97). The alkylation of a series of 5-mercapto-substituted bis-*O*-trimethylsilyl uracil derivatives by 2-deoxy-3,5-di-*O*-(*p*-chlorobenzoyl)-α-*O*-ribofuranosyl chloride afforded either or both of the two anomers of the nucleoside, depending on the reaction conditions (98) (eq. 103).

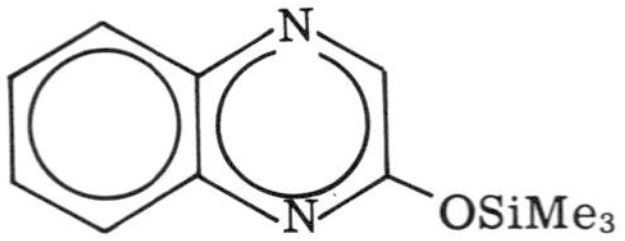

(13) OSiMe$_3$ (14) SiMe$_3$ CH$_3$

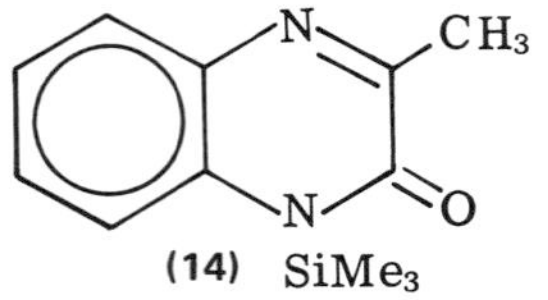

OSiMe$_3$ SR Me$_3$SiO + ArCOO ArCOO Cl

1. 80–110°C, neat or in PhH
2. hydrolysis

ArCOO ArCOO HN SR O and/or HN SR ArCOO ArCOO (103)

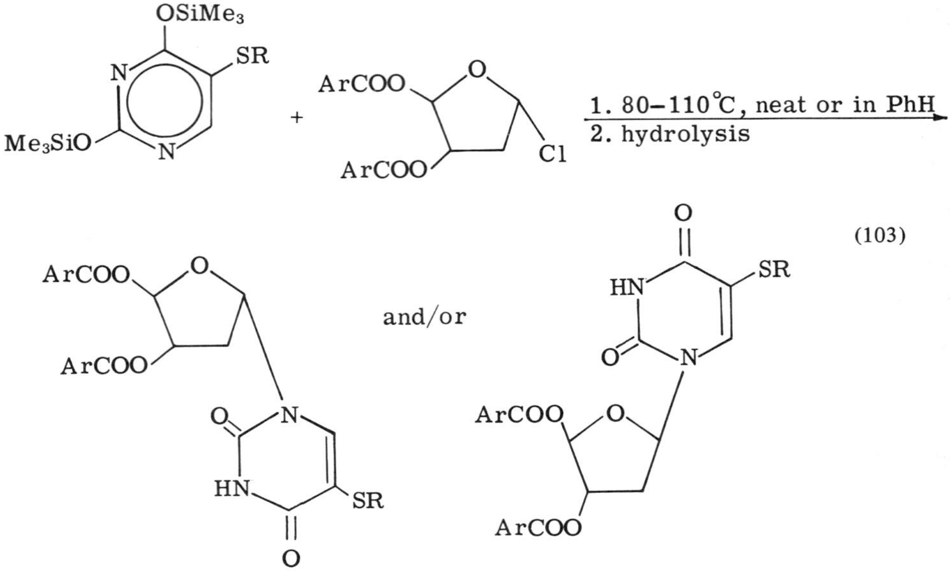

The β-nucleoside was the predominant product obtained in the absence of any added trimethylchlorosilane, as expected from an S_N2 mechanism involving a Walden inversion. In the presence of trimethylchlorosilane, the α-nucleoside was formed predominantly which was rationalized by rapid equilibration of the α-furanosyl chloride to the mixture of the α and β forms by chlorosilane and the predominant reaction of the less stable β anomer.

C. *ACYLATION AND ALKYLATION OF ENOLATES AND CARBOXYLATES*

Silyl enolates are accessible by reaction of enolizable ketones with sodium bis(trimethylsilyl)amide (100), or with a silyl amide in the presence of catalytic amounts of base (101), or by reaction of the ketone with sodium hydride followed by addition of trimethylchlorosilane/triethylamine (102); in many cases, silylation with bis(trimethylsilyl)acetamide without a catalyst is sufficient (37).

The reaction of 2-trimethylsiloxypent-2-ene-4-one with benzoyl chloride at 140°C resulted in acylation at C-3; trimethylchlorosilane distilled off, and the triketone was obtained in 75% yield by fractionation (103) (eq. 104).

$$CH_3COCH{=}C(CH_3)OSiMe_3 + PhCOCl \longrightarrow CH_3COCH(COPh)COCH_3 + Me_3SiCl \qquad (104)$$

β-Trimethylsiloxy crotonate gave the α-benzoyl derivative under the same conditions (eq. 105).

$$CH_3C(OSiMe_3){=}CHCOOEt \xrightarrow[-Me_3SiCl]{PhCOCl} CH_3COCH(COPh)COOEt \qquad (105)$$

An earlier report of the reaction of ethyl β-trimethylsiloxy crotonate with acetyl chloride at 135-150°C claimed the formation of β-acetoxy crotonate by oxygen-acylation (104). A reinvestigation of the two conflicting reports showed that the carbon-acylation product predominates in both cases, but that about 10% of *O*-acyl crotonate is formed with acetyl chloride as well as benzoyl chloride, in addition to much ethyl benzoate in the latter case (105).

Examination of a reasonable transition state for the condensation reaction shows that *C*- as well as *O*-acylation can be rationalized readily (eq. 105a). The condensation of silyl enolates with benzaldehyde proceeds similarly at reflux temperature of the mixture and leads to *C*-alkylated products (103) (eq. 106). The initial adduct cannot be isolated, presumably due to the reactivity of the acidic hydrogen on C_3, resulting in elimination of trimethylsilanol which in turn attacks the remaining siloxybenzyl pentanedione, leading to some of the parent carbinol.

$$\text{(RCO)(H)C{=}C(R')O\cdots SiMe_3 / R''CO\cdots Cl (transition state, } \delta^-, \delta^+) \xrightarrow{-Me_3SiCl} \begin{cases} (RCO)(H)C(COR')(COR'') \\ (RCO)(H)C{=}C(R')(OCOR'') \end{cases} \qquad (105a)$$

$$CH_3COCH{=}C(CH_3){-}O{-}SiMe_3 + PhCH{=}O \longrightarrow \left[\begin{array}{c} CH_3COCHCOCH_3 \\ | \\ HCOSiMe_3 \\ | \\ Ph \end{array}\right] \begin{array}{l} \xrightarrow{-\,Me_3SiOH} CH_3COC(=CHPh)COCH_3 \ (26\%) \\ \xrightarrow[-\,(Me_3Si)_2O]{Me_3SiOH} CH_3COCH(CH(OH)Ph)COCH_3 \ (32\%) \end{array} \quad (106)$$

In the absence of hydrogen on C_3, the addition product is thermally stable (103) (eq. 107).

$$(CH_3CO)(CH_3)C{=}C(OSiMe_3)(CH_3) \xrightarrow{PhCHO} CH_3COC(CH_3)(CH(OSiMe_3)Ph)COCH_3 \ (40\%) \quad (107)$$

When the mixed ethyl-trimethylsilyl ester of malonic acid is metalated and then reacted with two equivalents of valeryl chloride, ethyl divaleryl acetic ester is obtained (105,106) (eq. 108). The metalation is carried out at $-60°C$ and subsequent acylation at room temperature; a yield of about 40% of the divaleryl ester can be obtained. The monovaleryl ethyl acetate is formed in about 55% yield by reaction of the metalated malonic ester with one equivalent of valeryl chloride for 10 min at $0°C$ and subsequent hydrolysis.

$$H_2C(COOEt)(COOSiMe_3) \xrightarrow[2.\ C_4H_9COCl]{1.\ n\text{-}C_4H_9Li} C_4H_9COCH(COOEt)(COOSiMe_3) \xrightarrow[-Me_3SiCl,\ -CO_2]{C_4H_9COCl} (C_4H_9CO)_2CHCOOEt \quad (108)$$

The double acylation on carbon has not been observed with alkyl esters of malonic acid. If instead of the mixed alkyl-trimethylsilyl ester of malonic acid, bis(trimethylsilyl)malonate is subjected to acylation, divaleryl methane (eq. 109) and, in the case of metalation with isopropyl magnesium bromide, even trivaleryl methane can be obtained (105) (eq. 110).

$$H_2C(COOSiMe_3)_2 \xrightarrow[\text{2. } C_4H_9COOCOOEt]{\text{1. PhLi}} \underset{(60\%)}{C_4H_9COCH_2COC_4H_9} \quad (109)$$

$$H_2C(COOSiMe_3)_2 \xrightarrow[\text{2. } C_4H_9COOCOOEt]{\text{1. } i\text{-PrMgBr}} \underset{(40\%)}{(C_4H_9CO)_3CH} \quad (110)$$

This method has been applied to the synthesis of polycarbonyl compounds; for stabilization, some of the keto groups were protected with ethylene hemithioketal functions which can be removed with mercuric chloride after completion of the chain extension (105) (eqs. 111 and 112).

Ethyl or trimethylsilyl *N,N*-bis(trimethylsilyl)glycinate can be metalated with sodium bis(trimethylsilyl)amide or other bases; treatment with alkyl halides results in alkylation at C_1 and formation of higher amino acids whereas trimethylchlorosilane attacks the ester oxygen (107) (eq. 113). The *N,N*-bistrimethylsilylglycine esters are prepared by heating glycine or ethyl glycinate with excess trimethylsilyl diethylamine in the presence of a catalytic amount of $(NH_4)_2SO_4$ at 120°C until no more diethylamine is evolved. The metalation is

$$EtOCOCH_2-\underset{S\;\;\;\;O}{C}-COOH \xrightarrow[Et_3N]{ClCOOEt}$$

$$EtOCOCH_2-\underset{S\;\;\;\;O}{C}-COOCOOEt \xrightarrow[\text{2. hydrolysis; } -CO_2]{\text{1. } (CH_3CO\overset{\ominus}{C}HCOOSiMe_3)Li^{\oplus}} \quad (111)$$

$$EtOCOCH_2-\underset{S\;\;\;\;O}{C}-(COCH_2)_2CH_3$$

$$CH_3{-}\underset{\text{S--O ring}}{C}{-}CH_2COOCOOEt \xrightarrow[\text{2. hydrolysis; } -CO_2]{\text{1. } \overset{\ominus}{H}C(COOSiMe_3)_2} \tag{112}$$

$$CH_3{-}\underset{\text{S--O ring}}{C}{-}CH_2COCH_2COCH_2CH_2{-}\underset{\text{S--O ring}}{C}{-}CH_3$$

$$\underset{N(SiMe_3)_2}{CH_2COOEt} \xrightarrow{(Me_3Si)_2NNa} \underset{N(SiMe_3)_2}{CH}{\overset{\ominus}{=\!=}}C(O)OEt$$

$$\xrightarrow[-NaI]{C_2H_5I} \underset{N(SiMe_3)_2}{C_2H_5CHCOOEt}\ (65\%)$$

$$\xrightarrow[-NaCl]{Me_3SiCl} \underset{N(SiMe_3)_2}{CH}{=}C(OSiMe_3)OEt\ (77\%) \tag{113}$$

carried out at $-10°C$ in ether and the subsequent reaction with alkyl halide at reflux for 10-15 hr. Hydrolysis with dilute HCl leads to the amino acid hydrochlorides or their ethyl esters.

Aldehydes can be added to the metalated glycine esters; the resulting salts of the β-hydroxy amino acid esters have been converted to the β-trimethylsiloxy compounds; treatment with dilute acid leads to the β-substituted serine ester (107) (eq. 114). Reaction of the parent, nonmetalated *N,N*-bis(trimethylsilyl)glycine esters with aldehydes affords imines and hexamethyl disiloxane.

$$\underset{N(SiMe_3)_2}{{}^{\ominus}CHCOOEt} \xrightarrow[\text{2. } Me_3SiCl]{\text{1. PhCHO}} Ph\overset{OSiMe_3}{CH}\underset{N(SiMe_3)_2}{CH}COOEt \xrightarrow{HCl} Ph\underset{HO}{CH}\underset{NH_2}{CH}COOEt \tag{114}$$

D. PHOSPHONATION OF SILYL ACETYLENES

The Arbusov reaction of trialkyl phosphites with chloroethynyl trimethylsilane [prepared by reaction of dichloroacetylene first with phenyllithium, then trimethylchlorosilane (108)] allows the synthesis of ethynyl phosphonate esters in good yields which are otherwise not conveniently accessible (109) (eq. 115).

$$Me_3SiC{\equiv}CCl \xrightarrow[CH_3Cl]{(CH_3O)_3P} \underset{(70\%)}{(CH_3O)_2\overset{O}{\overset{\|}{P}}C{\equiv}CSiMe_3} \xrightarrow{aq.\ Na_2CO_3} \underset{(63\%)}{(CH_3O)_2\overset{O}{\overset{\|}{P}}C{\equiv}CH} \qquad (115)$$

The trimethylsilyl group has the twofold role of facilitating the halide displacement by way of electron withdrawal and protecting the molecule against side reactions by temporary removal of the acidic hydrogen. The reaction was carried out by refluxing the mixture for 5 hr. The diethyl and the diisopropyl trimethylsilylethynylphosphonate were prepared by the same procedure in 82 and 98% yield, respectively. Cleavage of the carbon-silicon bond could be achieved by treatment with 10% aqueous sodium carbonate solution at room temperature for 1-2 min. The final products could also be obtained without isolation of the intermediate trimethylsilyl derivatives in yields of about 60% based on the silane.

IV. Addition of Organosilanes to Reactive Double Bonds and Cyclic Compounds

This section describes the synthesis of compounds formed by an apparent insertion of a reactive organic moiety into silicon-nitrogen and silicon-carbon bonds. Little is known about the actual mechanism of most of these reactions; in some cases, such as the adduct formation from silyl amines and carbon dioxide, a four-center type mechanism may be operative and "insertion" may be an adequate description. Reactions such as the base-catalyzed additions to epoxides and aldehydes certainly proceed by a different mechanism, and the similarity to addition reactions with compounds containing cumulated double bonds is only a formal one.

A. ADDITION TO CUMULATED DOUBLE BONDS; FORMATION OF UREAS, IMIDES, AND ACETOACETAMIDES

Silylamine addition reactions with compounds containing cumulated double bonds such as carbon dioxide (110), carbon disulfide (111,112), sulfur trioxide (113), isocyanates (114-119), isothiocyanates (112), and ketene (120,121) have been reported. The literature up to 1967 has been reviewed (122).

The reaction of silylamines with carbon dioxide leads to insertion products which at higher temperatures lose CO_2 and revert to the starting material (110). In the special case of 1,2,2,3-tetramethyl silaimidazolidine, the insertion product with CO_2 can be converted thermally to a cyclic urea derivative (123) (eq. 116).

$$\underset{\substack{\diagup\ \ \diagdown \\ Me\qquad Me}}{CH_3N \underset{Si}{\diagdown\ \diagup} NCH_3} \xrightarrow{2\ CO_2} \underset{\substack{Me\qquad Me}}{\overset{CH_3N\!-\!CO\!-\!O\qquad O\!-\!CO\!-\!NCH_3}{\diagdown\ Si\ \diagup}} \longrightarrow CH_3N\underset{\substack{C \\ \| \\ O}}{\diagdown\ \diagup}NCH_3 + \text{polymer} \tag{116}$$

The addition of silylamines and silazanes to isocyanates or isothiocyanates leads in high yield to silyl-substituted urea or biuret derivatives (114-119) which are readily hydrolyzed to the parent compound (eqs. 117-120).

$$PhNCO + Me_2NSiMe_3 \longrightarrow Ph\underset{\substack{| \\ SiMe_3}}{N}CONMe_2 \tag{117}$$

$$MeNCO + MeN(SiMe_3)_2 \longrightarrow Me\underset{\substack{| \\ SiMe_3}}{N}CO\underset{\substack{\diagdown \\ SiMe_3}}{N}Me \xrightarrow{MeNCO} Me\underset{\substack{\diagup \\ SiMe_3}}{N}CO\underset{\substack{| \\ Me}}{N}CO\underset{\substack{\diagdown \\ SiMe_3}}{N}Me \tag{118}$$

$$PhCONCO + (Me_3Si)_2NMe \longrightarrow PhCO\underset{\substack{| \\ Me_3Si}}{N}CO\underset{\substack{| \\ SiMe_3}}{N}Me \xrightarrow{H_2O} \underset{(73\%)}{PhCONHCONHMe} \tag{119}$$

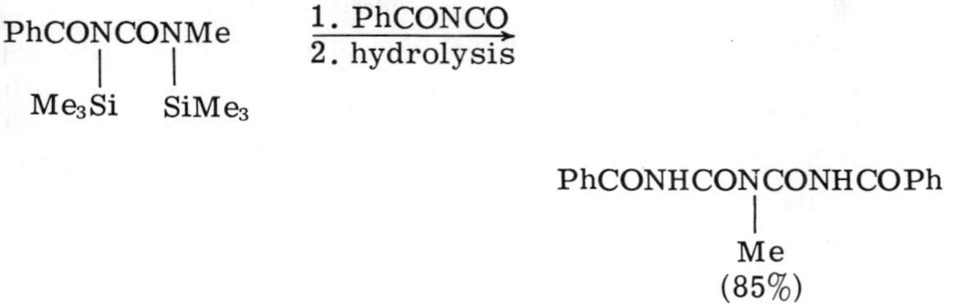

$$\underset{\displaystyle\quad Me_3Si\quad\; SiMe_3}{PhCON\,CONMe} \xrightarrow[\text{2. hydrolysis}]{\text{1. PhCONCO}} \underset{\displaystyle Me}{PhCONHCONCONHCOPh} \quad (120)$$

(85%)

If the 1:1 addition product of a silazane and an isocyanate (eq. 118) bears hydrogen instead of methyl on one of the nitrogens, reaction with a second molecule of isocyanate affords a symmetrical urea and Me_3SiNCO is eliminated (114). This reaction was utilized for the synthesis of the otherwise inaccessible divinylurea (124) (eq. 121).

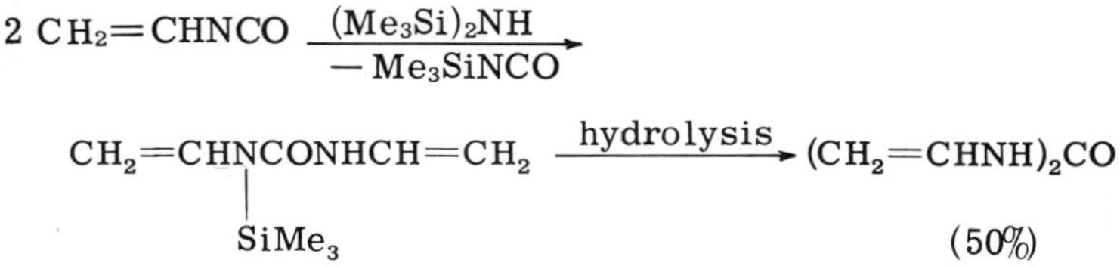

$$2\,CH_2{=}CHNCO \xrightarrow[-\,Me_3SiNCO]{(Me_3Si)_2NH} \underset{\displaystyle SiMe_3}{CH_2{=}CHNCONHCH{=}CH_2} \xrightarrow{\text{hydrolysis}} (CH_2{=}CHNH)_2CO \quad (121)$$

(50%)

The addition of silylated diamines to diisocyanates was used to prepare silyl-substituted ureas which could be handled in solvents of low polarity due to the less polar character of the polymers as compared with normal polyureas (125) (eq. 122).

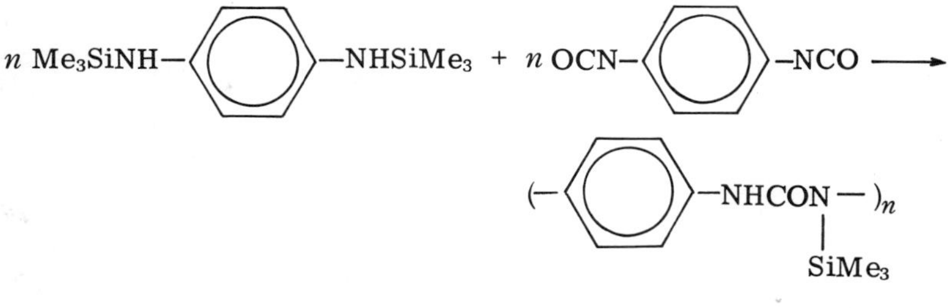

$$n\,Me_3SiNH{-}C_6H_4{-}NHSiMe_3 + n\,OCN{-}C_6H_4{-}NCO \longrightarrow ({-}C_6H_4{-}NHCON(SiMe_3){-})_n \quad (122)$$

Addition of dialkylsilylamines to cyanates affords silyl-substituted isoureas in quantitative yield (eq. 123) which may be hydrolyzed to the parent isoureas, acylated with acid chlorides, or reacted with further cyanate to form triazines (126) (eq. 124). Monoalkyl aminosilanes and cyanates in a ratio of 1:2 gave cyano-substituted isoureas in about 50% yield (126) (eq. 125). *N*-Alkyl-substituted silazanes are able to add two molecules of cyanate; reaction with more cyanate at higher temperatures leads to displacement of Me_3Si- by $-CN$ (eq. 126). Hexamethyldisilazane adds one molecule of cyanate; further cyanate again displaces the silyl group, yielding 24-62% cyano-substituted isoureas after hydrolysis (eq. 127). The interaction of ketene with silylated amines and amides has been studied in detail (121). Reaction at low temperature leads to 1:1 adducts which can be isolated with proper care. At room temperature or above, a second molecule of ketene is added and high yields of β-siloxyvinylacetic acid amides can be isolated. Hydrolysis afforts acetoacetic acid amide (eq. 128). Analogous products are obtained from ketene and the silylamides (**15** and **16**).

$$PhOC{\equiv}N + Et_2NSiMe_3 \longrightarrow PhOC(NEt_2){=}NSiMe_3 \quad (100\%) \tag{123}$$

$$PhOC(NEt_2){=}NSiMe_3 \xrightarrow{ROH} PhOC(NEt_2){=}NH + Me_3SiOR$$

$$PhOC(NEt_2){=}NSiMe_3 \xrightarrow{PhCOCl} PhOC(NEt_2){=}NCOPh + Me_3SiCl \tag{124}$$

$$PhOC(NEt_2){=}NSiMe_3 \xrightarrow{2\ PhOCN} \text{2-}NEt_2\text{-4,6-}(PhO)_2\text{-1,3,5-triazine} + Me_3SiOPh$$

$$PhOCN + EtNHSiMe_3 \longrightarrow \left[PhOC(NHEt){=}NSiMe_3 \rightleftharpoons PhOC({=}NEt)NHSiMe_3 \right] \xrightarrow{PhOCN} PhOC(NHEt){=}NCN + PhOSiMe_3 \tag{125}$$

$$CH_3N(SiMe_3)_2 \xrightarrow[25°C]{2\ PhOCN} \begin{matrix} PhOC{=}NSiMe_3 \\ | \\ CH_3N \\ | \\ PhOC{=}NSiMe_3 \end{matrix} \xrightarrow[-2\ PhOSiMe_3]{2\ PhOCN,\ 2\ hr,\ 80°C} \begin{matrix} PhOC{=}NCN \\ | \\ CH_3N \\ | \\ PhOC{=}NCN \\ (50\%) \end{matrix} \qquad (126)$$

$$2\ MeO\text{-}C_6H_4\text{-}OCN \xrightarrow[25°C]{(Me_3Si)_2NH} MeO\text{-}C_6H_4\text{-}\underset{\displaystyle NHSiMe_3}{OC{=}NCN} \xrightarrow{\text{hydrolysis}} \underset{(62\%)}{MeO\text{-}C_6H_4\text{-}\underset{\displaystyle NH_2}{OC{=}NCN}} \qquad (127)$$

$$Me_3SiNMe_2 \xrightarrow[-15°C]{CH_2{=}C{=}O} CH_2{=}\underset{\displaystyle OSiMe_3}{CNMe_2} \xrightarrow{CH_2{=}C{=}O} CH_2{=}\underset{\displaystyle OSiMe_3}{CCH_2CONMe_2} \xrightarrow{H_2O} CH_3COCH_2CONMe_2 \qquad (128)$$

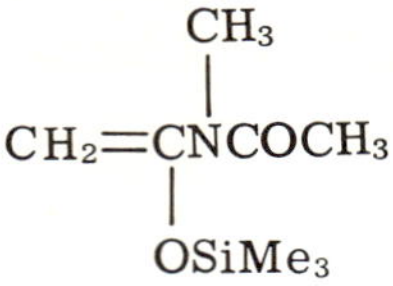

$$CH_2{=}C(OSiMe_3)N(CH_3)COCH_3 \quad \textbf{(15)}$$

$$CH_2{=}C(OSiMe_3)CH_2CON(CH_3)COCH_3 \quad \textbf{(16)}$$

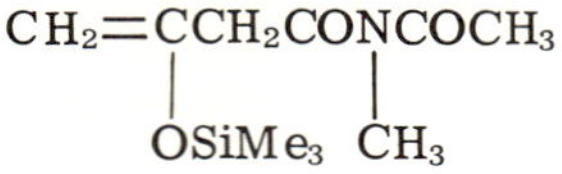

The addition of the second molecule of ketene appears to proceed via rupture of the silicon-oxygen bond rather than by insertion into the carbon-nitrogen bond as shown by addition to diphenylketene to the 1:1 adduct (eq. 129). The 1:1 adduct—α-trimethylsiloxyvinyldimethylamine—undergoes a slow rearrangement to α-trimethylsiloxyacetamide. This interesting silyl migration from oxygen to carbon is greatly accelerated by catalytic amounts of trimethylbromosilane (eq. 130).

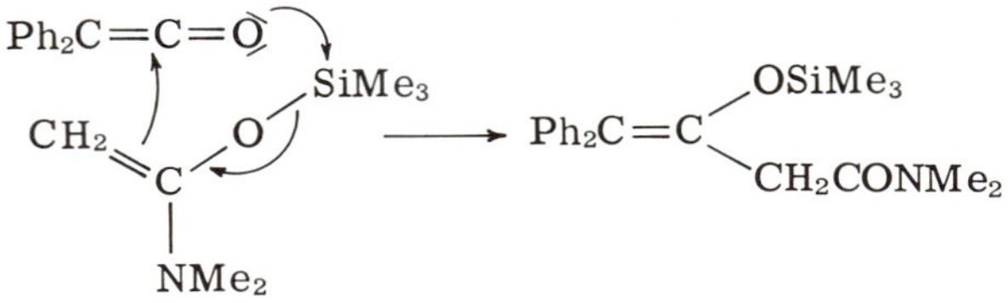

$$Ph_2C{=}C{=}O + CH_2{=}C(NMe_2)OSiMe_3 \longrightarrow Ph_2C{=}C(OSiMe_3)CH_2CONMe_2 \qquad (129)$$

$$CH_2{=}C(NMe_2)OSiMe_3 \xrightarrow[15\ \text{min}]{(Me_3SiBr)} Me_3SiCH_2CONMe_2 \qquad (130)$$

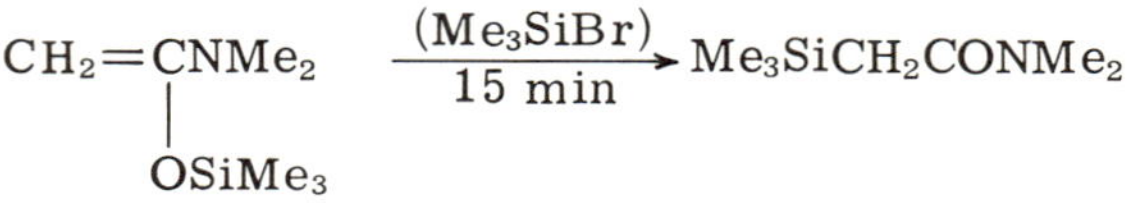

The addition of silylamines to diketene affords β-trimethylsiloxycrotonic acid amide, probably via a vinyl acetic acid precursor which was found to undergo this rearrangement in an independent reaction (eq. 131).

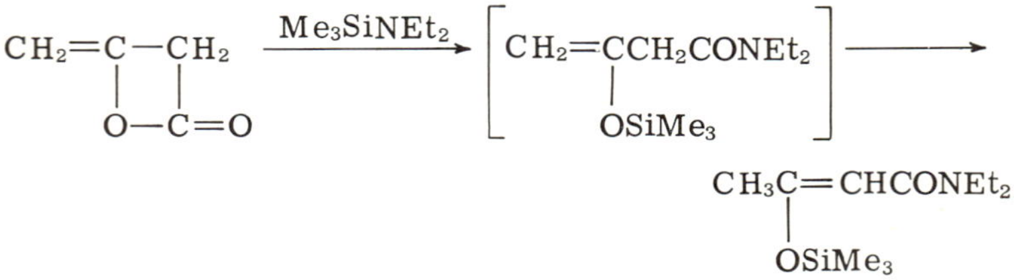

$$\underset{\text{(diketene: } CH_2{=}C{-}CH_2,\ O{-}C{=}O)}{CH_2{=}C{-}CH_2} \xrightarrow{Me_3SiNEt_2} \left[CH_2{=}C(OSiMe_3)CH_2CONEt_2\right] \longrightarrow CH_3C(OSiMe_3){=}CHCONEt_2 \qquad (131)$$

B. ADDITION TO LACTONES, ANHYDRIDES, AND EPOXIDES; FORMATION OF β-AMINO- AND AMIDO-ALCOHOLS; AMIC ACIDS

Reaction of silylamines with lactones could conceivably take two different courses: cleavage of the carbonyl-to-oxygen bond would result in formation of amides whereas cleavage of the methylene-to-oxygen bond would lead to aminocarboxylic acids. The latter reaction was observed exclusively and the silyl esters of *N,N*-dialkyl-β-alanine were formed in 64-88% yield (127) (eq. 132).

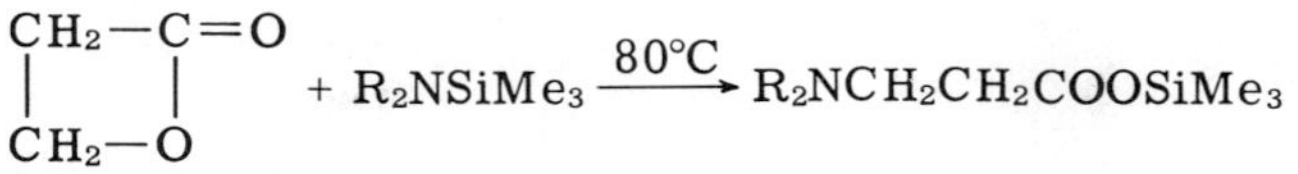

$$\begin{array}{l} CH_2-C=O \\ | \quad\quad | \\ CH_2-O \end{array} + R_2NSiMe_3 \xrightarrow{80°C} R_2NCH_2CH_2COOSiMe_3 \qquad (132)$$

In contrast, cyclic anhydrides are opened up with attack of the amine group at carbonyl (eq. 133).

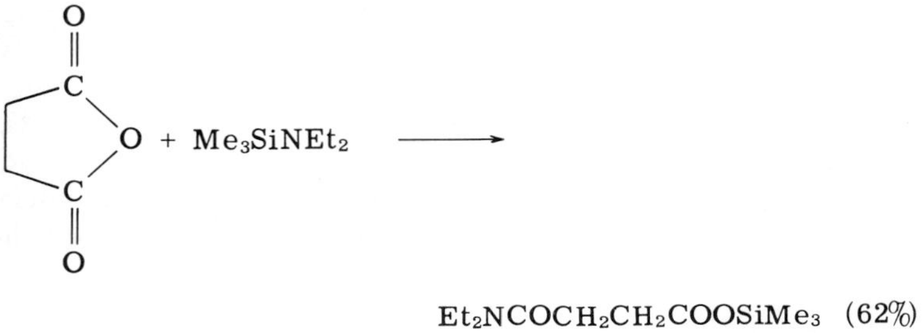

$$\text{(succinic anhydride)} + Me_3SiNEt_2 \longrightarrow Et_2NCOCH_2CH_2COOSiMe_3 \ (62\%) \qquad (133)$$

Aromatic anhydrides react in the same fashion; if a silylated primary amine is employed, the intermediate silylated "amic acid" can eliminate silanol at higher temperature with formation of an imide. These reactions are quantitative and have been employed for the preparation of polyimides; the presence of trimethylsilyl groups provides for solubility of the polymeric intermediates in nonpolar solvents; hydrocarbons can thus be used instead of the customary polar solvents such as dimethylformamide and *N*-methylpyrrolidone (128,129) (eq. 134). The reactions of phthalic anhydride with bis(trimethylsilyl)acetamide led to *N*-acetylphthalimide; the presumed intermediate adduct could not be isolated (130) (eq. 135).

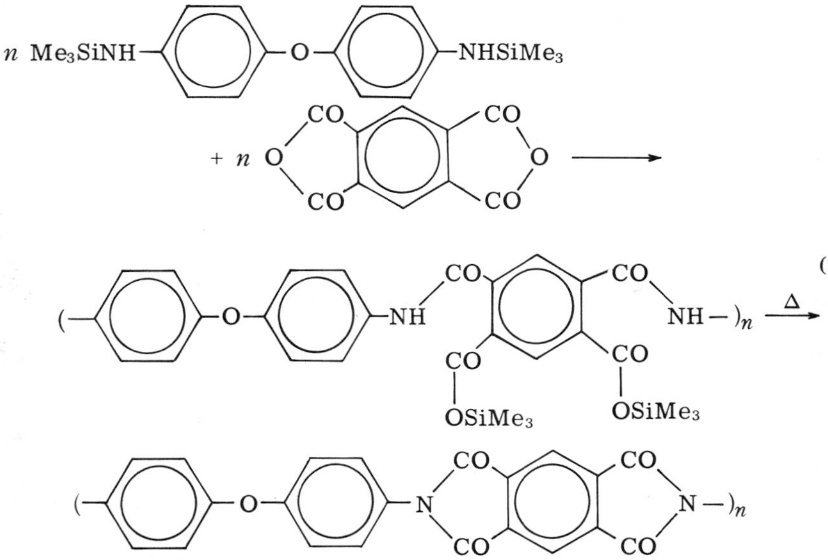

$$n\ Me_3SiNH-C_6H_4-O-C_6H_4-NHSiMe_3 + n\ O(CO)_2C_6H_2(CO)_2O \longrightarrow$$

$$(-C_6H_4-O-C_6H_4-NH-CO-C_6H_2(COOSiMe_3)_2-CO-NH-)_n \xrightarrow{\Delta} \qquad (134)$$

$$(-C_6H_4-O-C_6H_4-N(CO)_2C_6H_2(CO)_2N-)_n$$

$$\text{C}_6\text{H}_4(\text{CO})_2\text{O} \xrightarrow{\text{CH}_3\text{CON(SiMe}_3)_2} \left[\text{C}_6\text{H}_4(\text{COOSiMe}_3)(\text{CON(SiMe}_3)\text{COCH}_3)\right] \xrightarrow{-(\text{Me}_3\text{Si})_2\text{O}} \text{C}_6\text{H}_4(\text{CO})_2\text{NCOCH}_3 \qquad (135)$$

Ring opening of epoxides activated by an electron-withdrawing substituent with silylamines resulted in the expected β-aminoethyl trimethylsilyl ethers in 45-91% yield (121) (eq. 136).

$$\text{Cl}_3\text{CCH}\overset{\text{O}}{—}\text{CH}_2 + \text{Me}_3\text{SiNR}_2 \longrightarrow \text{Cl}_3\text{CCH(OSiMe}_3)\text{CH}_2\text{NR}_2 \qquad (136)$$

The silicon-free parent amino alcohols were obtained by hydrolysis.

The base-catalyzed reaction of silyl-substituted amides with epoxides does not require an electron-withdrawing substituent on the epoxide ring (132) (eq. 137). The additions to butene oxide and styrene oxide were carried out without a solvent and required temperatures of 95-150°C to afford the adducts in 66-87% yield. Attack of the amide group occurred at the unsubstituted carbon exclusively. Most adducts could be purified by distillation; acid hydrolysis led to the crystalline β-aminoalcohols.

$$\text{PhCH}\overset{\text{O}}{—}\text{CH}_2 + \text{CH}_3\text{CON(SiMe}_3)\text{CH}_3 \xrightarrow[100°\text{C}]{1\%\ \text{NaOSiMe}_3} \text{PhCH(OSiMe}_3)\text{CH}_2\text{N(CH}_3)\text{COCH}_3 \qquad (137)$$

Thermal instability was observed with the formyl-substituted adduct which suffered decarbonylation upon vacuum distillation (eq. 138).

$$\text{PhCH(OSiMe}_3)\text{CH}_2\text{N(CH}_3)\text{CHO} \xrightarrow[-\text{CO}]{170°\text{C}} \underset{(67\%)}{\text{PhCH(OSiMe}_3)\text{CH}_2\text{NHCH}_3} \xrightarrow{\text{hydrolysis}} \text{PhCH(OH)CH}_2\text{NHCH}_3 \qquad (138)$$

The mechanism of the base-catalyzed ring-opening reaction is believed to involve an amide-anion intermediate (eq. 139).

$$\underset{\substack{|\\ SiMe_3}}{CH_3CONCH_3} \xrightarrow[-(Me_3Si)_2O]{Me_3SiO^{\ominus}} \underset{\substack{|\\ CH_3}}{CH_3CON^{\ominus}} \xrightarrow{PhCH{-}CH_2\ (\text{via } O)}$$

$$\underset{\substack{|\quad\quad\quad\ \ |\\ O^{\ominus}\quad\quad CH_3}}{PhCHCH_2NCOCH_3} \xrightarrow[-\underset{\substack{|\\ CH_3}}{CH_3CON^{\ominus}}]{CH_3CONCH_3SiMe_3} \underset{\substack{|\quad\quad\quad\quad |\\ Me_3SiO\quad\quad CH_3}}{PhCHCH_2NCOCH_3} \quad (139)$$

The silanolate-catalyzed reaction of silyl amides with enolizable ketones results in a silyl proton exchange reaction rather than addition; the base presumably promotes enolate formation (132) (eq. 140).

$$RCOCH_2R' + \underset{\substack{|\\ SiMe_3}}{CH_3CONCH_3} \xrightarrow{(NaOSiMe_3)} \underset{\substack{|\\ OSiMe_3}}{RC{=}CHR'} + CH_3CONHCH_3 \quad (140)$$

C. *ADDITION TO ALDEHYDES*

1. *Silicon-Nitrogen Compounds; Formation of Amides*

Silylamides react with aldehydes by addition across the carbonyl bond; no catalyst is required in contrast to the reaction with epoxides and ketones. The parent aldehyde and amide are produced upon hydrolysis and the reaction is thus not useful from an organic synthetic point of view (133,134) (eq. 141).

$$\underset{\substack{|\\ SiMe_3}}{RCONR'} + R''CHO \longrightarrow \underset{\substack{|\\ RCONR'}}{R''CHOSiMe_3} \xrightarrow{H_3O^+} R''CHO + RCONHR' \quad (141)$$

Aliphatic aldehydes and silylazides form adducts in the presence of zinc chloride (135); the α-trimethylsiloxyalkyl azides are formed in 65-97% yield. They are hydrolytically relatively stable; pyrolysis leads to elimination and amide formation (eq. 142). Analogous trimethylsiloxy-*n*-alkyl azides pyrolyze to a mixture of two amides resulting from proton and alkyl migration, respectively (eq. 143).

$$\mathrm{Me_2CHCHO + Me_3SiN_3 \xrightarrow{ZnCl_2} Me_2CH\underset{\displaystyle N_3}{\underset{|}{C}}HOSiMe_3 \xrightarrow[-N_2]{250-300^\circ C} Me_2CHCONHSiMe_3} \quad (142)$$

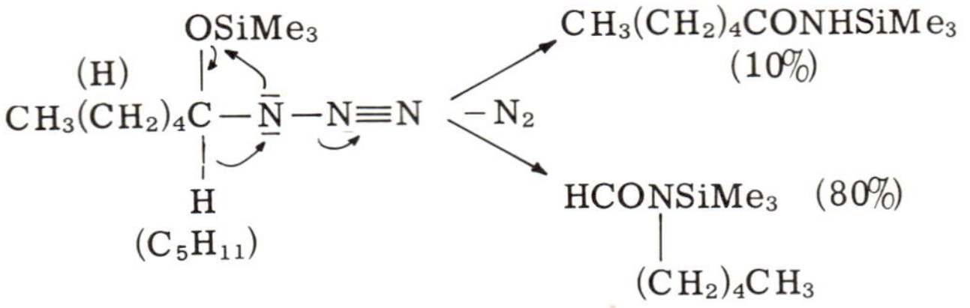

$$\mathrm{CH_3(CH_2)_4\overset{\displaystyle OSiMe_3}{\overset{|}{\underset{\displaystyle H}{\underset{|}{C}}}}-\bar{N}-N{\equiv}N \xrightarrow{-N_2} CH_3(CH_2)_4CONHSiMe_3\ (10\%) \;+\; HCO\underset{\displaystyle (CH_2)_4CH_3}{\underset{|}{N}}SiMe_3\ (80\%)} \quad (143)$$

(H) ... (C_5H_{11})

N,N-Dialkyl-substituted formamides were obtained by addition of silylamines with bulky substituents on nitrogen to chloral and a subsequent β elimination (136) (eq. 144). The yields increased with increasing size of the alkyl substituents. In addition, varying amounts of the initial adducts were obtained; the adduct was the only product in the case of trimethylsilyl dimethylamine (eq. 145).

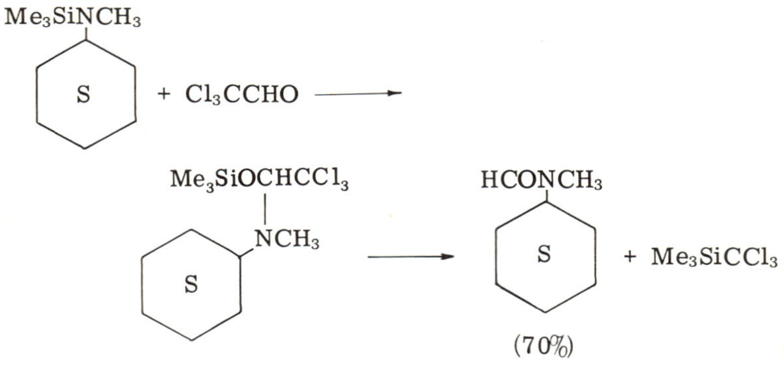

$\mathrm{Me_3SiNCH_3}$ (on ring S) + $\mathrm{Cl_3CCHO}$ ⟶

$\mathrm{Me_3SiOCHCCl_3}$ / $\mathrm{NCH_3}$ (on ring S) ⟶ $\mathrm{HCONCH_3}$ (on ring S) + $\mathrm{Me_3SiCCl_3}$ (144)

(70%)

$$\mathrm{Me_3SiNMe_2 + Cl_3CCHO \longrightarrow Cl_3C\underset{\displaystyle NMe_2}{\underset{|}{C}}HOSiMe_3} \quad (145)$$

(66%)

2. *Silicon-Carbon Compounds with Electron-Withdrawing β Substituents; Chain Extension Reactions*

Ethyl trimethylsilylacetate and similar compounds containing a carbonyl group β to silicon have been reported to undergo facile cleavage reactions of the Si-CH_2 bond with dilute acid and alkali and even with ethanol (137,138).

Activation of the silicon-carbon bond by ester or cyano substituents allows the addition of such compounds to aldehydes; catalytic amounts of base are required for the reactions (139) (eq. 146).

$$O_2N\text{-}C_6H_4\text{-}CHO + Me_3SiCH_2COOEt \xrightarrow[160-180^\circ C]{KAl(OCH_3)_4} O_2N\text{-}C_6H_4\text{-}CH(OSiMe_3)CH_2COOEt \quad (81\%) \tag{146}$$

Cyanomethyltrimethylsilane gave a similar yield; benzaldehyde and anisaldehyde afforded the adducts in 30-40% yield and aliphatic aldehydes formed small amounts of the corresponding products (eq. 147).

$$CH_3CH_2C(CH_3)_2CHO + Me_3SiCH_2CN \xrightarrow{NaOH} CH_3CH_2C(CH_3)_2\text{—}CH(OSiMe_3)CH_2CN \quad (10\%) \tag{147}$$

Hydrolysis under mild conditions (heating with aqueous ethanol) leads to the β-hydroxy esters or β-hydroxy nitriles. The method is valuable where the products cannot be prepared by the Reformatsky reaction, such as in the case of *p*-nitrobenzaldehyde which is unreactive under Reformatsky conditions or where the latter reaction leads to unsaturated products, for example, formation of ethyl *p*-methoxycinnamate from anisaldehyde.

The mechanism of the addition is thought to involve initial anion formation by abstraction of a proton or a silyl cation (eqs. 148 and 149).

$$Me_3SiCH_2X \xrightarrow[-BH]{B^\ominus} Me_3Si\overset{\ominus}{C}HX \xrightarrow{RCHO} RCH(O^\ominus)CH(SiMe_3)X \longrightarrow$$

$$RCH(OSiMe_3)\overset{\ominus}{C}HX \xrightarrow{BH} RCH(OSiMe_3)CH_2X + B^\ominus \tag{148}$$

$$\underset{\displaystyle\text{SiMe}_3}{\text{Me}_2\text{CCOOEt}} \xrightarrow[-\text{Me}_3\text{SiB}]{\text{B}^{\ominus}} \text{Me}_2\overset{\ominus}{\text{C}}\text{COOEt} \xrightarrow{\text{PhCHO}}$$

$$\underset{\displaystyle\text{O}^{\ominus}}{\text{PhCHC(CH}_3)_2\text{COOEt}} \xrightarrow{\text{Me}_3\text{SiB}} \tag{149}$$

$$\underset{\displaystyle\text{OSiMe}_3}{\text{PhCHC(CH}_3)_2\text{COOEt}} + \text{B}^{\ominus}$$

D. *ADDITION REACTIONS OF SILYLATED MULTIPLE-BOND COMPOUNDS*

1. *Formation of Triazoles, Tetrazoles, and Aziridines*

Azides add to trimethylsilyl-substituted acetylenes with formation of triazoles (79). Analogous products are also obtained from acetylenes without silyl substituents; the bulky trimethylsilyl group has a directional effect on the addition (eq. 150). The addition of phenylacetylene to phenyl azide proceeds with little orientation (140).

$$\text{PhN}_3 + \text{PhC}{\equiv}\text{CSiMe}_3 \longrightarrow$$

$$\underset{(78\%)}{\text{1-Ph-5-Ph-4-SiMe}_3\text{-1,2,3-triazole}} + \underset{(3\%)}{\text{1-Ph-5-SiMe}_3\text{-4-Ph-1,2,3-triazole}} \tag{150}$$

The carbon-silicon bond of the adducts can be cleaved with aqueous methanolic HCl.

Of greater synthetic value are addition reactions with trimethylsilyl azide which is accessible by a number of methods (141). The addition leads, after hydrolysis, to 1,2,3-triazoles with a free NH group, which are otherwise only accessible in generally low yields by the addition of the explosive HN_3 to acetylenes.

A number of *N*-trimethylsilyl-substituted triazoles were prepared in yields of 40-90% by heating the components in an autoclave at 120-150°C for 10 to 20 hr (142) (eq. 151). The symmetrical position of the trimethylsilyl group was assigned on the basis of ir and nmr spectra and dipole moments.

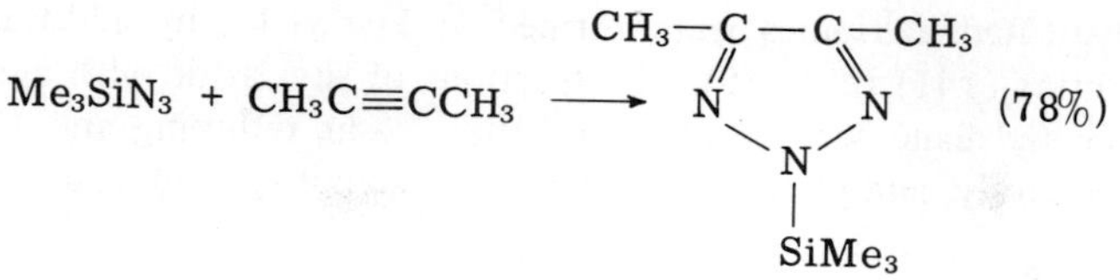

$$Me_3SiN_3 + CH_3C{\equiv}CCH_3 \longrightarrow \text{(1-trimethylsilyl-4,5-dimethyl-1,2,3-triazole)}\ (78\%) \qquad (151)$$

Diethynyl compounds afford bis(triazolyl)alkanes (eq. 152). Hydrolysis or alcoholysis leads to the parent triazoles.

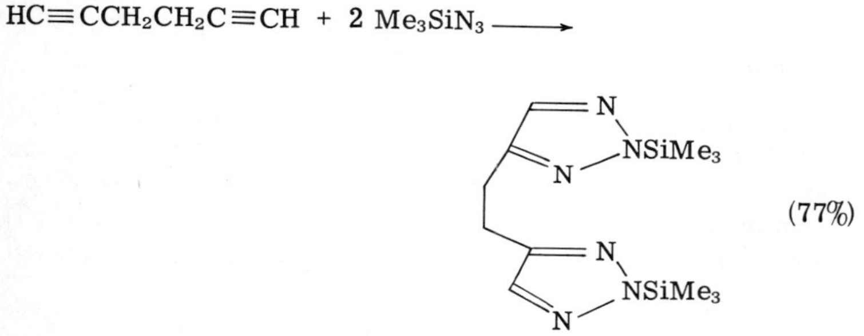

$$HC{\equiv}CCH_2CH_2C{\equiv}CH + 2\ Me_3SiN_3 \longrightarrow \text{1,2-bis(2-trimethylsilyl-1,2,3-triazol-4-yl)ethane (NSiMe}_3\text{)}\ (77\%) \qquad (152)$$

The reaction of acetyl chloride with 2-trimethylsilyl-1,2,3-triazoles leads to the formation of the 1-acetyl derivative and trimethylchlorosilane (142) (eq. 153).

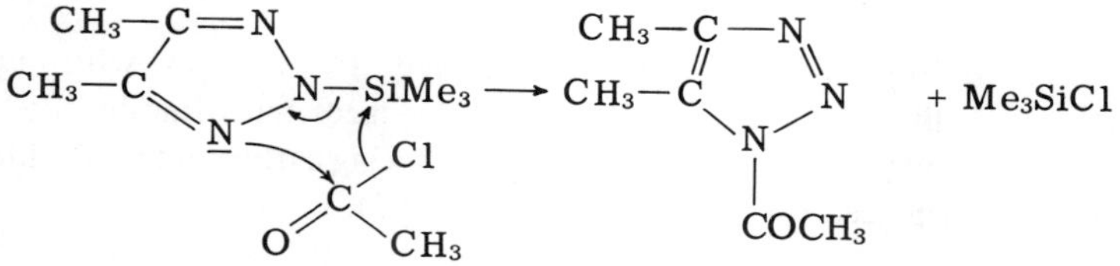

$$\text{4,5-dimethyl-2-(SiMe}_3\text{)-1,2,3-triazole} + CH_3COCl \longrightarrow \text{4,5-dimethyl-1-(COCH}_3\text{)-1,2,3-triazole} + Me_3SiCl \qquad (153)$$

The addition of silylazide to the carbon-nitrogen triple bond has also been carried out. Heating a mixture of benzonitrile and trimethylsilyl azide at 180°C afforded a tetrazole derivative in 80% yield (141) (eq. 154).

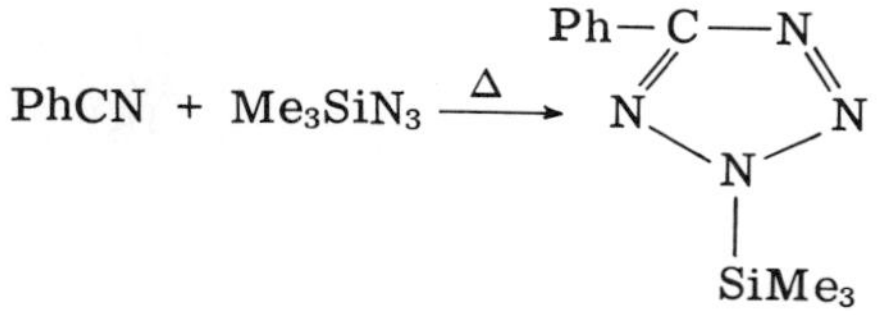

$$PhCN + Me_3SiN_3 \xrightarrow{\Delta} \text{5-phenyl-1-(SiMe}_3\text{)tetrazole} \qquad (154)$$

The corresponding benzyl and ethyl derivatives were obtained in 40 and 26% yield, respectively. Reaction with acetyl chloride led to displacement of the silyl group in these compounds by acetyl.

Silyl-substituted aziridines were formed in low yields by addition of silyl azide to olefins (141) (eq. 155). The reactions of silyl azide with cyclohexene and triethylvinylsilane were incomplete after 24 hr refluxing and 10 days at 125°C, respectively, and yielded 13 and 20% of the corresponding aziridines.

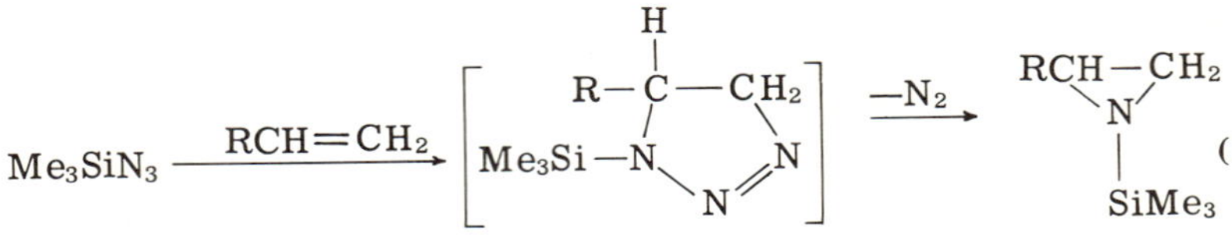

$$\mathrm{Me_3SiN_3} \xrightarrow{\mathrm{RCH{=}CH_2}} \left[\begin{array}{c} \text{H} \\ | \\ \mathrm{R{-}C{-}CH_2} \\ \mathrm{Me_3Si{-}N \quad N} \\ \mathrm{N} \end{array}\right] \xrightarrow{-\mathrm{N_2}} \begin{array}{c} \mathrm{RCH{-}CH_2} \\ \diagdown\ \diagup \\ \mathrm{N} \\ | \\ \mathrm{SiMe_3} \end{array} \qquad (155)$$

2. *Formation of Pyrazoles*

The 1,3-addition of diazo compounds to silyl acetylenes proceeds in the same way as the addition of azides; yields of 95% of pyrazole derivatives were obtained when bis(trimethylsilyl)acetylene was allowed to stand with diazomethane at room temperature or heated at 100°C with diazoacetic ester (143) (eq. 156). Trimethylsilyl phenylacetylene afforded 3-trimethylsilyl-4-phenyl pyrazole in 35% yield.

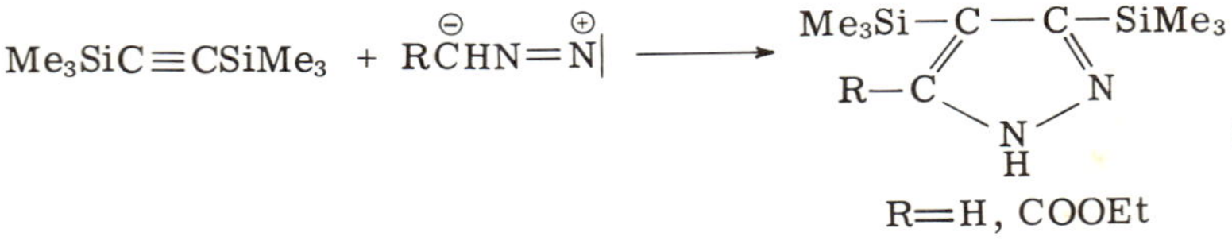

$$\mathrm{Me_3SiC{\equiv}CSiMe_3} + \mathrm{R\overset{\ominus}{C}HN{=}\overset{\oplus}{N}|} \longrightarrow \begin{array}{c} \mathrm{Me_3Si{-}C{-}C{-}SiMe_3} \\ \mathrm{R{-}C \qquad N} \\ \mathrm{N} \\ \mathrm{H} \end{array} \qquad (156)$$

$$\mathrm{R{=}H,\ COOEt}$$

The trimethylsilyl groups can be removed by heating with aqueous methanolic HCl. Bromine at low temperature displaces the trimethylsilyl group on the carbon of highest electron density; in refluxing carbon tetrachloride both silyl groups are displaced by bromine (143) (eq. 157).

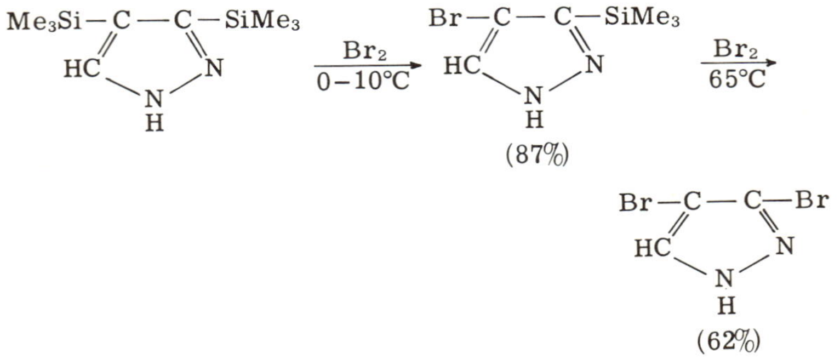

$$\begin{array}{c} \mathrm{Me_3Si{-}C{-}C{-}SiMe_3} \\ \mathrm{HC \qquad N} \\ \mathrm{N} \\ \mathrm{H} \end{array} \xrightarrow[0-10^\circ\mathrm{C}]{\mathrm{Br_2}} \begin{array}{c} \mathrm{Br{-}C{-}C{-}SiMe_3} \\ \mathrm{HC \qquad N} \\ \mathrm{N} \\ \mathrm{H} \\ (87\%) \end{array} \xrightarrow[65^\circ\mathrm{C}]{\mathrm{Br_2}} \begin{array}{c} \mathrm{Br{-}C{-}C{-}Br} \\ \mathrm{HC \qquad N} \\ \mathrm{N} \\ \mathrm{H} \\ (62\%) \end{array} \qquad (157)$$

3. *Self-Condensation and Addition of Silyl-aci-nitromethane and Silyl-aci-malononitrile*

Silylation can be expected in certain cases not only to activate the molecule toward attack by other reagents, but also to facilitate self-condensation or addition reactions. This is observed in the silylation of nitromethane with *N*-trimethylsilyl diphenylurea at room temperature; the presumed initial product, *O*-trimethylsilyl-aci-nitromethane, which could not be isolated, undergoes a condensation reaction with a second molecule of nitromethane to the trimethylsilyl ether of nitroacetaldoxime which is further silylated to *syn*- and *anti*-bis(trimethylsilyl)methazonic acid (144) (eq. 158).

$$CH_3NO_2 \xrightarrow[-\,PhNHCONHPh]{PhN(SiMe_3)CONHPh} \left[CH_2{=}N(\rightarrow O)OSiMe_3\right] \xrightarrow[-\,H_2O]{CH_3NO_2}$$

$$O_2NCH_2CH{=}NOSiMe_3 \xrightarrow[-\,PhNHCONHPh]{PhN(SiMe_3)CONHPh} Me_3SiN(\rightarrow O){=}CHCH{=}NOSiMe_3 \qquad (158)$$

syn and *anti* (95%)

Similarly, the initial silylation product of malononitrile was not isolated; trimerization of the presumed unstable silylated "aci"-malononitrile (silyl ketene imine) at 5°C led to tricyanotri(silylamino)benzene which afforded the parent *sym*-tricyanotriaminobenzene upon hydrolysis (144) (eq. 159).

$$3\ NCCH_2CN \xrightarrow[-\,3\ PhNHCONHPh]{3\ PhN(SiMe_3)CONHPh} \left[3NCCH{=}C{=}NSiMe_3\right] \longrightarrow$$

$$\left[\text{cyclo-}\{CH(CN)\text{–}C({=}NSiMe_3)\text{–}CH(CN)\text{–}C({=}NSiMe_3)\text{–}CH(CN)\text{–}C({=}NSiMe_3)\}\right] \longrightarrow \qquad (159)$$

$$1,3,5\text{-}(CN)_3\text{-}2,4,6\text{-}(Me_3SiNH)_3C_6 \xrightarrow{H_2O} 1,3,5\text{-}(CN)_3\text{-}2,4,6\text{-}(H_2N)_3C_6$$

V. Elimination Reactions of Organosilanes

A. *FORMATION OF ISOCYANATES AND ISOCYANATO CARBOXYLIC ACIDS*

A number of thermal cleavage reactions of compounds containing *N*-silyl carbonamide moieties (**17**) have been found in which isocyanates and organosilanes $Me_3Si\text{-}X$ are formed. These elimination reactions can be regarded as the reverse of addition reactions of silylamines with isocyanates where the equilibrium generally is shifted toward the adduct, with the exception of one case of an elimination of trimethylsilylisocyanate from the adduct of a bis(trimethylsilyl)urea and an alkyl isocyanate (see Section IV, eq. 121) (eq. 160). The 1:2 adduct, which is presumably formed prior to the elimination, was not isolated in this case.

$$-\underset{\substack{| \\ SiMe_3}}{N}\overset{\substack{O \\ \|}}{C}X$$

(17)

$$(Me_3Si)_2NH + RNCO \longrightarrow \underset{\substack{| \\ Me_3Si}}{HN}CO\underset{\substack{| \\ SiMe_3}}{N}R \xrightarrow{RNCO}$$

$$\left[Me_3SiNHCO\underset{\substack{| \\ R}}{N}CO\underset{\substack{| \\ SiMe_3}}{N}R \right] \longrightarrow R\underset{\substack{| \\ H}}{N}CO\underset{\substack{| \\ SiMe_3}}{N}R + Me_3SiNCO \qquad (160)$$

The reversible adduct formation between silylamines and carbon dioxide, carbon disulfide, and isothiocyanates (110-112,114) assumes an intermediate position between the addition reactions, described in Section IV, and a series of elimination reactions which represent a new route to isocyanates and related compounds.

When *N*-silylated chlorides, anhydrides, or esters of carbamic acids are heated to temperatures up to 120°C, in some cases as low as –20°C, isocyanates are formed in yields of 70-95% (145,146) (eq. 161). The syntheses start with amines which are first acylated with a chloroformic or carbonic ester and subsequently silylated with chlorosilane/*t*-amine or trimethylsilyl acetamide; the isocyanate is produced by heating of the mixture without isolation of intermediates. If hydroxyl, thiol, or amino groups are present elsewhere in the molecules, the corresponding silylated isocyanato compounds are formed (eq. 162). Similarly, silyl esters of ω-isocyanato carboxylic acids are obtained by thermolysis of silylated *N*-phenoxycarbonyl β-, γ-, or ϵ-amino acids (147) (eq. 163).

$$R'N(SiMe_3)C(=O)X \xrightarrow{\Delta} R'N{=}C{=}O + Me_3SiX \qquad (161)$$

$$X = Cl,\ OCOR,\ OR''$$
$$R',\ R'' = \text{Alkyl, cycloalkyl, aryl}$$

$$HOCH_2CH_2NH_2 \longrightarrow Me_3SiOCH_2CH_2N(SiMe_3)COX \longrightarrow \underset{(70\%)}{Me_3SiOCH_2CH_2NCO} \qquad (162)$$

$$PhOCON(SiMe_3)(CH_2)_nCOOSiMe_3 \longrightarrow OCN(CH_2)_nCOOSiMe_3 + PhOSiMe_3 \qquad (163)$$

$$n = 2,\ 3,\ 5$$

The corresponding elimination products from α-amino acids exist in a temperature-dependent equilibrium with silyl α-*N*-carboxyamino acid anhydrides; the isocyanates predominate at high temperature, the anhydrides are favored by low temperature. Both compounds have been identified by reaction with a primary or secondary amine (eq. 164).

$$O{=}C{=}NCH_2C(=O)OSiMe_3 \underset{\Delta}{\rightleftharpoons} \text{Me}_3\text{SiN—CH}_2\text{ (cyclic: } O{=}C\text{—O—}C{=}O) \qquad (164)$$

$$O{=}C{=}NCH_2COOSiMe_3 \xrightarrow{R_2NH} R_2NCONHCH_2COOSiMe_3$$

$$\text{anhydride} \xrightarrow{R_2NH} R_2NCOCH_2NHCOOSiMe_3$$

Formation of trimethylsilyl α-isocyanato carboxylates had been reported earlier by reaction of carbobenzoxylated amino acids with excess trimethylchlorosilane in pyridine (148) (eq. 165). The intermediate formation of a carbamoyl chloride was proposed for this reaction (**18**).

$$\underset{\displaystyle \text{HNCOOCH}_2\text{Ph}}{\overset{\displaystyle\text{RCHCOOH}}{|}} \xrightarrow[\text{pyridine}]{\text{Me}_3\text{SiCl}} \underset{\displaystyle \text{NCO}}{\overset{\displaystyle\text{RCHCOOSiMe}_3}{|}} + \text{Me}_3\text{SiOCH}_2\text{Ph} \qquad (165)$$

$$\underset{\displaystyle \text{HNCOCl}}{\overset{\displaystyle\text{RCHCOOSiMe}_3}{|}}$$

(18)

However, the similarity to the later discovered cleavage reactions (eqs. 161-163) suggests that silylation of the carbobenzoxy amino acid at the carboxyl group and at the nitrogen precedes the isocyanate formation.

B. THERMOLYSIS OF SILYL-SUBSTITUTED TETRAZOLES; FORMATION OF 1,3-DIPOLAR FRAGMENTS

Heating of 1-trimethylsilyl tetrazole at 134-140°C results in expulsion of nitrogen and formation of bis(trimethylsilyl)carbodiimide and polymeric cyanamide; the same mixture of products is obtained when the silylation of tetrazole is attempted in refluxing hexamethyldisilazane (127°C) (149) (eq. 166). The ability of the trimethylsilyl group to accommodate electrons by $d\pi$-p_π bonding thus facilitates the elimination of nitrogen similarly to a phenyl substituent in the same position (150).

$$\text{(1-trimethylsilyltetrazole: CH—NSiMe}_3\text{, N, N, N ring)} \xrightarrow[-\text{N}_2]{\Delta} \left[\text{Me}_3\text{Si}\overset{\oplus}{\underline{\text{N}}}\text{CH}=\overset{\ominus}{\text{N}}\right] \longrightarrow \qquad (166)$$

$$1/2\ \text{Me}_3\text{SiN}=\text{C}=\text{NSiMe}_3 + 1/2\ (\text{H}_2\text{NCN})_n$$

A 1,3-dipolar fragment is also formed upon heating of 2-trimethylsilyl-5-phenyltetrazole (obtained by silylation of 5-phenyltetrazole or addition of benzonitrile to silylazide; see Section IV.D) at 200°C; stabilization occurs by dimerization and rearrangement to a triazole derivative which is formed in 95% yield (149) (eq. 167). An isomeric structure was claimed for this product by other workers (**19**); however, no support is given for this structure (141).

The intermediacy of the symmetrical tetrazine in reaction (167), which could not be isolated, is supported by the fact that silylation of 3,6-diphenyldihydro-

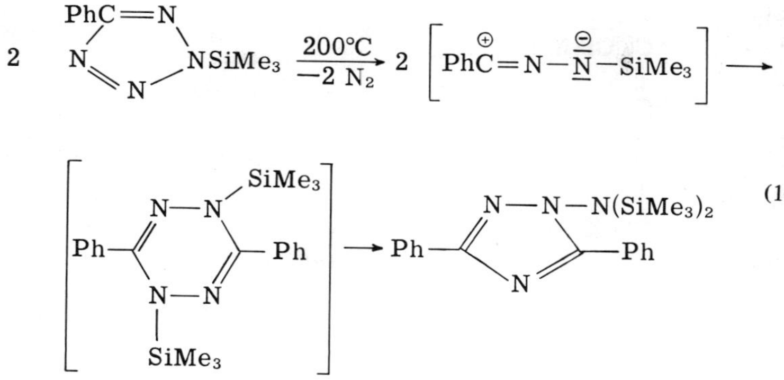

$$2\ \text{PhC}{=}\text{N}{-}\text{N}(\text{SiMe}_3){-}\text{N}{=}\text{N}\ (\text{ring}) \xrightarrow[-2\,\text{N}_2]{200°\text{C}} 2\left[\text{Ph}\overset{\oplus}{\text{C}}{=}\text{N}{-}\overset{\ominus}{\underline{\overline{\text{N}}}}{-}\text{SiMe}_3\right] \longrightarrow$$

$$\left[\text{3,6-diphenyl-1,4-bis(SiMe}_3\text{)-dihydro-1,2,4,5-tetrazine}\right] \longrightarrow \text{Ph}{-}\text{C}_2\text{N}_3(\text{Ph}){-}\text{N}(\text{SiMe}_3)_2 \quad (167)$$

$$\text{3,5-Ph}_2\text{-4-N(SiMe}_3)_2\text{-1,2,4-triazole}$$

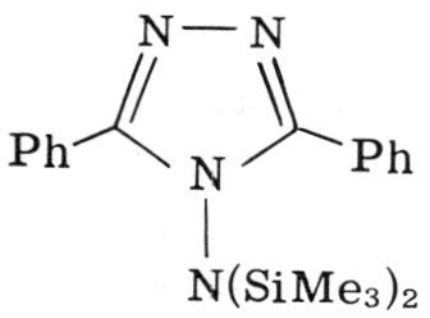

(19)

1,2,4,5-tetrazine in refluxing hexamethyldisilazane also afforded the bis(trimethylsilyl)amino diphenyltriazole.

It was possible to capture the primary fragment in eq. (167) by 1,3-dipolar addition to ethyl cinnamate and bis(trimethylsilyl)fumarate (149) (eqs. 168 and 169).

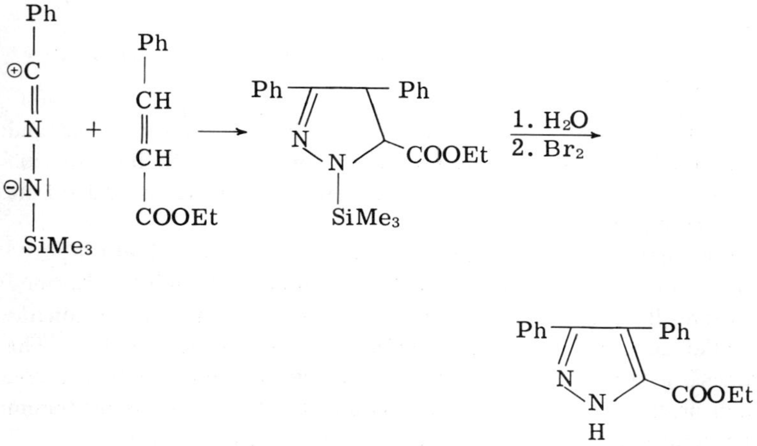

$$\text{Ph}\overset{\oplus}{\text{C}}{=}\text{N}{-}\overset{\ominus}{\text{N}}{-}\text{SiMe}_3 + \text{PhCH}{=}\text{CHCOOEt} \longrightarrow \text{3,4-Ph}_2\text{-5-COOEt-1-SiMe}_3\text{-pyrazoline} \xrightarrow[2.\ \text{Br}_2]{1.\ \text{H}_2\text{O}} \quad (168)$$

$$\text{3,4-Ph}_2\text{-5-COOEt-1}H\text{-pyrazole (N–H)}$$

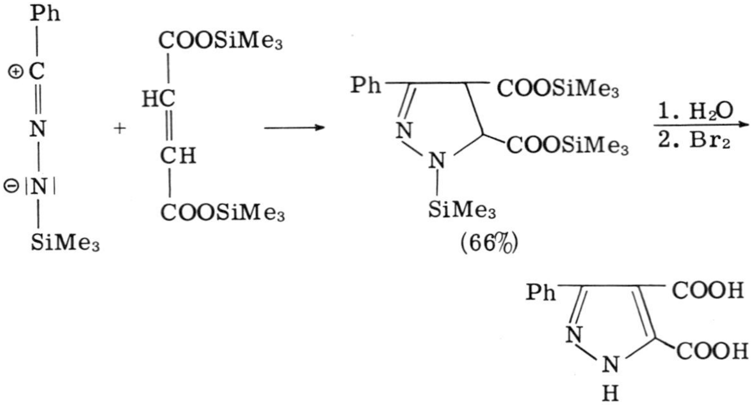

$$\mathrm{Ph{-}\overset{\oplus}{C}{=}N{-}\overset{\ominus}{N}{-}SiMe_3} + \mathrm{Me_3SiOOC{-}CH{=}CH{-}COOSiMe_3} \longrightarrow$$

Ph, $COOSiMe_3$, N, $COOSiMe_3$, N, $SiMe_3$ (66%) $\xrightarrow[2.\ Br_2]{1.\ H_2O}$ Ph, COOH, N, COOH, N, H (169)

A third mode of thermal decomposition of a silyl tetrazole derivative was observed when 1-trimethylsilyl-5-trimethylsilylaminotetrazole—prepared by refluxing 5-aminotetrazole in hexamethyldisilazane—was heated at 150°C; trimethylsilylazide was formed in addition to bis(trimethylsilyl)carbodiimide and polymeric cyanamide (149) (eq. 170).

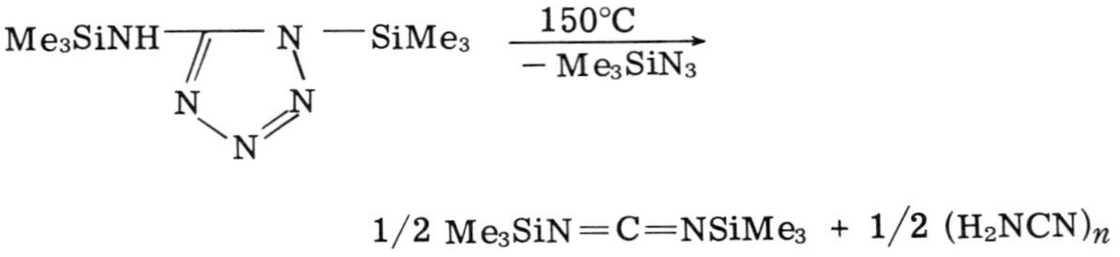

Me_3SiNH–tetrazole (N, N, N, N)–$SiMe_3$ $\xrightarrow[-\ Me_3SiN_3]{150°C}$ (170)

$$1/2\ Me_3SiN{=}C{=}NSiMe_3 + 1/2\ (H_2NCN)_n$$

C. *FORMATION OF OLEFINS FROM CARBONYL COMPOUNDS WITH SILYL ORGANOMETALLICS*

Certain similarities of phosphorus and silicon in their affinity to oxygen and sensitivity to attack by alkoxide suggested a new approach to olefin formation from carbonyl compounds using appropriate metalated silanes instead of the known Wittig reagents (151).

Silanes containing activated methylene groups were metalated with *n*-butyllithium or its complex with tetramethylethylenediamine and added to carbonyl compounds. Hydrolysis of the reaction mixtures and extraction of the aqueous phase with ether afforded the desired olefins in 50-80% yields (eq. 171). The corresponding Grignard adducts (**20**) did not undergo the elimination reaction; however, hydrolysis to the alcohol and treatment with sodium or potassium hydride again resulted in formation of the olefins in comparable yield.

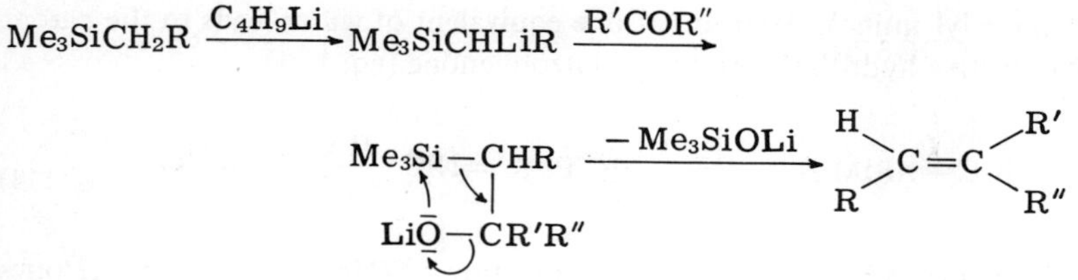

$$Me_3SiCH_2R \xrightarrow{C_4H_9Li} Me_3SiCHLiR \xrightarrow{R'COR''}$$

$$\begin{array}{c} Me_3Si-CHR \\ | \quad\ | \\ Li\bar{O}-CR'R'' \end{array} \xrightarrow{-Me_3SiOLi} \begin{array}{c} H \\ \\ R \end{array}\!\!>C=C<\!\!\begin{array}{c} R' \\ \\ R'' \end{array} \qquad (171)$$

R = Ph, CH_3S, Ph_2P

R′, R″ = H, CH_3, Ph, $-C_5H_{10}-$

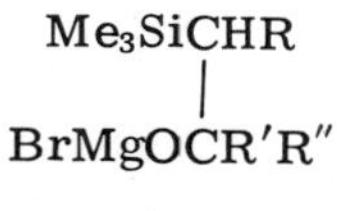

$$\begin{array}{c} Me_3SiCHR \\ | \\ BrMgOCR'R'' \end{array}$$

(20)

VI. Condensation of Carbonyl Compounds with Sodium Bis(trimethylsilyl)amide

A. *FORMATION OF IMINES*

Sodium bis(trimethylsilyl)amide (152) has found use as a reagent for the introduction of trimethylsilyl amino groups, particularly in reaction with halides of sulfur, silicon, and phosphorus (153-155) (e.g., eq. 172). The applicability of this reagent was widened when it was found that it undergoes condensation reactions with nonenolizable aromatic ketones, affording yellow *N*-silyl-substituted imines (156) (eq. 173).

$$2\ (Me_3Si)_2NNa + SCl_2 \longrightarrow (Me_3Si)_2NSN(SiMe_3)_2 + 2\ NaCl \qquad (172)$$

$$PhCOPh + (Me_3Si)_2NNa \xrightarrow{80°C} Ph_2C{=}NSiMe_3 + NaOSiMe_3 \qquad (173)$$

The *N*-silyl benzophenone imine is obtained in 84% yield after 7 hr refluxing of the components in benzene, addition of trimethylchlorosilane to convert the silanolate into hexamethyldisiloxane and sodium chloride, and subsequent fractionation (removal of the silanolate in this fashion improves the isolated yield of the silyl imine). Addition of one equivalent of water leads to the parent imine; exhaustive hydrolysis produces benzophenone (eq. 174).

$$Ph_2C{=}NSiMe_3 \xrightarrow[-\,Me_3SiOH]{H_2O} Ph_2C{=}NH \xrightarrow[-\,NH_3]{H_2O} Ph_2CO \quad (174)$$

No intermediates of the condensation reaction of $(Me_3Si)_2NNa$ with ketones could be detected; the reaction presumably proceeds by an addition-elimination mechanism involving a silyl migration (eq. 175).

$$PhCOPh \xrightarrow{(Me_3Si)_2N^{\ominus}} \left[PhC(O^{\ominus})(Ph)N(SiMe_3)_2 \rightarrow PhC(OSiMe_3)(Ph)N^{\ominus}SiMe_3 \right] \xrightarrow{-Me_3SiO^{\ominus}} Ph_2C{=}N{-}SiMe_3 \quad (175)$$

The hydrolytically very sensitive silyl imines of benzaldehyde and *p*-benzoquinone were obtained by the same procedure (**21** and **22**).

$Ph(H)C{=}NSiMe_3$ (61%) **(21)**

$Me_3SiN{=}C_6H_4{=}NSiMe_3$ (20%) **(22)**

The same reaction conditions were found applicable to the 1,2-diketones benzil and phenanthrene quinone, leading to the corresponding bis(*N*-silyl)imines (157) (**23** and **24**).

$PhC(=NSiMe_3){-}C(=NSiMe_3)Ph$ (91%) **(23)**

9,10-bis($NSiMe_3$)phenanthrene (34%) **(24)**

Alcoholysis of the two compounds afforded the previously unknown parent 1,2-diimines.

Ketones with α-methylene groups react with sodium bis(trimethysilyl)amide with formation of an enolate which can be stabilized by subsequent silylation (156) (eq. 176).

$$RCH_2COR \xrightarrow[-(Me_3Si)_2NH]{(Me_3Si)_2NNa} RCH{=}\underset{\displaystyle O^{\ominus}}{\underset{|}{C}}R \xrightarrow[-NaCl]{Me_3SiCl} RCH{=}\underset{\displaystyle OSiMe_3}{\underset{|}{C}}R \qquad (176)$$

B. CONVERSION OF BIS(TRIMETHYLSILYL)DIIMINES INTO HETEROCYCLIC COMPOUNDS

The silicon-nitrogen bond in bis(trimethylsilyl)imines was found to possess the high reactivity toward acid chlorides typical for silyl amines. Thus, a number of *N,N'*-diacyl derivatives of benzil diimine were prepared by heating its bis(trimethylsilyl) derivative with the acyl chloride until the calculated amount of trimethylchlorosilane had distilled over (157) (eq. 177).

$$\begin{array}{c} PhC{=}NSiMe_3 \\ | \\ PhC{=}NSiMe_3 \end{array} \xrightarrow{2\ PhCOCl} \begin{array}{c} PhC{=}NCOPh \\ | \\ PhC{=}NCOPh \end{array}\ (85\%) + 2\ Me_3SiCl$$

$$\begin{array}{c} PhC{=}NSiMe_3 \\ | \\ PhC{=}NSiMe_3 \end{array} \xrightarrow{2\ ClCOOEt} \begin{array}{c} PhC{=}NCOOEt \\ | \\ PhC{=}NCOOEt \end{array}\ (65\%) + 2\ Me_3SiCl \qquad (177)$$

$$\begin{array}{c} PhC{=}NSiMe_3 \\ | \\ PhC{=}NSiMe_3 \end{array} \xrightarrow{2\ p\text{-}TolSO_2Cl} \begin{array}{c} PhC{=}NSO_2{-}p\text{-}Tol \\ | \\ PhC{=}NSO_2{-}p\text{-}Tol \end{array}\ (20\%) + Me_3SiCl$$

Of particular interest is the formation of a number of new heterocyclic compounds under similar conditions (eqs. 178 and 179).

$$\mathrm{PhC{=}NSiMe_3 / PhC{=}NSiMe_3} \xrightarrow{SCl_2} \text{3,4-diphenyl-1,2,5-thiadiazole (96\%)}$$

$$\xrightarrow{SOCl_2} \text{3,4-diphenyl-1,2,5-thiadiazole 1-oxide (92\%)}$$

$$\xrightarrow{PhPOCl_2} \text{2,4,5-triphenyl-1,3,2-diazaphosphole 2-oxide (82\%)}$$

$$\xrightarrow{COCl_2} \text{4,5-diphenyl-2H-imidazol-2-one (37\%)} \qquad (178)$$

$$\text{9,10-bis(trimethylsilylimino)phenanthrene} \xrightarrow{SCl_2} \text{phenanthro[9,10-c][1,2,5]thiadiazole (78\%)}$$

$$\xrightarrow{SOCl_2} \text{phenanthro[9,10-c][1,2,5]thiadiazole 2-oxide (80\%)} \qquad (179)$$

The reaction of benzil-bis(trimethylsilyl)imine with sulfuryl chloride afforded the bischloroimine in addition to the expected cyclic sulfone (eq. 180).

$$\begin{array}{c} PhC{=}NSiMe_3 \\ | \\ PhC{=}NSiMe_3 \end{array} + SO_2Cl \begin{cases} \nearrow \text{3,4-diphenyl-1,2,5-thiadiazole 1,1-dioxide } (Ph\text{-}C{=}N\text{-}SO_2\text{-}N{=}C\text{-}Ph) + 2\ Me_3SiCl \quad (6\%) \\ \searrow \begin{array}{c} PhC{-}CPh \\ \| \quad \| \\ N \quad N \\ | \quad | \\ Cl \quad Cl \end{array} + 2\ Me_3SiCl,\ SO_2 \quad (60\%) \end{cases} \tag{180}$$

Condensation with aromatic aldehydes at temperatures above 100°C in the presence of catalytic amounts of $(NH_4)_2SO_4$ resulted in imidazole derivatives such as lophine (eq. 181).

$$\begin{array}{c} PhC{=}NSiMe_3 \\ | \\ PhC{=}NSiMe_3 \end{array} + ArCHO \longrightarrow \text{4,5-diphenyl-2-Ar-2H-imidazole (Ph\text{-}C{=}N\text{-}C(H)(Ar)\text{-}N{=}C\text{-}Ph)} \tag{181}$$

$$Ar = Ph\ (90\%)$$
$$p\text{-}CH_3OPh\ (79\%)$$

C. FORMATION OF BIS(TRIMETHYLSILYL)BENZIMIDATE

The condensation of benzoic esters with sodium bis(trimethylsilyl)amide involves an addition-elimination reaction similar to that of benzophenone; however, sodium alkoxide is eliminated in this case and both silyl groups are retained in the product (158) (eq. 182).

$$PhCOOCH_3 \xrightarrow{(Me_3Si)_2N^{\ominus}} \left[\begin{array}{c} O^{\ominus} \\ | \\ PhCN(SiMe_3)_2 \\ | \\ OCH_3 \end{array} \longrightarrow \begin{array}{c} OSiMe_3 \\ | \ \ominus \\ PhCNSiMe_3 \\ | \\ OCH_3 \end{array} \right] \xrightarrow{-CH_3O^{\ominus}} PhC\begin{array}{l} \diagup OSiMe_3 \\ \diagdown\!\!\diagdown NSiMe_3 \end{array} \tag{182}$$

The isolation of the product is facilitated if the alkoxide is reacted with trimethylchlorosilane prior to fractionation; a yield of 51% was obtained. The same product is obtained by reaction of sodium bis(trimethylsilyl)amide with benzoic anhydride or benzamide. The aliphatic members of this class of compounds are accessible by reaction of amides with trimethylchlorosilane/base (37,159,160) (eq. 183).

$$CH_3CONH_2 + 2\ Me_3SiCl \xrightarrow[-Et_3N \cdot HCl]{Et_3N} CH_3C\begin{matrix} \diagup OSiMe_3 \\ \diagdown\!\!\diagdown NSiMe_3 \end{matrix} \tag{183}$$

The "bis(trimethylsilyl)amides" are powerful silylating agents (1,37,161); proton sources displace the two silyl groups in successive steps, and the monosilylated intermediate can be isolated (eq. 184).

$$CH_3C\begin{matrix} \diagup OSiMe_3 \\ \diagdown\!\!\diagdown NSiMe_3 \end{matrix} \xrightarrow[-ROSiMe_3]{ROH} CH_3CONHSiMe_3 \xrightarrow[-ROSiMe_3]{ROH} CH_3CONH_2 \tag{184}$$

Thermolysis of *N*-silyl-silylimidates affords disiloxane and nitrile (158) (eq. 185).

$$PhC\begin{matrix} \diagup OSiMe_3 \\ \diagdown\!\!\diagdown NSiMe_3 \end{matrix} \xrightarrow{200°C} PhCN + (Me_3Si)_2O \tag{185}$$

VII. Silyl Substituents as Protective Groups

A. *PROTECTION OF AMINE AND IMINE GROUPS*

The amino hydrogen atoms of anilines can be replaced by silyl groups by reaction with two equivalents of a Grignard reagent or phenyllithium followed by trimethylchlorosilane (162,163). The same method is applicable to ring-halogenated anilines; the halogen atom of the resulting *N,N*-bis(trimethylsilyl)-bromo(chloro)aniline can then be replaced by lithium using the free metal or butyllithium and the metalated intermediates can be used for a variety of syntheses.

Carbonation leads to the *p*-aminobenzoic acids; reaction with diethyl carbonate affords the corresponding triaryl carbinols (164) (eq. 186). Treatment of the *p*-lithio derivative with benzaldehyde and benzophenone yields monoaminocarbinols (165) (eq. 187).

$$\text{Br-C}_6\text{H}_4\text{-NH}_2 \xrightarrow[\text{2. 2 Me}_3\text{SiCl}]{\text{1. 2 PhLi}} \text{Br-C}_6\text{H}_4\text{-N(SiMe}_3)_2 \quad (80\%)$$

$$\text{Br-C}_6\text{H}_4\text{-N(SiMe}_3)_2 \xrightarrow{\text{1. } n\text{-BuLi; 2. (EtO)}_2\text{CO; 3. hydrol.}} \text{HOC(-C}_6\text{H}_4\text{-NH}_2)_3$$

$$\text{Br-C}_6\text{H}_4\text{-N(SiMe}_3)_2 \xrightarrow{\text{1. } n\text{-BuLi; 2. CO}_2\text{; 3. hydrol.}} \text{HOOC-C}_6\text{H}_4\text{-NH}_2 \tag{186}$$

$$\text{Br-C}_6\text{H}_4\text{-N(SiMe}_3)_2 \xrightarrow{\text{1. Li; 2. PhCHO; 3. hydrol.}} \text{H}_2\text{N-C}_6\text{H}_4\text{-CH(OH)Ph} \quad (76\%)$$

$$\text{Br-C}_6\text{H}_4\text{-N(SiMe}_3)_2 \xrightarrow{\text{1. Li; 2. PhCOPh; 3. hydrol.}} \text{H}_2\text{N-C}_6\text{H}_4\text{-C(OH)Ph}_2 \quad (80\%) \tag{187}$$

Trialkylsilyl and -germyl groups can be introduced into the aromatic ring by the same method (165); the reaction can be carried out under Wurtz-Fittig conditions without isolation of the metalated intermediate (eq. 188). Hydrolysis of the silyl anilines was readily effected in all cases by refluxing in methanol containing a small amount of sodium hydroxide or by treatment with dilute hydrochloric acid at 40-60°C.

$$3\text{-ClC}_6\text{H}_4\text{N(SiMe}_3)_2 + \text{Me}_3\text{SiCl} \xrightarrow[\text{PhCH}_3,\ 115°\text{C}]{\text{Na}} 3\text{-Me}_3\text{SiC}_6\text{H}_4\text{N(SiMe}_3)_2 \qquad (188)$$

The reduction of the carboxyl group in L-phenylglycine was accomplished after silylation (166) (eq. 189).

$$\text{L}-\text{PhCH(NH}_2)\text{COOH} \xrightarrow[\text{pyridine}]{\text{Me}_3\text{SiCl}} \text{L}-\text{PhCH(NHSiMe}_3)\text{COOSiMe}_3 \xrightarrow[\text{2. hydrolysis}]{\text{1. LiAlH}_4} \text{L}-\text{PhCH(NH}_2)\text{CH}_2\text{OH} \qquad (189)$$

A simple synthesis for the hydrolytically unstable phosphine imines is provided by the reaction of phosphines with trimethylsilyl azide leading to *N*-trimethylsilyl phosphine imines which afford the free imines by acid-catalyzed methanolysis at low temperature. The reaction is applicable to aromatic as well as aliphatic phosphines (167,168) (eq. 190). Analogous reactions of trimethylsilyl azide with phosphorus-nitrogen compounds have also been reported (169-172).

$$(\text{C}_3\text{H}_7)_3\text{P} + \text{Me}_3\text{SiN}_3 \xrightarrow[-\text{N}_2]{80°\text{C},\ 6\ \text{hr}} (\text{C}_3\text{H}_7)_3\text{P}{=}\text{NSiMe}_3 \xrightarrow[-20°\text{C}]{\text{CH}_3\text{OH},\ \text{H}^+} (\text{C}_3\text{H}_7)_3\text{P}{=}\text{NH} \qquad (190)$$

B. PROTECTION OF HYDROXYL AND THIOL GROUPS

A series of silyl-substituted phenols was synthesized in high yield from the corresponding chlorophenols by first refluxing with trimethylchlorosilane to prepare the trimethylsiloxy derivative and then refluxing with a second equivalent of halosilane and sodium in toluene solution (173) (eq. 191).

The introduction of silyl groups into the ring of trimethylsiloxy halobenzenes was also accomplished by preparation of the intermediate Grignard or lithium derivative (174-176). The free phenols could be obtained by acid-catalyzed alcoholysis of the trimethylsiloxy derivatives.

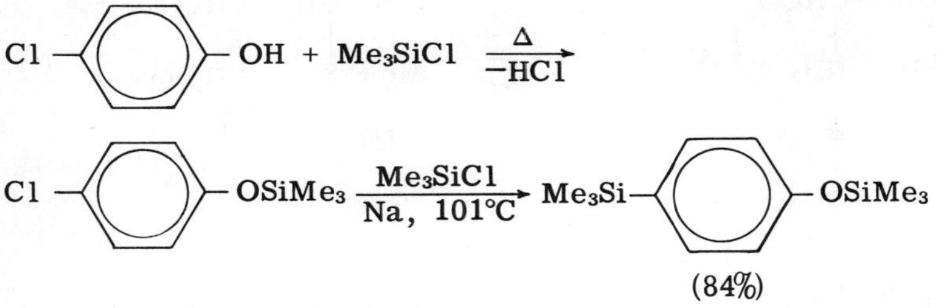

$$\text{Cl-C}_6\text{H}_4\text{-OH} + \text{Me}_3\text{SiCl} \xrightarrow[-\text{HCl}]{\Delta}$$

$$\text{Cl-C}_6\text{H}_4\text{-OSiMe}_3 \xrightarrow[\text{Na, 101°C}]{\text{Me}_3\text{SiCl}} \text{Me}_3\text{Si-C}_6\text{H}_4\text{-OSiMe}_3 \quad (84\%) \tag{191}$$

The isolation of oligomeric aryloxyphenols prepared by a redistribution reaction of mesitol with poly(2,6-dimethylphenylene oxide) proved to be difficult because the equilibrium shifted toward the side of starting materials upon distiliation (eq. 192). Silylation of the hydroxy groups with bistrimethylsilyl acetamide arrested the equilibrium, allowing the separation of the oligomers by distillation (37).

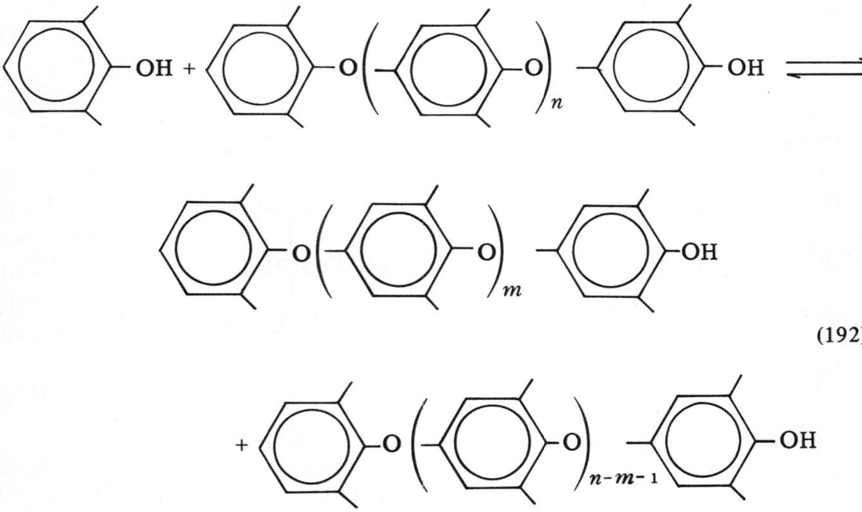

$$\text{ArOH} + \text{Ar-O}\left(\text{Ar-O}\right)_n\text{-Ar-OH} \rightleftharpoons \text{Ar-O}\left(\text{Ar-O}\right)_m\text{-Ar-OH} + \text{Ar-O}\left(\text{Ar-O}\right)_{n-m-1}\text{-Ar-OH} \tag{192}$$

$$m = 0, 1, 2, 3, \ldots; \; n = 1, 2, 3, \ldots$$

Hydroxyl and thiol groups in amino acids used for peptide syntheses have to be protected, for example, by benzylation. This operation and the subsequent removal of the protective group involves losses of material. The peptide synthesis discussed in Section III.A.2 using *N*-carbobenzoxy amino acids, *N,N'*-carbonyl diimidazole, and *N*-silyl amino acid silyl esters avoids a separate protective operation for OH and SH functions. These groups are silylated along with the amino and carboxyl group, and the silyl groups are subsequently hydrolyzed in the workup of the peptide (177) (eqs. 193 and 194). *N*-Glycyl serine was obtained in 80% yield by this procedure.

$$\text{L}-\underset{\displaystyle \text{SH}}{\text{CH}_2}-\underset{\displaystyle \text{NH}_2\cdot\text{HCl}}{\text{CHCOOH}} \xrightarrow[127^\circ\text{C, 3 hr}]{(\text{Me}_3\text{Si})_2\text{NH}} \text{L}-\underset{\displaystyle \text{Me}_3\text{SiS}}{\text{CH}_2}-\underset{\displaystyle \text{NHSiMe}_3}{\text{CHCOOSiMe}_3} \quad (73\%) \tag{193}$$

$$\text{CboNHCH}_2\text{COOH} + \left(\text{Im}-\right)_2\text{CO} \longrightarrow \text{CboNHCH}_2\text{COIm}$$

$$\xrightarrow[\text{2. hydrolysis}]{\text{1. Me}_3\text{SiSCH}_2\underset{\displaystyle \text{NHSiMe}_3}{\text{CHCOOSiMe}_3}} \text{CboNHCH}_2\text{CONH}\underset{\displaystyle \text{CH}_2\text{SH}}{\text{CHCOOH}} \quad (40\%) \tag{194}$$

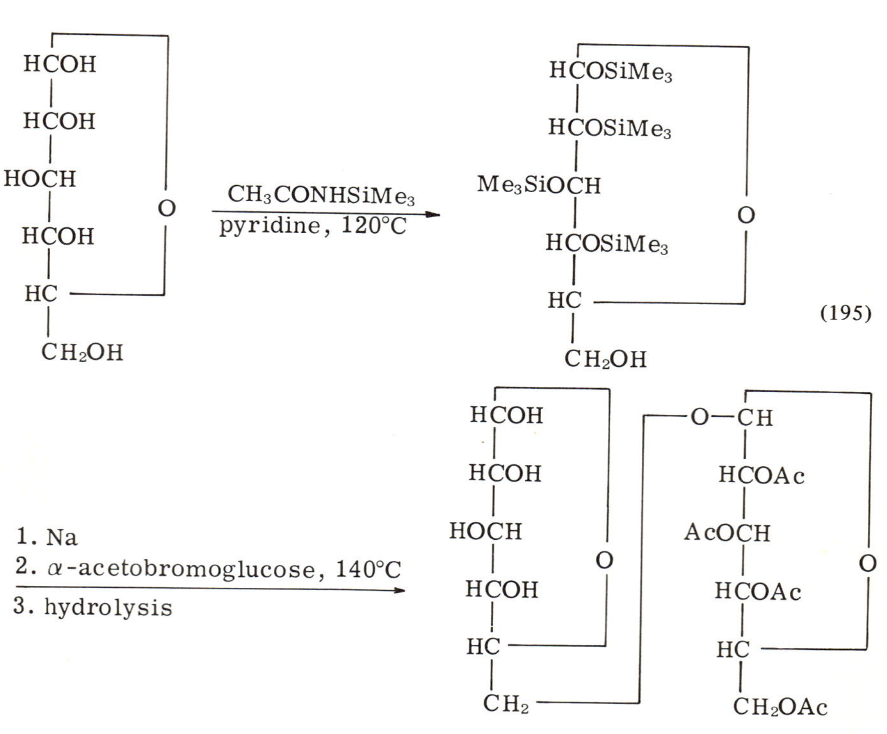

A partially silylated glucose with an open 6 position could be obtained by refluxing glucose with 4.4 equivalents of *N*-trimethylsilyl acetamide in pyridine solution. Only the secondary hydroxyl groups are silylated under these conditions; replacement of the primary hydroxyl group requires a larger excess of the silyl donor, or more readily, treatment of glucose with the silyl amide in the melt at 180°C. 1,2,3,4-Tetrakis(trimethylsilyl)-D-glucose was then converted to the alcoholate with sodium in ether, reacted with acetobromoglucose at 140°C, and hydrolyzed to 2′,3′,4′,6′-tetraacetylgentiobiose (178) (eq. 195).

The yield of cyclic α-hydroxy ketones by the acyloin condensation method has not been satisfactory for 4-, 5-, and 6-membered rings. The enolates formed in the condensations apparently are particularly prone in these cases to undergo further condensation and polymerization reactions. The capture of the enolate intermediates as siloxy ethers by addition of trimethylchlorosilane to the reaction mixtures largely inhibits such side reactions; the bis(trimethylsilyl)-cycloalkanes can be isolated in high yield (179) (eq. 196).

$$EtOOC{-}(CH_2)_3{-}COOEt \xrightarrow{4\ Na} \left[(CH_2)_3\begin{matrix} \diagup CONa \\ \| \\ \diagdown CONa \end{matrix} + 2\ EtONa \right] \xrightarrow[-4\ NaCl,\ -2\ EtOSiMe_3]{4\ Me_3SiCl} (CH_2)_3\begin{matrix} \diagup C{-}OSiMe_3 \\ \| \\ \diagdown C{-}OSiMe_3 \end{matrix} \quad (93\%) \qquad (196)$$

The hydrolysis was carried out in 1*N* HCl in tetrahydrofuran and the α-hydroxyketone distilled after neutralization; some polymerization of the 5- and 6-membered products was observed during this procedure (eq. 197).

$$(CH_2)_3\begin{matrix} \diagup COSiMe_3 \\ \| \\ \diagdown COSiMe_3 \end{matrix} \xrightarrow[-(Me_3Si)_2O]{H_3O^+} (CH_2)_3\begin{matrix} \diagup C{=}O \\ | \\ \diagdown CHOH \end{matrix} \quad (78\%) \qquad (197)$$

The method has been found successful in the synthesis of a series of cyclobutane derivatives which were obtained in 80-90% yield (180) (eq. 198).

$$C_6H_4(COOEt)_2 \xrightarrow[-4\ NaCl,\ -2\ EtOSiMe_3]{4\ Na,\ 4\ Me_3SiCl} \text{benzocyclobutene-1,2-bis}(OSiMe_3) \quad (89\%) \qquad (198)$$

The same method has been used to prepare acyloins from carboxylic esters with functional groups in the β position. Satisfactory yields were obtained since the reaction medium remains neutral whereas the strongly basic reaction medium of the Bouveault-Blanc reaction generally leads to β eliminations (181) (eqs. 199 and 200).

$$2\ X(CH_2)_2COOEt \xrightarrow[-4\ NaCl,\ -4\ EtOSiMe_3]{4\ Na,\ 4\ Me_3SiCl}$$

$$X(CH_2)_2C(OSiMe_3)\!=\!C(OSiMe_3)(CH_2)_2X \xrightarrow{\text{hydrolysis}} X(CH_2)_2C(=O)CH(OH)(CH_2)_2X \qquad (199)$$

X = EtO (47%), EtS (69%), Et_2N (62%)

X = EtO (70%), EtS (68%)

$$EtOOC(CH_2)_2X(CH_2)_2COOEt \longrightarrow \text{cyclo-}[(CH_2)_2C(OSiMe_3)\!=\!C(OSiMe_3)(CH_2)_2X] \qquad (200)$$

X = S (26%), NPh (74%), NMe (67%)

The Simmons-Smith reaction with 1,2-bis(trimethylsiloxy)alkenes allowed the preparation of bis(trimethylsiloxy)cyclopropanes in 20-55% yield; the corresponding bisacetoxycyclopropanes were obtained by reaction with acetyl bromide. The parent cyclopropanediols are not accessible; hydrolysis yields α-hydroxyketones (182) (eq. 201).

C. PROTECTION OF TERMINAL ETHYNYL GROUPS

The synthesis of internal polyacetylenes is readily accomplished by oxidative coupling; however, this technique cannot be adapted to the controlled synthesis of oligoacetylenes with terminal ethynyl groups.

$$EtC{=}CEt(OSiMe_3)_2 \xrightarrow{CH_2I_2/Zn} \text{Et(Me}_3\text{SiO)C—C(OSiMe}_3\text{)Et (cyclopropane)}$$

$$\xrightarrow{AcBr} \text{Et(AcO)C—C(OAc)Et (cyclopropane)} \qquad \xrightarrow{aq.\ CH_3OH/PhH} EtCOC(Et)(OH)CH_3 \tag{201}$$

Introduction of a trialkylsilyl group into acetylene or a diacetylene allows the oxidative coupling of the unprotected end of the molecule. The trialkylsilyl group can be removed quantitatively by treatment with methanolic alkali (183) (eqs. 202 and 203).

$$2Et_3SiC{\equiv}CH \xrightarrow[R_3N]{O_2} Et_3SiC{\equiv}CC{\equiv}CSiEt_3 \tag{202}$$

$$PhC{\equiv}CH + Br(C{\equiv}C)_2SiEt_3 \longrightarrow Ph(C{\equiv}C)_3SiEt_3 \xrightarrow{OH^-} Ph(C{\equiv}C)_3H \tag{203}$$

m- and *p*-Bromophenylacetylenes were converted into the corresponding ethynylsilanes via the ethynyl Grignard reagents (eq. 204). The protected acetylenic compounds afforded a number of ring-substituted derivatives by Grignard synthesis; in the last step the silyl group was removed by dilute alkali to give terminal acetylene derivatives which otherwise were not, or less conveniently, accessible (184) (eq. 205).

The corresponding meta derivatives as well as *m*- and *p*-trimethylsilylethynyl benzene were prepared by the same method.

$$Br{-}C_6H_4{-}C{\equiv}CH \xrightarrow[2.\ Me_3SiCl]{1.\ EtMgBr} Br{-}C_6H_4{-}C{\equiv}CSiMe_3 \tag{204}$$

$$Me_3SiC{\equiv}C\text{-}C_6H_4\text{-}Br \xrightarrow[3.\ n/10\ OH^-]{1.\ Mg;\ 2.\ CO_2} HC{\equiv}C\text{-}C_6H_4\text{-}COOH\ (55\%)$$

$$Me_3SiC{\equiv}C\text{-}C_6H_4\text{-}Br \xrightarrow[3.\ n/10\ OH^-]{1.\ Mg;\ 2.\ Me_3SnBr} HC{\equiv}C\text{-}C_6H_4\text{-}SnMe_3\ (29\%) \qquad (205)$$

The selective reduction of a triyne to the enediyne was possible after silylation of the terminal ethynyl groups. The latter are not reduced by hydrogen in the presence of Pd-$BaSO_4$. The silyl groups were then removed with silver nitrate and potassium cyanide (185) (eq. 206).

$$HC{\equiv}CCH{=}C(CH_3)CH_2C{\equiv}CCH_2C(CH_3){=}CHC{\equiv}CH$$

$$\xrightarrow[3.\ H_2/Pd\text{-}BaSO_4]{1.\ EtMgBr;\ 2.\ Me_3SiCl} \xrightarrow[2.\ KCN\ in\ H_2O]{1.\ AgNO_3\ in\ aq.\ EtOH} \qquad (206)$$

$$HC{\equiv}CCH{=}C(CH_3)CH_2CH{=}CHCH_2C(CH_3){=}CHC{\equiv}CH$$

l-Trimethylsilylpropyne was prepared from propynyllithium and chlorosilane; metalation with butyllithium afforded lithio-l-trimethylsilylpropyne which proved a valuable reagent for chain extension reactions (186) (eq. 207). The trimethylsilyl group allows the metalation of the propargyl moiety in C_3 specifically and screens the molecule from attack at C_1 during the alkylation reaction, thus essentially preventing the formation of allenic by-products.

$$CH_3C{\equiv}CLi + Me_3SiCl \xrightarrow{30\ hr,\ 35°C} CH_3C{\equiv}CSiMe_3 \xrightarrow{n\text{-}BuLi}$$

$$(\overset{\ominus}{C}H_2C{\equiv}CSiMe_3)\ Li^{\oplus} \xrightarrow{CH_3(CH_2)_5I} CH_3(CH_2)_6C{\equiv}CSiMe_3\ (55\%) \qquad (207)$$

$$(\overset{\ominus}{C}H_2C{\equiv}CSiMe_3)\ Li^{\oplus} \xrightarrow{PhCH_2I} Ph(CH_2)_2C{\equiv}CSiMe_3\ (50\%)$$

The chain extension reaction was applied in the preparation of an essential intermediate in the synthesis of cecropia juvenile hormone (187) (eq. 208).

$$\xrightarrow[\text{Et}_2\text{O}]{(\overset{\ominus}{\text{C}}\text{H}_2\text{C}\equiv\text{CSiMe}_3)\text{Li}^{\oplus}}$$

CH_2Br

$CH_2CH_2C\equiv CSiMe_3$ $\xrightarrow[\text{2. KCN}]{\text{1. AgNO}_3/\text{EtOH}}$ (208)

$CH_2CH_2C\equiv CH$ (80%)

VIII. Experimental Procedures

(a) N-Acetylimidazole (16) (Section II.B.1). *N*-Trimethylsilylimidazole was prepared by refluxing 13.6 g of imidazole with 24.6 g of hexamethyldisilazane for 10 hr under anhydrous conditions. Fractionation afforded the product in 85% yield, bp 91°C/12 mm, $n_D^{20} = 1.4756$. Acetylchloride, 0.1 mole in 25 ml of anhydrous benzene, was added dropwise with stirring at 0°C to a solution of 0.1 mole of *N*-trimethylsilylimidazole in 25 ml of anhydrous benzene. Stirring was continued for 15 min and the benzene removed *in vacuo*. Nearly pure *N*-acetylimidazole remained, yield 96%, mp 105°C.

(b) S-Acetylthiolacetic Acid (33) (Section II.C.1). *S*-Trimethylsilylthiolacetic acid trimethylsilyl ester was obtained by adding 0.48 mole of trimethylchlorosilane dissolved in 50 ml of benzene during 1 hr to the ice-cooled solution of 0.2 mole of freshly distilled thiolacetic acid and 0.48 mole of triethylamine in 150 ml of benzene. Triethylamine hydrochloride was filtered off after an additional 30 min stirring, washed with benzene, and the combined filtrate and washing were fractionated; the colorless liquid boiled at 100°C/10 mm, $n_D^{20} = 1.4515$, yield 73%. Addition of 50 mmoles of acetylchloride to 50 mmoles of *S*-trimethylsilylthiolacetic acid trimethylsilyl ester in 50 ml of anhydrous benzene was followed by 1 hr heating at reflux and fractionation. *S*-Acetylthiolacetic acid trimethylsilyl ester was obtained in 62% yield, bp 102°C/10 mm, $n_D^{20} = 1.4529$. Warming of 5.5 g of the acetyl ester for 10 min in 40 ml of moist ether afforded *S*-acetylthiolacetic acid, bp 103°C/0.8 mm.

(c) Cyclohexanone-2,6-dicarboxylic Acid Dimethyl Ester (37) (Section II.D.1). To a slurry of 10 mmoles of cyclohexanone-2,6-dicarboxylic acid in 5 ml of anhydrous ether was added 40 mmoles of bis(trimethylsilyl)acetamide and the mixture stirred under anhydrous conditions until a clear solution was obtained. The solution was added to 10 ml of 2*M* methanolic hydrogen chloride at 5°C. The reaction mixture was allowed to stand for 10 min and was then poured into water and extracted several times with ether. The ether extracts were washed and dried, and the solvent evaporated. The dimethyl ester was obtained in 85% yield, mp 144-146°C.

(d) Phenyl Sulfide (73) (Section II.E.3). An ether solution containing phenyl sulfoxide and trichlorosilane in a molar ratio of 1:2.5 was stirred at room temperature under anhydrous conditions for 1 hr. The mixture was worked up by hydrolysis and the product purified by column chromatography. Phenyl sulfide was obtained in 98% yield. Benzylic and aliphatic sulfoxides afforded mercaptals $RCH(SCH_2R)_2$ in addition to sulfides.

(e) 1-Trimethylsilylbutyn-3-one (79) (Section II.F.2). A mixture of 15.2 g of acetylchloride and 33 g of bis(trimethylsilyl)acetylene in a small amount of carbon disulfide or nitrobenzene was added dropwise to a solution of 26 g of anhydrous aluminum chloride in CS_2 or $PhNO_2$ cooled to 2-8°C. After stirring for 30 min, the mixture was decomposed with ice/hydrochloric acid, the aqueous phase extracted with CS_2, and the residue fractionated after drying and removal of solvent. Yield, 24 g (90%); bp 156°C/760 mm; $n_D^{20} = 1.4415$.

(f) Tetrakis(trimethylsilyl)uric Acid (82) (Section III.A.1). Tetrakis(trimethylsilyl)uracil was prepared by refluxing 10 g of 4,5-diaminouracil with 30 g of hexamethyldisilazane for 10 hr. Fractionation afforded 29 g (90%), bp 105°C/0.001 mm; mp 60°C. A solution containing 16.5 g of tetrakis(trimethylsilyl)uracil and 8.5 g of triethylamine in 100 ml of toluene was cooled to 0°C and 4.0 g of phosgene dissolved in 50 ml of toluene was added dropwise. After several hours the triethylamine hydrochloride was filtered and the filtrate fractionated. The product was obtained in 80% yield (14 g), bp 135-140°C/0.02 mm, mp 114°C (from acetonitrile). Hydrolysis with aqueous EtOH gave uric acid in quantitative yield.

(g) DL-*Valyl-glycl-β-alanine* (86) (Section III.A.2). *N*-Trimethylsilyl-glycyl-β-alanine trimethylsilyl ester was prepared by refluxing 2.5 g of glycyl-β-alanine in 20 g of hexamethyldisilazane for 5 hr and distillation; bp 138°C/0.025 mm; yield, 3.35 g (68%). Eleven millimoles of the silylated dipeptide was added to a solution of *N*-carbobenzoxy-DL-valylimidazolide, prepared by warming 10 mmoles of *N*-carbobenzoxy-DL-valine with 1.75 g of *N,N'*-carbonyldiimidazole in 50-100 ml of THF at 50°C for 1 hr. The mixture was heated at reflux for 3 hr, THF removed *in vacuo*, and the residue treated with 40 ml of water.

Hexamethyldisiloxane was removed and the aqueous solution acidified with 6*N* HCl after filtration. The N-Cbo derivative of the tripeptide was obtained crystalline; yield, 3.3 g (87%); mp 175°C from water.

(h) 3,7-Dimethylxanthine (84) (Section III.B.1). 2,6,7-Tris(triethylsilyl)-xanthine was obtained by refluxing 10.0 g of xanthine with 51.7 g of triethylsilylamine for 10 hr, and distillation of the mixture; bp 160°C/0.02 mm; mp 34°C; 25.8 g (85%). A mixture of 5 g of the silylated product and 15 g of methyliodide was heated at 50-60°C for 12 hr. The reaction mixture was then diluted with acetone and hydrolyzed by addition of a little methanol. The precipitate was filtered off, suspended in water, and neutralized with ammonia. The solid product was filtered off and recrystallized from water. A yield of 1.0 g (55%) of colorless needles was obtained, mp 351°C.

(i) Trivalerylmethane (105) (Section III.C). Bis(trimethylsilyl)malonate was prepared by adding 217 g of trimethylchlorosilane at 0°C to the stirred solution of 104 g of malonic acid and 158 g of pyridine in 1 liter of anhydrous ether. After stirring for 2 hr, 1 liter of petroleum ether was added, the mixture filtered, and the filtrate fractionated. A yield of 280 g of the bissilyl ester was obtained, bp 48°C/0.01 mm, $n_D^{20} = 1.4140$.

A solution of 0.2 mole of isopropyl magnesium bromide in ether was added below −20°C to the solution of 49.6 g of bis(trimethylsilyl)malonate in a mixture of 150 ml of anhydrous THF and 150 ml of anhydrous ether. Subsequently, 24 g of valeryl chloride was slowly added and the mixture was allowed to warm to room temperature whereupon a vigorous evolution of CO_2 occurred from 0°C on, and the initially formed precipitate dissolved. The solution was stirred for 24 hr, the solvent removed *in vacuo*, and the residue dissolved in 300 ml of anhydrous THF. Valeryl chloride, 24 g, was added with stirring at room temperature, stirring was continued for 3 hr, and the reaction mixture then shaken with 100 ml of dilute hydrochloric acid. The organic phase was washed, dried, and the solvent removed. Fractionation of the residue yielded 27 g of trivalerylmethane, bp 106°C/0.001 mm; $n_D^{20} = 1.4758$. The product crystallized slowly, mp 73°C, after recrystallization from petroleum ether at −30°C.

(j) Dimethyl Ethynylphosphonate (109) (Section III.D). Chloroethynyltrimethylsilane (108) was prepared by dropwise addition of 20 ml of a 3.47*M* solution of dichloroacetylene in ether to 56 ml of a 1.17*M* phenyllithium solution in ether and stirring for 1.5 hr, followed by addition of 6.1 g of trimethylchlorosilane in 30 ml of ether at a rate to maintain reflux. Stirring was continued for 30 min at room temperature, and an additional 30 min at reflux; the salts were filtered off and the filtrate fractionated, giving 4 g; bp 98-105°C; refractionation afforded pure product; bp 101.5°C/762 mm. A mixture of 8.64 g (0.065 mole) of chloroethynyltrimethylsilane and 9.21 g (0.075 mole) of

trimethylphosphite was heated at reflux under dry nitrogen for 5 hr, and the methyl chloride evolved was collected in a trap. The reaction mixture was then fractionated to give dimethyl trimethylsilylethynylphosphonate; bp 66°C/0.2 mm; $n_D^{25} = 1.4305$. The phosphonate, 7.6 g, was shaken with 24 ml of 10% aqueous sodium carbonate solution for 1-2 min. When the mildly exothermic reaction was complete, the mixture was extracted with ether. The dried extracts were fractionated to give 3.31 g (63%) of dimethyl ethynylphosphonate; bp 50-51°C/0.35; $n_D^{19} = 1.4241$.

(k) Divinylurea (124) (Section IV.A). A solution of vinyl isocyanate and hexamethyldisilazane in a 2:1 molar ratio in THF was heated at reflux. The solvent and trimethylsilyl isocyanate were evaporated *in vacuo* and the residue treated with methanol in benzene solution. The solid product was recrystallized twice from ethyl acetate, affording divinylurea in 50% yield; mp 155°C (decomp).

(l) N-*Methyl*-N-*(β-phenylethyl)acetamide* (132) (Section IV.B). A mixture of equimolar amounts of styrene oxide and *N*-methyl-*N*-trimethylsilylacetamide was heated with stirring in the presence of 1 mole % of sodium trimethylsilanolate at 95°C for 8 hr, or at 150°C for 1 hr, and then fractionated. The adduct distilled at 118.5-119.5°C/0.02 mm; yield 74%; $n_D^{20} = 1.4961$. The siloxy compound (38.4 g) was hydrolyzed by stirring with 40 ml of dilute hydrochloric acid; the crystalline product was filtered off and recrystallized from 140 ml of water; yield 35 g (81%); mp 84°C.

*(m) Ethyl β-hydroxy-β-(*p*-nitrophenyl)propionate* (139) (Section IV.C.2). Ethyl trimethylsilylacetate was prepared according to Gold, Sommer, and Whitmore (137). A solution of 325 g, 3 moles, of freshly distilled ethyl chloroformate in 1000 ml of anhydrous ether was placed in a 5-liter, three-necked flask fitted with a mercury-sealed, electrically driven Hershberg stirrer, a reflux condenser leading to a sulfuric acid trap via a Dry Ice-acetone trap, and a dropping funnel containing the Grignard reagent prepared in 95% yield from 2 moles of chloromethyltrimethylsilane in 500 ml of dry ether. While stirring, addition of the Grignard reagent at a rate sufficient to maintain reasonable reflux was followed by heating with a warm water-bath for 5 hr. Copious precipitation of magnesium chloride etherate necessitated addition of 1000 ml of ether in order to maintain stirring, but after another 12 hr at room temperature the reaction mixture had virtually solidified.

After hydrolysis with 1000 ml of 3% hydrochloric acid, washing of the ether solution with water and dilute sodium carbonate, and drying over anhydrous sodium sulfate, the ether was distilled and the residue fractionated through a column of 15 theoretical plates to give 113.5 g, 1.05 moles, 35% recovery, of ethyl chloroformate, bp 43.5°C at 120 mm; 239.5 g, 1.50 moles of ethyl trimethylsilylacetate, bp 75.5°C at 42 mm (157°C at 730 mm); $n_D^{20} = 1.4149$; 74.5% yield based on chloromethyltrimethylsilane; and 10 g of residue.

A mixture of 22.7 g of *p*-nitrobenzaldehyde, 30.0 g of ethyl trimethylsilylacetate, and 330 mg of potassium tetramethylalanate was heated in an atmosphere of dry nitrogen for 9 hr at 160-180°C. The catalyst was then filtered off and the product fractionated on a spinning band column. A yield of 37.6 g (81%) of ethyl β-trimethylsiloxy-β-(*p*-nitrophenyl)propionate was obtained, bp 126°C/0.1 mm; mp 43-44°C (from EtOH/water). The hydrolysis was carried out by refluxing 7 g of the silyl ether in 15 ml of ethanol containing 0.2 ml of concd HCl for 1 hr. The solvent was evaporated and the residue recrystallized; mp 46°C.

(n) 4,5-Dimethyl-1,2,3-triazole (142) (Section IV.D.1). Trimethylsilyl azide was prepared (141) by dissolving benzazide in an equimolar amount of trimethylsilyl-*n*-butylamine, adding a catalytic amount of anhydrous aluminum chloride, and allowing the mixture to stand at room temperature for 3 hr. Distillation afforded a 93% yield of Me_3SiN_3; bp 96°C; $n_D^{20} = 1.4161$.

A mixture of 0.5 mole of dimethylacetylene and 0.6 mole of trimethylsilylazide was heated in an autoclave at 150°C for 20 hr. The 2-trimethylsilyl-4,5-dimethyl-1,2,3-triazole was dissolved in anhydrous ether and fractioned, bp 65-66°C/13 mm; $n_D^{25} = 1.4537$; yield 78-87%. The parent triazole was obtained by treatment with water or ethanol.

(o) Ethyl 3,4-diphenylpyrazole-5-carboxylate (149) (Section V.B). 2-Trimethylsilyl-5-phenyltetrazole was obtained by refluxing 14.6 g of 5-phenyltetrazole in 50 g of hexamethyldisilazane for 5 hr. The clear solution was fractionated, yielding 88% of the colorless product, bp 104-106°C/0.05 mm; mp 25-27°C.

Heating a mixture of 7.3 g of 2-trimethylsilyl-5-phenyltetrazole and 25 g of ethyl cinnamate in an oil bath at 190-195°C for 2 hr led to the evolution of 628 cm^3 of nitrogen. Fractionation of the residue afforded, in addition to ethyl cinnamate, an oil boiling at 156-160°C/0.05 mm which was predominantly ethyl 1-trimethylsilyl-3,4-diphenyl-Δ^2-pyrazoline-5-carboxylate from which by alcoholysis and crystallization from ethanol 5.5 g of the parent ester was obtained. Treatment of 1.8 g of the ester in 100 ml of chloroform with 0.96 g of bromine at room temperature for 2 days, shaking with 50 ml of water containing 2 g of $NaHCO_3$, drying, and removal of the solvent afforded an 80% yield of ethyl 3,4-diphenylpyrazole-5-carboxylate; mp 156°C after recrystallization from 50% ethanol.

(p) 3,4-Diphenyl-1,2,5-thiadiazole (157) (Section VI.B). Benzil-bistrimethylsilylimine was prepared by adding dropwise 36.6 g (0.19 mole) of sodium bis(trimethylsilyl)amide in 400 ml of benzene to a solution of 16.8 g (0.08 mole) of benzil in 80 ml of benzene, stirring of the deep red solution for 3 hr at 20°C and 7 hr at 70°C, and then adding slowly 20.8 g (0.19 mole) of trimethylchlorosilane at room temperature, followed by stirring at 60°C for 5 hr. Filtration and fractionation of the filtrate yielded 25.6 g (91%) of the yellow compound; bp 150°C/0.1 mm; mp 66-71°C.

A solution of 0.02 mole of benzil-bistrimethylsilylimine in anhydrous ether or benzene was mixed with 0.02 mole of sulfur dichloride in the same solvent. The solvent and trimethylchlorosilane were distilled off and the residue recrystallized from ethanol. Colorless crystals; mp 86-87°C; yield 96%.

(q) (p-Aminophenyl)diphenylmethanol (165) (Section VII.A). *p*-Bromo-*N,N*-bis(trimethylsilyl)aniline was prepared by the following procedure: Ethylmagnesium bromide (1.3 moles) in ether (600 ml) was added to *p*-bromoaniline (103 g, 0.6 mole) during 30 min. The mixture was refluxed for 4 hr to complete the metallation, treated with chlorotrimethylsilane (140 g, 1.3 moles), and refluxed for a further 8 hr. Precipitated magnesium salts were filtered off and washed with ether (150 ml) and the combined filtrates evaporated to remove ether and ethyltrimethylsilane. Distillation afforded *p*-bromo-*N,N*-bis(trimethylsilyl)aniline (43% based on *p*-bromoaniline); bp 106°C/1.2 mm; n_D^{25} = 1.5140.

To *p*-lithio-*N,N*-bis(trimethylsilyl)aniline, from *n*-butyllithium (0.027 mole) and *p*-bromo-*N,N*-bis(trimethylsilyl)aniline (0.027 mole) in ether/hexane (30 ml), was added benzophenone (4.5 g, 0.025 mole) in ether (20 ml). The mixture was refluxed for 30 min and hydrolyzed with dilute hydrochloric acid (2*M*). After standing overnight the mixture was made alkaline and the ether layer separated, dried (Na_2SO_4), and evaporated to dryness. Recrystallization of the residue from toluene gave (*p*-aminophenyl)diphenylmethanol (80%); mp 116°C.

(r) Glutaroin (179) (Section VII.B). Sodium, 46 g, was dispersed in 600 ml of hot toluene in an atmosphere of dry nitrogen by rapid stirring in a flask fitted with mechanical stirrer, reflux condenser, and addition funnel. After cooling to room temperature, 2 moles of trimethylchlorosilane and subsequently 0.5 mole of diethyl glutarate were added to the suspension and the mixture then heated at reflux. After all the sodium was consumed (about 22 hr), the sodium chloride was filtered off and the filtrate fractionated. 1,3-Bis(trimethylsiloxy)cyclopentene-1 boiled at 93-94°C/10-12 mm; n_D^{20} = 1,4426; yield 93%. The hydrolysis was carried out by mixing 0.33 mole of the disiloxypentene with 18.6 g of *N*-hydrochloric acid (~1 mole H_2O) in 100 ml of THF. The mixture was heated at reflux on a steam bath for 15 min and, after cooling, treated for 15 min with 12 g of precipitated calcium carbonate, and fractionated after filtration. Glutaroin boiled at 78-80°C/10-12 mm; n_D^{20} = 1.4868; yield 78%.

REFERENCES

1. A. E. Pierce, *Silylation of Organic Compounds,* Pierce Chemical Co., Rockford, Illinois, 1968.
2. L. Birkofer and A. Ritter, *Angew. Chem.,* **77**, 414 (1965).
3. K. Rühlmann, *Z. Chem.,* **5**, 130 (1965).
4. C. Eaborn, *Organosilicon Compounds,* Butterworths, London, 1960.
5. (a) L. H. Sommer, *Stereochemistry, Mechanism and Silicon,* McGraw-Hill, New York, 1965; (b) L. H. Sommer and K. T. Rosborough, *J. Amer. Chem. Soc.,* **91**, 7067 (1969), and earlier papers by L. H. Sommer et al. cited therein.

6. M. Kumada and H. Hattori, *J. Chem. Soc. Japan,* **56,** 269 (1953).
7. R. O. Sauer, U.S. Pat. 2,534,149 (1950).
8. N. I. Shuikin, I. F. Bel'ski, and I. E. Grushko, *J. Gen. Chem. USSR,* **31,** 748 (1961).
9. K. A. Andrianov, G. A. Kurakov, and L. M. Khananashvili, *J. Gen. Chem. USSR,* **33,** 2567 (1963).
10. U. Krüerke, *Chem. Ber.,* **95,** 174 (1962).
11. R. Calas, J. Dunogues, C. Biran, and N. Duffaut, *J. Organometal. Chem.,* **20,** 22 (1969).
12. J. Dunogues, R. Calas, and N. Duffaut, *J. Organometal. Chem.,* **20,** 20 (1969).
13. R. Fessenden and J. S. Fessenden, *Chem. Rev.,* **61,** 361 (1961).
14. B. J. Aylett, in *Preparative Inorganic Reactions,* Vol. 2 (W. L. Jolly, ed.), Wiley-Interscience, New York, 1965.
15. U. Wannagat, *Adv. Inorg. Radiochem.,* **6,** 225 (1964).
16. L. Birkofer, P. Richter, and A. Ritter, *Chem. Ber.,* **93,** 2804 (1960).
17. T. J. Pullukat and G. Urry, *Tetrahedron Lett.,* **1967,** 1953.
18. K. Rühlmann, *Chem. Ber.,* **94,** 2311 (1961).
19. K. Rühlmann, A. Reiche, and M. Becker, *Chem. Ber.,* **98,** 1814 (1965).
20. O. J. Scherer, *Organometal. Chem. Rev.,* **3A,** 281 (1968).
21. H. Nöth, *Angew. Chem.,* **72,** 638 (1960).
22. C. Summerford and K. Wade, *J. Chem. Soc., A,* **1969,** 1487.
23. M. Becke-Goehring and H. Krill, *Chem. Ber.,* **94,** 1059 (1961).
24. E. W. Abel, D. Brady, and B. R. Lerwill, *Chem. Ind.,* **1962,** 1333.
25. E. W. Abel, D. A. Armitage, and G. R. Willey, *J. Chem. Soc.,* **1965,** 57.
26. O. J. Scherer and P. Klusmann, *Angew. Chem. Int. Ed.,* **8,** 752 (1969).
27. L. Birkofer, W. Gilgenberg, and A. Ritter, *Angew. Chem.,* **73,** 143 (1961).
28. E. W. Abel and D. A. Armitage, *J. Chem. Soc.,* **1964,** 3122.
29. J. F. Klebe, *J. Polym. Sci., A,* **2,** 2673 (1964).
30. K. Itoh, A. Nozawa, and Y. Ishii, *Tetrahedron Lett.,* **1969,** 1421.
31. L. Birkofer, H. Dickopp, and S. K. Majlis, *Chem. Ber.,* **102,** 3094 (1969).
32. E. W. Abel and D. A. Armitage, *Adv. Organometal. Chem.,* **5,** 2 (1967).
33. M. Rimpler, *Chem. Ber.,* **99,** 1528 (1966).
34. E. W. Abel, D. A. Armitage, and R. P. Bush, *J. Chem. Soc.,* **1964,** 2455.
35. L. Birkofer, A. Ritter, and H. Goller, *Chem. Ber.,* **96,** 3289 (1963).
36. L. Birkofer and H. Krebs, *Tetrahedron Lett.,* **1968,** 885.
37. J. F. Klebe, H. Finkbeiner, and D. M. White, *J. Amer. Chem. Soc.,* **88,** 3390 (1966).
38. K. Rühlmann, *J. Prakt. Chem.,* **16,** 172 (1962).
39. L. H. Edwards, C. E. Turner, C. E. Lane, Jr., and S. F. Thames, *Tetrahedron Lett.,* **1969,** 4225.
40. L. Birkofer, A. Ritter, and J. Schramm, *Chem. Ber.,* **95,** 426 (1962).
41. K. D. Petrov, *J. Gen. Chem. USSR,* **17,** 1099 (1947).
42. K. D. Petrov and E. S. Lagutscheva, *J. Gen. Chem. USSR,* **18,** 1150 (1948).
43. Y. K. Jurjev and G. B. Eljakov, *J. Gen. Chem. USSR,* **26,** 2350 (1956).
44. A. N. Nesmejanov, E. G. Perewalova, R. W. Golovnja, and O. A. Nesmejanova, *Ber. Akad. Wiss. USSR,* **97,** 459 (1954).
45. N. C. Cook and J. E. Lyons, *J. Amer. Chem. Soc.,* **87,** 3283 (1965).
46. N. C. Cook and J. E. Lyons, *J. Amer. Chem. Soc.,* **83,** 3396 (1966).
47. E. J. Moriconi and R. E. Misner, *J. Org. Chem.,* **34,** 3672 (1969).
48. H.-P. Becker and W. P. Neumann, *J. Organometal. Chem.,* **20,** P3 (1969).
49. J. F. Klebe, unpublished results.
50. L. Birkofer, E. Bierwirth, and A. Ritter, *Chem. Ber.,* **94,** 1263 (1961).
51. W. Fink, *Helv. Chim. Acta,* **49,** 1408 (1966).
52. R. Calas, *Pure Appl. Chem.,* **13,** 61 (1966).
53. E. Frainnet, *Pure Appl. Chem.,* **19,** 489 (1969).

54. J. W. Jenkins and H. W. Post, *J. Org. Chem.*, **15**, 556 (1950).
55. S. I. Sergejeva, D. D. Zitovich, and M. G. Voronkov, *Ber. Acad. Wiss. USSR*, **134**, 1371 (1960).
56. E. Frainnet, A. Bazouin, and R. Calas, *C. R. Acad. Sci., Paris*, **257**, 1304 (1963).
57. R. Calas, E. Frainnet, and A. Bazouin, *C. R. Acad. Sci. Paris*, **254**, 2357 (1962).
58. R. Calas, E. Frainnet, and A. Bazouin, *C. R. Acad. Sci., Paris*, **252**, 420 (1961).
59. E. Frainnet, R. Calas, and A. Berthault, *C. R. Acad. Sci., Paris*, **258**, 613 (1964).
60. H. Kautsky, H. Keck, and H. Kunze, *Z. Naturforsch.*, **9b**, 165 (1954).
61. S. Nitzsche and M. Wick, *Angew. Chem.*, **69**, 96 (1957).
62. R. Calas and N. Duffaut, *C. R. Acad. Sci., Paris*, **245**, 906 (1957).
63. R. Calas, E. Frainnet, and J. Bonastre, *C. R. Acad. Sci., Paris*, **251**, 2987 (1960).
64. A. D. Petrov and S. I. Sadyk-Sade, *Bull. Soc. Chim. Fr.*, **1959**, 1932.
65. E. Frainnet and R. Bourhis, *Bull. Soc. Chim. Fr.*, **1965**, 574
66. E. Frainnet and C. Esclamadon, *C. R. Acad. Sci., Paris*, **254**, 1814 (1962).
67. E. Frainnet and C. Fritsch, *Bull. Soc. Chim. Fr.*, **1960**, 596.
68. F. C. Whitmore, E. W. Pietrusza, and L. H. Sommer, *J. Amer. Chem. Soc.*, **69**, 2108 (1947).
69. D. N. Dolgov, S. N. Borisov, and M. G. Voronkov, *J. Gen. Chem. USSR*, **27**, 716 (1957).
70. H. Fritzsche, U. Hasserodt, and F. Korte, *Chem. Ber.*, **97**, 1988 (1964).
71. H. Fritzsche, U. Hasserodt, and F. Korte, *Chem. Ber.*, **98**, 171 (1965).
72. H. Fritzsche, U. Hasserodt, and F. Korte, *Chem. Ber.*, **98**, 1681 (1965).
73. T. H. Chan, A. Melnyk, and D. N. Harpp, *Tetrahedron Lett.*, **1969**, 201.
74. R. A. Benkeser, R. A. Hickner, D. I. Hoke, and O. H. Thomas, *J. Amer. Chem. Soc.*, **80**, 5289 (1958).
75. R. A. Benkeser, D. I. Hoke, and R. A. Hickner, *J. Amer. Chem. Soc.*, **80**, 5294 (1958).
76. L. Birkofer and E. Krämer, *Chem. Ber.*, **100**, 2776 (1967).
77. S. N. Battacharya, C. Eaborn, and D. R. M. Walton, *J. Chem. Soc., C*, **1969**, 1367.
78. R. West and L. Quass, *J. Organometal. Chem.*, **18**, 55 (1969).
79. L. Birkofer, A. Ritter, and H. Uhlenbrauck, *Chem. Ber.*, **96**, 3280 (1963).
80. R. Calas and P. Bourgeois, *C. R. Acad. Sci., Paris*, **268**, 1525 (1969).
81. L. H. Sommer, R. E. Van Strien, and F. C. Whitmore, *J. Amer. Chem. Soc.*, **71**, 3056 (1949).
82. L. Birkofer, H. P. Kühltau, and A. Ritter, *Chem. Ber.*, **93**, 2810 (1960).
83. H. P. Kühltau, Diploma Thesis, University of Cologne, Germany, 1960.
84. L. Birkofer, A. Ritter, and H. P. Kühltau, *Chem. Ber.*, **97**, 934 (1964).
85. W. Giessler, Doctoral Thesis, University of Cologne, Germany, 1963.
86. L. Birkofer, A. Ritter, and P. Neuhausen, *Justus Liebigs Ann. Chem.*, **659**, 190 (1962).
87. L. Birkofer, W. Konkol, and A. Ritter, *Chem. Ber.*, **94**, 1263 (1961).
88. G. E. Hilbert and T. B. Johnson, *J. Amer. Chem. Soc.*, **52**, 4489 (1930).
89. E. Wittenburg, *Chem. Ber.*, **99**, 2380 (1966).
90. H. Kaye, *J. Polym. Sci., Part B*, **7**, 1 (1969).
91. L. Birkofer, A. Ritter, and H. P. Kühltau, *Angew. Chem.*, **75**, 209 (1963).
92. T. Nishimura and I. Iwai, *Chem. Pharm. Bull.* (Tokyo), **12**, 357 (1964).
93. T. Nishimura, B. Shimizu, and I. Iwai, *Chem. Pharm. Bull.* (Tokyo), **12**, 1471 (1964).
94. T. Nishimura and B. Shimizu, *Chem. Pharm. Bull.* (Tokyo), **13**, 803 (1965).
95. E. Wittenburg, *Chem. Ber.*, **101**, 1095 (1968).
96. E. Wittenburg, *Z. Chem.*, **4**, 303 (1964).
97. F. Reisser and W. Pfleiderer, *Chem. Ber.*, **99**, 547 (1966).
98. M. Kotick, C. Szantay, and T. Bardos, *J. Org. Chem.*, **34**, 3806 (1969).
99. L. Pichat, B. Massé, J. Deschamps, and P. Dufy, *C. R. Acad. Sci., Paris, Ser. C.*, **268**, 197 (1969).

100. C. Krüger, E. G. Rochau, and U. Wannagat, *Chem. Ber.*, **96**, 2132 (1963).
101. L. Birkofer and H. Dickopp, *Chem. Ber.*, **102**, 14 (1969).
102. G. Stork and P. F. Hudrlik, *J. Amer. Chem. Soc.*, **90**, 4462 (1968).
103. L. Birkofer, A. Ritter, and H. Vernaleken, *Chem. Ber.*, **99**, 2518 (1966).
104. H. Gilman and R. N. Clark, *J. Amer. Chem. Soc.*, **69**, 967 (1947).
105. U. Schmidt and M. Schwochau, *Monatsh, Chem.*, **98**, 1492 (1967).
106. U. Schmidt and M. Schwochau, *Tetrahedron Lett.*, **1967**, 4491.
107. K. Rühlmann and G. Kuhrt, *Angew. Chem.*, **80**, 797 (1968).
108. H. G. Viehe, *Chem. Ber.*, **92**, 3064 (1959).
109. D. W. Burt and P. Simpson, *J. Chem. Soc., C,* **1969**, 2273.
110. H. Breederveld, *Rec. Trav. Chim. Pays-Bas,* **81**, 276 (1962).
111. H. Breederveld, *Rec. Trav. Chim. Pays-Bas,* **79**, 1126 (1960).
112. K. Itoh, I. R. Lee, I. Matsuda, S. Sakai, and Y. Ishii, *Tetrahedron Lett.*, **1967**, 2667.
113. M. Schmidt and H. Schmidbaur, *Angew. Chem.*, **70**, 657 (1958).
114. J. F. Klebe, J. B. Bush, Jr., and J. E. Lyons, *J. Amer. Chem. Soc.*, **86**, 4400 (1964).
115. G. Oertel, H. Malz, and H. Holtschmidt, *Chem. Ber.*, **97**, 891 (1964).
116. W. Fink, *Chem. Ber.*, **97**, 1433 (1964).
117. I. Matsuda, K. Itoh, and Y. Ishii, *J. Organometal. Chem.*, **19**, 339 (1969).
118. D. Y. Zhinkin, M. M. Morgunova, K. K. Popkov, and K. A. Andrianov, *Dokl. Akad. Nauk USSR,* **158**, 641 (1964).
119. K. Itoh, N. Kato, S. Sakai, and Y. Ishii, *J. Chem. Soc., C.*, **1969**, 2005.
120. W. W. Limburg and H. W. Post, *Rec. Trav. Chim. Pays-Bas,* **81**, 430 (1962).
121. I. F. Lutsenko, Y. I. Baukov, A. S. Kostyuk, N. I. Savelyeva, and V. K. Krysina, *J. Organometal. Chem.*, **17**, 241 (1969).
122. M. L. Lappert and P. Prokai, *Adv. Organometal. Chem.*, **5**, 225 (1967).
123. C. H. Yoder and J. J. Zuckerman, *Chem. Commun.*, **1966**, 694.
124. C. G. Overberger and S. Ishida, *J. Polym. Sci., Part B,* **3**, 789 (1965).
125. J. F. Klebe, *J. Polym. Sci., Part B,* **2**, 1079 (1964).
126. D. Martin, K. Witke, P. Reich, and K. Nadolski, *Chem. Ber.*, **101**, 3185 (1968).
127. K. Itoh, S. Sakai, and Y. Ishii, *J. Org. Chem.*, **31**, 3948 (1966).
128. E. Boldebuck and J. F. Klebe, Fr. Pat. 1,437,751 (1966).
129. G. Greber and D. Lohmann, *Meeting of the German Chemical Society, Abstracts,* Sept. 1969, p. 97.
130. G. Nelson and J. F. Klebe, unpublished results.
131. K. Itoh, S. Sakai, and Y. Ishii, *J. Org. Chem.*, **32**, 2210 (1967).
132. L. Birkofer and H. Dickopp, *Chem. Ber.*, **102**, 14 (1969).
133. L. Birkofer and H. Dickopp, *Angew. Chem.*, **76**, 648 (1964).
134. L. Birkofer and H. Dickopp, *Chem. Ber.*, **101**, 3579 (1968).
135. L. Birkofer, F. Müller, and W. Kaiser, *Tetrahedron Lett.*, **1967**, 2781.
136. K. Itoh, M. Fukui, and Y. Ishii, *Tetrahedron Lett.*, **1968**, 3867.
137. J. R. Gold, L. H. Sommer, and F. C. Whitmore, *J. Amer. Chem. Soc.*, **70**, 2874 (1948).
138. L. S. Sommer and N. S. Maras, *J. Amer. Chem. Soc.*, **72**, 1935 (1950).
139. L. Birkofer, A. Ritter, and H. Wieden, *Chem. Ber.*, **95**, 971 (1962).
140. W. Kirmse and L. Horner, *Justus Liebigs Ann. Chem.*, **614**, 1 (1958).
141. E. Ettenhuber and K. Rühlmann, *Chem. Ber.*, **101**, 743 (1968).
142. L. Birkofer and P. Wegner, *Chem. Ber.*, **99**, 2512 (1966).
143. L. Birkofer and M. Franz, *Chem. Ber.*, **100**, 2681 (1967).
144. J. F. Klebe, *J. Amer. Chem. Soc.*, **86**, 3399 (1964).
145. G. Greber and H. R. Kricheldorf, *Angew. Chem. Int. Ed.*, **7**, 941 (1968).
146. V. F. Mironov, V. D. Sheludyakov, V. P. Kozyukov, and G. D. Khatuntzev, *Doklad. Akad. Nauk USSR,* **181**, 115 (1968); *Chem. Abstr.*, **70**, 398 (1969).
147. G. Greber and H. Kricheldorf, *Angew. Chem. Int. Ed.*, **7**, 942 (1968).

148. L. Birkofer, W. Knipprath, and A. Ritter, *Angew. Chem.*, **70,** 404 (1958).
149. L. Birkofer, A. Ritter, and P. Richter, *Chem. Ber.*, **96,** 2750 (1963).
150. R. Huisgen, J. Sauer, and M. Seidel, *Chem. Ber.*, **94,** 2503 (1961).
151. D. J. Peterson, *J. Org. Chem.*, **33,** 780 (1968).
152. U. Wannagat and H. Niederprüm, *Chem. Ber.*, **94,** 1540 (1961).
153. U. Wannagat, *Angew. Chem.*, **75,** 173 (1963).
154. U. Wannagat and H. Niederprüm, *Z. Anorg. Allg. Chem.*, **308,** 337 (1961).
155. U. Wannagat and H. Kuckertz, *Angew. Chem.*, **74,** 117 (1962).
156. C. Krüger, E. G. Rochow, and U. Wannagat, *Chem. Ber.*, **96,** 2132 (1963).
157. G. Tuchtenhagen and H. Rühlmann, *Justus Liebigs Ann. Chem.*, **711,** 174 (1965).
158. C. Krüger, E. G. Rochow, and U. Wannagat, *Chem. Ber.*, **96,** 2138 (1963).
159. L. Birkofer, A. Ritter, and W. Giessler, *Angew. Chem.*, **75,** 93 (1963).
160. J. Pump and E. G. Rochow, *Chem. Ber.*, **97,** 627 (1964).
161. J. F. Klebe and J. B. Bush, Jr., First International Symposium on Organosilicon Chemistry, Prague, 1965.
162. E. W. Abel and G. R. Willey, *J. Chem. Soc.*, **1964,** 1528.
163. U. Wannagat, C. Krüger, and H. Niederprüm, *Z. Anorg. Allg. Chem.*, **314,** 82 (1962).
164. W. Broser and W. Harrer, *Angew. Chem.*, **77,** 1139 (1965).
165. D. R. M. Walton, *J. Chem. Soc., C,* **1966,** 1706.
166. P. S. Venkateswaran and T. J. Bardos, *J. Org. Chem.*, **32,** 1256 (1967).
167. L. Birkofer, A. Ritter, and S. M. Kim, *Chem. Ber.*, **96,** 3099 (1963).
168. L. Birkofer and S. M. Kim, *Chem. Ber.*, **97,** 2100 (1964).
169. K. L. Placiorek and R. H. Kratzer, *J. Org. Chem.*, **31,** 2426 (1966).
170. O. J. Scherer and J. Wokulat, *Z. Anorg. Allg. Chem.*, **361,** 296 (1968).
171. O. J. Scherer and G. Schieder, *Chem. Ber.*, **101,** 4184 (1968).
172. O. J. Scherer and P. Klusmann, *Angew. Chem.*, **80,** 560 (1968).
173. J. L. Speier, *J. Amer. Chem. Soc.*, **74,** 1003 (1952).
174. K. C. Frisch and P. D. Schroff, *J. Amer. Chem. Soc.*, **75,** 1249 (1953).
175. R. G. Neville, *J. Org. Chem.*, **25,** 1063 (1960).
176. R. G. Neville, *J. Org. Chem.*, **26,** 3031 (1961).
177. L. Birkofer, W. Konkol, and A. Ritter, *Chem. Ber.*, **94,** 1263 (1961).
178. L. Birkofer, A. Ritter, and F. Bentz, *Chem. Ber.*, **97,** 2196 (1964).
179. U. Schräpler and K. Ruhlmann, *Chem. Ber.*, **97,** 1383 (1964).
180. J. J. Bloomfield, *Tetrahedron Lett.*, **1968,** 587.
181. K. Rühlmann, H. Seefluth, and H. Becker, *Chem. Ber.*, **100,** 3820 (1967).
182. R. LeGoaller and P. Arnaud, *C. R. Acad. Sci., Paris, Ser. C,* **268,** 2322 (1969).
183. R. Eastmond and D. R. M. Walton, *Chem. Commun.*, **1968,** 204.
184. C. Eaborn, A. R. Thompson, and D. R. M. Walton, *J. Chem. Soc., C,* **1967,** 1364.
185. H. M. Schmidt and J. F. Arens, *Rec. Trav. Chim. Pays-Bas,* **86,** 1138 (1967).
186. E. J. Corey and H. A. Kirst, *Tetrahedron Lett.*, **1968,** 5041.
187. E. J. Corey, J. A. Katzenellenbogen, N. W. Gilman, S. A. Roman, and B. W. Erickson, *J. Amer. Chem. Soc.*, **90,** 5618 (1968).

LATENT FUNCTIONALITY IN ORGANIC SYNTHESIS

By DANIEL LEDNICER, *Research Laboratories, The Upjohn Company, Kalamazoo, Michigan*

CONTENTS

I. Introduction and Scope

The use of protecting groups in organic synthesis is so well established that the origin of the concept would be difficult to find. We would probably not be too far wrong if we surmised that the protecting group originated not long after the first synthetic scheme involving a polyfunctional molecule. As synthetic organic chemists brought within their scope ever more complex and delicate molecules, protecting groups underwent a commensurate increase in sophistication. There are today, for example, probably no fewer than half a dozen protecting groups for carboxylic acids; each of these has different stabilities and each in turn can be removed under slightly different conditions. The current state of the art has recently been discussed by McOmie in an excellent review (1).

Even when properly employed, the use of protecting groups carries with it inherent flaws. To begin with, these groups necessarily involve the labor and expense of prolonging a given synthesis by two steps—that is, the steps in which the protecting group is attached and later removed. These further operations also mean that the possibly sensitive intermediate needs to be exposed to reagents that many more times, with the additional possibilities for degradation. Finally, the juggling of protecting groups is an esthetically inelegant process. By way of illustration one need only bring to mind some steroid intermediate fairly bristling with a bismethylenedioxy ether, a ketal, and several acetates.

Starting about 20 years ago, there evolved a concept complimentary to that of the protecting group. This idea, which we have chosen to call *latent functionality,* carries some necessary function through one or more steps of a synthesis in a precursor form; at the proper stage the precursor is converted to the needed group. The use of an anisole ring as a potential cyclohexenone is perhaps the most familiar example of this concept. Because of its aromatic character the anisole can be carried through many steps of a synthesis without any particular precautions; lithium in ammonia reduction at the proper moment reveals the enone. A well-designed scheme will also take advantage of the stereochemical consequences of both the method of formation of the latent function (e.g., the *cis* fusion resulting from a Diels-Alder reaction) as well as its unmasking (e.g., retention of configuration in a cyclic rearrangement).

Though the concept of latent functionality is inherent in all of organic chemistry, it has only recently been applied to the preparation of complicated molecules. This chapter is intended to cover some of these applications. Since concepts are almost nonretrievable from the bulk of the literature, no claims will be made to the comprehensiveness of this work. Examples of what we mean by latent functionality were selected for their illustrative value. We choose this time to apologize to those whose work we have not included. Many of the examples cited come from syntheses of very complex molecules; in the interests of clarity only those steps which pertain to the theme of this chapter have been shown.

II. Cleavage of Carbon-Carbon Double Bonds

A. MONOCARBONYL COMPOUNDS FROM TERMINAL OLEFINS

The two and three carbon units C—C—O and C—CO—C have many applications in organic synthesis, particularly in the construction of rings. Though few methods exist for the direct introduction of these units by alkylation or condensation reactions, the olefinic counterparts can be readily incorporated into molecules; allyl halides can be easily added to the α position of ketones by alkylation of the enolate. Oxidation of the double bond by either ozone or a combination of a hydroxylation-cleavage sequence uncovers the latent carbonyl group. The inertness of the olefinic function to all but these reagents makes it possible to carry such a group through many steps of synthesis.

1. Methallyl Groups as Latent Acetone Residues

A convenient method for the introduction of what will finally be the acetone moiety consists in first incorporation of a methallyl residue by either alkylation (methallyl halide) or condensation (methallyl Grignard). Oxidative cleavage of the double bond either with ozone or by a hydroxylation-glycol cleavage sequence unmasks the ketone (eq. 1).

Z → Z—CH₂C(CH₃)=CH₂ → Z—CH₂C(CH₃)=O (1)

In an early approach to the total synthesis of steroids, the tetralone **1** was alkylated with methallyl bromide; ozonization followed by treatment with methanol gave the internal ketal (**5**) of the desired product (**4**). An alternate oxidation scheme (performic acid followed by periodate) led to the same product. The great stability of that ketal precluded its use as a synthetic intermediate (2).

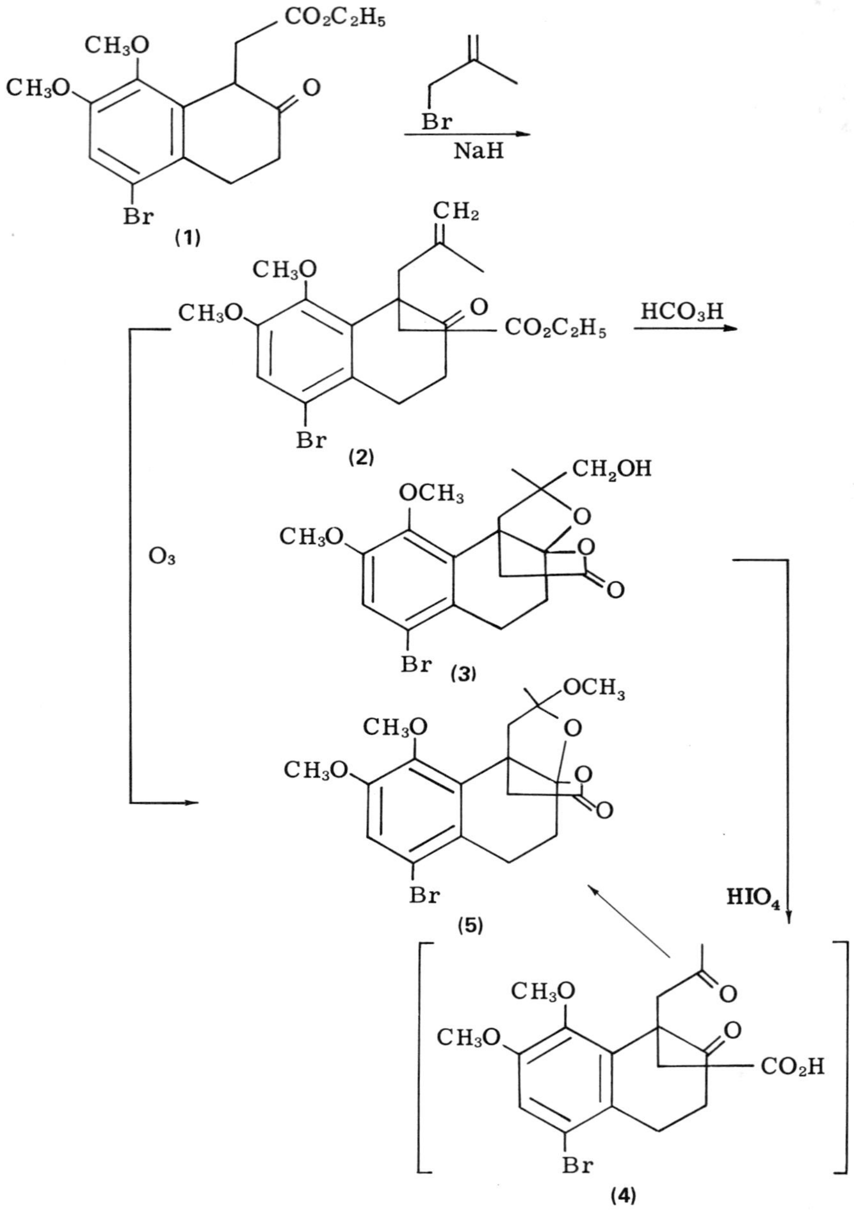

$CO_2C_2H_5$
CH_3O
CH_3O
O
Br
Br
NaH
(1)
CH_2
CH_3O
CH_3O
O
$CO_2C_2H_5$
HCO_3H
Br
(2)
CH_2OH
OCH_3
CH_3O
O
O
O
O_3
Br (3)
OCH_3
CH_3O
O
CH_3O
O
O
Br
HIO_4
(5)
CH_3O
O
CH_3O
O
CO_2H
Br
(4)

A similar scheme for forming an acetone side chain was employed by the Merck group in their total synthesis of cortisone (3). Alkylation of the tricyclic ketone (**6**) with methallyl chloride gave compound **7**; the ketone function was elaborated in a series of transformations to **8**. Cleavage of the double bond by the hydroxylation-glycol cleavage sequence gave the methyl ketone **9**. This last was subjected to an internal alkylation reaction to form the 5-membered ring D. An analogous scheme for forming the D ring was later used in the preparation of 18,19-bisnorprogesterone **(11-13)** (4).

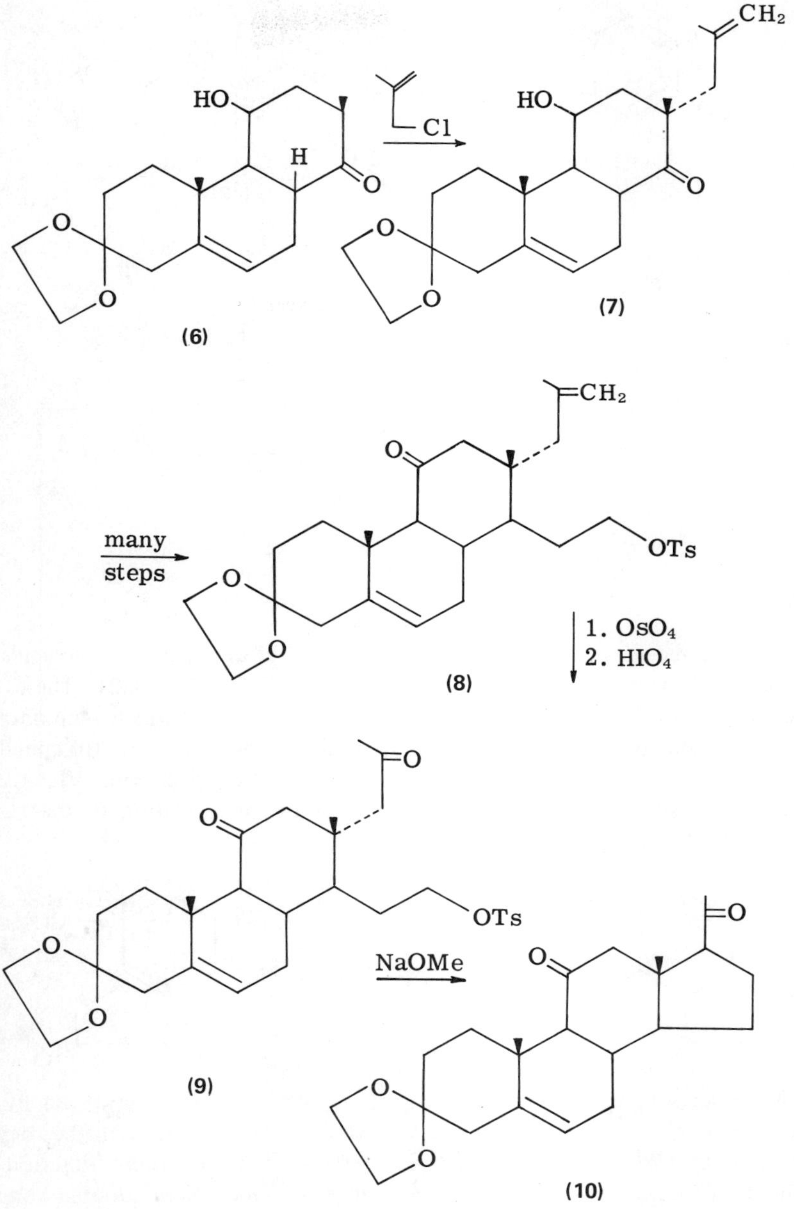

HO H O O O Cl HO O O O CH_2

(6) **(7)**

many steps

O O O CH_2 OTs

(8)

1. OsO_4
2. HIO_4

O O O O OTs

NaOMe

O O O O

(9) **(10)**

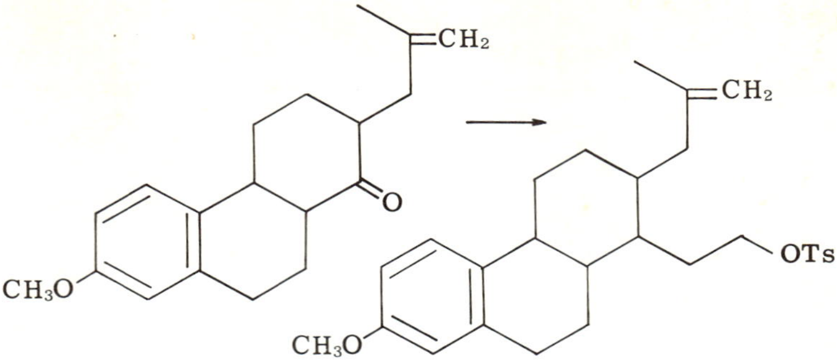

CH_2 O CH_3O CH_2 OTs CH_3O

(11) (12)

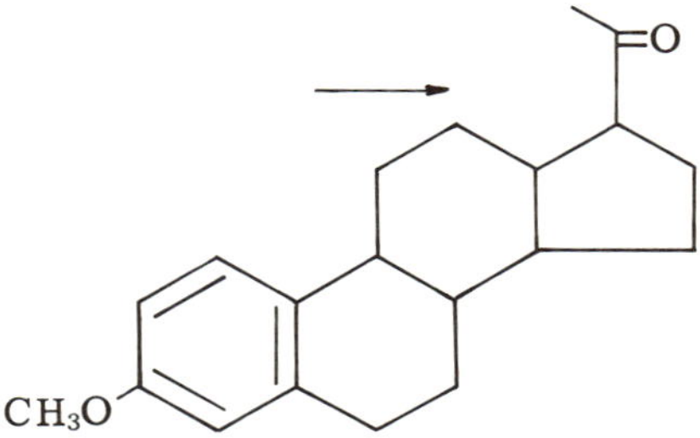

O CH_3O

(13)

In some recent work relating to the synthesis of deoxypodocarpic acids, the preparative route led through hydrindanones such as **17** and **21**. These were prepared by cyclization of the acetyl ketones (**16** and **20**) which were obtained by oxidation of the methallyl compounds. Isomer **15** was obtained by methylation of the Claisen rearrangement product **14** (see Section VI.A), while the epimer **19** was the predominant product of the allylation of the methyl compound **18** (5,6).

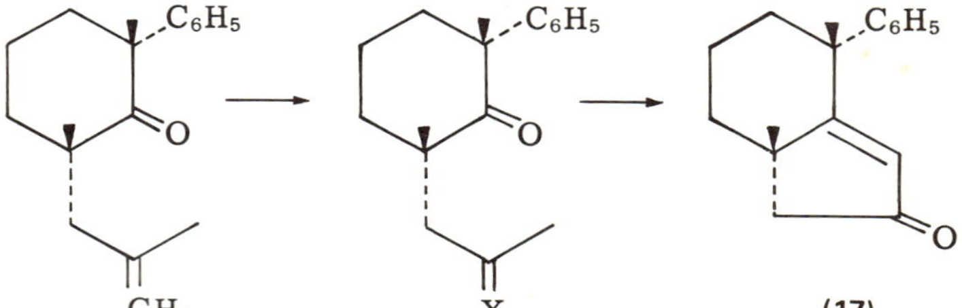

C_6H_5 O CH_2 C_6H_5 O X C_6H_5 O

(14) (17)

(15) X = CH_2
(16) X = O

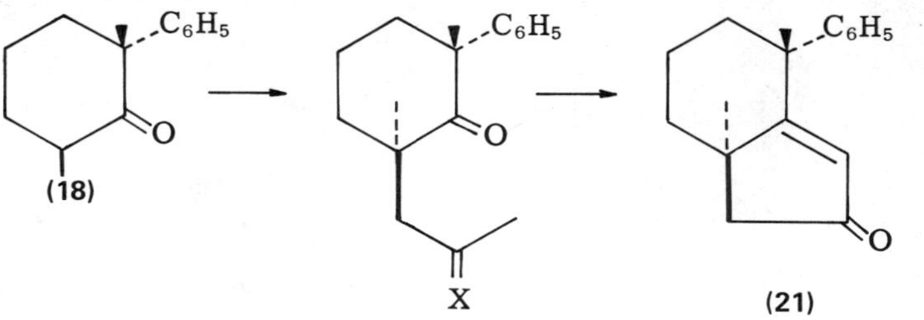

(18)

(19) X = CH_2
(20) X = O

(21)

The use of this method in the formation of complex furans is illustrated in the total synthesis of furopellargone (7). Elaboration of the aldehyde group of photocitral **(22)** leads by a series of transformations to the keto-diene **(23)** which contains one actual and two latent carbonyl groups. Ozonolysis gave the tricarbonyl compound **(24)** (an attempted cleavage with osmium tetroxide-periodate was in this case unsatisfactory); treatment of this product with acetic anhydride gave the furan **(25)**, cyclization taking place to afford the less crowded furan.

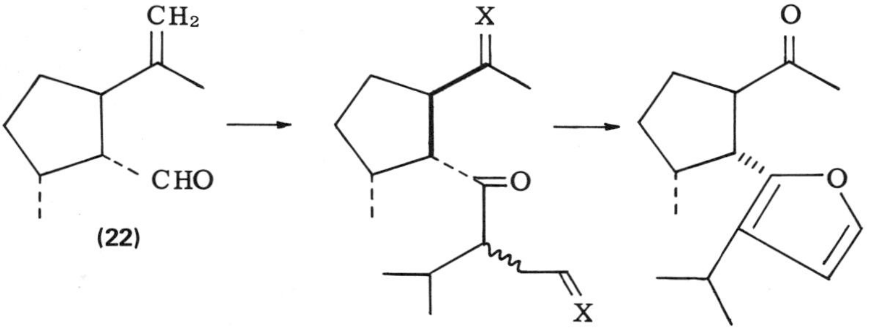

(22)

(23) X = CH_2
(24) X = O

(25)

Reaction of enol lactone **(26)** with the Grignard reagent from 2-methyl-5-bromopentene-1, followed by treatment with strong base, led to a steroid intermediate containing rings B, C, and D, and the requisite carbon atoms for ring A. Angular methylation of the enolate followed by ozonolysis and cyclization gave the tetracyclic ketone **(28)** (8).

2. Allyl Groups as Latent Acetic Acids

In an approach to the erythrina bases, the diallyl compound **(29)** (9) was oxidized with potassium permanganate—the accompanying oxidation to the lactam could not be circumvented. Esterification to **30** followed by acyloin condensation of the diester gave the desired product (10). The allyl group was

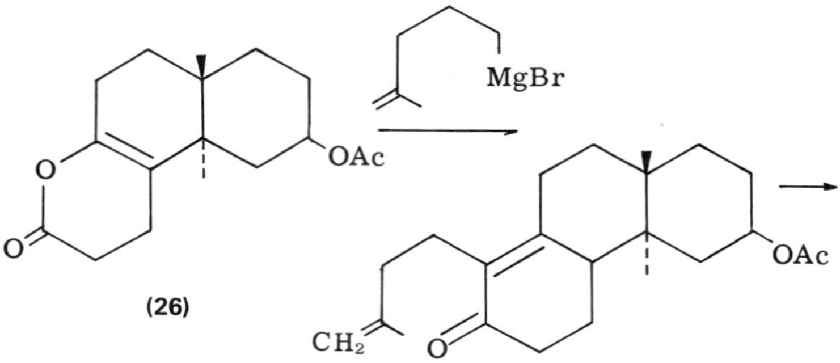

MgBr

OAc

(26)

OAc

CH_2

(27)

OAc

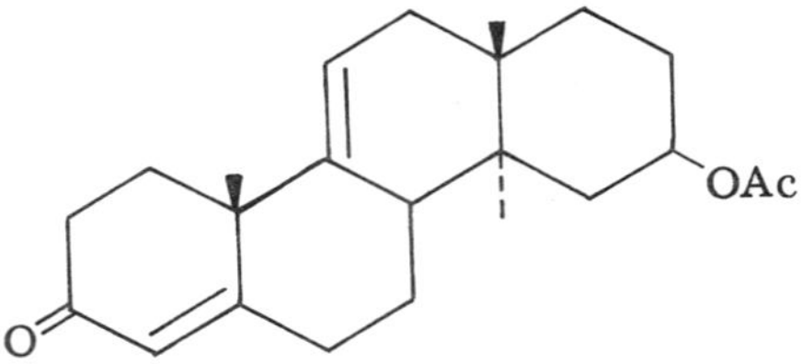

(28)

similarly used to introduce the acetic acid residue in the overall conversion of **32** to **33**. The olefin was in this case oxidized by means of osmium tetroxide/ periodic acid (11), followed by silver oxide.

CH_3O CH_3O N CH_2 CH_2

(29)

1. $KMnO_4$
2. CH_2N_2

CH_3O CH_3O N O CO_2CH_3 CO_2CH_3

(30)

Na / NH_3

CH_3O CH_3O N O O OH

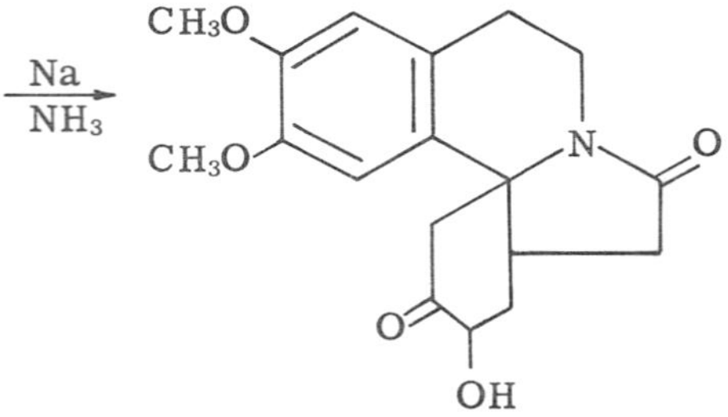

(31)

3. *The Allyl Group as a Latent Acetaldehyde: Formation of Bridged Bicyclic Rings*

The allylation-oxidation sequence has found particularly widespread use in the preparation of bridged-ring natural products. In broad outline a cyclohexanone, usually blocked at the more active position, is alkylated with an allyl halide. The blocking group is then removed and the allyl group oxidized to the aldehyde, usually by means of the catalytic osmium tetroxide/periodate procedure (12). The last ring is then closed with acid or base (eq.2).

This scheme was used to prepare the intermediate **35** in the total synthesis of veatchine and garryine (13,14), the intermediates **38** and **39** in the preparation of aconitine degradation products (15,16), and kaure-16-en-19-oic acid **(41)** (17). When applied to the keto-aldehyde **42** in a synthesis aimed at atisine, cyclization took place in the undesired sense to **43**. Blocking this position prior to the cleavage step gave the desired bicyclic skeleton (18). Subsequent work showed that the use of different cyclization conditions led to a mixture from which the ketol **(44)** could be obtained directly (19).

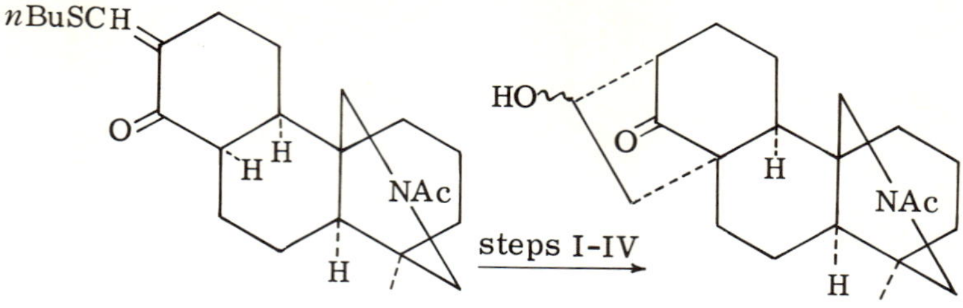

nBuSCH

O

H

H

NAc

H

steps I–IV

(34)

HO

O

H

NAc

H

(35)

OCH_3

R

O

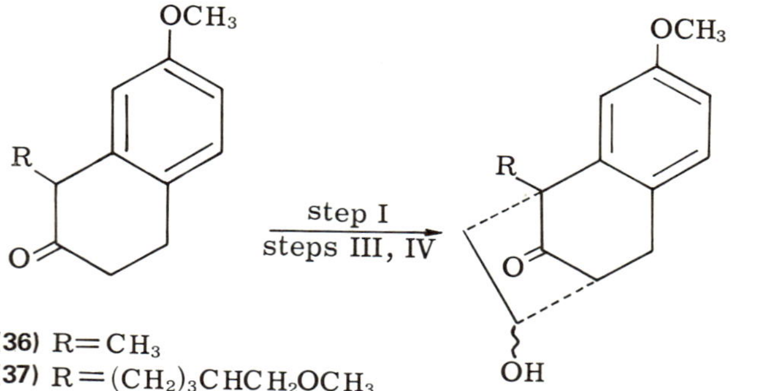

(36) $R=CH_3$

(37) $R=(CH_2)_3CHCH_2OCH_3$ (CH bearing OBz)

step I

steps III, IV

OCH_3

R

O

OH

(38) $R=CH_3$

(39) $R=(CH_2)_3CHCH_2OCH_3$ (CH bearing OBz)

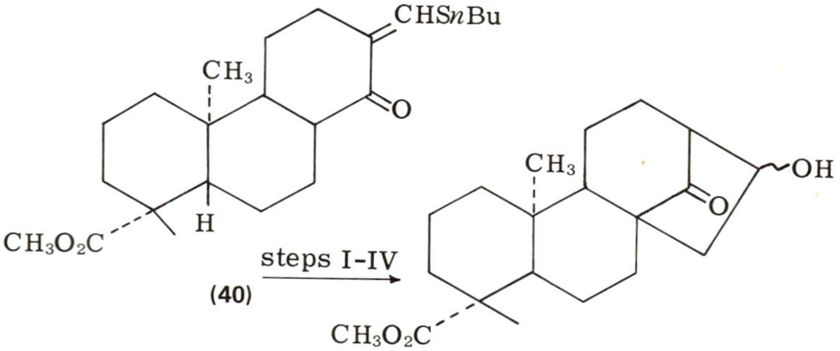

CHSnBu

CH_3

O

H

CH_3O_2C

steps I–IV

(40)

CH_3

OH

O

CH_3O_2C

(41)

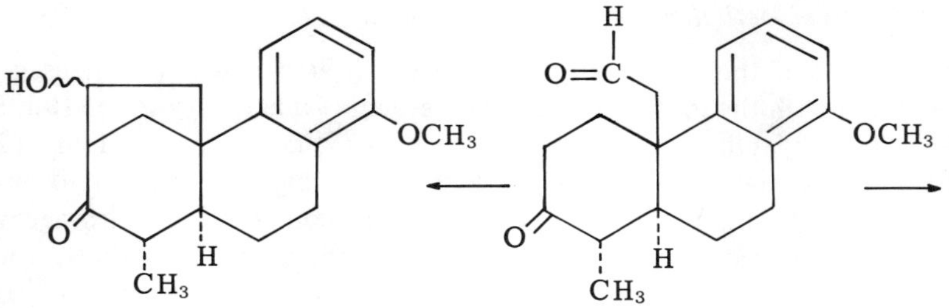

HO OCH$_3$ O CH$_3$ H (43) ← H O=C OCH$_3$ O H CH$_3$ (42) →

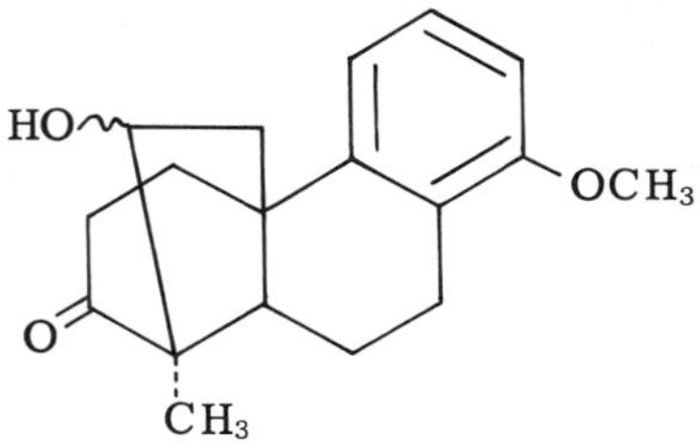

HO OCH$_3$ O CH$_3$ (44)

In an interesting variation on this theme, keto acid **45** was condensed with homoallylmagnesium bromide; the resulting hydroxy acid was lactonized to **46**. Oxidation to the aldehyde followed by condensation gave hydroxy lactone **(47a)**; this was then oxidized to the model for the A ring of giberellic acid **(47b)** (20).

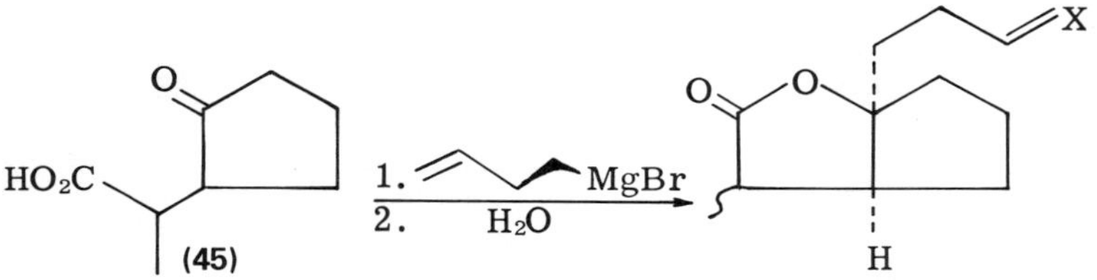

O HO_2C (45) 1. MgBr 2. H_2O → O O X H

(46a) X=CH_2
(46b) X=O

→ O O R

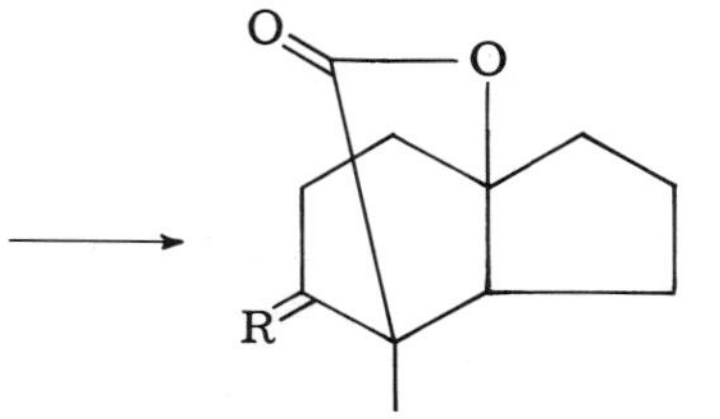

(47a) R=H, OH
(47b) R=O

4. The Terminal Methylene in Alkaloid Synthesis

In the field of the alkaloids, generation of an aldehyde has been used for condensation with the indole nitrogen. This scheme gained entry to the Hunteria alkaloids (and by subsequent rearrangement, to aspidospermia alkaloids) (21). Thus reaction of the ester-aldehyde **48** with tryptamine followed by hydroxylation by means of osmium tetroxide gave tetracyclic compound **49**. Cleavage with periodic acid formed the last ring (21,22) to afford **50**. It was later found possible to preform glycol **52** by oxidation of **51** with potassium permanganate. Condensation with tryptamine led to the diol **49** (23).

Finally, it was found that treatment of the glycol (**49**) with dicyclohexylcarbodiimide in dimethylsulfoxide led to the cyclized glycol (**53**) (22).

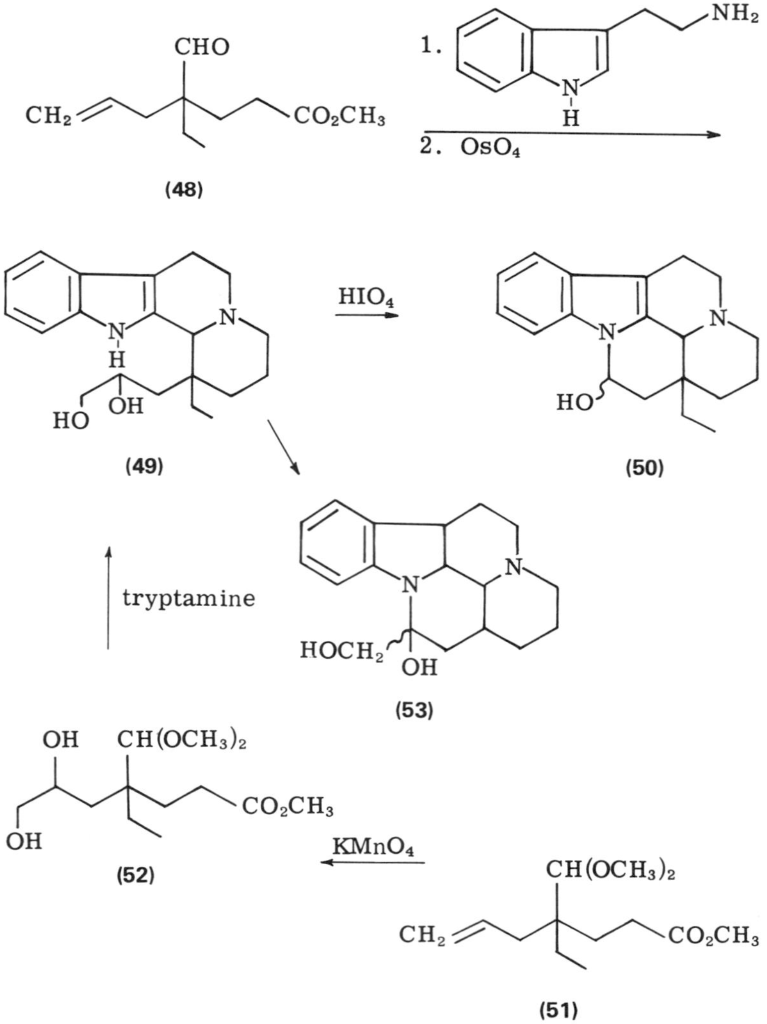

CHO CH2 CO2CH3 (48)

1. tryptamine (NH2, N-H) 2. OsO_4

(49) HO OH N-H N

HIO_4

(50) HO N N

(53) HOCH2 OH N N

tryptamine

(52) OH CH(OCH3)2 CO2CH3 OH

$KMnO_4$

(51) CH(OCH3)2 CH2 CO2CH3

In a related sequence, allylation of the ester **54** followed by oxidation gave the aldehyde **56**. Condensation with tryptamine gave the tricyclic lactam **57**, an intermediate in the synthesis of vinca alkaloids (24).

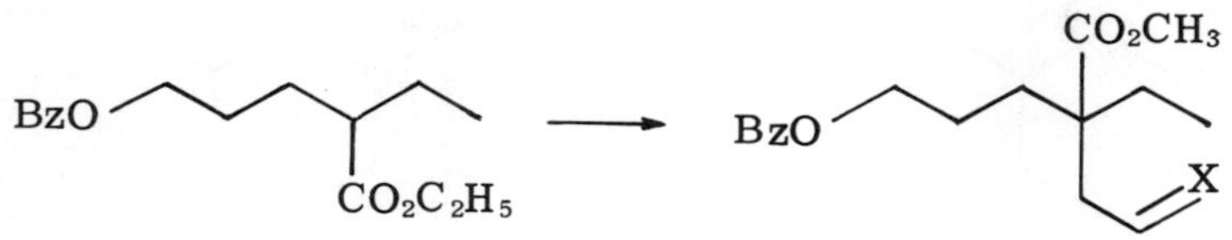

BzO ... $CO_2C_2H_5$ → BzO ... CO_2CH_3, X

(54) (55) X=CH_2
(56) X=O

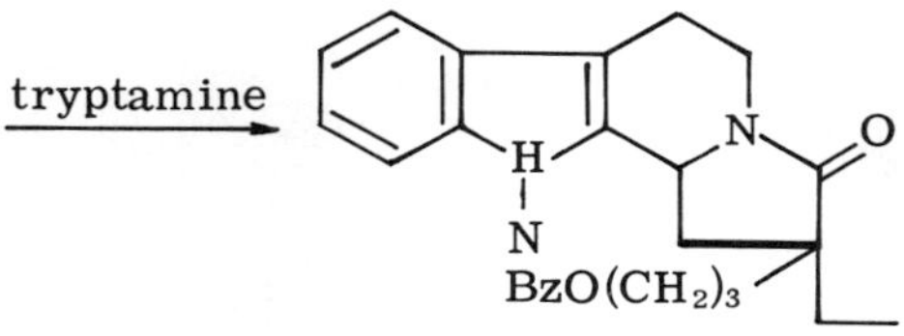

tryptamine → H, N, N, O, $BzO(CH_2)_3$

(57)

5. *Oxidation of Terminal Allyl Groups without Cleavage*

Several workers have used a modification of this general scheme in which all carbon atoms are retained. Thus the allyl alcohol **58** (obtained by reaction of homocamphor with allylmagnesium chloride) gave on hydroboration, followed by oxidation, the diol **59**. Selective dehydration of the tertiary alcohol followed by chromic acid oxidation gave the unsaturated acid **60**, used as an intermediate in the total synthesis of patchouli alcohol (25).

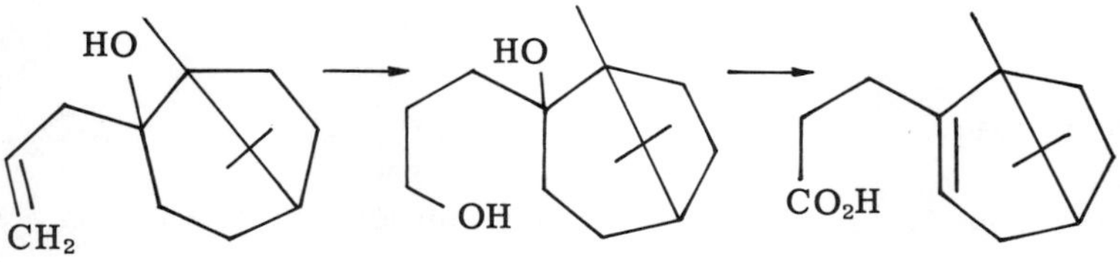

HO, CH_2 → HO, OH → CO_2H

(58) (59) (60)

In an analogous sequence, the tricyclic imminium salt **(61)** was condensed with methallylmagnesium chloride. Hydroboration followed by treatment with hydrogen peroxide gave the lycopodium alkaloid **(62)** (25a).

In an alternate scheme for the formation of bridged bicyclic systems such as that in **44**, the vinyl enone **(63a)** was first epoxidized to **63b**. Treatment with base effected intramolecular alkylation to **64**. The hydroxymethyl group was then removed in a series of steps which led ultimately to **(65)** (26).

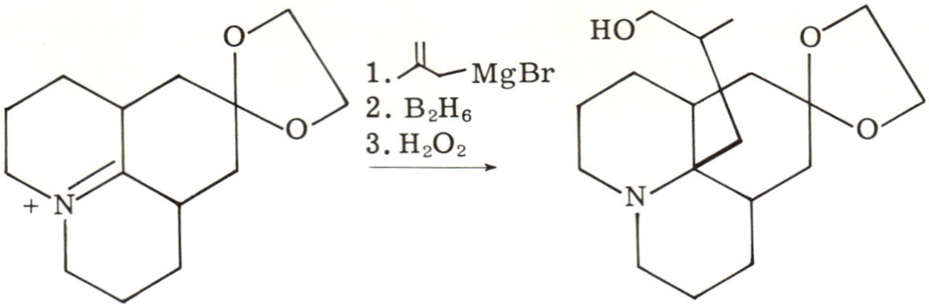

1. MgBr
2. B_2H_6
3. H_2O_2

(61) (62)

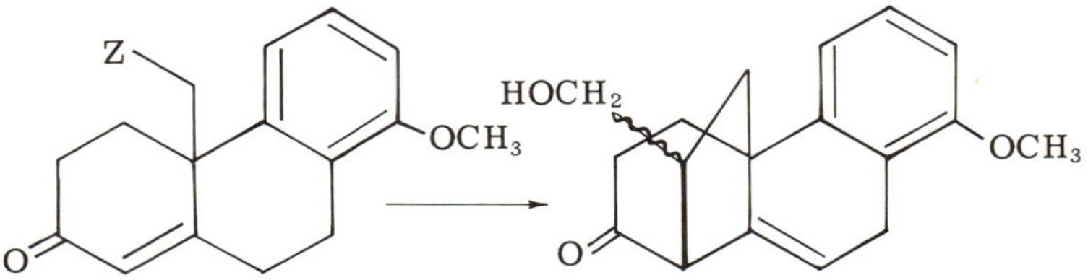

(63a) $Z{=}CH{=}CH_2$

(63b) $Z{=}CH{-}CH_2$ (epoxide, O)

(64)

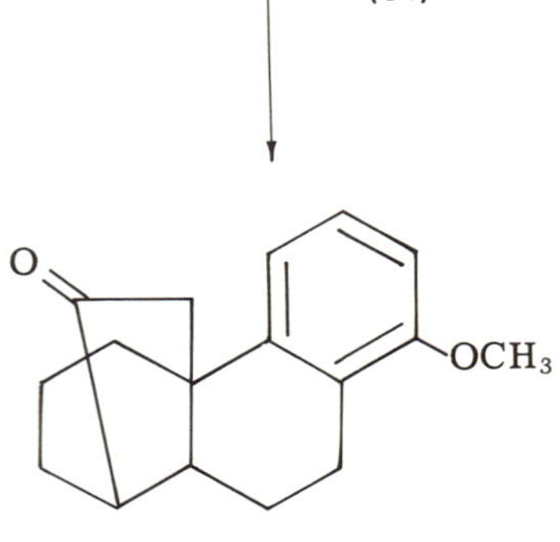

(65)

B. DICARBONYL COMPOUNDS FROM INTERNAL OLEFINS

1. Subsequent Cyclization to Carbocyclic Compounds

Internal olefins have found widespread use as latent carbonyl compounds for much the same reasons which commend the use of the terminal olefins. In addition, however, since many of the starting materials can be obtained by a Diels-Alder reaction, two centers of asymmetry in the final molecule can be readily manipulated. These centers may be maintained as obtained from the initial condensation (Scheme 1, eq. 3). It is often the case that the condensation product is more stable in the alternate configuration (e.g., if R^1 and R^2 are joined in a 6-membered ring); if one of the groups adjacent to the ring junction contains an enolizing function, equilibration of the adduct prior to cleavage (Scheme 2, eq. 3) will then give the other isomer.

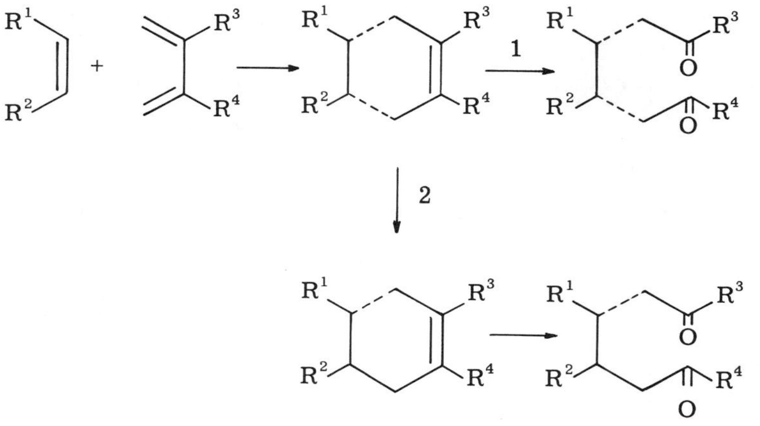

(3)

(a) The Cyclohexene to Cyclopentenecarboxaldehyde Sequence. In its simplest form, this scheme starts with a substituted cyclohexene; the double bond is hydroxylated by means of osmium tetroxide. In some cases the resulting glycol has been isolated and carried through several synthetic steps in either this form or protected as an acetonide. At a suitable point in the synthesis, the glycol is cleaved to the dicarbonyl compound with either periodate or lead tetraacetate. Alternately, the dicarbonyl compound can be obtained directly from the olefin by ozonization, or by cleavage with periodate containing a catalytic amount of osmium tetroxide (12). Both acid and base have been used to effect the cyclization step (eq. 4).

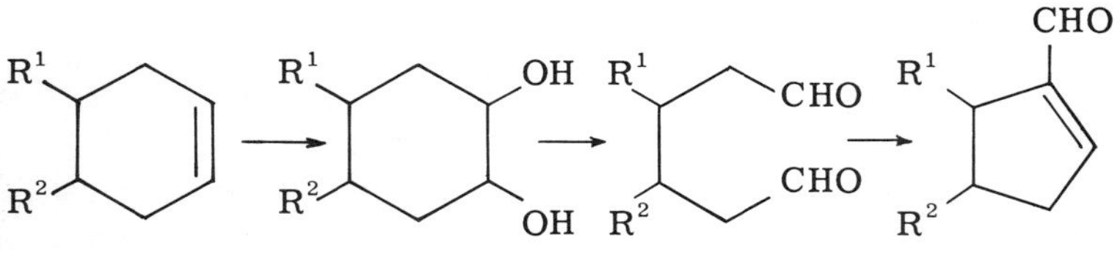

(4)

This sequence was crucial to the design for the total synthesis of cantharidin (27,28). Diels-Alder reaction of the oxide **(66)** (itself a product of the condensation of furan with dimethyl maleate) with butadiene established the *cis* stereochemistry at the quaternary carbon atoms. Subsequent steps were aimed at converting the carboxyl groups to methyls, and the cyclohexene bridge to a carboxylic anhydride. Thus **67** was converted by a series of transformations to the bisthioether **(68)**; hydroxylation and then desulfurization followed by glycol cleavage led to the aldehyde (**70**). This last was then cyclized to the cyclopentenecarboxaldehyde (**71**); further oxidation led to the target molecule.

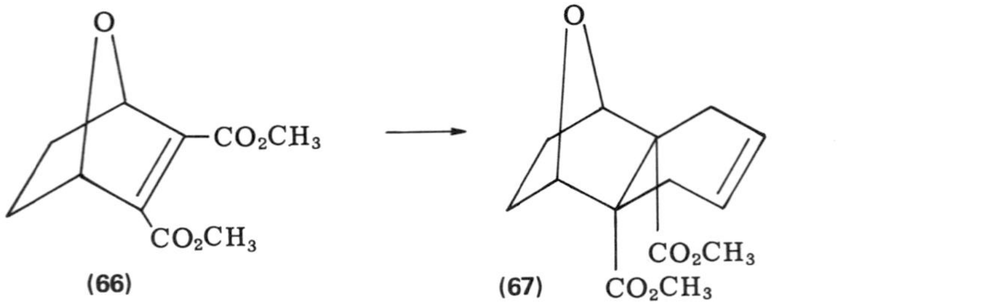

CO_2CH_3
CO_2CH_3
(66)
CO_2CH_3
CO_2CH_3
(67) 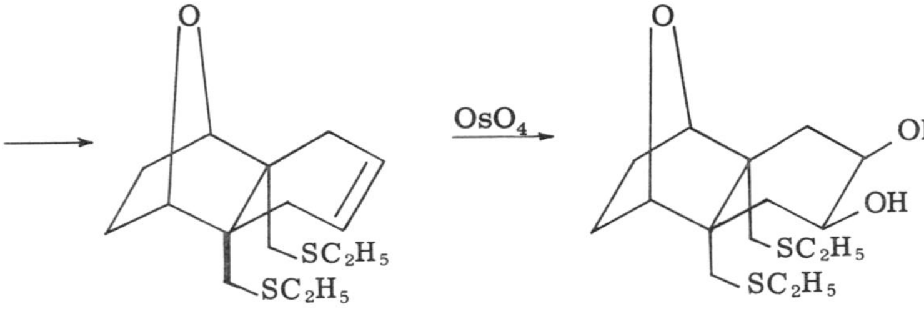
OsO_4
OH
OH
SC_2H_5
SC_2H_5
(68)
SC_2H_5
SC_2H_5
(69) 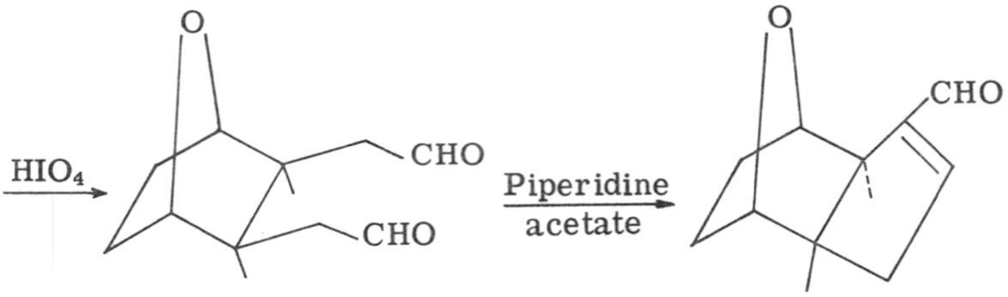
HIO_4
CHO
CHO
Piperidine acetate
CHO
(70)
(71)

The fact that a *cis* hydrindane is more stable than the *trans* counterpart adds a measure of complication to the total synthesis of steroids since the C-D ring junction in those compounds is *trans*. Many of the classical total synthesis, however, have taken advantage of the fact that the situation is reversed in reduced naphthalene derivatives. Thus equilibration of the Diels-Alder adduct (**72a**) leads to the more stable *trans* isomer (**72b**). This was then transformed through a series of reactions to the tricyclic intermediate (**73**). The electron-rich isolated double bond was selectively hydroxylated and the resulting glycol protected as the acetonide (**74**). This in turn was elaborated to the glycol (**75**). Application of the ring contraction sequence led to the steroid (**76**) (29).

In parallel work resulting in the total synthesis of cortisone, hydroxylation was postponed until the tetracyclic compound (**77**) had been reached—the stereochemical arguments are the same as those above. Hydroxylation of that compound was found to proceed selectively to **75** (30).

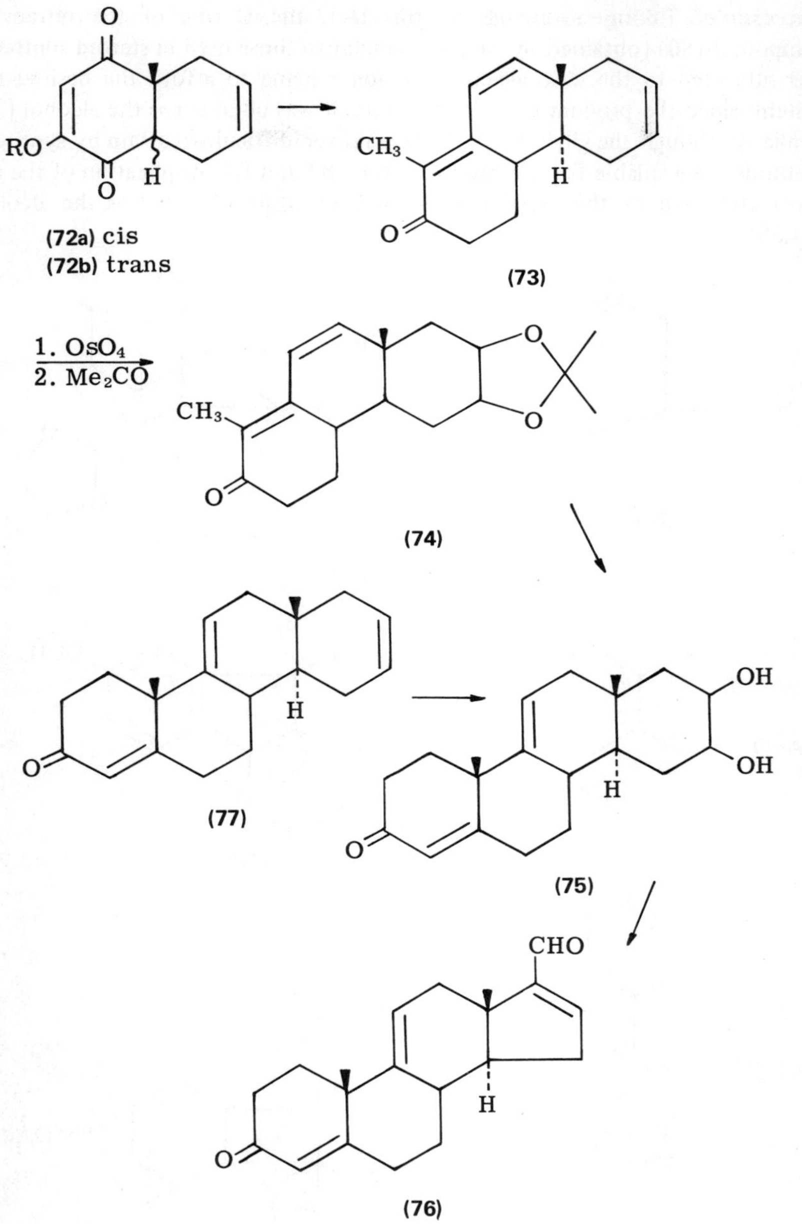

(72a) cis
(72b) trans

(73)

1. OsO_4
2. Me_2CO

(74)

(77)

(75)

(76)

Application of the ring contraction to the bridged bicyclic derivative **(78)** of carvone led to the carbon skeleton **(79)** of helminthosporal in a total synthesis of that compound (31,32). Though a host of schemes have been elaborated for the total synthesis of steroids, the veratrum alkaloid nucleus is relatively inaccessible. Taking advantage of this fact, the C ring of the tetracyclic compound (**80**) (obtained by methods similar to those used in steroid synthesis) was subjected to the cleavage-condensation scheme to afford the desired ring system; since the product cannot dehydrate, it was obtained as the alcohol (33). Similarly, though the giberellic acids have proven difficult to attain by synthesis, methods are available for the preparation of **82** and **83**. Application of the ring contraction led to the 7α-gibanes **84** and **85**, again obtained as the alcohols (34,35).

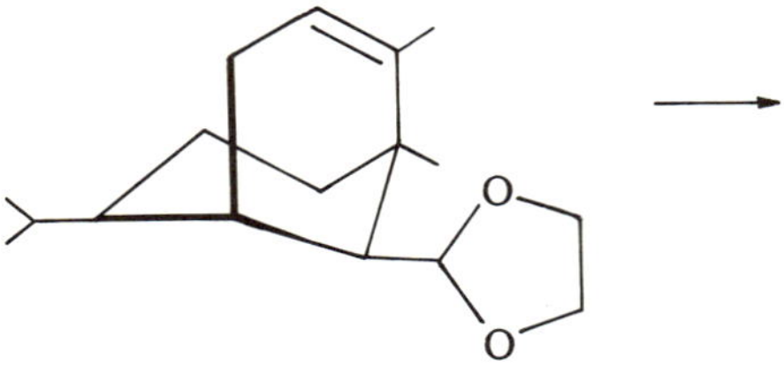

CHO

(**78**)

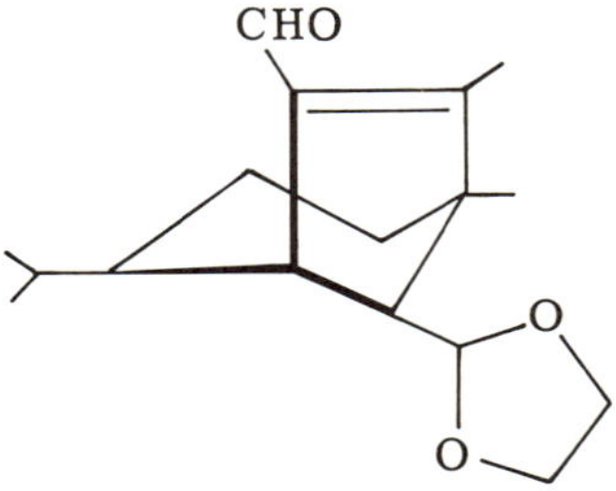

(**79**)

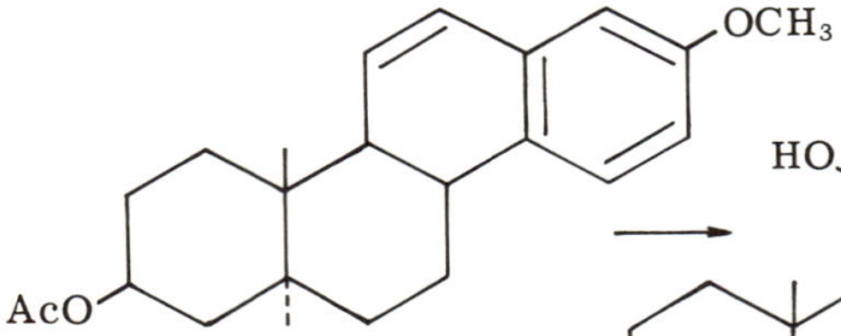

OCH_3

HO OCH_3

AcO

CHO

(**80**)

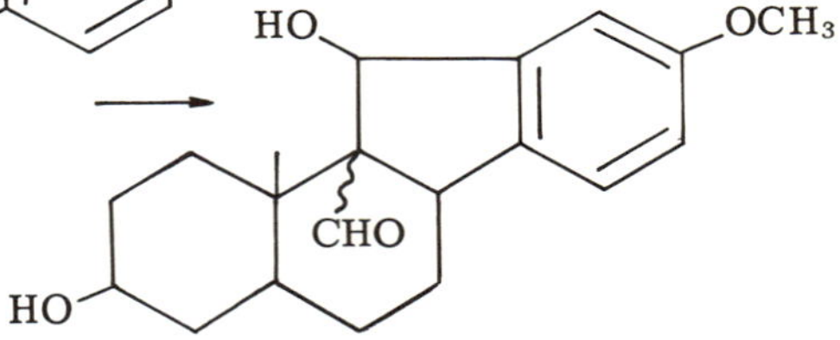

HO

(**81**)

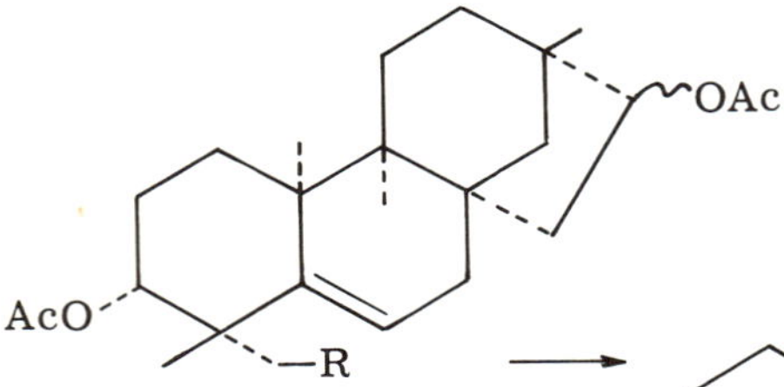

OAc

AcO

R

(**82**) R=H
(**83**) R=OAc

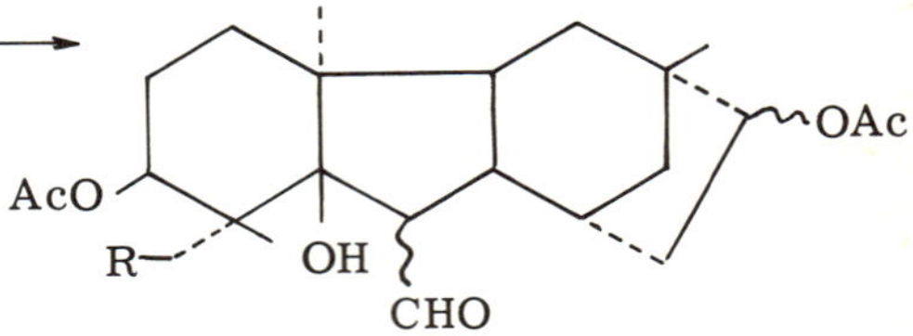

OAc

AcO

R OH

CHO

(**84**) R=H
(**85**) R=OAc

In the course of the total synthesis of the natural product genipin, diene **86** was oxidized to the tetrahydroxy compound; selective opening of the 5-membered ring followed by lactonization gave the diol **(87)**. Treatment of the diol with periodic acid followed by piperidine acetate gave the desired intermediate **(88)** (36). In the synthesis of the related compound fulvoplumierin, diene **89** was selectively hydroxylated. The diol was carried as such through several stages in which the oxygenated ring was elaborated. Application of the ring contraction scheme to the diol **(90)** gave the desired fulvene **(91)** (37).

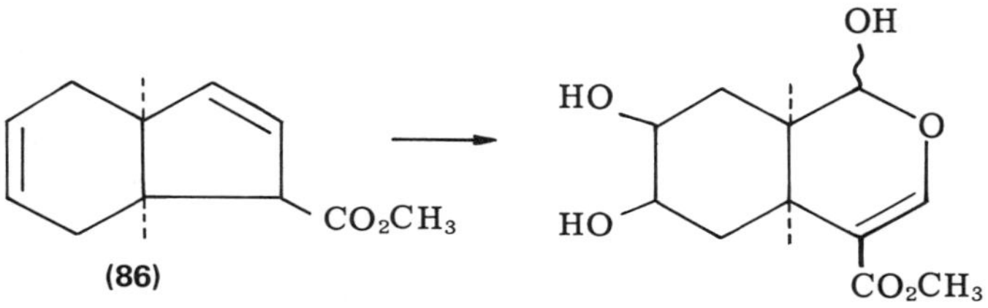

CO_2CH_3 HO OH O HO CO_2CH_3

(86) **(87)**

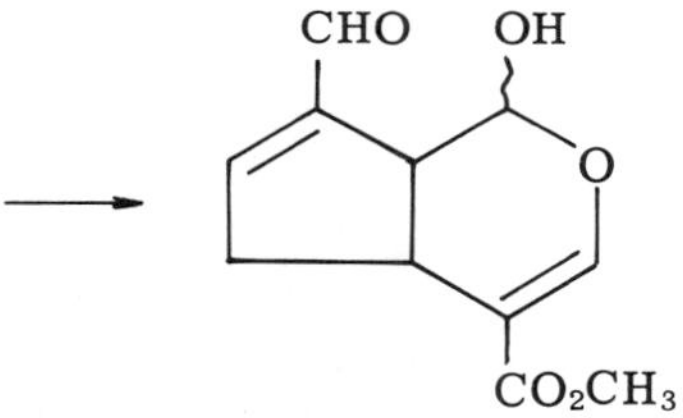

CHO OH O CO_2CH_3

(88)

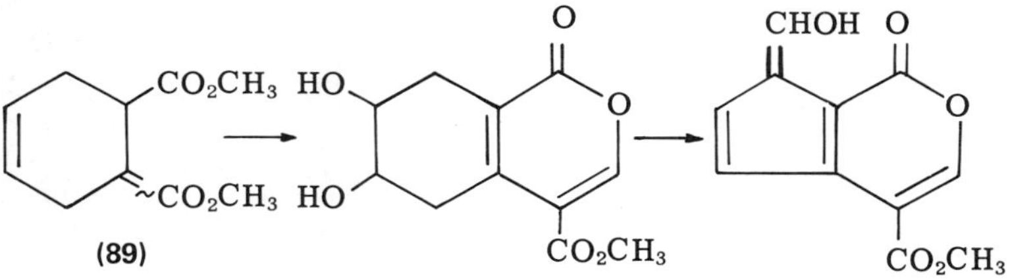

CO_2CH_3 HO O CHOH O O CO_2CH_3 HO

(89) CO_2CH_3 CO_2CH_3

(90) **(91)**

In the total synthesis of tetracyclines, tricyclic compound **92** was prepared in order to fix the stereochemistry of the oxygen-bearing carbon atoms. In order to proceed with the synthesis, two carbons of the cyclohexene ring needed to be eliminated. Application of the ring contraction sequence led to the cyclopentene **(93)**; this was then again opened by oxidation. Basic cleavage of the β-diketone served to shed the superfluous carbon atoms (38,39).

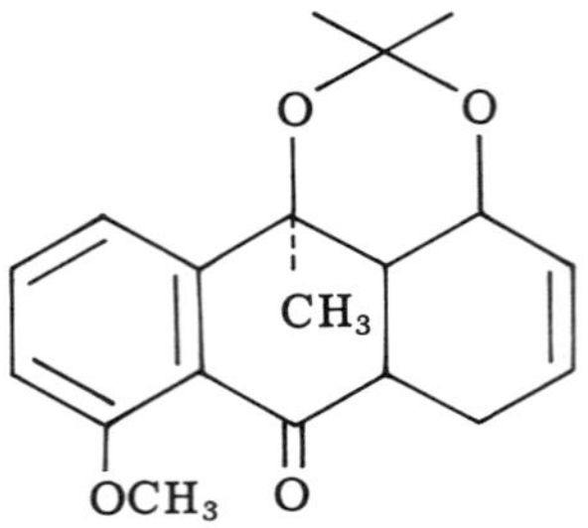

(b) The Methylcyclohexene to Acetylcyclopentene Sequence. In a modification of the above ring contraction, one carbon of the olefin bears a methyl group. Cleavage of the double bond will thus result in the formation of a methyl ketone from one terminal of the double bond. Cyclization as above leads to conjugated acetyl-cyclopentenes, or in those cases where a 6-membered ring results, a cyclohexenone (eq. 5).

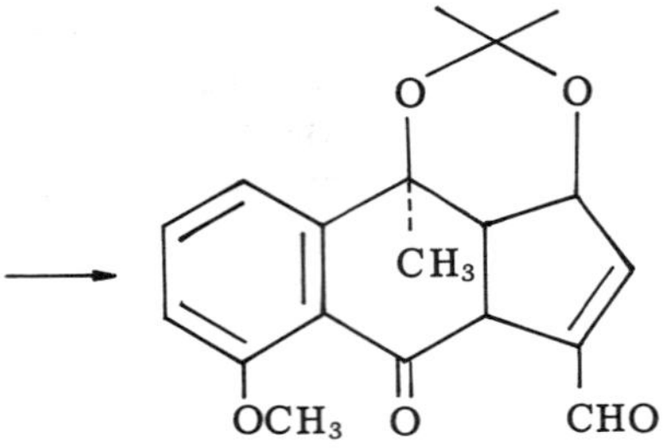

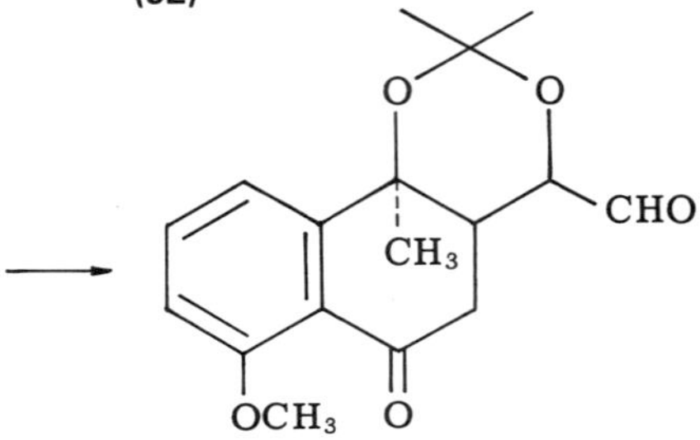

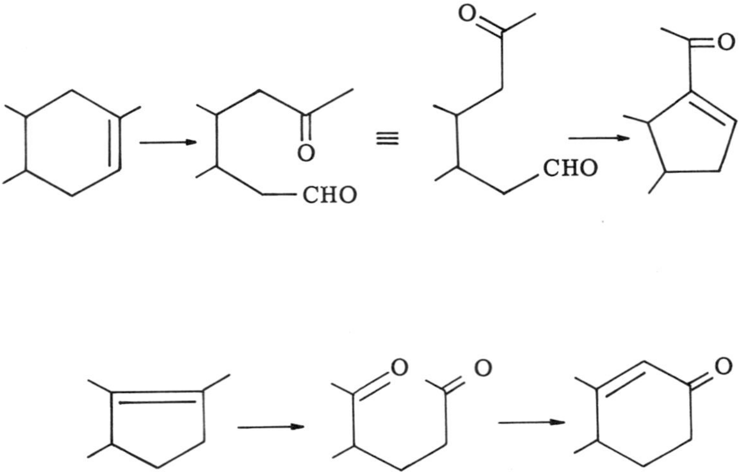

The preparation of 18,19-bisnorprogesterone from estradiol methyl ether illustrates both these processes. Rearrangement of the starting material affords the olefin **(97)**. Ozonolysis followed by cyclization leads to the enone **(98)**. This was converted to olefin **(99)** by consecutive reduction, addition of methylmagnesium bromide, and dehydration. A second application of the olefin cleavage-condensation sequence afforded the methyl ketone **(100)** (40,41).

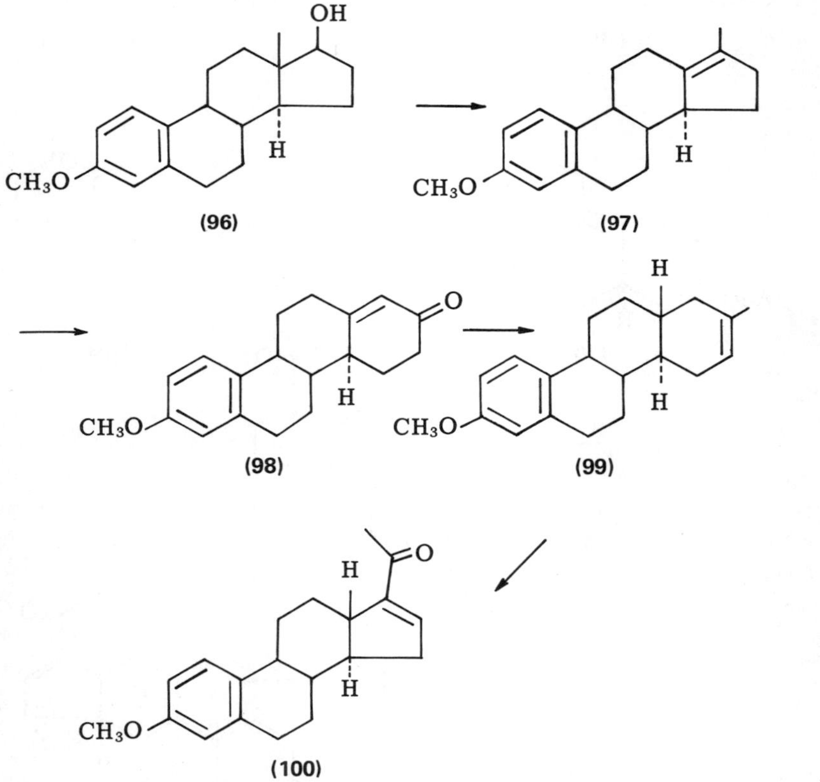

(96) **(97)**

(98) **(99)**

(100)

A similar sequence was used in the preparation of the intermediate **(102)** for the total synthesis of conessine; the carboxyl group was ultimately transformed to the nitrogen-bearing angular methyl group of the natural product (42). In another approach to conessine the tetracyclic intermediate **(103)** was subjected to the contraction sequence; the product contained an enone system which readily added cyanide in a Michael reaction; this nitrile group was eventually converted to the nitrogen-substituted angular methyl group (43). The same intermediate was also used for a progesterone synthesis (44). The product of a biogenetic-type polyolefin cyclization **(106)** was converted to 16-dehydroprogesterone by simultaneous cleavage-reclosure of both rings A and D (45).

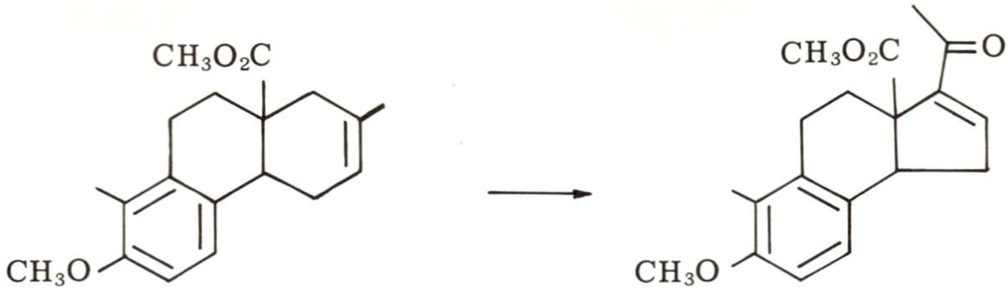

CH_3O_2C CH_3O **(101)** → CH_3O_2C O CH_3O **(102)**

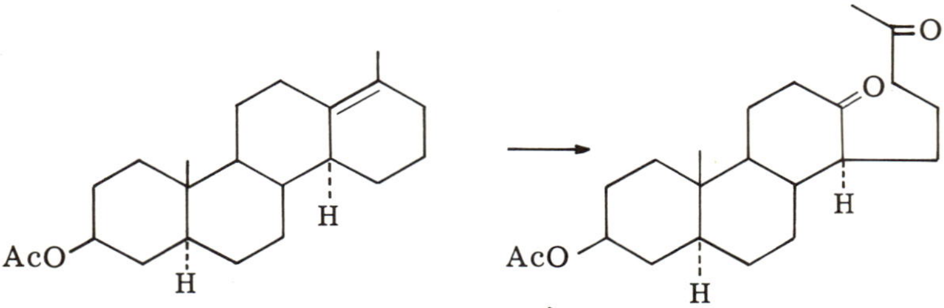

AcO H H **(103)** → O O AcO H H **(104)**

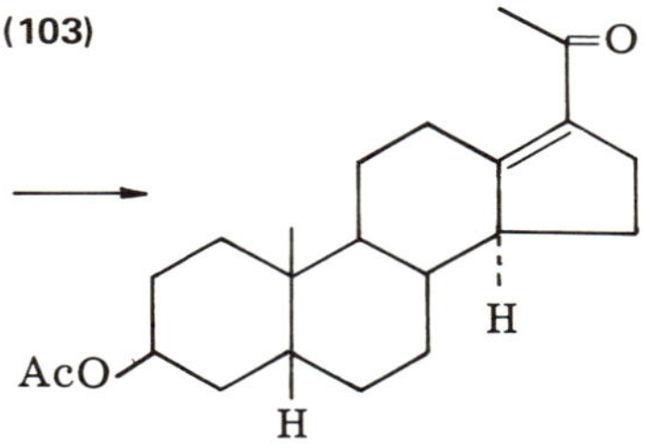

→ O AcO H H **(105)**

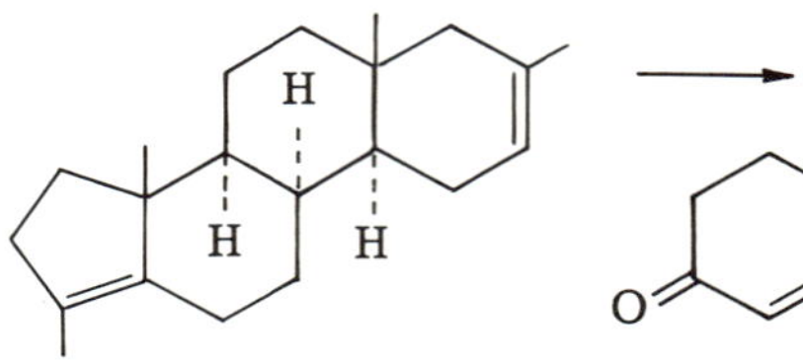

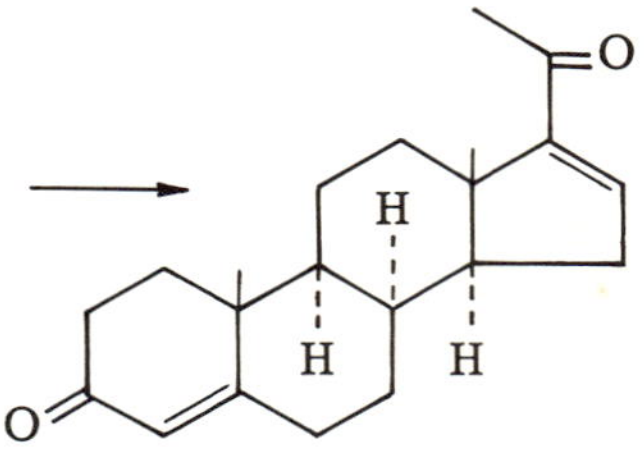

H H H **(106)** → O H O H H **(107)**

(c) Miscellaneous Cleavages. In the total synthesis of α-onocerin, the tricyclic enone **(108)** was used to establish the substitution pattern and stereochemistry of the two-ring unit. Ozonization cleaved the enone ring to give the keto-acid. The acid side-chain was then shortened by one carbon atom and that product **(109)** coupled by electrolysis to an α-onocerin degradation product (46,47). In an interesting method for "reversing" the stereochemistry at an angular carbon atom, the acid **(110)** was cleaved by ozonization and converted to the ester anhydride **(111)**; Dieckmann condensation in this case occurs preferentially to give, after hydrolysis and decarboxylation, **112**, with a net "inversion" (48).

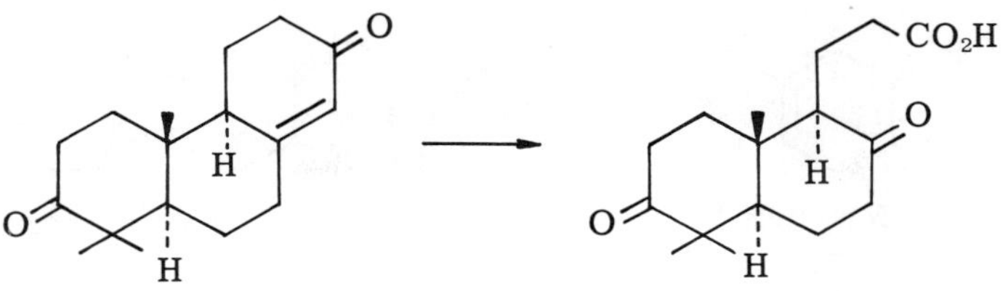

(108) (109)

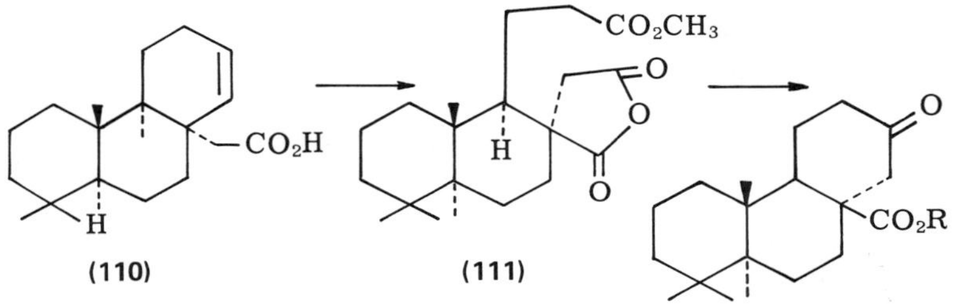

(110) (111) (112)

In a variation on the cleavage-condensation theme, the glycol **(113a)** from hydroxylation of the corresponding olefin was isolated and converted to the methanesulfonate **(113b)**. Treatment of this with strong base led to ring contraction to afford in this case the saturated cyclopentane ring in **114** (49). Recently in one of a series of prostaglandin syntheses, the ring contraction was used to good advantage. Olefin **115**, elaborated from what was originally a Diels-Alder adduct, was cleaved with osmium tetroxide followed by lead tetraacetate **(116)**; cyclization was accomplished by treatment with diazabicyclononane. It is of note that the aldol reaction in this case stops at the ketol stage **(117)**, necessary in the prostaglandins (50).

In the recently reported preparation of the boll weevil pheromone, the methylcyclohexenone:ethylene adduct **(118)** was elaborated to allyl alcohol **(119)** by a series of straightforward reactions. Oxidative cleavage by means of osmium tetroxide/periodate gave keto acid **(120)** which was then taken on to the natural product (51). It is of note that the first-formed hydroxyaldehyde undergoes further cleavage with the reagent with loss of a carbon atom.

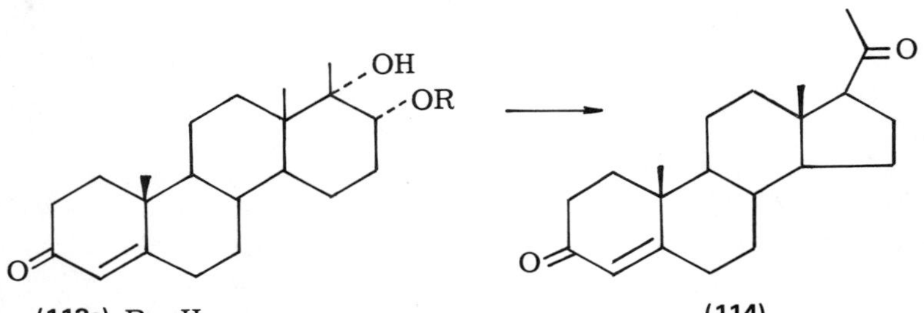

(113a) R=H
(113b) $R{=}SO_2CH_3$

(114)

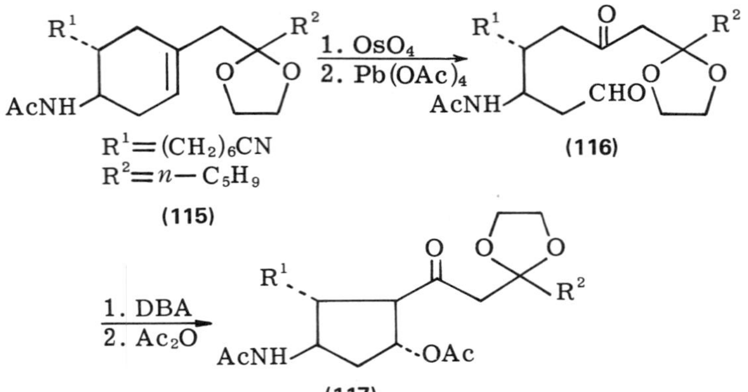

$R^1 = (CH_2)_6CN$
$R^2 = n-C_5H_9$

(115) → 1. OsO_4 2. $Pb(OAc)_4$ → (116)

1. DBA 2. Ac_2O → (117)

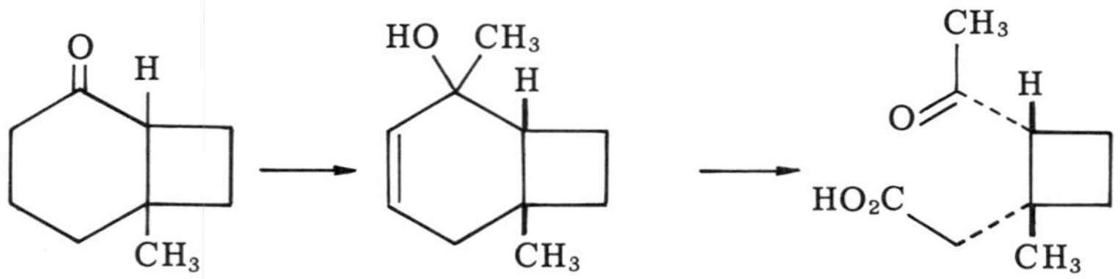

(118) → (119) → (120)

2. *Subsequent Cyclization to Heterocycles*

The ease with which aldehydes react with amines (see also Section II.A.4) makes these attractive partners in the formation of piperidine rings. This very reactivity, however, makes it mandatory that the aldehyde be carried through the synthesis in either protected or latent form. In practice this has been accomplished by generation of the reactive intermediate from either a cyclohexenone (eq. 6) or a cyclopentene (eq. 7). In the case of the latter only one of the carbonyls may be used in the ring formation, the other being reserved for some later use.

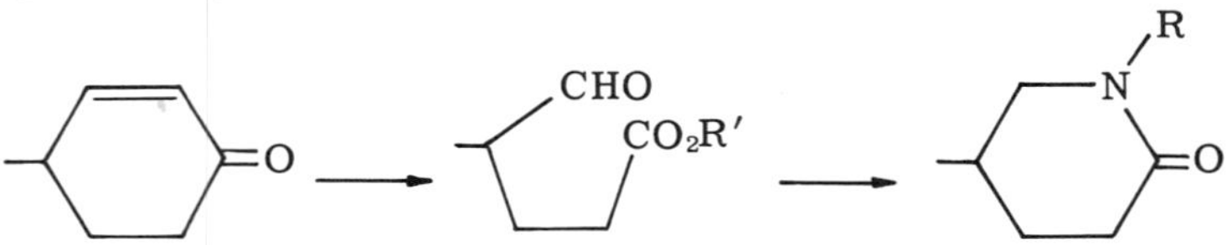

(6)

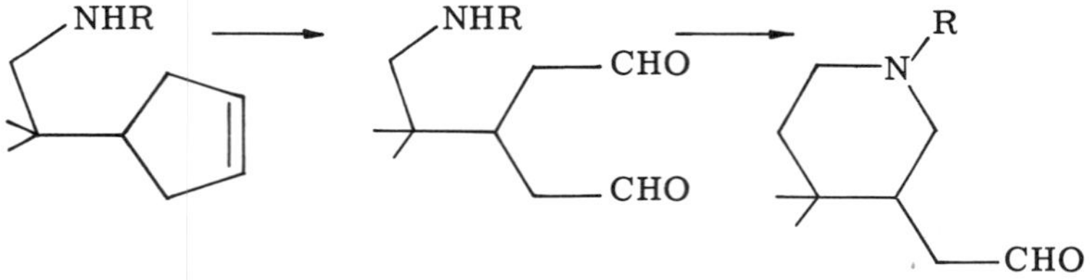

(7)

Thus Woodward's total synthesis of reserpine started with the Diels-Alder adduct **121**. In a series of elegant steps the proper functions of the future E ring were introduced in their proper steric relationships. Particular care was taken to preserve the *cis* ring junction introduced in the Diels-Alder reaction since this was to be the *cis* D-E junction in the product. Hydroxylation with osmium tetroxide followed by cleavage of the glycol and then esterification gave the ester aldehyde **(123)**. Condensation with tryptamine led to piperidone **(124)**; acid-catalyzed cyclization gave the pentacyclic indole **(125)**, which now has all but one of the asymmetric centers of reserpine in place (52-54).

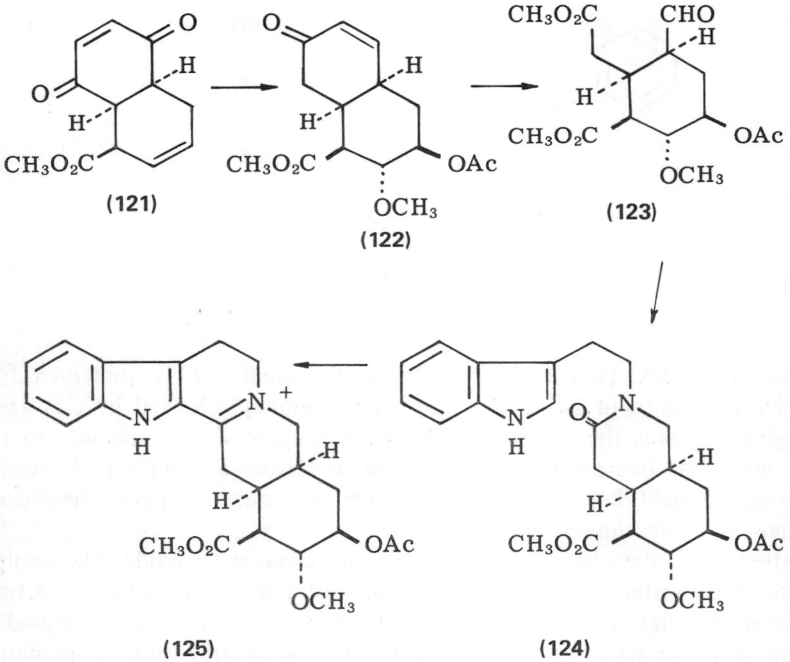

(121) **(122)** **(123)**

(125) **(124)**

The D-E ring fusion of yohimbine, on the other hand, has the *trans* configuration. Thus, in an approach similar to that employed for reserpine, a compound derived from a Diels-Alder adduct was chosen as starting material, in this case as the more stable *trans* isomer **(126)**. This ester was reacted with tryptamine to give amide **127**. Ring opening with osmium tetroxide followed by periodate and then cyclization gave the tricyclic compound **(128)**. Treatment with acid closed the remaining ring; a further series of reactions was used to shorten the acetaldehyde chain to a carboxylic acid and thus afford pseudoyohimbine (55).

The 3-cyclopentenyl grouping has proven very useful as a latent glutardialdehyde grouping. This difunctional aldehyde in conjunction with basic derivatives of indoles in various stages of reduction has led to some complex polycyclic

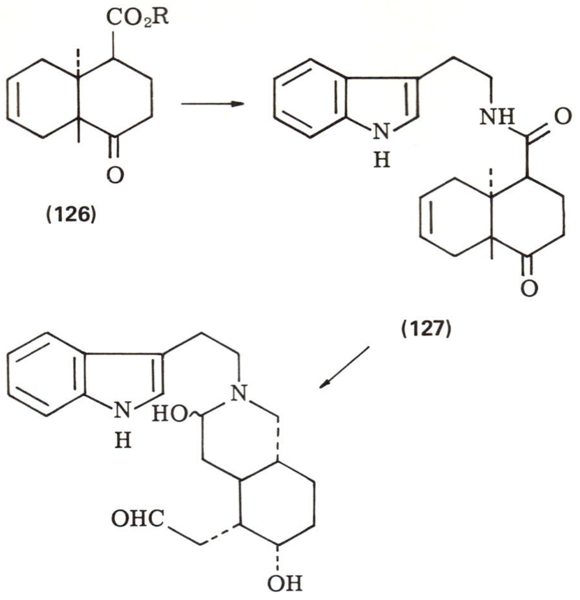

(126)

(127)

(128)

natural products. Thus hydroxylation of the amide **(129)** [obtained from indoleacetic acid and 2-(3-cyclopentenyl)butylamine] followed by cleavage of the glycol affords the dialdehyde **(130)** which cyclizes spontaneously to **131**. The latter on treatment with mild base undergoes a complicated series of reactions to yield directly a polycyclic compound which "contains the essential framework of strychnine" (56).

Alternately, the olefin is first treated with osmium tetroxide. The resulting glycol is converted to the oxindole amine by a series of standard reactions. Cleavage of that product affords the dialdehyde **(132)**. This compound on cyclization with acid and then reduction with borohydride leads to the natural product rhyncophyllol **(133)** (57).

A similar set of transformations was employed in the initial steps of the recently reported total synthesis of ajmaline. Cleavage of the cyclopentene ring of the aminoalcohol **(134)** with osmium tetroxide and periodate gave a dialdehyde which cyclized spontaneously to the piperidine **(135)**. Treatment of this with acid gave the tetracyclic intermediate **(136)** (58).

In one of the earlier approaches to the veatchine and garryine alkaloids, the heterocyclic rings were generated by means of dialdehydes. A series of these compounds **(138)** was obtained from the cleavage of the corresponding bridged olefins. It is of note that ammonia failed to accomplish the desired reaction **(139 →140)**, probably because of the crowding at the reaction site. It was, however, found that reduction of the bisoxime over platinum in acetic acid gave the bridged piperidines (14,59).

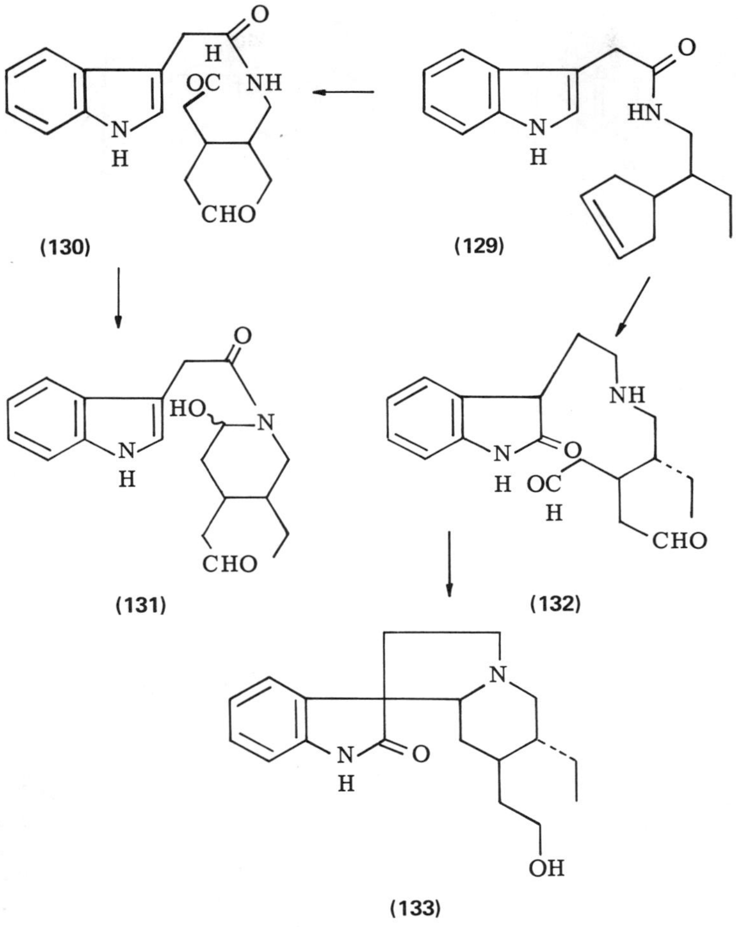

(130) (129)

(131) (132)

(133)

C. ESTERS BY CLEAVAGE OF ETHERS OF PHENOLS AND DIHYDROPHENOLS

As the concept of the inviolability of the aromatic ring slowly died, the use of ethers of phenols as sources of six carbon atoms became established. This general method has the particular virtue of carrying these carbon atoms through a sequence of steps in very nonreactive form. A fuller discussion of this method as applied to derivatives of cyclohexane will be found in Section V. We are presently concerned only with those few cases where the aromatic ring is actually cleaved in the course of the synthesis.

The first and perhaps most elegant use of this method comes from the Woodward group's synthesis of strychnine. A crucial aspect of this synthesis was the elaboration of the spirane ring in the 3 position of the indole at an early stage. Since the 2 position is the more reactive, it was necessary to block that position. Rather than use some grouping which would have to be removed at

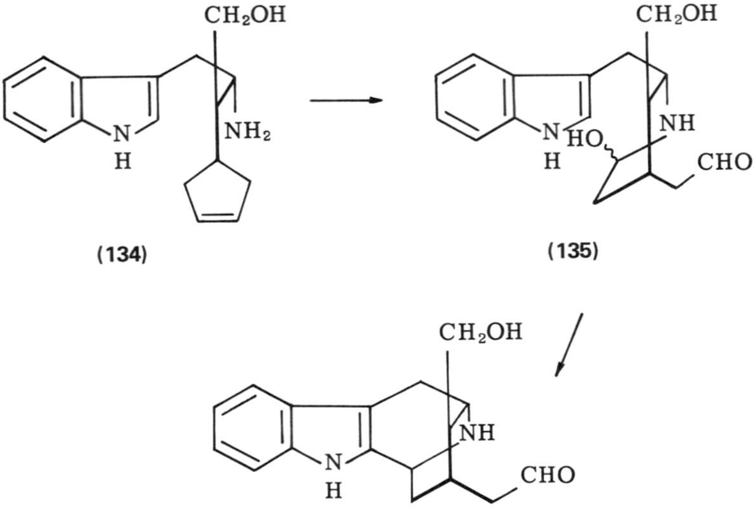

CH_2OH NH$_2$ N H

(134)

CH_2OH NH CHO N H HO

(135)

CH_2OH NH CHO N H

(136)

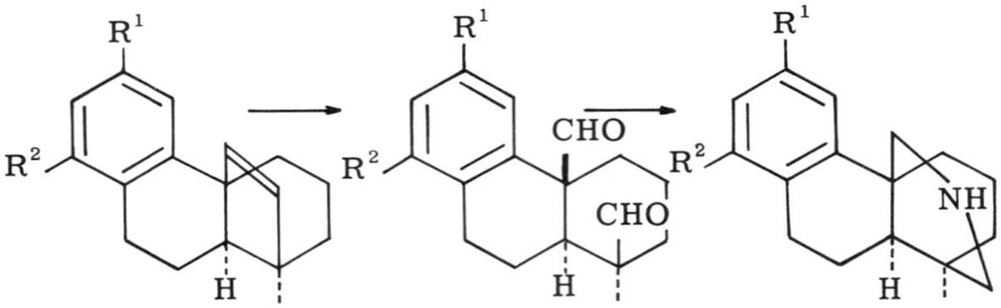

R^1 R^2 H

$R^1 = R^2 = H$
$R^1 = OCH_3$, $R^2 = H$
$R^1 = H$, $R^2 = OCH_3$

(138)

R^1 R^2 CHO CHO H

(139)

R^1 R^2 NH H

(140)

some later stage in the preparation, these workers employed the veratrole ring as the blocking group, choosing to regard this as a latent hexadienedicarboxylic acid. Thus the veratryl indole **(141)** was elaborated to the indoline **(142)**. Ozonization afforded on work-up the diester **(143)**. Three of the aromatic ring's carbon atoms were now utilized in the formation of a new ring; heating the compound with acid gave the lactam **(144)**. Following an additional series of reactions, the remaining atoms of the veratrole ring were used in a Dieckmann condensation to form yet another new ring to yield **145** (60,61).

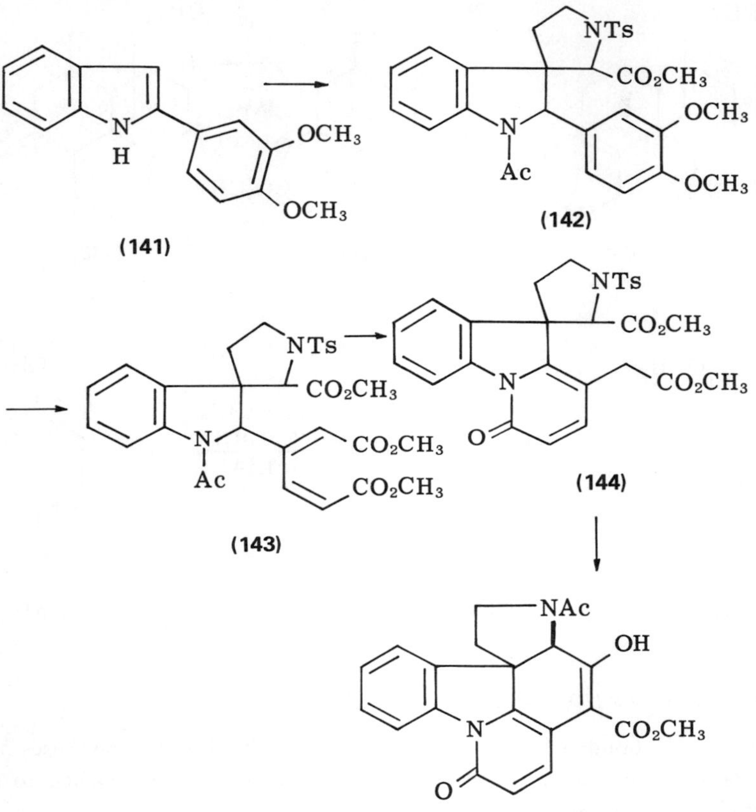

(141)

(142)

(143)

(144)

(145)

In a recent total synthesis of lycopodine, the dihydroanisole (**146**) was obtained from the corresponding aromatic compound by reduction with lithium in ammonia followed by isomerization to the conjugated diene with strong base. Ozonization of this diene followed by oxidative cleavage of the carbonyl of the resulting unsaturated aldehyde with selenium dioxide and hydrogen peroxide gave the keto ester (**147**). At this time all but one of the carbon atoms obtained from the aromatic ring were utilized in forming the final ring necessary for lycopodine (62) by cyclization to **148**.

A 7-membered moiety with stringent requirements for stereochemistry about the double bond was required as an intermediate in a recent elegant total synthesis of the cecropia juvenile hormone. In order to obtain this compound, the workers called into play the cisoid nature of cyclohexenes. Ozonization of the Birch reduction product (**149**) went preferentially at the enol-ether double bond. The aldehyde-ester (**150**) was then taken on to the homoallylic alcohol (**151**) (63).

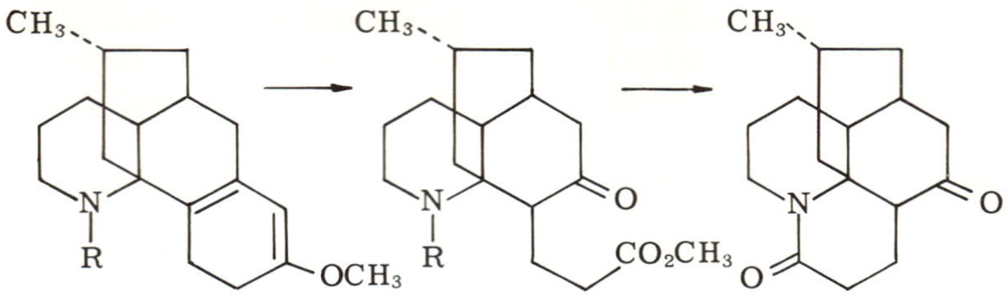

CH_3 CH_3 CH_3

N N N O O

R OCH_3 R CO_2CH_3 O

(146) (147) (148)

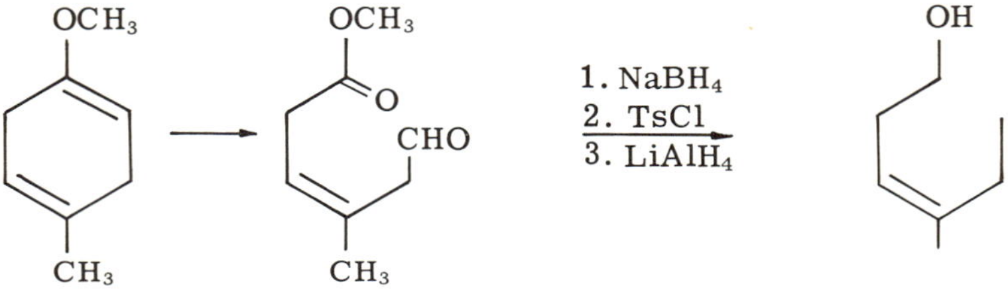

OCH_3 OCH_3 OH

O CHO

1. $NaBH_4$
2. TsCl
3. $LiAlH_4$

CH_3 CH_3

(149) (150) (151)

D. NITRENE INSERTION IN DOUBLE BONDS

The double bond has been used several times in complex syntheses as the acceptor of a nitrene to form an aziridine which was then opened to some bridged bicyclic ring system.

This general scheme was exploited in an approach to the delphinine-type diterpene alkaloids. Addition of benzenesulfonyl azide to the olefin **(152)** (obtained from addition of benzyne to the cyclopentadienecarboxylic ester) gave the bridged aziridine **(153)**. Reaction of this with water results in opening of the aziridine ring with concomitant rearrangement of the bicycloheptene ring to give **154**. The net result is the apparent shift of the amine function to the methylene bridge (63,64).

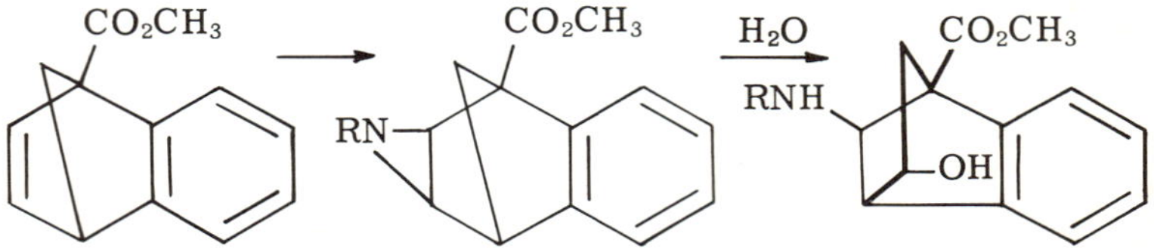

CO_2CH_3 CO_2CH_3 H_2O CO_2CH_3

RN RNH OH

$R{=}C_6H_5SO_2$

(152) (153) (154)

In one of the recent syntheses of the iboga alkaloids, internal nitrene addition was used as a link in a sequence leading to the preparation of the isoquinuclidine moiety. Thus treatment of the primary amines (**155**) with lead tetraacetate leads, via the nitrene, to the bridged aziridines (**156**). These aziridines on opening with indoleacetic anhydride afford the quinuclidine intermediates (**157**) as their indoleacetate esters. These were elaborated to desethylibogamine (65), ibogamine, and *epi*-ibogamine (66).

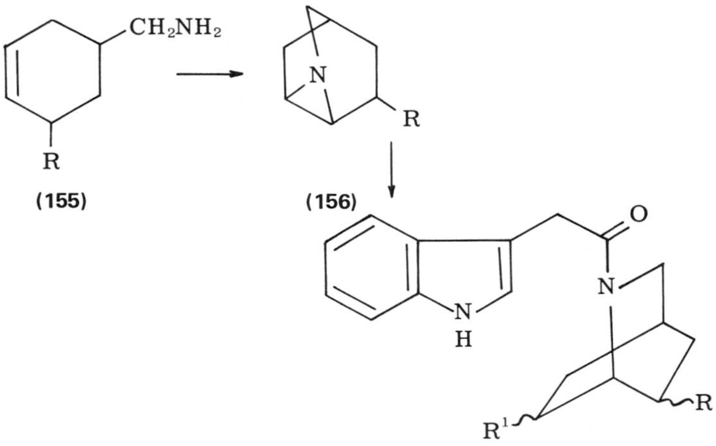

CH_2NH_2

R

(155)

N

R

(156)

O

N

N

H

R

R^1

R^1 = indoleacetate

(157)

III. Cleavage of Ketones

The carbonyl group is not, in the strict sense of the term, a latent function. There are, however, enough cases where a given carbonyl function is not the goal of a synthetic scheme nor transformable to the goal by a simple manipulation. Rather, the carbonyl group in question is used in much the same manner as was the double bond in the preceding section for the construction of a molecule. Since latency in the case of a carbonyl function is somewhat difficult to define, the examples chosen below are of necessity somewhat arbitrary. We have tried to include cases where the desired end products could not be easily obtained by other methods.

A. OXIDATIVE CLEAVAGE TO DICARBONYL COMPOUNDS

1. The Cyclohexanone to Methylcyclopentanone Sequence

Cyclohexanones are readily available from several reaction sequences. Perhaps no single reaction has proved more fruitful in this respect than the Birch reduction of anisoles (see Section V). This availability, in fact, led Johnson to plan his several schemes for the total synthesis of steroids so as to carry the D ring for several stages as a 6-membered ring; at an appropriate point this would be contracted to the desired 5-membered ring (67). The general scheme chosen to effect this overall transformation was first applied to 1-decalone **(158)** (68). The more reactive position towards alkylation was blocked as the benzylidene derivative **(159)**. Alkylation with base and methyl iodide gave a mixture of the *cis* and *trans* ketones (this lack of steric control would increase the difficulty of application of this sequence to steroid syntheses). In the original work the benzal ketone was then oxidized to the dicarboxylic acid **(161)** by means of potassium permanganate. When applied to steroid syntheses, both ozonization and alkaline hydrogen peroxide were at times employed to effect cleavage. Finally, pyrolysis of a heavy metal salt of the dicarboxylic acid (barium, lead) gave the desired cyclopentanone **(162)**. The earlier observation that D-homoestrone methyl ether would undergo part of the sequence (69) **163→165** encouraged its application to steroid synthesis.

Application of this scheme to the A ring aromatic totally synthetic tetracyclic compound **(166)** afforded synthetic *d,l*-estrone methyl ether **(167)** (70). By appropriate modification of prior reductions, epimers of **166** were obtained. These were then taken on as above to afford eventually all possible epimers of estrone (71,72).

The general scheme was subsequently found to be equally applicable to the synthesis of steroids possessing a saturated A ring. In all this later work, the furfurylidene group was used to block the α position rather than the benzylidene group. Thus the tetracyclic compound **(168)** was taken on to *epi*-androsterone (73,74) **(169)**, and the 11-hydroxy analog **(170)** was taken on to 11-hydroxy*epi*-androsterone (75,76) **(171)**. (In this last case cyclization of the acid was accomplished by Dieckmann cyclization of the diester.) Testosterone was obtained as its ethylene ketal **(175)** by application of the contraction scheme to the intermediate **(174)** (77,78). Interestingly, in the case of the unsaturated compound **(172)**, methylation went almost exclusively to afford the desired stereoisomer (18β-methyl); completion of that sequence gave $\Delta^{9,11}$-dehydro*epi*-androsterone **(173)**, (79,80).

An application to nonsteroids illustrates the use of this sequence without alkylation. Ketone **177**, an intermediate to the synthesis of cuparene, proved unexpectedly difficult to prepare due to overcrowding about the ring. The cyclohexanone **(176)** was, however, fairly readily accessible. Ring contraction of this last compound via the furfurylidene derivative readily led to **177** (81).

2. *Cleavage of α-Hydroxyketones and α-Diketones*

Conditons required to effect cleavage of cycloalkanones tend to be fairly strenuous. By first preparing an alkylidene derivative, as described in the foregoing section, somewhat milder oxidants may be used to effect the

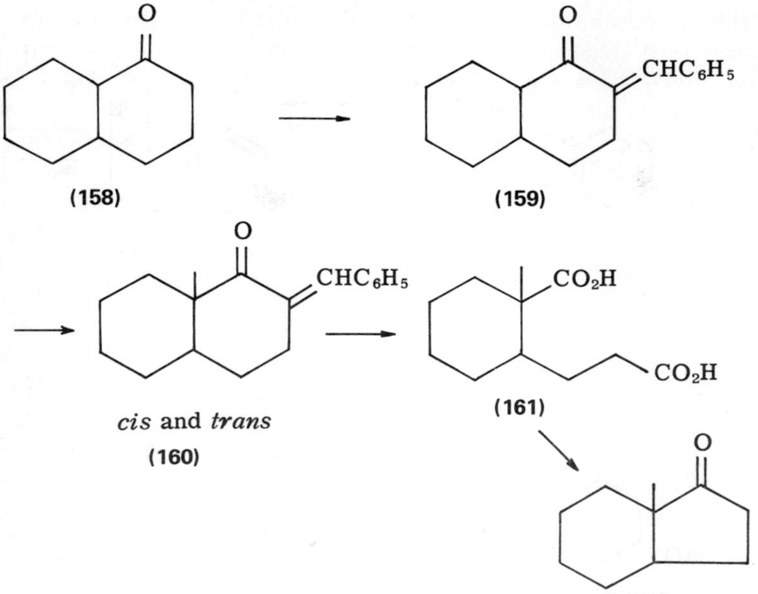

O
CHC_6H_5
(158) (159)

O
CHC_6H_5
CO_2H
CO_2H
cis and *trans*
(160)
(161)
O
(162)

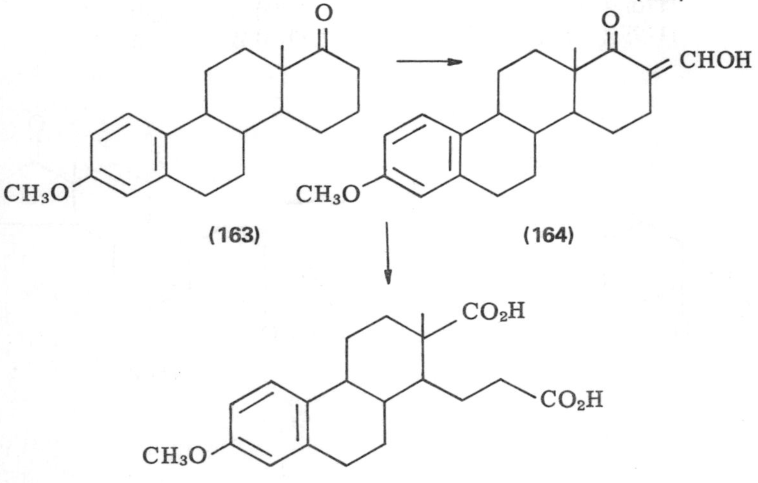

O
O
CHOH
CH_3O
CH_3O
(163) (164)

CO_2H
CO_2H
CH_3O
(165)

transformation. Another approach to the use of relatively mild reagents to effect ring opening depends on α-hydroxyketone scission; this functional group is known to be cleaved readily by periodate or lead tetraacetate. This method is thus more consistent with the presence of easily oxidized functions than is, say, permanganate oxidation. It should be further noted that this scheme leads to acid aldehydes rather than diacids.

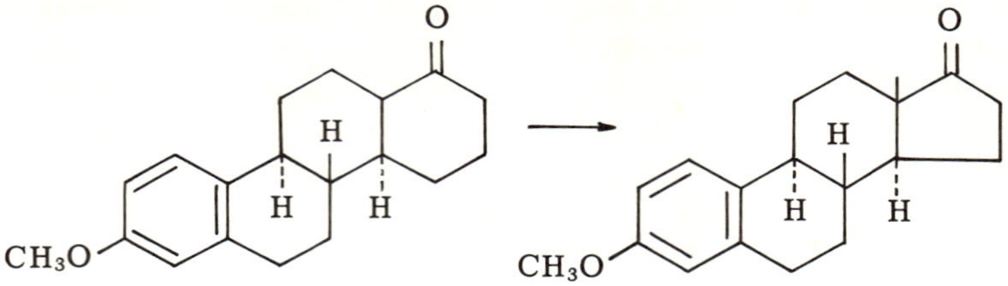

(166) **(167)**

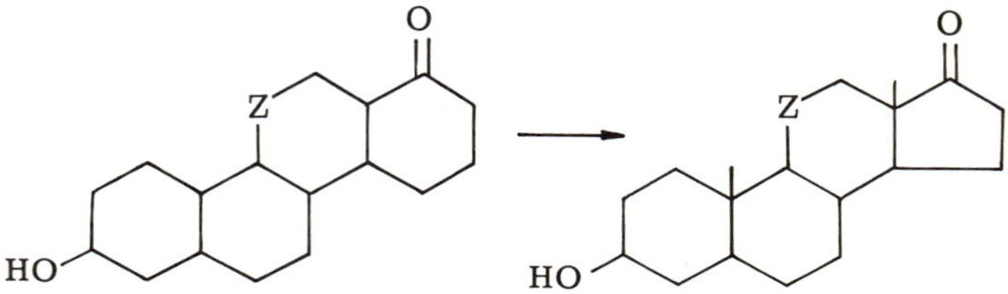

(168) Z=CH_2
(170) Z=CHOH
(172) Z=CH, Δ^{9-11}

(169) Z=CH_2
(171) Z=CHOH
(173) Z=CH, Δ^{9-11}

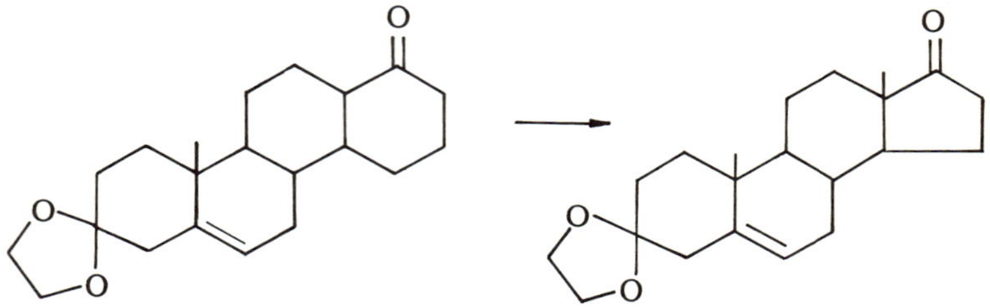

(174) **(175)**

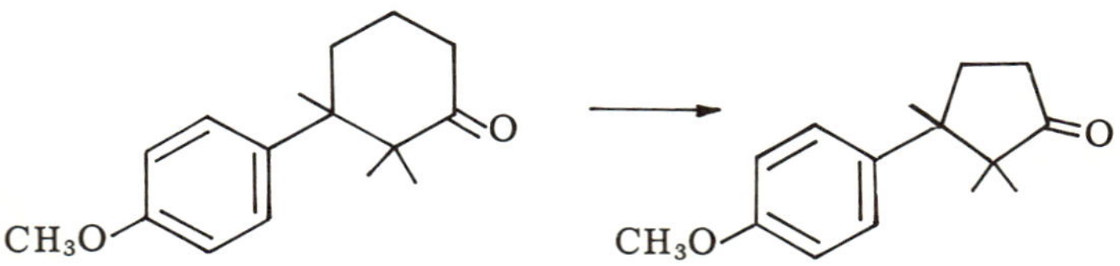

(176) **(177)**

A very real drawback to the alkylation-ring-contraction scheme used in the preparation of steroids was the fact that introduction of the 18-methyl group was not only nonstereospecific, but usually took place to give largely the undesired isomer. In a model scheme to circumvent this bottleneck, the same workers attempted to take advantage of the tendency for alkylation to occur from the α side. Thus the nitrile **178** (obtained by Michael reaction of the corresponding ketone with methacrylonitrile) was converted to the corresponding α-hydroxyketone **(179)** by successive reduction with sodium borohydride, protection as the tetrahydropyranyl ether, ozonization, and finally hydrolysis. Cleavage with sodium periodate gave the acid aldehyde **180**. By a series of reactions, the aldehyde was converted to the angular methyl group, the nitrile side-chain to an acetone moiety, and the carboxylic acid to an aldehyde with one less carbon atom. This last compound **(181)** was then cyclized with base to the model for rings C and D of a steroid **(182)** (82).

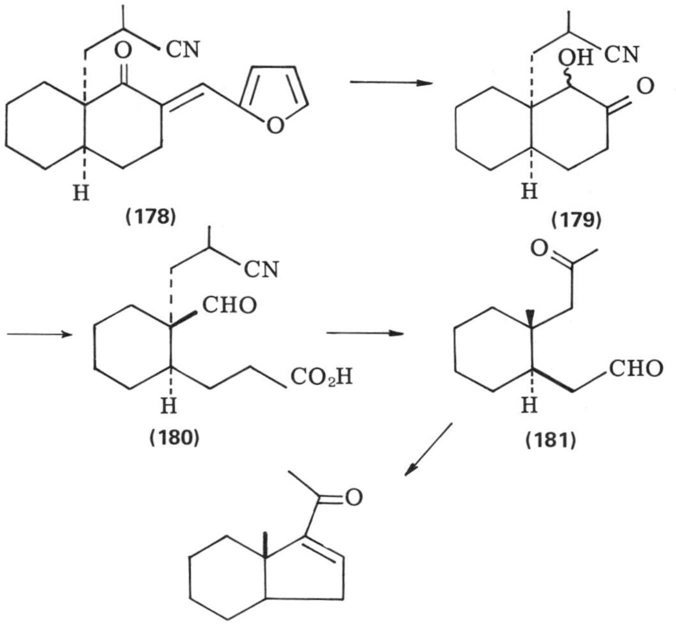

(178) **(179)**

(180) **(181)**

(182)

The procedure was somewhat modified when actually applied to the successful total synthesis of aldosterone. The starting 11-hydroxylated tetracyclic ketone **(184)** was converted to the cyanoethylated diketone **(186)** by a reaction sequence much like that described above. Cleavage with hydrogen peroxide gave a dicarboxylic acid which was saponified to a triacid. This last readily formed the diacid lactone **(188)**, which now possesses an oxidized version of the 18-oxygenated methyl group of aldosterone. The 5-membered ring D was constructed by degradation of the two acid side-chains to afford **189**, and then cyclization (83,84).

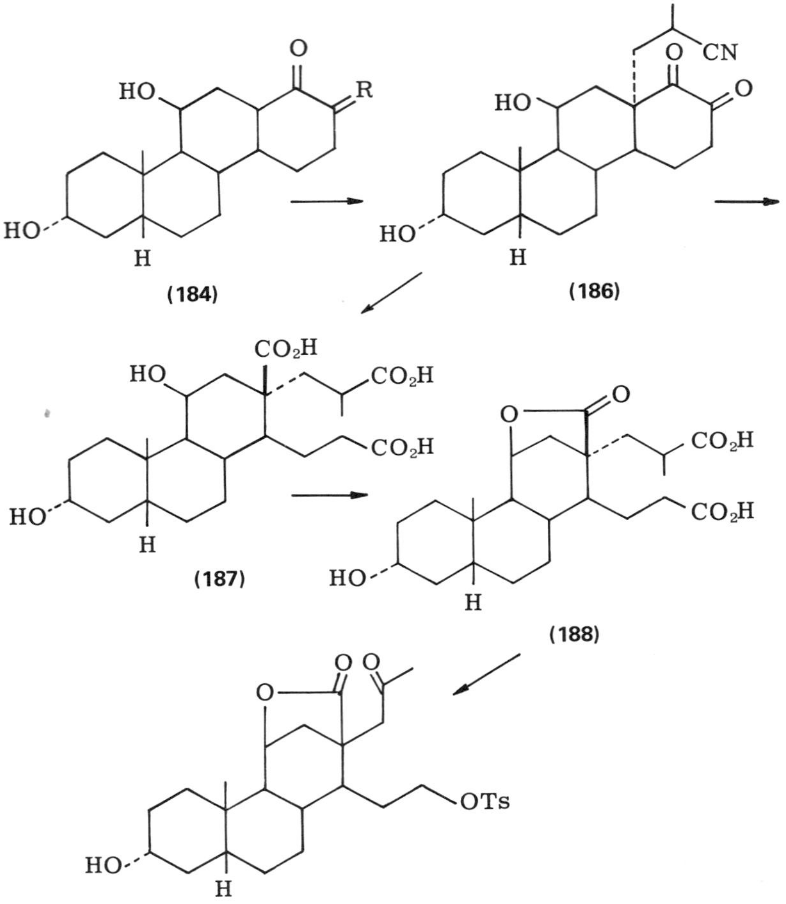

(184)

(186)

(187)

(188)

(189)

A similar sequence was employed in the total synthesis of *d,l*-nepetalactone **(194)**. Bicyclic compound **190** was prepared as a means of fixing the stereochemistry of the three groups attached to the cyclopentane ring. There remained then the task of converting the cyclopentanone ring to an enol lactone. Thus the ketone was converted to the benzal derivative **(191)**. This was then reduced to the alcohol and acetylated. Ozonization followed by cleavage of the thus obtained α-hydroxy ketone **(192)** with periodic acid gave the acid aldehyde **(193)** which was dehydrated to the enol lactone **(194)** (85).

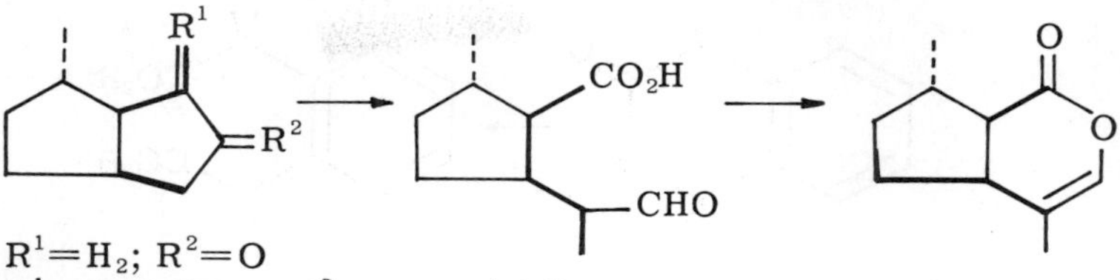

(**190**) $R^1 = H_2$; $R^2 = O$
(**191**) $R^1 = C_6H_5CH=$; $R^2 = O$
(**192**) $R^1 = O$; $R^2 = OH$

(**193**)

(**194**)

A modification of this general procedure was used to obtain the enantiomer of nepetalactone. The unsaturated ketone (**195**) was obtained in optically active form in several steps from *trans*-pulegenic acid. Epoxidation of the double bond followed by acid-catalyzed rearrangement and catalytic reduction gave the α-hydroxyketone (**197**). This was then cleaved by means of lead tetraacetate to the acid aldehyde–obtained in this case as the lactol (**198**). Pyrolysis gave nepetalactone (the absolute configuration of the product is opposite to that shown for **194**) (86).

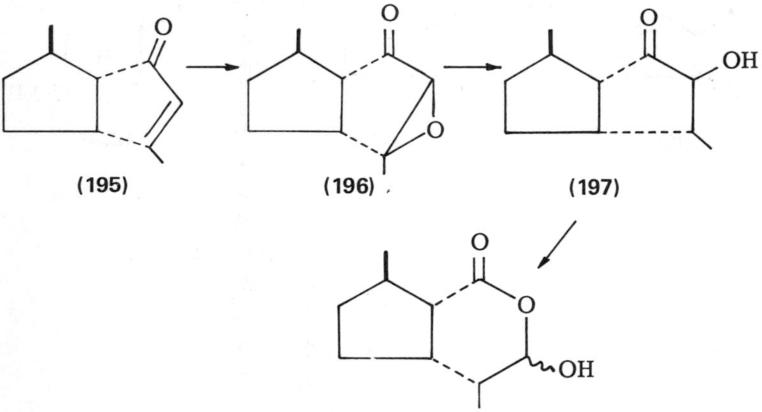

(**195**)

(**196**)

(**197**)

(**198**)

Many attempts at the preparation of di-*t*-butylbenzene failed because the bulk of the steric strain was developed late in the synthesis. In an interesting approach, the tetralone (**199**), which in effect contains the necessary carbon skeleton, was the starting material. Treatment with selenium dioxide gave the α-diketone (**200**). This last was cleaved with periodic acid to diacid **201**. The synthesis was completed by conversion of the carboxyl groups to methyls by standard methods (87).

In work aimed at a derivative of podocarpic acid, the stereochemistry was fixed by taking advantage of the properties of the hydrindane system. Thus reduction of the α-diketone (**202**), first with sodium borohydride and then with hydrogen, gave the hydroxyketone (**203**). It is of interest that only one of the

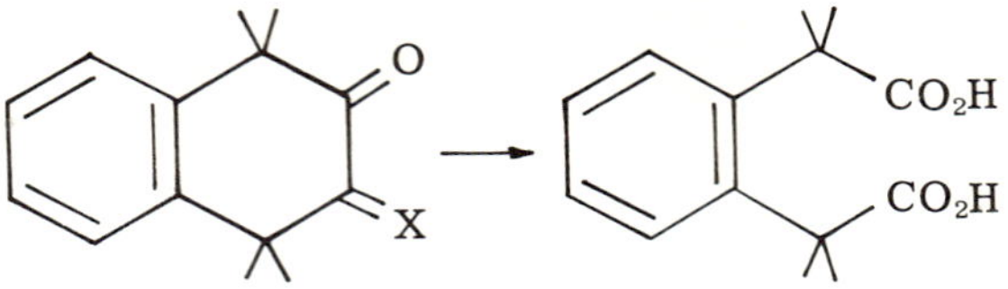

(199) X=H_2
(200) X=O

(201)

two carbonyl groups is reduced by the borohydride. Since the diacid was in this case the goal, simultaneous cleavage and oxidation to **204** was accomplished by treatment with chromic acid. The diacid on cyclization gave ketoacid (**205**), which possesses the desired carbon skeleton in the proper stereochemical arrangement (5).

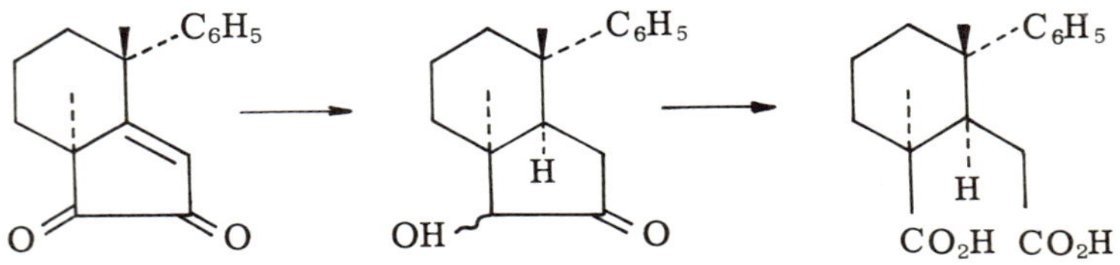

(202) (203) (204)

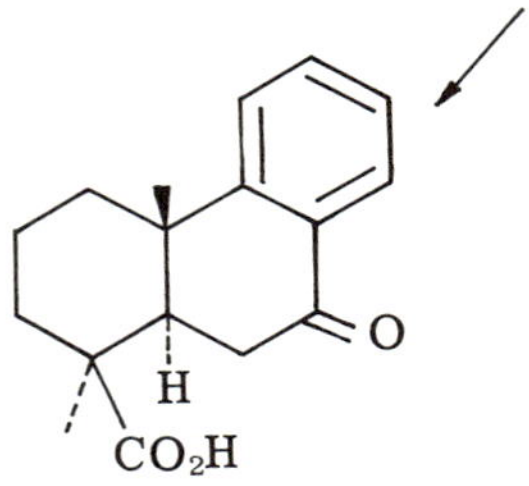

(205)

3. *Other Cleavages*

A sequence similar to that which has been used to accomplish the contraction of steroid D rings was used to effect "inversion" of ring C in a terpene synthesis. In order to construct the desired bicyclic ring system, it was necessary first to change the configuration of the existing ring of **206**. Reaction with furfural served in this case both to block the more active position to alkylation and "activate" the ring to the subsequent cleavage. In fact, ozonization of the allylated compound (**207**) followed by esterification gave the triester (**208**). Dieckmann cyclization followed by hydrolysis, decarboxylation, and re-esterification gave the inverted product (**209**) (88).

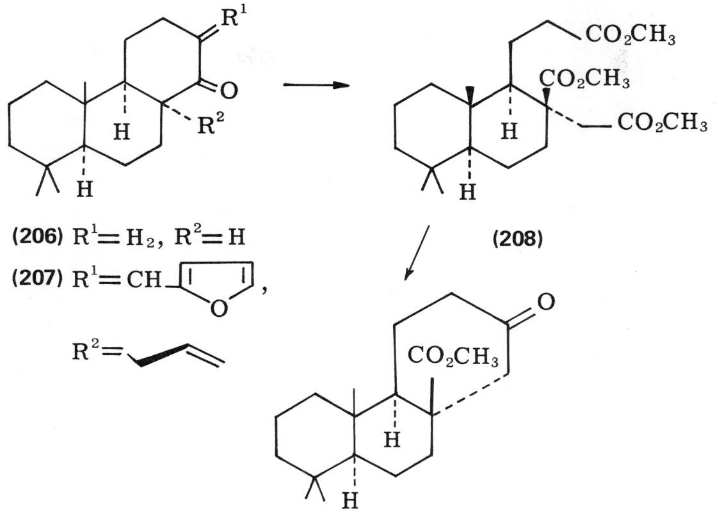

(206) $R^1 = H_2$, $R^2 = H$

(207) $R^1 = CH$–(3-furyl), $R^2 =$ (allyl)

(208)

(209)

In the preparation of compounds related to the terpene acids, the stereochemical properties of decalins were exploited in order to fix the steric relationships in the final product. Thus lithium in ammonia reduction of the enone (**210**) gave the more stable isomer (**211**). It was now necessary to cleave the cyclohexanone ring in order to provide the carbon atoms for construction of ring B and for the carboxyl function of the product. Formylation of the ketone to **212** followed by ozonization gave the key dicarboxylic acid (**213**), which was converted to **214** by standard methods (89,90).

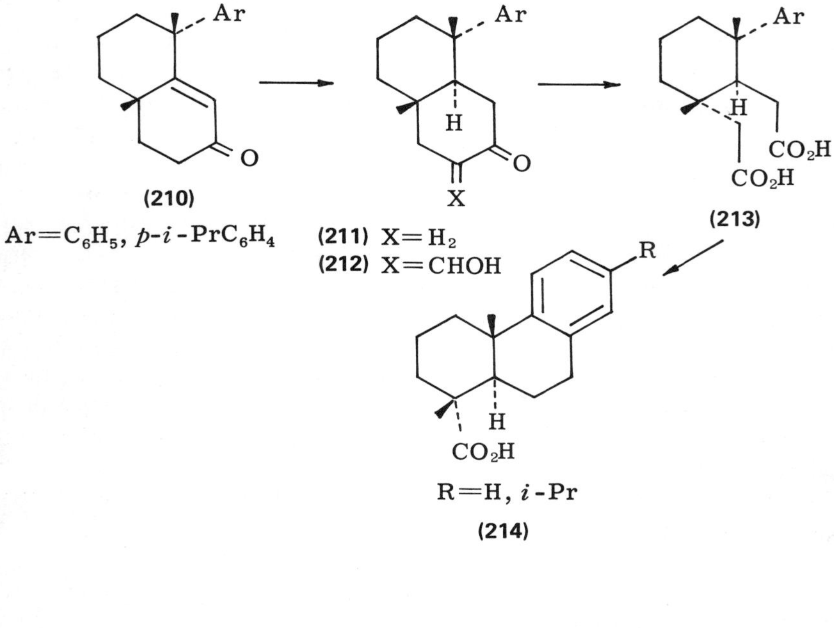

(210)
$Ar = C_6H_5$, p-i-PrC_6H_4

(211) $X = H_2$
(212) $X = CHOH$

(213)

$R = H$, i-Pr
(214)

B. CLEAVAGE TO HYDROXYKETONE DERIVATIVES

1. Alkaloid Syntheses

It is far more difficult to control stereochemistry in acyclic compounds than in ring systems. In the latter the centers are often locked with respect to each other and such factors as ring stabilities and neighboring group participation can be exploited. We have already seen these factors employed in the preparation of complex polycyclic alkaloids in Section II.B.2, where the key reaction was the opening of an internal olefin to a dialdehyde. It frequently happens that the key intermediate is a cycloalkanone rather than an olefin. In such cases the Baeyer-Villiger reaction has been used to open that ketone to the lactone. The latter is then opened with alcoholic hydrogen bromide to give a bidentate bromoester. Reaction of that intermediate with a primary amine leads to a lactam, for the net insertion of nitrogen (eq. 8). It is of note that the reactions of the scheme leave the centers of asymmetry quite unaffected.

$(CH_2)_n$ ... C=O ... $(CH_2)_n$ → $(CH_2)_n$–O–C(=O)–$(CH_2)_{n'}$

→ $(CH_2)_nBr$ / $(CH_2)_{n'}-CO_2R$ → $(CH_2)_n$–N(R)–C(=O)–$(CH_2)_{n'}$ (8)

Thus the total synthesis of *d,l*-yohimbane started with the *trans* hydrindanone (**215a**). Cleavage with perbenzoic acid gave the lactone (**216**) (the symmetrical nature of the ketone makes the direction of cleavage irrelevant). Treatment with alcoholic hydrogen bromide gave the bromoester (**217**), which in turn afforded the tricyclic amide (**218**) with tryptamine. Cyclization with phosphorus oxychloride followed by reduction gave yohimbane (**219**) (91,92). Application of the same sequence to the *cis* hydrindanone (**215b**) afforded a tricyclic compound isomeric with **218** at the ring junction. That intermediate was taken on to both *d,l*-alloyohimbane and *d,l*-epialloyohimbane (93,94).

In an analogous application of this reaction sequence, the cyclopentanone (**220**) was converted in two steps to the bromoester. Reaction with tryptamine followed by cyclization and reduction afforded dihydrocorynanthene (**222**) (95); the same sequence starting with 3,4-dimethoxyphenethylamine led to the model compound (**224**) used to determine the stereochemistry of the ipecac alkaloids (96).

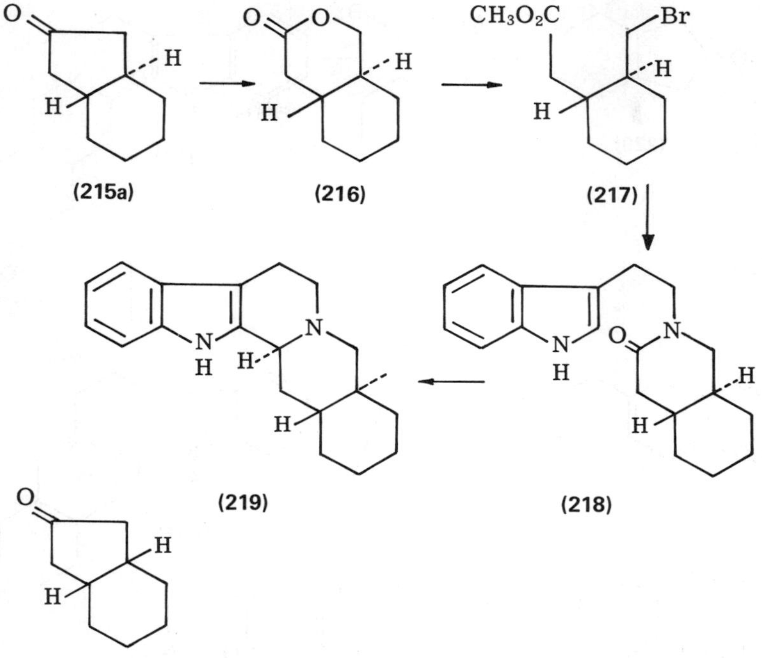

(215a) (216) (217)

(219) (218)

(215b)

2. Other Applications of Hydroxyacids

The ring hydroxyls of the prostaglandins have posed a serious problem in synthetic design of these molecules. In several of his more recent approaches to this molecule Corey exploited the Baeyer-Villiger cleavage of a ketone in order to fix the relationship of one of those hydroxyl groups relative to the adjacent side chain.

In the first of these syntheses the bicyclic ketone (**225**) was cleaved by means of *m*-chloroperbenzoic acid. The resulting lactone (**226**) of necessity establishes the *cis* relation of the 11-hydroxyl and the side chain bearing the ester at 8. In brief, the acetic acid residue of **226** was used to introduce the remaining hydroxyl by means of the iodolactone; the methoxymethyl group was converted to the aldehyde of (**227**) and then this used as a means of introducing one of the side chains, giving **228**. Elaboration of the acetic acid side chain to seven carbon atoms completed the synthesis (97).

An interesting variation of this method used the ring junction of the bicyclic photoaddition product (**229**) to fix the *cis* relationship of the side chain at the 8 position and the 9-hydroxyl. Peracid-catalyzed ring opening went as expected, with retention of configuration, to give the lactone (**230**). It proved possible

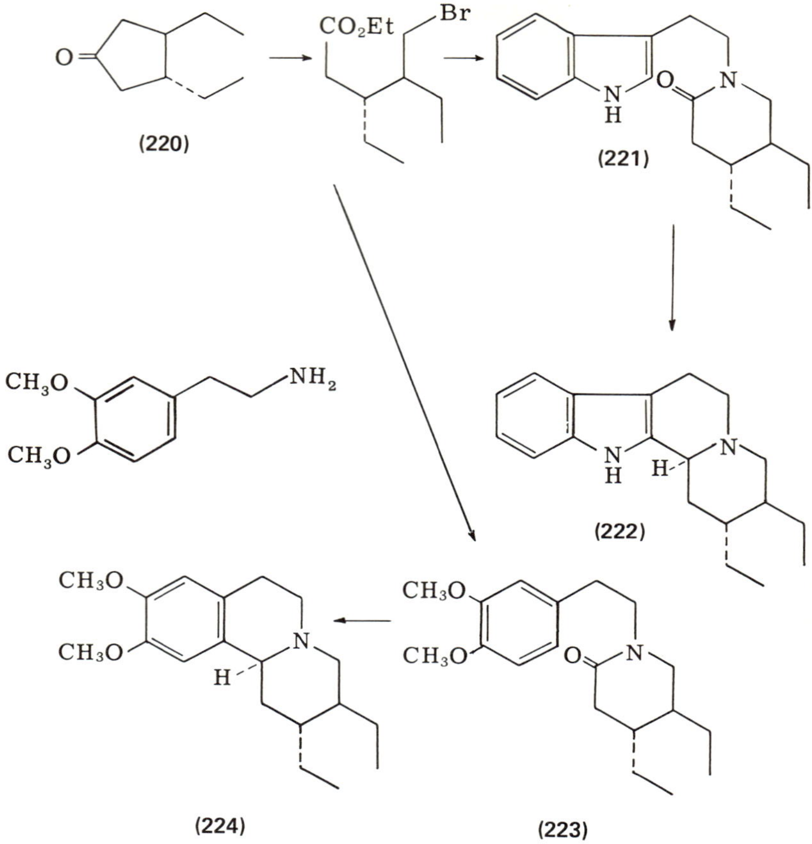

(220) (221) (222) (223) (224)

under very special conditions to transform this to the epoxide (**231**). Reaction of the hemiacetal obtained from the epoxide with the anion of the dimethyl ether of 1,3-dimercaptopropene gave a mixture of alkylation products from which it was possible to isolate **233**. This on hydrolysis gave **234**, which was taken on to $PGF_{2\alpha}$ by previously developed methods (98).

In yet another variation on this theme, the bicyclic ether (**235**) (obtained by reaction of methoxycarbene with cyclopentadiene) was condensed with dichloroketene. Reduction (zinc in acetic acid) gave the tricyclic compound (**236**). This was then taken on to the lactone (**237**) by hydrogen peroxide in acetic acid. Sequential cleavage of the methyl ether to the cyclopropanol and oxidation gave the hydroxyaldehyde (**238**). The poor yield of this last reaction detracts from the attractiveness of this scheme. The hydroxyaldehyde was then convertible to prostaglandins by a sequence similar to that by which **227** was taken on (99).

A modification of this cleavage was used in a simple synthesis of latia luciferin. Thus the aldehyde (**239**), on treatment with selenium dioxide and

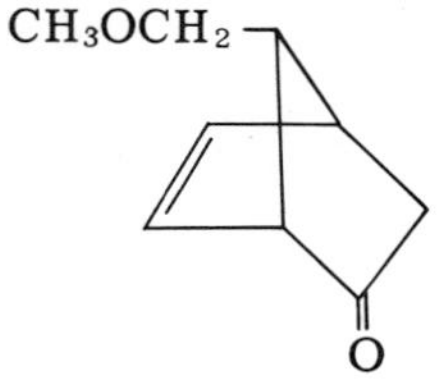

(225) 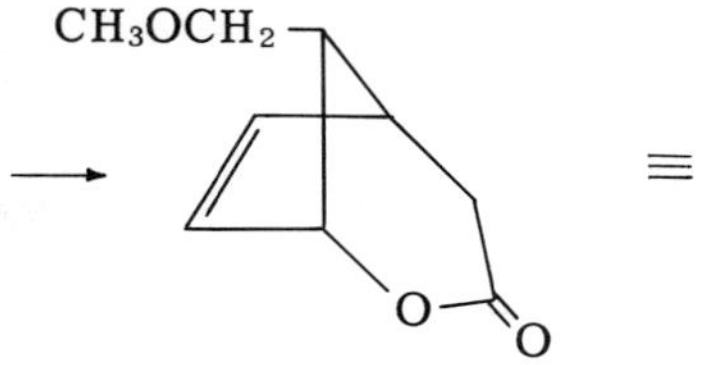 (226)

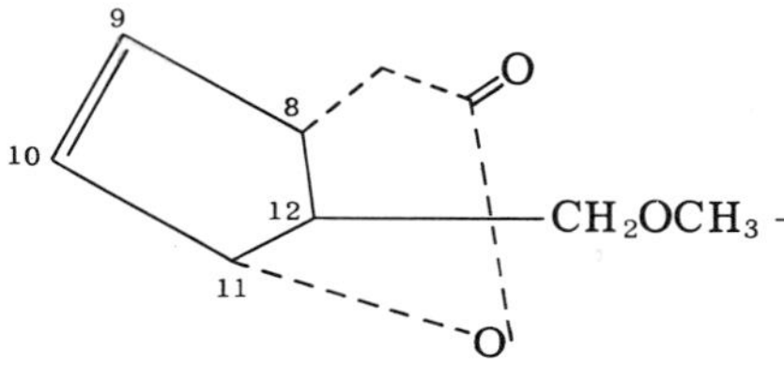

(226)

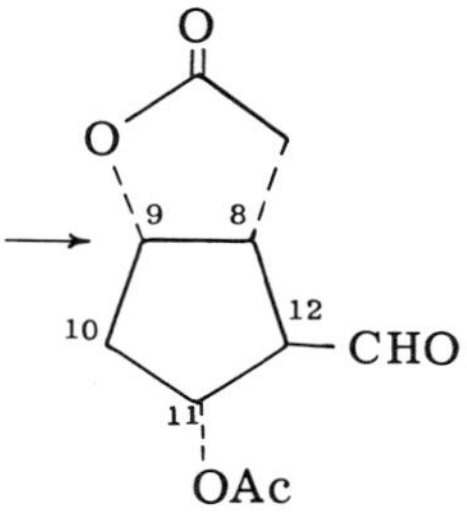

(227)

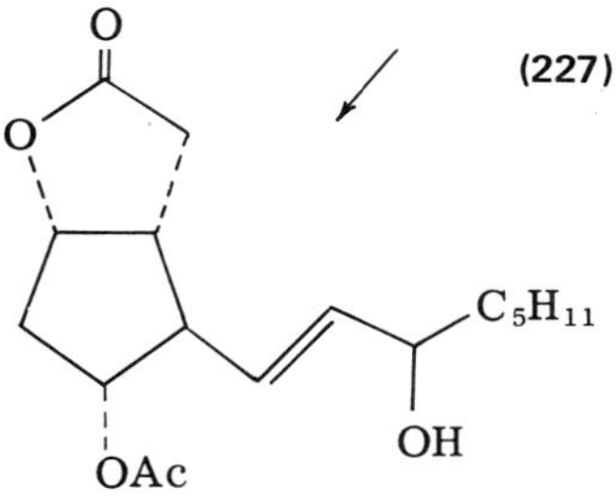

(228)

hydrogen peroxide, afforded the otherwise difficulty accessible enol formate **(240)**. It is of note that the aldehyde isomeric about the double bond gave an isomeric enol formate (100).

C. REVERSE ALDOL REACTION

Allene has been shown to add to enones photolytically so as to place the methylene group β to the carbonyl. The products of this reaction, though readily available, at first sight do not offer much promise as synthetic intermediates. A clever scheme was developed, however, by which either two or all three of the carbon atoms of the added allene are incorporated in the final molecule. The overall conversion amounts to the as yet unrealized Michael addition of acetaldehyde or acetone to the enone.

The sequence starts by protection of the carbonyl group of the adduct as its ketal. In the first scheme the methylene group is cleaved to the ketone by means of osmium tetroxide and periodic acid (see Section II.A.4); the resulting

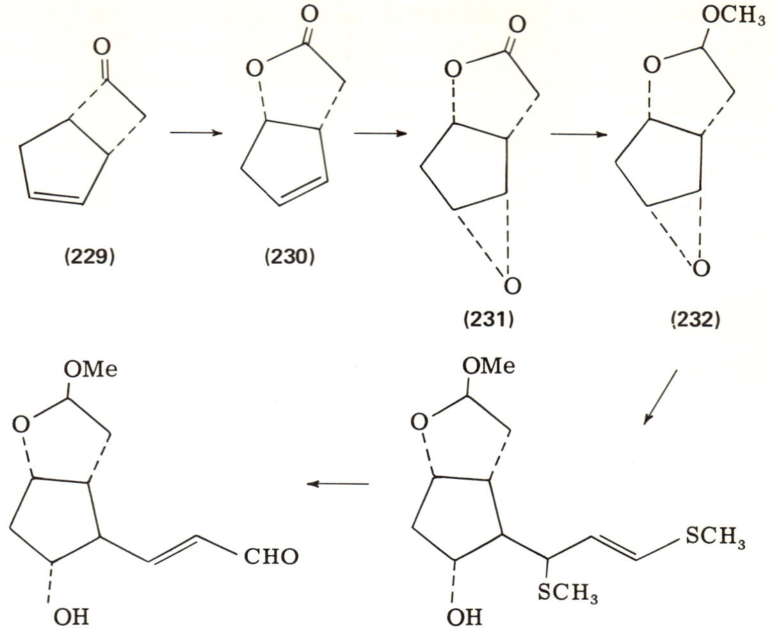

O

O

O

OCH$_3$

(229)

(230)

(231)

(232)

OMe

OMe

CHO

SCH$_3$

SCH$_3$

OH

OH

(234)

(233)

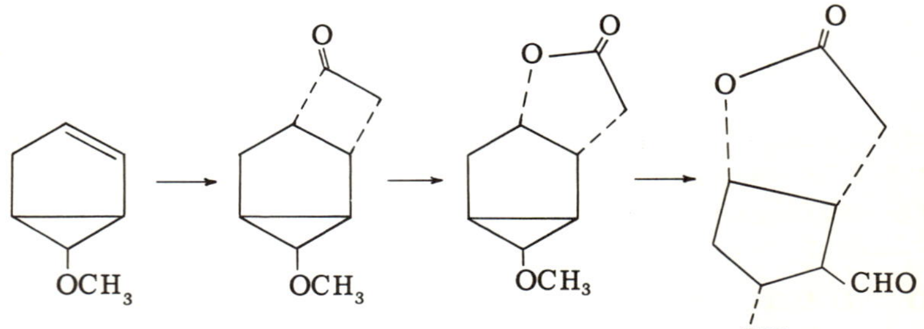

(235) OCH$_3$

(236) OCH$_3$

(237) OCH$_3$

(238) CHO OH

CHO

OCHO

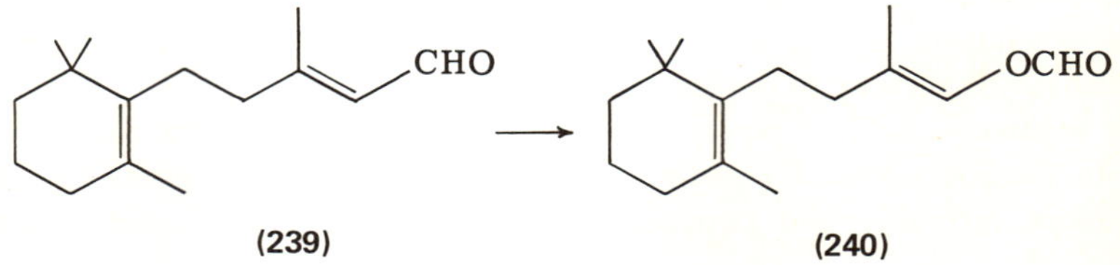

(239)

(240)

cyclobutanone is reduced to the alcohol. In the crucial step of the scheme, the ketal is removed by hydrolysis to reveal a β ketol. This highly strained "product" of an unlikely aldol reaction opens under the deketalization conditions to afford the ketoaldehyde (eq. 9).

In a variation of the sequence, the ketalized adduct is converted to the epoxide and this converted to the alcohol by means of borohydride. Hydrolysis again reveals a highly strained β ketol. This opens as above to yield in this case a diketone.

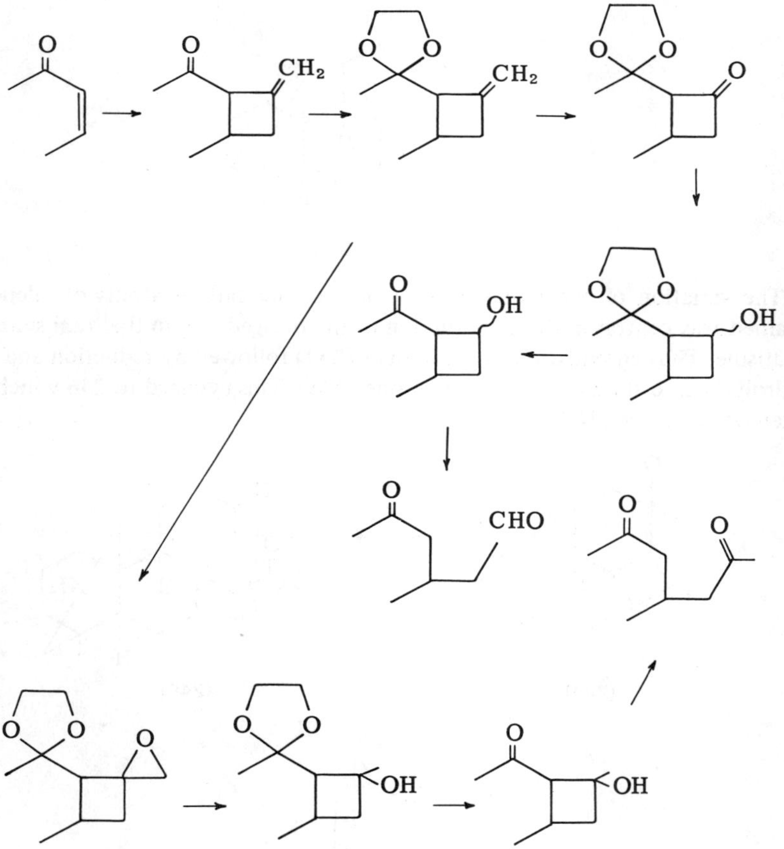

(9)

The general method was first used in a model study towards the synthesis of atisine. The allene photoadduct **(241)** was subjected in sequence to the transformations detailed above to afford finally the ketoaldehyde **(242)**; this last was not isolated but treated in turn with silica gel and chromic acid to afford the 2,2,2-bicyclooctyl system of **243** (101).

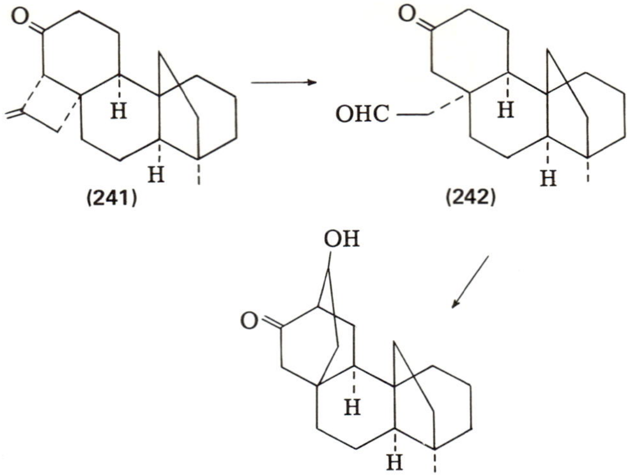

(241)

(242)

(243)

The variation of the sequence in which all the carbon atoms of allene are retained was central in the construction of the bridged ring in the total synthesis of atisine. Thus epoxidation of the adduct **(244)** followed by reduction and then hydrolysis gave the intermediate diketone **(245)**. This cyclized to **246** which was taken on to atisine (102).

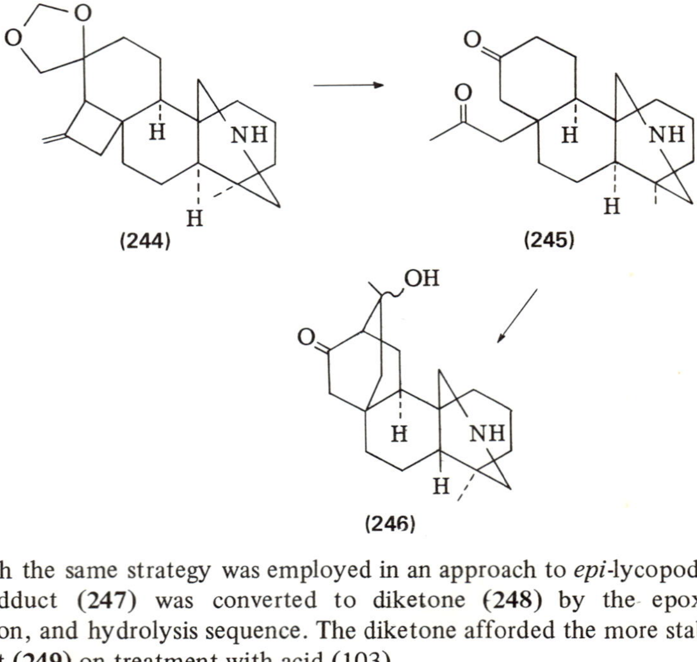

(244)

(245)

(246)

Much the same strategy was employed in an approach to *epi*-lycopodine. The photoadduct **(247)** was converted to diketone **(248)** by the epoxidation, reduction, and hydrolysis sequence. The diketone afforded the more stable aldol product **(249)** on treatment with acid (103).

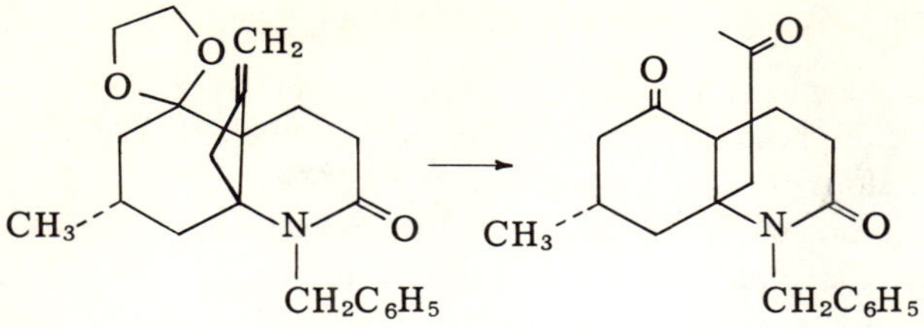

(247) (248)

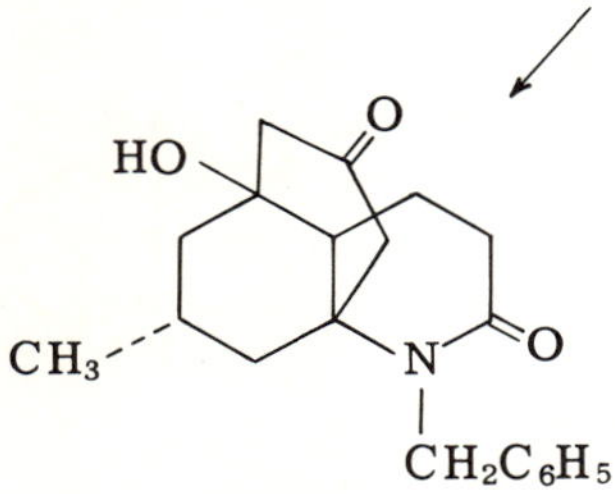

(249)

In what is probably one of the most elegant total syntheses of *epi*-lycopodine, the allene photoaddition-reverse-aldol reaction sequence was used in order to introduce both bridging rings of the target molecule. Irradiation of the cleverly designed allene **(250)** led to internal addition to form both a cyclohexane and cyclobutane ring **(251)**. The latter was converted to a methylketone **(252)** by the epoxidation scheme. Aldol condensation of that intermediate furnished the ring system of *epi*-lycopodine (104).

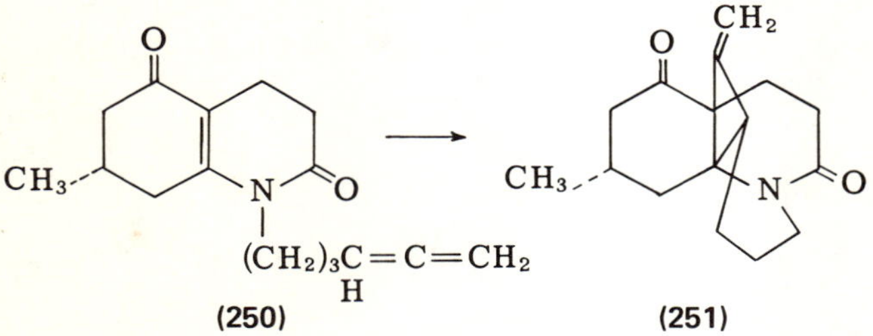

(250) (251)

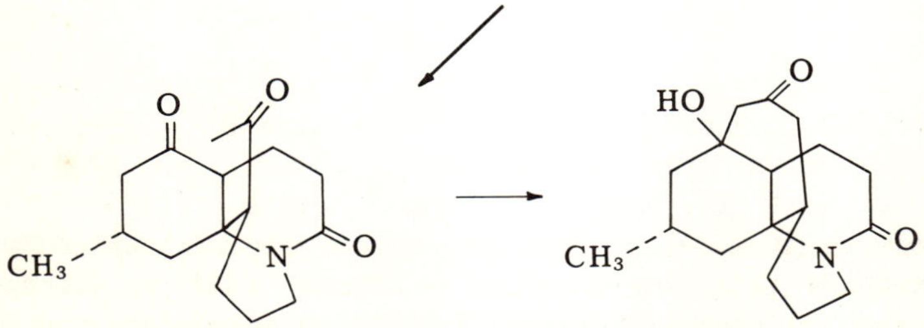

(252) (253)

IV. Nonoxidative Ring-Opening Reactions

It is a well-recognized principle in organic chemistry that the more rigid a given molecule, the better the control that can be exerted over the stereochemistry of its reactions. The regiospecificity of organic reactions similarly tends to increase in rigid molecules. The goal of synthesis cannot, of course, always be a rigid molecule; the organic chemist is frequently faced with the problem of designing a synthesis of some large, relatively flexible, ring. It has often been found advantageous to approach such molecules by a circuitous route. Thus the synthesis in its initial stages takes full advantage of the known chemistry of some bicyclic system which contains the necessary carbon atoms for the final product. After the appropriate manipulations have been carried out, one of the bonds of the bicyclic system is broken to lead to the target molecule or to some suitable precursor. Though this principle has been used most frequently for the formation of medium rings, this is not the limit of its applicability.

A. BASE-INITIATED FRAGMENTATIONS

1. Compounds Related to 1,3-Glycols

This fragmentation reaction is one of a large class of elimination reactions which have been formulated according to the general scheme (105):

$$\text{A—B—C—D—X} \rightarrow \text{A—B} + \text{C=D} + \text{X:}$$

As used in synthesis, the groups A and D are usually connected in some bicyclic system. Most frequently the starting material contains an alcohol and some such leaving group as mesylate in a 1,3 relationship; formation of the alkoxide initiates the reaction and leads to concerted ketone formation, bond cleavage, and loss of the leaving group (eq. 10).

H O OSO_2CH_3 → $O^{\ominus}$ OSO_2CH_3 → O (10)

Though this reaction had been recognized earlier (106), its first application to a rigid polycyclic system was in the steroid series. Thus, treatment of the hydroxy-tosylate (**253**) with strong base gave the keto-olefin (**254**). The isomeric hydroxy-tosylate (**255**), on the other hand, afforded a mixture of the cleavage product (**254**) along with the product of internal displacement (**256**) (107). the 1α-hydroxy-3β-tosylate (**257**) similarly underwent cleavage to afford in this case an aldehyde. It was at this time that the requirement for a *trans* antiparallel arrangement for the reactive centers in this transformation was enunciated. That is, that olefin will form from that conformation of the alkoxide which places the CO bond of the alkoxide and that of the leaving sulfonate in a *trans* antiparallel arrangement (108).

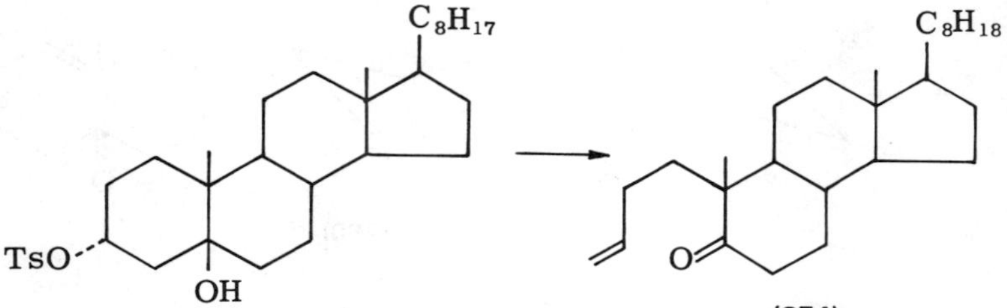

C_8H_{17} C_8H_{18}

TsO OH O

(253) → **(254)**

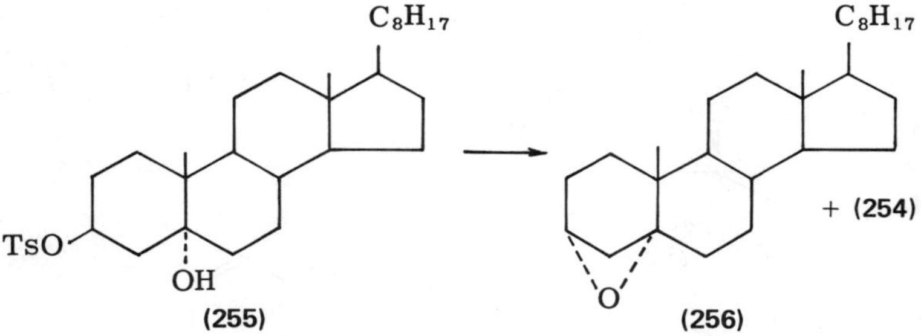

C_8H_{17} C_8H_{17}

TsO OH O + **(254)**

(255) → **(256)**

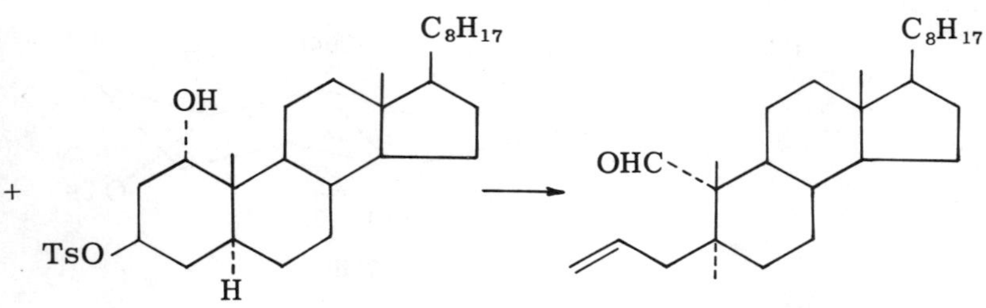

C_8H_{17} C_8H_{17}

+ OH TsO H OHC

(257) → **(258)**

The stereochemical consequences of these predictions were borne out in a series of systematic experiments. Rearrangement of the tetralin **(259)** thus went in good yield to afford the cyclodecenone **(260)** (109). The simpler analog **(261)** went as predicted to afford the *trans* double bond **(262)**. The isomer **(263)**, on the other hand, rearranges to the isomer with a *cis* double bond **(264)**. The isomer **(265)** in which the two groups cannot assume the required orientation gives none of the ring-opening reaction product (110,111).

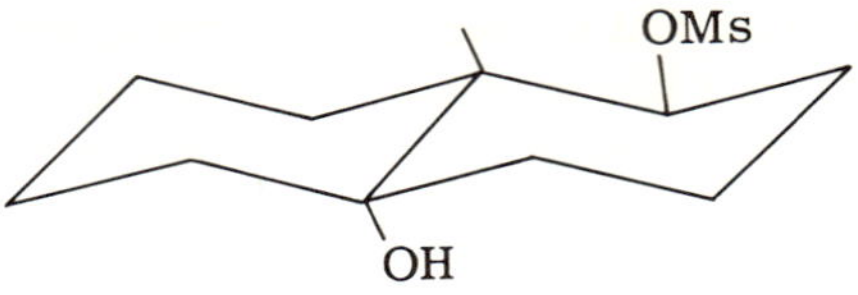

OMs

OH

(259)

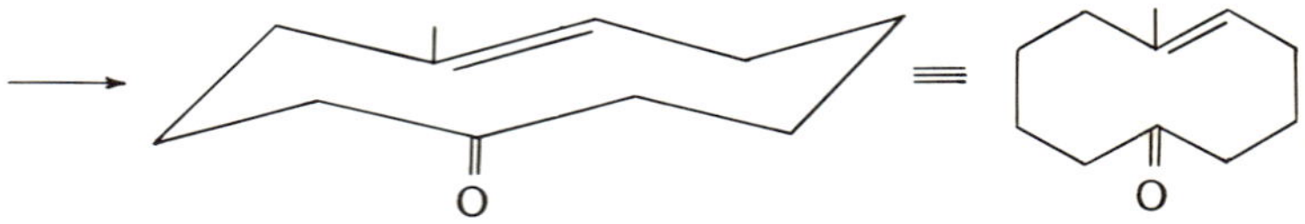

O O

(260)

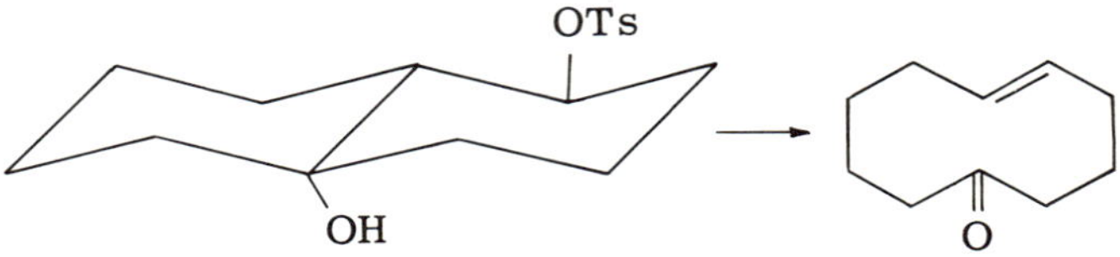

OTs

OH

(261)

O

(262)

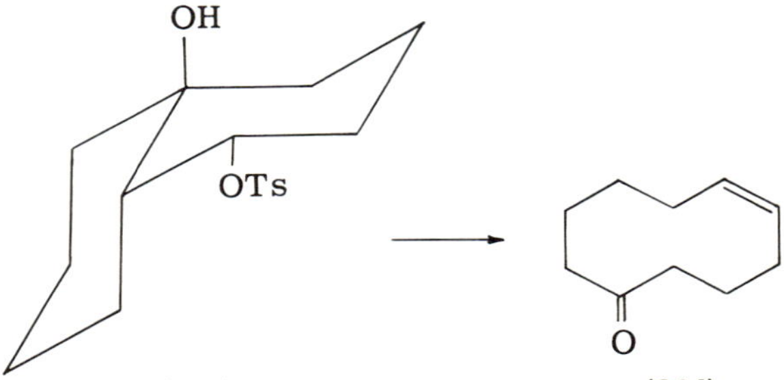

OH

OTs

(263)

O

(264)

H

OTs

OH

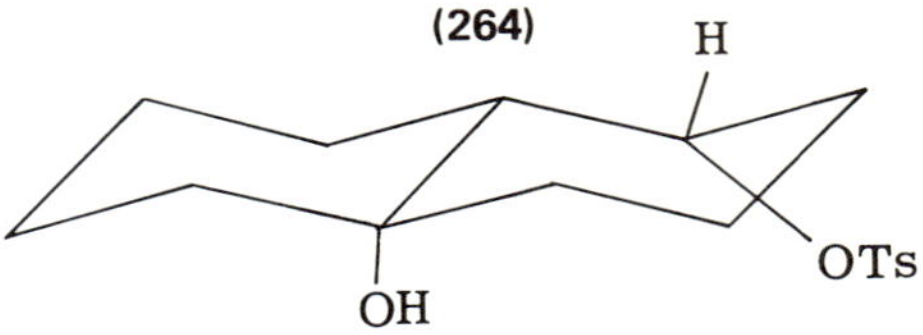

(265)

Caryophyllene and isocaryophyllene represent synthetic challenges in that each possesses an unconjugated cyclononene ring. A synthetic plan which depended on *de novo* construction of such a ring would have but small chance of success. A much more attractive scheme would be one in which an equivalent bicyclic system was first constructed equipped with the proper functionality for subsequent ring opening to the medium ring. Just such an approach was used by Corey and his collaborators in the preparation of these molecules. The tricyclic ketone **(266)** which contains all but one of the carbon atoms of the target molecule was prepared by an elegant series of transformations. Stereoselective reduction followed by tosylation led from that ketone to the intermediates **267** and **269**. When subjected to the conditions of the rearrangement, each of these fragmented to the corresponding enone. Thus **267** gave the ketone with the double bond in the *cis* configuration of isocaryophyllene, while the isomer **(269)** led to a compound which contained the *trans* olefin of caryophyllene **(270)** (112,113).

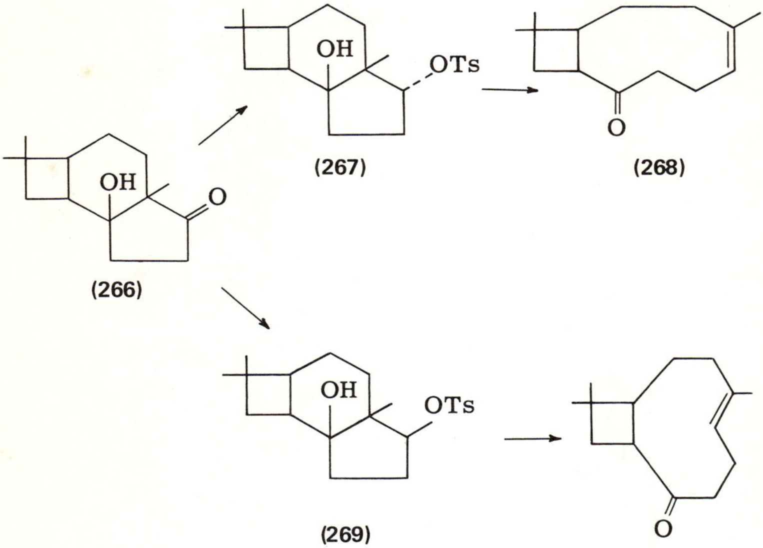

OH OTs

(266) (267) (268) (269) (270)

A fragmentation reaction was central in the conversion of the 3,2,1-bicylo-octyl system of **272** to the required 2,2,2-bicyclooctyl system in a total synthesis of atisine. Thus the diketone **(271)** was closed to the bicyclic system by aldol reaction and reduced to a diol. The product, as its acetate mesylate **(272)**, fragmented to the enone **(273)**. The terminal vinyl grouping was then converted to the tosylate **(274)** by hydroboration followed by tosylate formation. Internal alkylation of the ketone proceeded to give the bridged bicyclic compound **(275)** (114).

The total synthesis of L-3-oxovaleranone illustrates the utility of the cleavage reaction even in those cases where no ring enlargement is necessary; the reaction, in this case, is used to take advantage of the availability of a starting material of known configuration. Application of the fragmentation to the decalin **(276)** (prepared by homoannelation of tetrahydrocarvone followed by reduction and mesylation), gave largely the enone **(277)**. The ketone was then treated with methyllithium and the mixture of alcohols subjected to formolysis. Oxidation of the bicyclic alcohols thus produced gave the target molecule **(279)** (115).

Among the many cyclization products of terpenes are some medium ring compounds. Consideration of the fragmentation reaction leads to an interesting

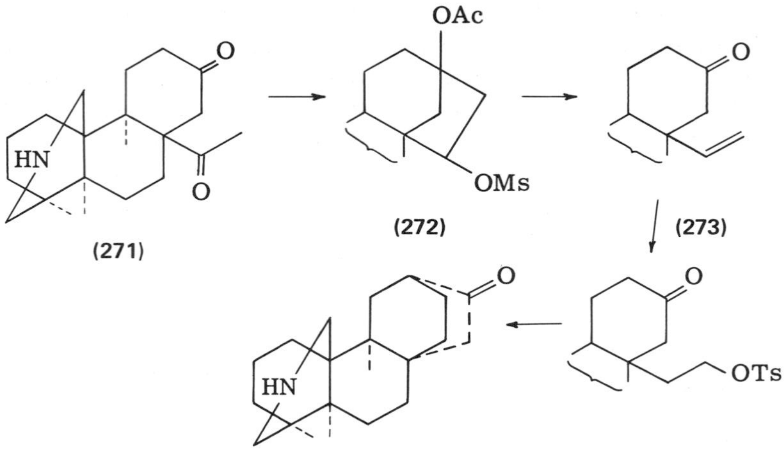

(271) (272) (273) (274) (275)

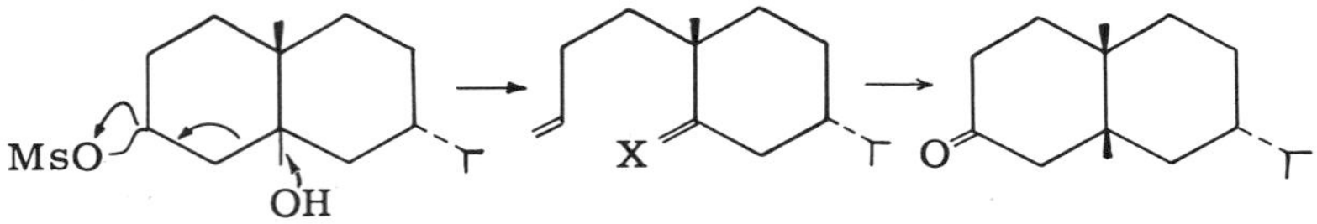

(276) (277) X=O (278) X=C(CH$_3$)(OH) (279)

scheme for stereoselective construction of such rings. Bicyclic compounds such as **281** are fairly readily available in two steps from the enamine of cyclodecanone and methyl acrylate. Reduction of the diketone followed by tosylation leads to selective formation of a monotosylate **(282)** (apparently due to hindrance about the hydroxyl group on the bridge). Treatment of that alcohol with strong base leads to the cyclodecene **(283)**. Alternately, the diketone can be alkylated prior to reduction **(284)**; at the end of the sequence the alkyl group will appear on the double bond as in **286** (116).

In an extension of this scheme, the keto ester **(287)** was condensed with methyl vinyl ketone in the straightforward homoannelation reaction. The resulting keto ester **(288)** was reduced to the diol and converted to the monotosylate **(289)**. Treatment of this exocyclic tosylate with strong base gave the cyclodecene **(290)** (117).

2. Related Fragmentations

A full negative charge is not a necessary requirement for initiation of the fragmentation sequence; in a properly designed molecule the unshared pair of

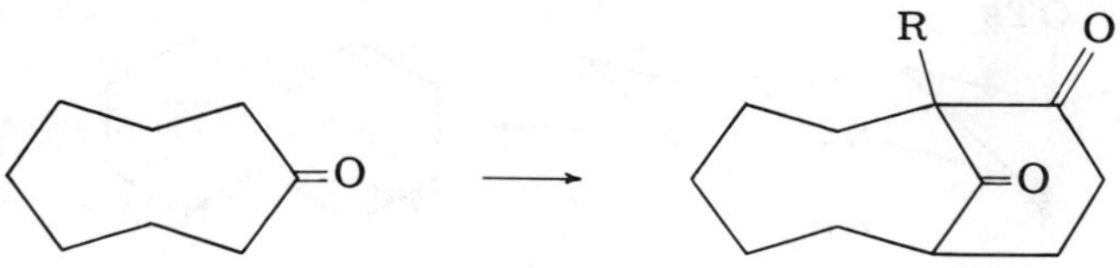

(280)

(281) R=H
(284) $R=CH_3$

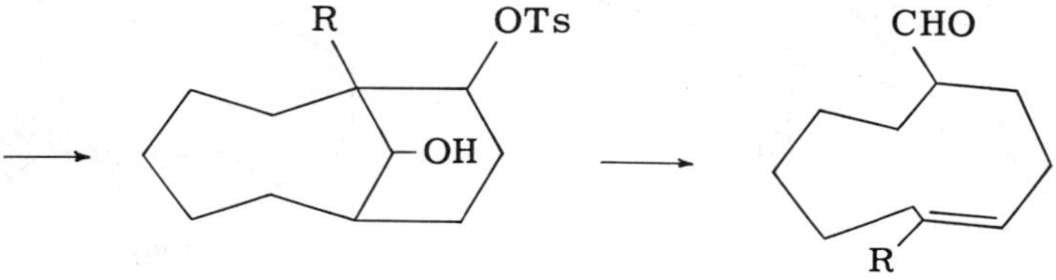

(282) R=H
(285) $R=CH_3$

(283) R=H
(286) $R=CH_3$

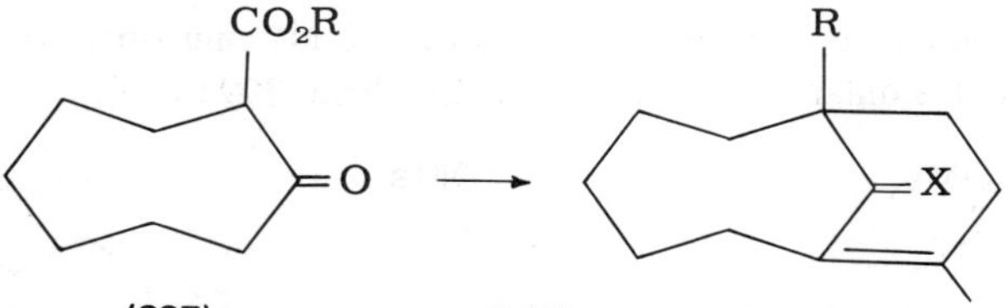

(287)

(288) X=O; $R=CO_2Et$
(289) X=<H, OH, $R=CH_2OTs$

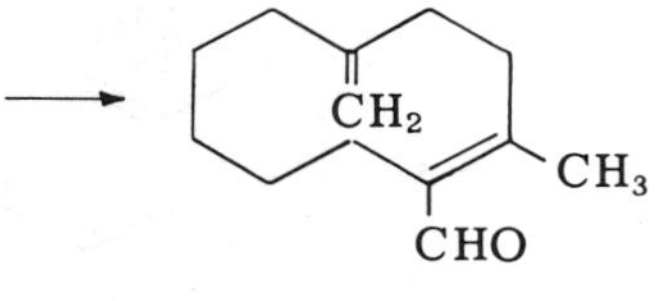

(290)

electrons of nitrogen will accomplish the same task. The steric requirements for this modification are essentially the same as for the prototypical reaction. The pi orbital of the unshared pair is treated as one of the reacting centers. Thus the tosylate of the *trans* fused perhydroquinoline **(291)** leads to the *trans* olefin **(293)** on solvolysis. The *cis* fused isomer **(294)**, on the other hand, affords the *cis* olefin **(295)**. Compounds in which the tosyl group is axial give only substitution or elimination products (118).

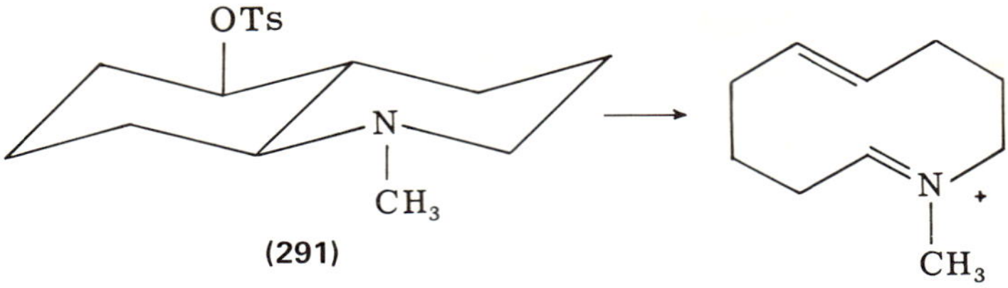

OTs

N—CH$_3$

(291)

N$^+$—CH$_3$

(293)

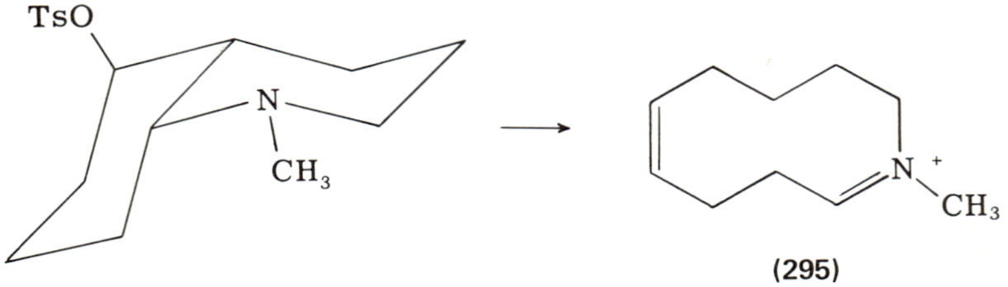

TsO

N—CH$_3$

(294)

N$^+$—CH$_3$

(295)

A reaction which at least formally fits the classification of a 1,4 elimination has been used for the stereoselective preparation of medium ring dienes. Hydroboration of the unsaturated mesylate **(296)** gave the intermediate addition product. Treatment of this with sodium hydroxide leads to the 1,4 elimination of the boron and mesylate group with concomitant ring opening (119). The isomeric olefin on similar treatment affords the diene **299** (120).

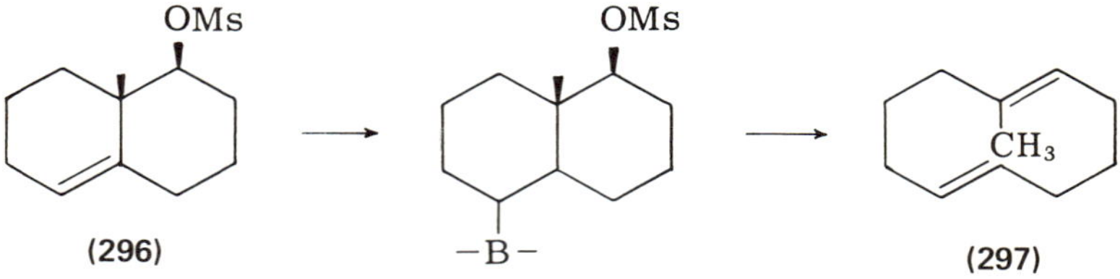

OMs

(296)

OMs

—B—

CH$_3$

(297)

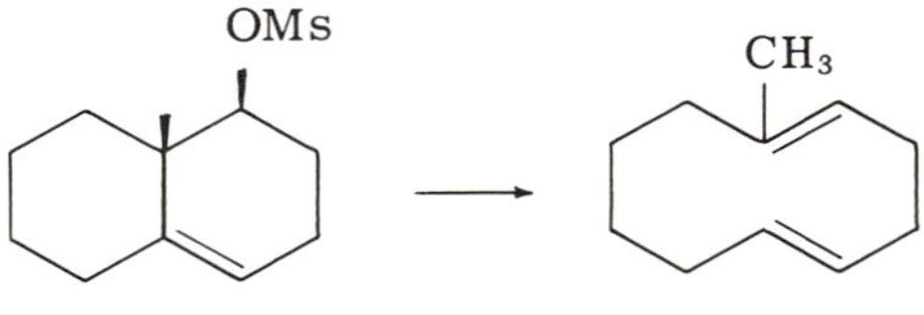

OMs

(298)

CH$_3$

(299)

Cleavage of systems where the anionic center and leaving group are in a 1,3 relation to each other has been employed in an interesting method for the addition of a terminally functionalized two carbon fragment to the α position of a cyclohexenone. The first application of this was to the synthesis of the carbon skeleton of the ormosia alkaloids. Thus photoaddition of vinyl acetate to the tricyclic enone **(300)** afforded the cyclobutane **(301)**. Halogenation of this ketone went on the 4-6 bridgehead position to afford **302**. Treatment with mild base leads, via the fragmentation reaction (and loss of bromine as the leaving group), to the keto-aldehyde **303** (121).

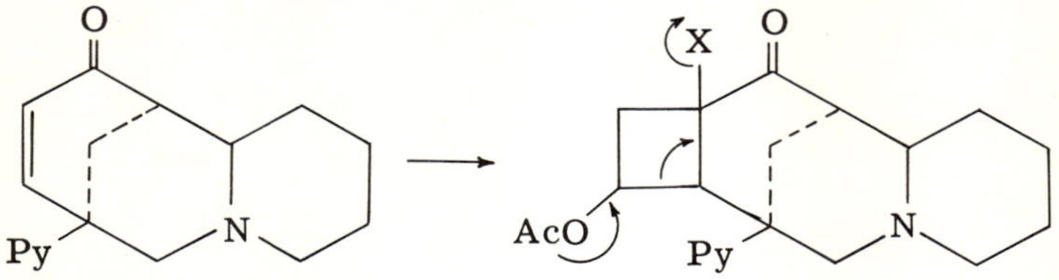

Py = 2 − pyridyl

(300)

(301) X=H
(302) X=Br

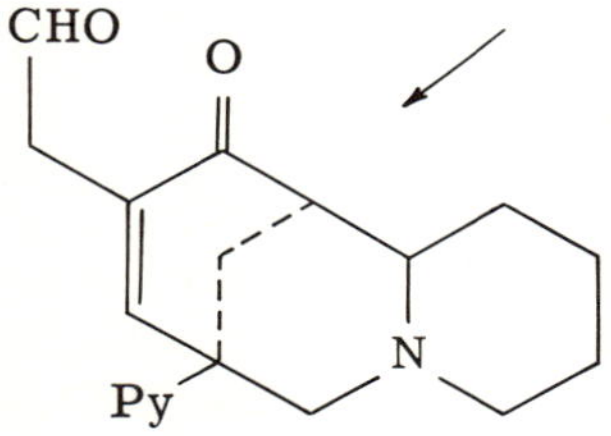

(303)

B. SECOND-ORDER BECKMANN REARRANGEMENT

Certain ketoximes will, when subjected to the conditions of the Beckmann rearrangement, yield a fragmented molecule rather than the expected amide. This seems to be especially true for oximes which are substituted with electron donating groups in the α position (e.g., carbonyl, alkoxy, amino), or those in which the potential migrating group forms an especially stable carbonium ion. Such a reaction of an oxime of a cyclic ketone will, of course, lead to the ring-opened product (eq. 11).

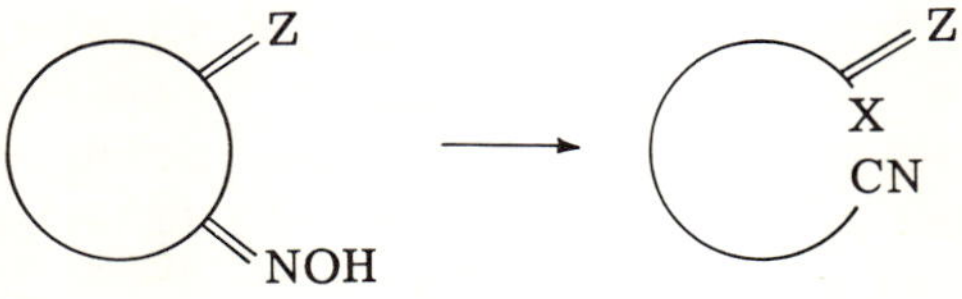

(11)

This procedure was the key in the ring contraction scheme in one of the early steroid syntheses. Nitrosation of the tricyclic ketone intermediate **(304)** afforded the α ketoxime **(305)**. This last, on treatment with toluenesulfonyl chloride and then base, gave the ring-opened acid nitrile **(306)**. That product was saponified to the diacid, esterified, and taken on to the ring-contracted product by means of a Dieckmann cyclization followed by hydrolysis (122).

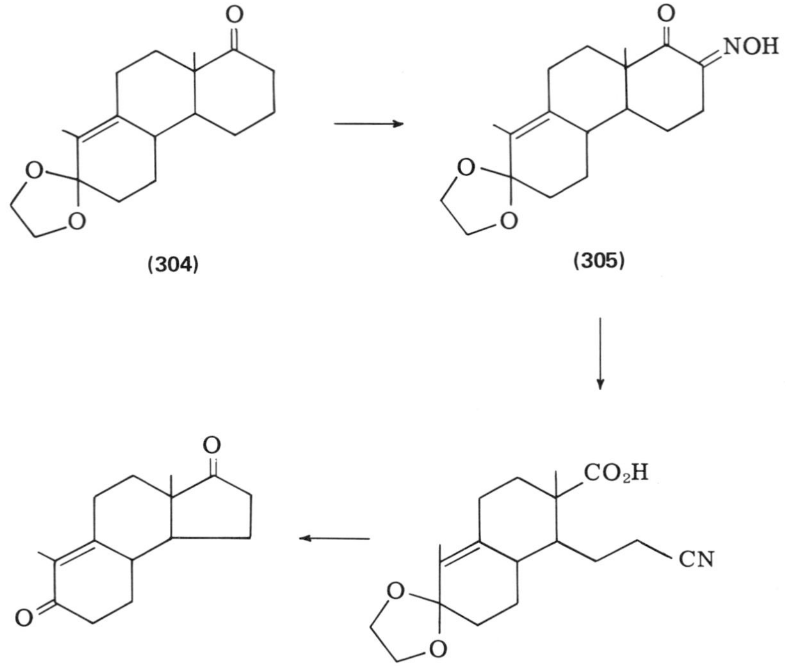

O NOH

(304) (305)

O CO_2H

CN

(306)

Various methods have been used to build the heterocyclic bridge of the atisine alkaloids. The obvious procedure, that is, Beckmann rearrangement of the ketone **(307)**, proved inapplicable to this transformation. Instead, starting much as above, the ketone was nitrosated to **308**. Application of the second-order Beckmann rearrangement followed by esterification gave the nitrile ester **(309)**. This last on reduction (diborane or controlled lithium aluminum hydride) afforded the lactam **(310)** (19,123). This proved to be isomeric with the lactam obtained on rearrangement of the oxime of **307** (123).

Fragmentations brought about by the presence of a stable potential migrating carbonium ion have been used to good purpose for the preparation of bicyclic cycloheptanols. Abnormal Beckmann rearrangement of the starting oxime **(311)** affords in this case the unsaturated nitrile **(312)**; the presumably intermediate carbonium ion at the *gem*-dimethyl carbon stabilizes by loss of a proton to form the olefin. Reduction of the nitrile by means of lithium aluminum tritertiary-butoxyhydride gave the unsaturated aldehyde **(313)**. This last on treatment with stannic chloride recyclized to the ring-expanded alcohol **(314)** (124). Proceeding in the same manner, oximes **315** and **317** were converted to the corresponding fused cycloheptanols. It is of note that this sequence leaves the ring fusion unaffected since those centers are not involved in the transformation (125).

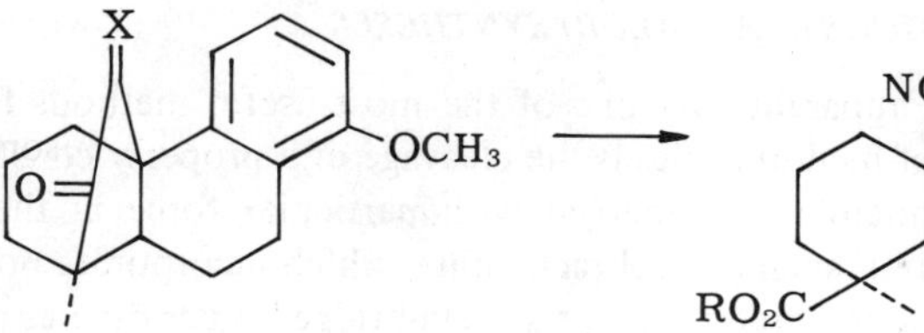

(307) X=H_2
(308) X=NOH

(309)

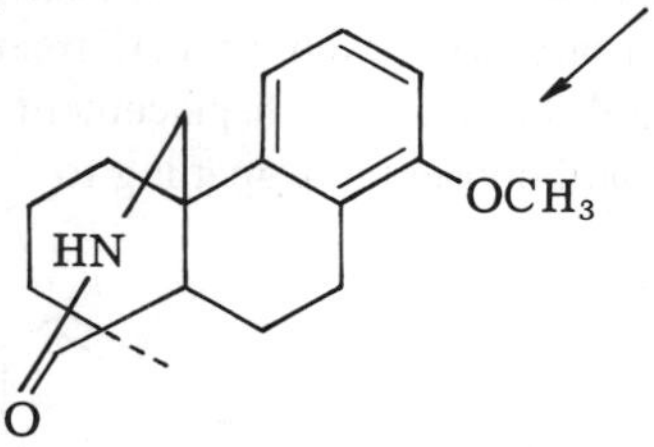

(310)

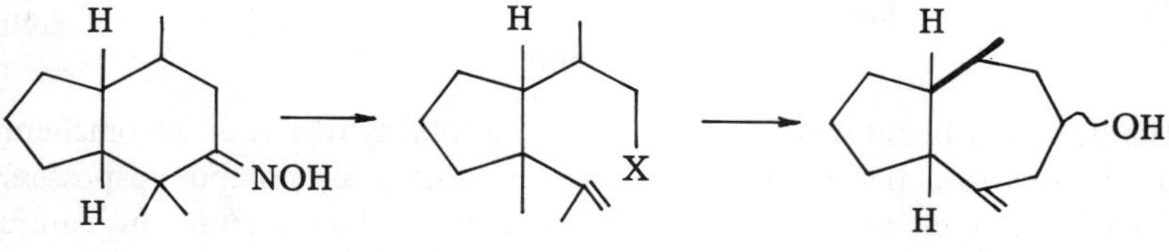

(311)

(312) X=CN
(313) X=CHO

(314)

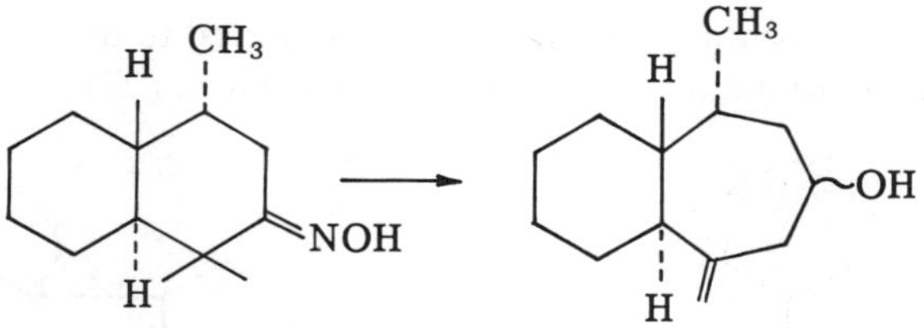

(315)

(316)

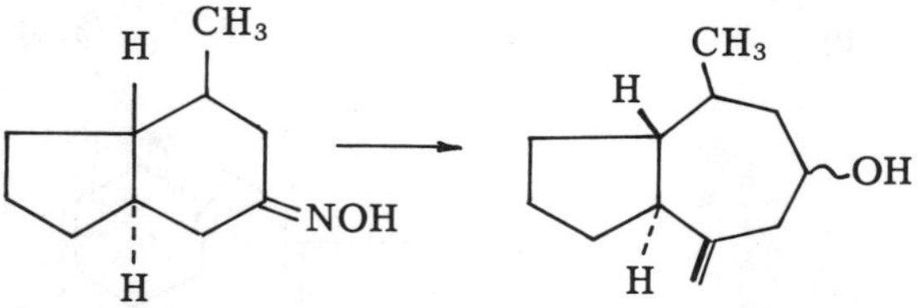

(317)

(318)

C. *RING-OPENING REACTIONS IN ALKALOID SYNTHESES*

It should by now be apparent that one of the most useful methods for the controlled construction of medium rings is the cleavage of a properly constituted bicyclic system. This principle has emerged with particular force in the total syntheses of indole alkaloids such as quebrachamine, which incorporate not only a difficultly accessible 9-membered ring, but a meta-bridged piperidine as part of that ring. This difficult synthetic problem was solved neatly by resort to a fragmentation reaction. In all but the first work, the synthetic schemes aim first for a polycyclic quaternary salt such as (a); treatment of this salt with a nucleophile (including hydride) leads to displacement on the carbon adjacent the 2 position of the indole and consequent opening to the desired medium ring (eq. 12).

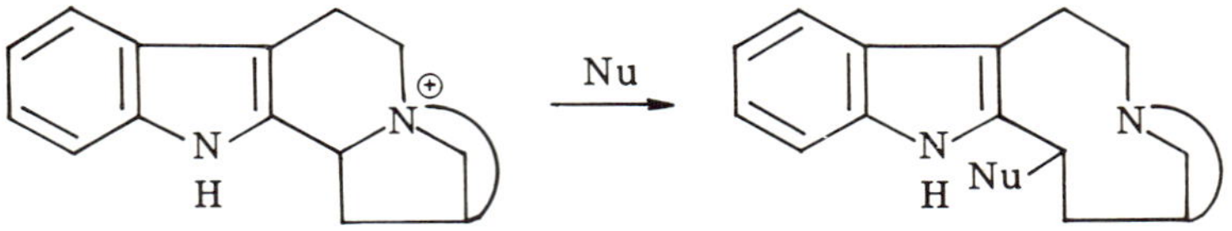

Nu

(a) (b) (12)

Scheme 5

The general scheme was first applied to the total synthesis of quebrachamine (**321**). In this case the initial goal of the synthesis was a compound possessing the aspidospermine nucleus (**320**). This last was obtained from the carefully directed "abnormal" Fischer indole condensation of the tricyclic aminoketone (**319**) with phenylhydrazine. (The sensitivity of this reaction to steric factors is shown by the fact that a similar reaction of the ketoamide fails to give the indolenine.) The pentacyclic amine, when subjected to the known (126) cleavage with sodium borohydride, afforded quebrachamine (127).

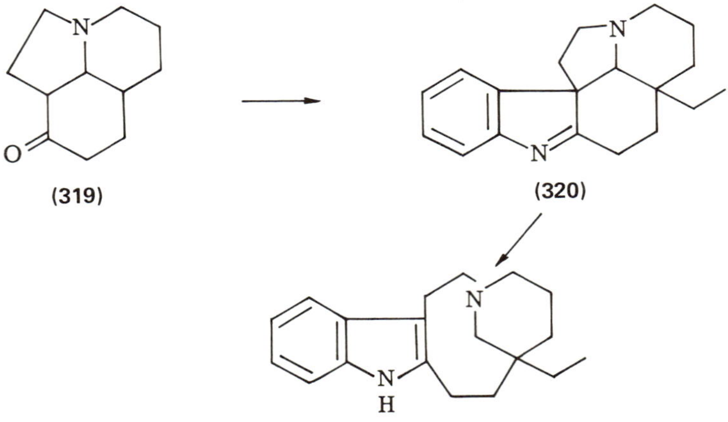

(319) **(320)**

(321)

Alkaloids of this type have proven useful as intermediates in the synthesis of the oncolytic vinca alkaloids. In recent work the key reaction was again a ring opening of a polycyclic system. The immediate precursor was, however, the more readily accessible pentacyclic compound such as **323**. These intermediates were prepared by the now fairly standard methods detailed in Sections II.A.4 and II.B.2. Mesylation of the primary alcohol of **322** followed by heating gave the required quaternary salts. Treatment of **323** with sodium in liquid ammonia gave quebrachamine (**321**) (128); the isomeric salt (**326**) gave dihydrocleavamine (129,130). Ring cleavage with potassium cyanide was used to prepare the nitrile (**327**) which on saponification gave a degradation product of cantharanthine (131). The salt (**323**) was similarly cleaved by cyanide to afford in this case an intermediate for synthesis of vinca alkaloids (23).

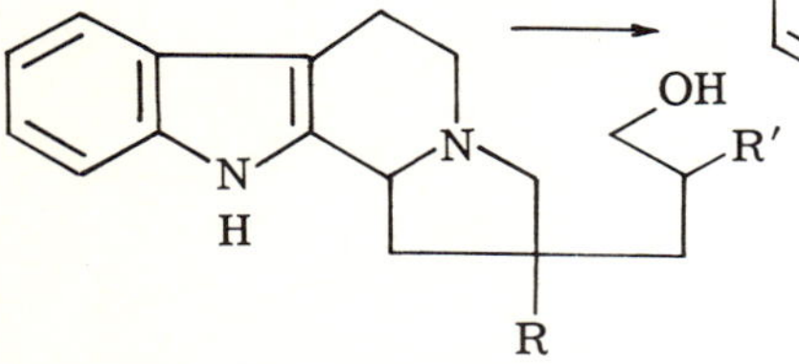

(**322**) R=Et, R′=H
(**325**) R=H, R′=Et

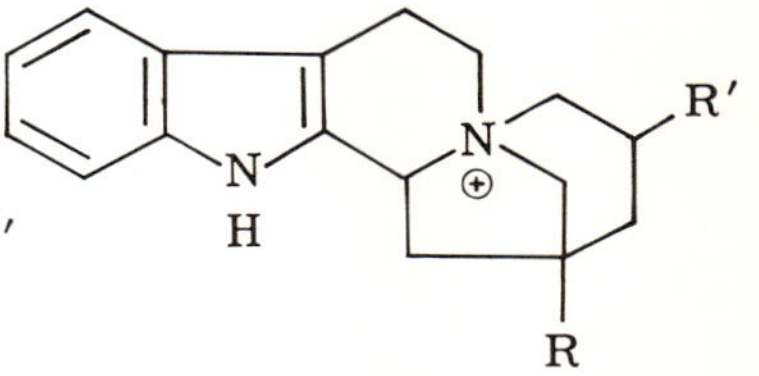

(**323**) R=Et, R′=H
(**326**) R=H, R′=Et

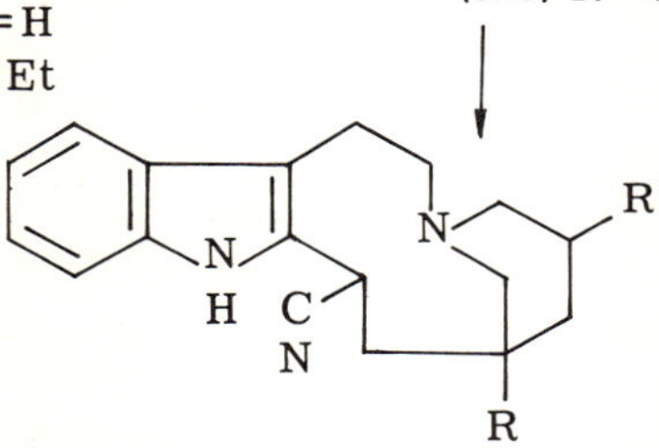

(**324**) R=Et, R′=H
(**327**) R=H, R′=Et

Added versatility was lent to this general reaction scheme by the finding that the piperidine ring need not be preformed. Thus treatment of the tetracyclic indole (**328**) with sodium acetate in acetic anhydride leads to simultaneous formation of the piperidone and cleavage of the bicyclic ring. Although no mechanistic work on this transformation has been carried out, some scheme such as that shown serves to explain the overall result (132). In yet a further simplication of the synthetic scheme these same workers found that the simple unsubstituted tetracyclic indole (**330**) affords the ring enlargement product (**331**) on treatment with α-chlorobutyric anhydride; a scheme analogous to that above can be formulated for this reaction. The product was in this case saponified and the resulting alcohol oxidized to the ketone. Treatment with base resulted in formation of **334** by internal alkylation; this intermediate was taken on to tubifoline and related alkaloids (133).

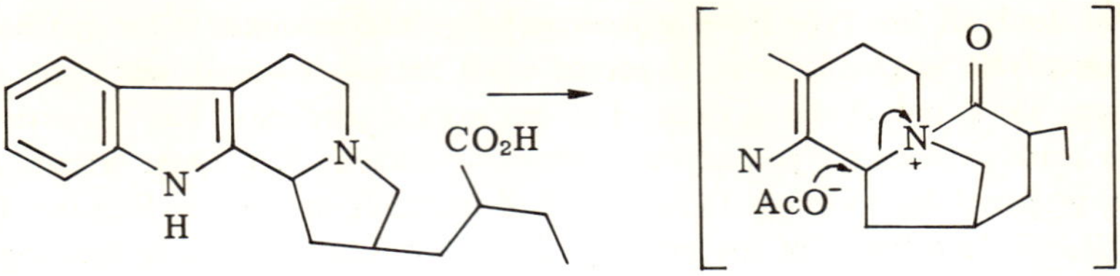

CO_2H

N, N, H, AcO⁻, O

(328)

N, H AcO, O

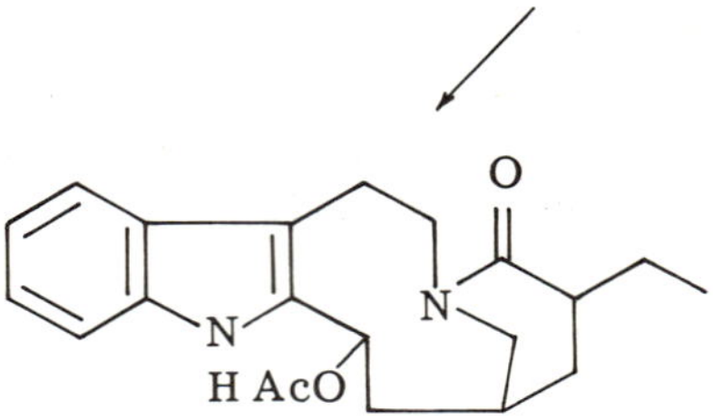

(329)

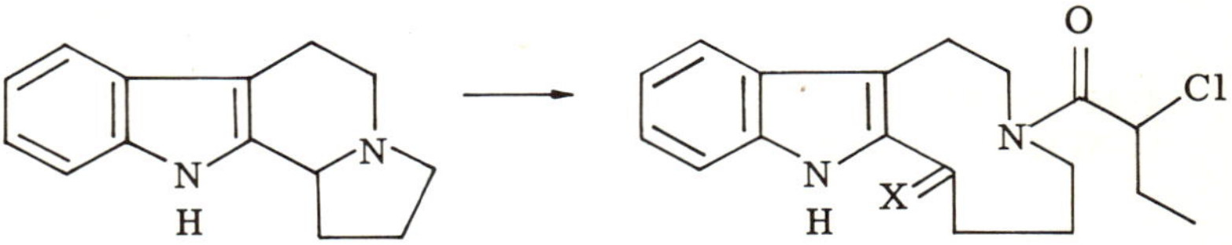

(330)

(331) X=<H, OCOCHClEt

(332) X=<H, OH

(333) X=O

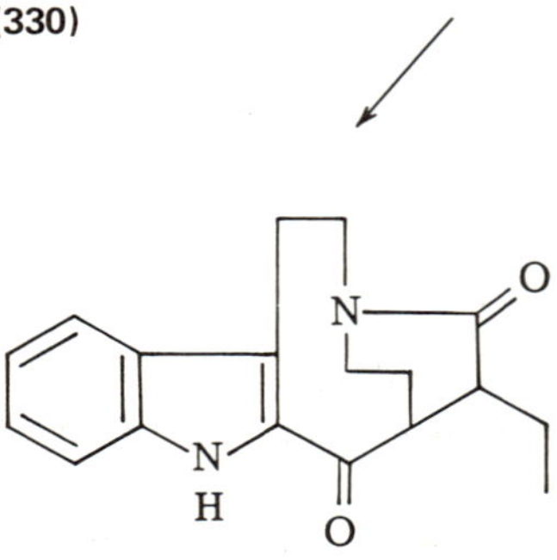

(334)

D. THE CYCLOPROPYLCARBINYL ION IN SYNTHESIS

The intraconversion of homoallylic alcohols and cyclopropyl carbinols or some derivative of each has been used extensively in steroid chemistry where the scheme is known as the *i*-steroid rearrangement. The scheme in fact is one of the methods of protecting the homoallylic system. The reaction has been used occasionally in steroids so as to take advantage of some of the reactivity of part of that system (eq. 13).

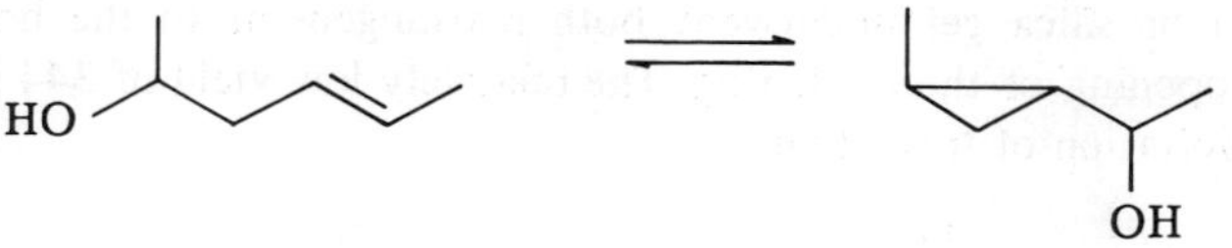

HO ⇌ OH (13)

The 7 position of a steroid is not easily activated towards alkylation without resorting to elaborate functional group manipulations. The *i*-steroid rearrangement provided a fairly direct method for achieving that aim. Thus Julia took advantage of the presence of oxygen in the 6 position of the *i*-steroid from cholesterol **(335)**. Oxidation of the alcohol gave ketone **(336)** and consequent activation of the 7 position. Alkylation with base and methyl iodide gave the dimethyl derivative **(337)**; reduction of the ketone back to the alcohol followed by rearrangement gave ultimately 7,7-dimethylcholesterol **(340)**. In an alternate procedure, the intermediate ketone was reacted with methyllithium. Rearrangement of that alcohol **(339)** afforded 6,7,7-trimethylcholesterol (134,135).

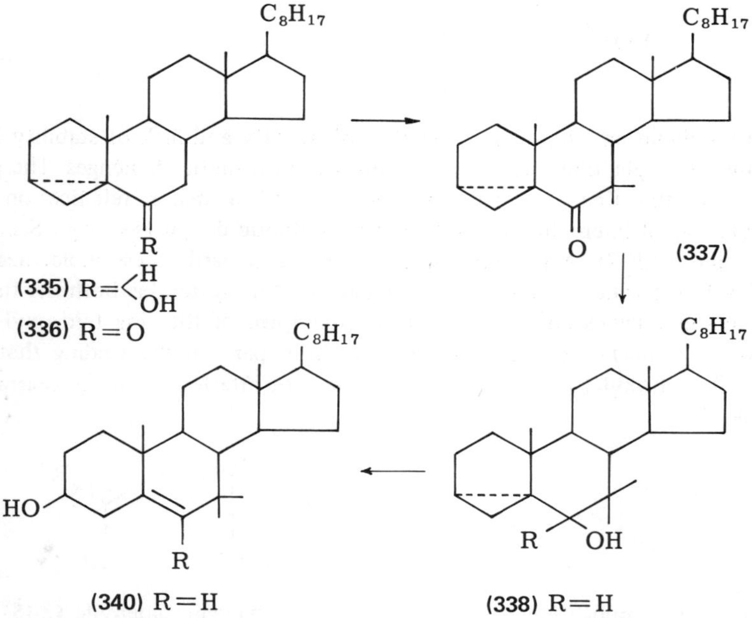

C_8H_{17} C_8H_{17}

R O **(337)**

(335) R=<H, OH

(336) R=O

C_8H_{17} C_8H_{17}

HO R R OH

(340) R=H
(341) $R{=}CH_3$

(338) R=H
(339) $R{=}CH_3$

The functionalization of the unactivated angular methyl group at the 19 position of a steroid has remained a major stumbling block towards a relatively simple preparation of the 19-nor compounds. An interesting scheme which took advantage of the presence of a hydroxyl group at 6 (and its proximity to 19) was recorded by several groups (136-139). The key reaction of this sequence was treatment of the *i*-steroid of the easily available dehydro*epi*androsterone with lead tetraacetate to afford the cyclic ether **(343)**. This last on treatment with aqueous acid or silica gel underwent both rearrangement to the homoallylic alcohol and opening of the oxide ring. The relatively low yield of **344** precluded efficient exploitation of this scheme.

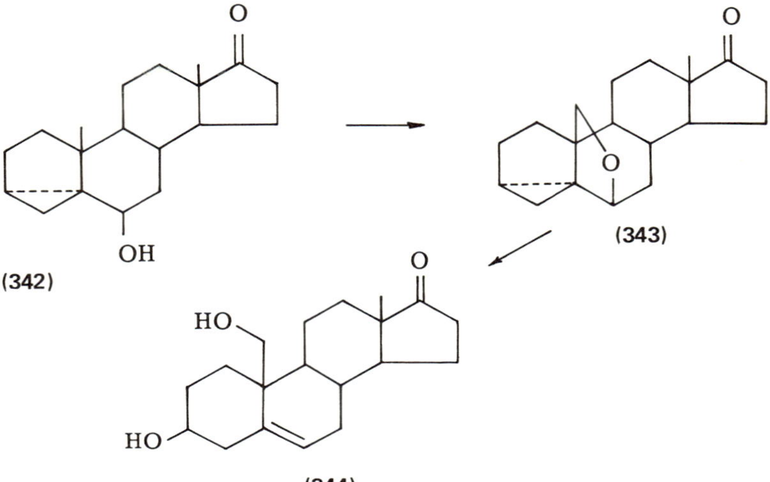

(342)

(343)

(344)

The combination of potent physiological activity and lack of stability have made the prostaglandins one of the current major synthetic challenges. The poor stability of the final molecule has also forced a major reliance on the incorporation of latent functionality in the synthetic design (see, e.g., Sections II.B.2.c and III.B.2). A general synthetic scheme towards these molecules has been developed which uses the [3.1.0] bicyclohexane system as both the future cyclopentane nucleus and also as a protected form of the ring hydroxyl and side-chain olefin. The transformation depends in part on the finding that the additional hydroxyl (or mesylate) group does not interfere with the rearrangement (eq. 14).

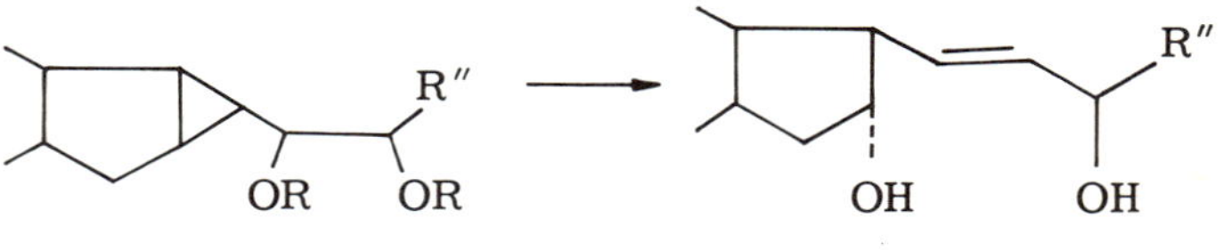

(14)

In the first application of that sequence, the bicyclic aldehyde (**345**) was converted by a series of standard steps to the ketone (**346**) which now carries the first of the two side-chains. The fatty acid side-chain was introduced by means of an alkylation reaction on the symmetrical ketone (**346**) and that reduced to **347**. The olefin was then treated with hydrogen peroxide and formic acid. The major product of the sequence was reported to be **348**, prostaglandin $F_{1\alpha}$ (140).

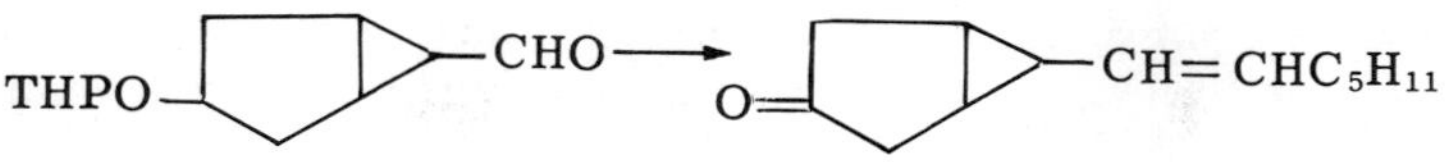

THPO— —CHO → O= —CH=CHC$_5$H$_{11}$

(345) **(346)**

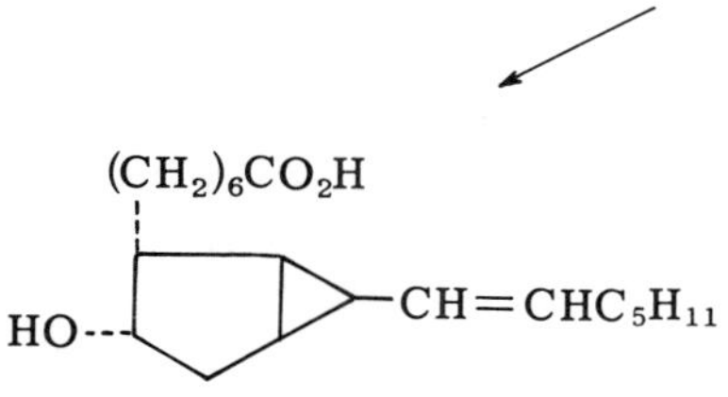

$(CH_2)_6CO_2H$

HO--- —CH=CHC$_5$H$_{11}$

(347)

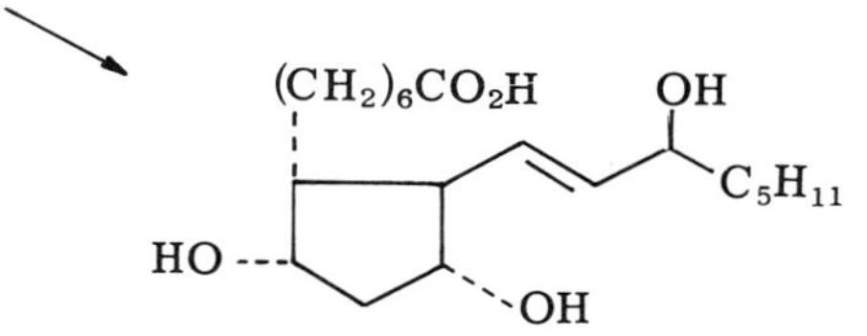

$(CH_2)_6CO_2H$ OH

C_5H_{11}

HO --- OH

(348)

The Upjohn group in one of their total syntheses of prostaglandin PGE_1 found that far more satisfactory results were obtained by working with a preformed glycol. Thus hydroxylation of the intermediate **(349)** (obtained by a scheme similar to that above) followed by mesylation afforded the bismesylate **(351)**. Solvolysis of that intermediate in aqueous acetone afforded the target molecule **(352)** (141). The best results were obtained by carrying out the synthesis with that isomer derived from the *endo* aldehyde as starting material (142). The same overall scheme was found widely applicable to the preparation of prostaglandins, such as, for example, **353**, which contains unsaturation in both side chains (143).

E. RING EXPANSION AS A SYNTHETIC METHOD

Ring expansions are, of course, some of the better known organic reactions and as such have been discussed at length elsewhere. Since this chapter deals with the strategy of synthesis, this is hardly the place to include an exhaustive discussion of such reactions. Rather we have, as before, chosen those applications of this reaction to synthesis which fit our thesis.

We have included ring expansion as a method involving "latency," since it is by this means that a difficultly accessible ring system is often established from a readily prepared precursor. This very principle was, in fact, invoked in order to prepare the tropone ring in a total synthesis of colchicine. Diels-Alder reaction

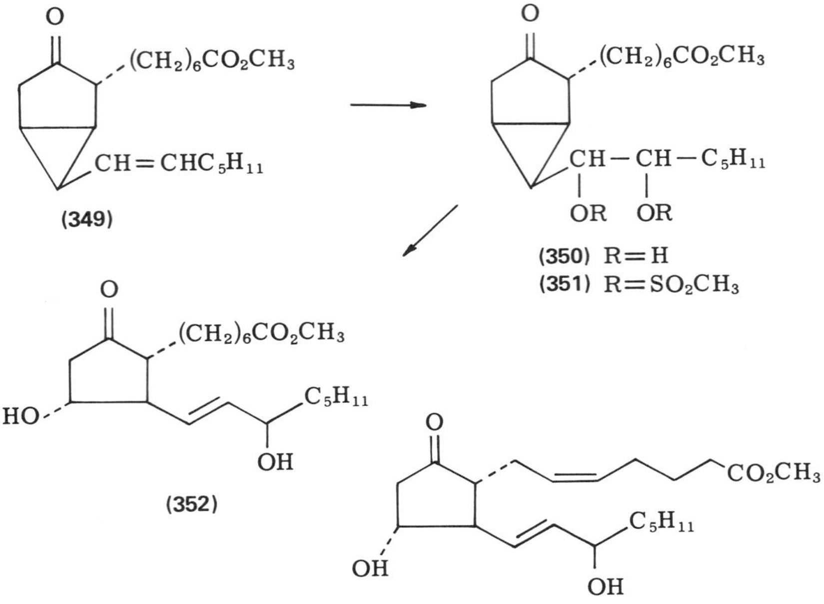

(349)

(350) R=H
(351) R=SO_2CH_3

(352)

(353)

of the pyrone **(354)** with the chloromethyl derivative of maleic anhydride gave (following the loss of carbon dioxide) the anhydride **(355)**. This last was converted to the diester. Treatment with strong base resulted in formation of an anion at the carbon bearing the carboxyl group. The anion then displaces the halogen from the chloromethyl group. Opening of the norcaradiene-like intermediate gives the desired cycloheptatriene **(357)**; that intermediate was then taken on to colchicine (144,145).

A considerable lore has been accumulated concerning the chemistry of octalone **358**. In the total synthesis of longifoline, Corey and co-workers took advantage of this for the preparation of the homologous intermediate **(361)**. Reaction of the octalone with ethylidene triphenylphosphorane gave the diene **(359)**. This was selectively hydroxylated by means of osmium tetroxide and converted to tosylate **(360)**. Treatment with lithium perchlorate and calcium carbonate led to the desired product **(361)** (146,147).

In a similar vein, a Wagner-Meerwein rearrangement was used to gain access to α-caryophyllene alcohol **(364)**. The tricyclic ketone **(362)** (obtained by photoaddition of 4,4-dimethylcyclopentene to 3-methylcyclohexenone) gave on reaction with methyllithium the tertiary alcohol **(363)**. This last on treatment with aqueous sulfuric acid underwent a simple 1,2 shift to afford the new ring system. Thus resort to photoaddition and ring expansion led to a simple elegant preparation of a complicated molecule (148,149).

Base-catalyzed ring expansion was the key reaction in a sequence for the formation of the dihydrothiazine system of Cephalosporin C. Thus the initial

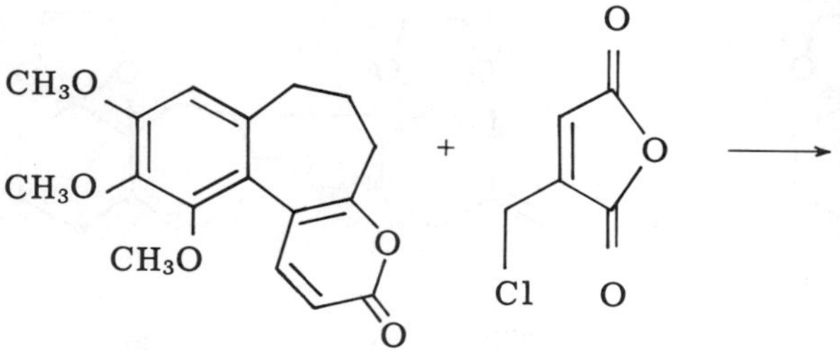

CH_3O CH_3O CH_3O + Cl O O O →

(354)

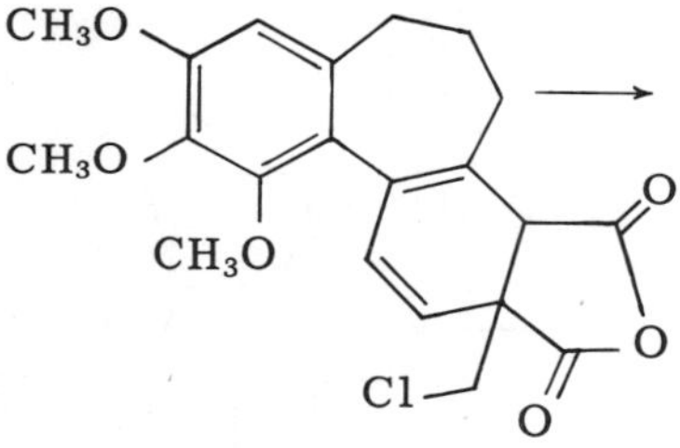

CH_3O CH_3O CH_3O Cl O O O → CH_3O CH_3O CO_2CH_3 Cl CO_2CH_3

(355) 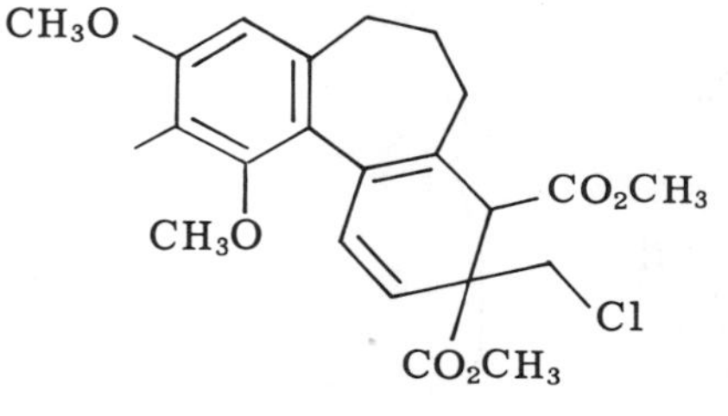 **(356)**

CH_3O CH_3O CH_3O CO_2CH_3 CO_2CH_3

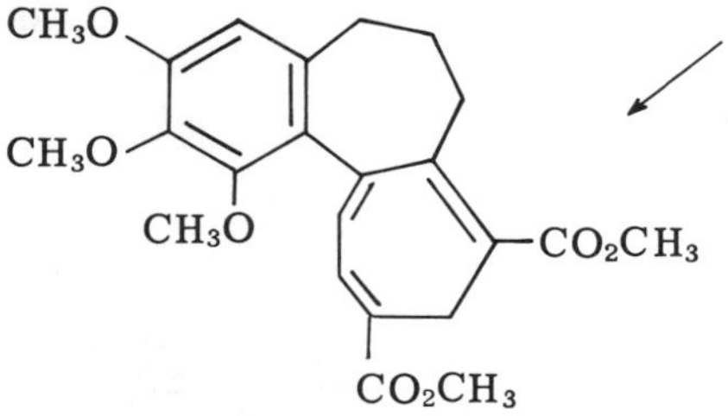

(357)

goal of the synthesis was the bicyclic thiazine **(365)**. When this was subjected to treatment with sodium acetate, ring expansion, presumably via the cyclopropyl compound, occurred. It is of note that this particular expansion not only affords a ring enlargement but provides, *inter alia,* the unconjugated lactone (150).

Pyrrolines such as **371** have proven useful intermediates in the preparation of compounds related to mesembrine (see Section VII.A). Those intermediates are conveniently prepared by a short sequence which terminates in a reaction which at least in the formal sense can be considered a ring enlargement. Reaction of the cyclopropyl nitrile **(367)** with an organometallic affords on careful work-up the imine **(369)**. Alternately, the nitrile is used as starting material for the aldehyde **(368)**; condensation of that aldehyde with a primary amine affords the *N*-substituted imine **(370)**. As the final step in the sequence the imines are heated with acid to afford the corresponding pyrrolines **(371)** (151,152).

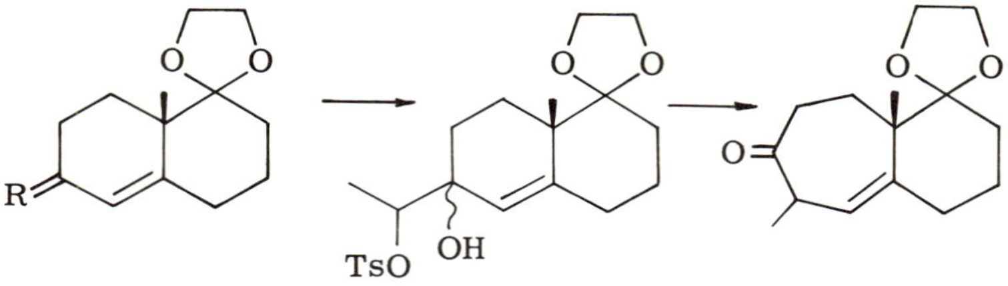

(358) R=O **(360)** **(361)**

(359) $R{=}CHCH_3$

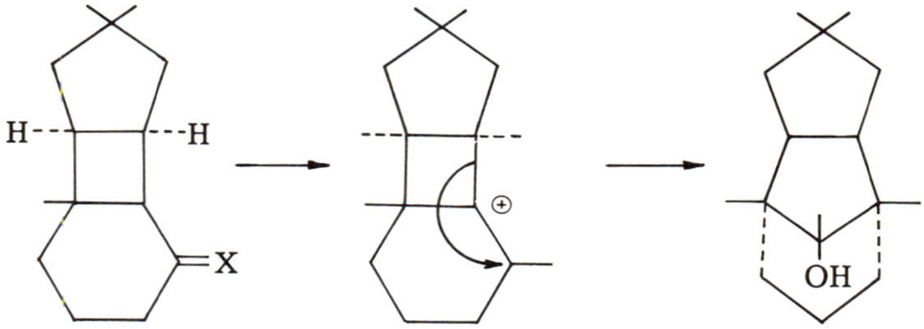

(362) X=O **(364)**

(363) $X{=}C(CH_3)(OH)$

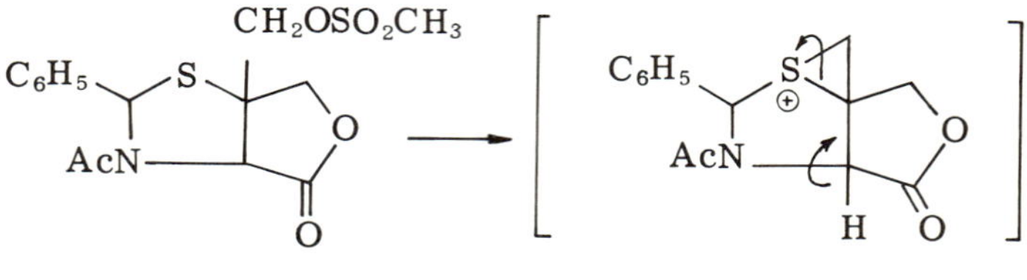

(365)

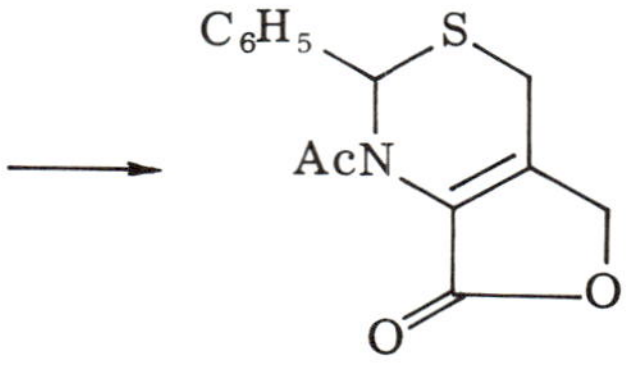

(366)

Ar CN
(367)

Ar CHO
(368)

(369) R=H, R′=CH_3
(370) R=CH_3, R′=H

(371)

V. Ethers of Phenols as Latent Cyclohexanes

It was all too easy for the organic chemist of but a few years ago, brought up on the concept of the inviolability of the aromatic ring, to forget that phenol is but a molecule of hydrogen removed from cyclohexenone. Once methods were developed to take advantage of this latency, however, the use of phenolic groups as latent cyclohexanones became almost routine. Syntheses designed around this concept permitted the use of "aromatic" chemistry in order to construct the carbon skeleton. In addition, the inertness of the aromatic ring toward many reaction conditions widened the scope of operations permitted in other parts of the molecule.

A. CYCLOHEXENONES BY REDUCTION

The latency alluded to above could not be fruitfully exploited until some laboratory method became available for the addition of a molecule of hydrogen to a phenolic ether. The development of the metal in ammonia reduction by Birch afforded the necessary method (153). The immediate product of that reduction is actually the enol ether; mild hydrolysis affords the unconjugated enone while treatment under more strenuous conditions gives the conjugated isomer (eq. 15). Occasional use has been made of the isomerization of the first-obtained diene to the conjugated isomer by means of strong base.

1. Reduction of Phenolic Ethers in Steroid Synthesis

(a) 19-Norsteroids by Reduction of Ring A Aromatic Precursors. Soon after the discovery of the metal in ammonia reduction, the reaction was applied to estrone (as its glycerol ether to enhance solubility in ammonia) by Birch to provide a facile preparation of the 19-nor compound. The initial product, the dihydro derivative (**373**), was hydrolyzed to the unconjugated ketone (**374**). Treatment of the last with base afforded the conjugated ketone (**375**) (154). Essentially the same scheme was used later for reduction of the A ring in a

synthesis of 19-norprogesterone **(376)** starting from aromatic A ring precursors (155). The general scheme gained special significance with the development of the "pill"; the ketone obtained by careful oxidation of the enol ether **(373)** (as the methyl ether) was reacted with the anion from acetylene. Hydrolysis under appropriate conditions afforded either **(377)** (156) or the unconjugated isomer **(378)** (157). These steroids are progestational constituents of some oral contraceptives. The bisnor-*D*-homosteroid derivative **(379)** was similarly prepared by Birch reduction of an aromatic A-ring precursor (158).

Several schemes for the preparation of steroids by total synthesis are now exploited commercially. One of these depends on the presence of an aromatic ring not only for the future A ring, but also to provide the proper activiation during construction of the skeleton. Thus the presence of the aromatic ring serves to stabilize the ambient cation probably intermediate in the "displacement" reaction on **380**. Stabilization of a cation similarly enhances the cyclization reaction to **382**. The sequence is completed by first catalytic reduction of the double bond in the D ring, then Birch reduction, in one or two stages (159-161).

(b) Preparation of Tricyclic Intermediates (BCD) for Total Synthesis. One of the early total syntheses of nonaromatic steroids had as its first goal the preparation of a tricyclic compound **(387)** incorporating the rings A, B, and C of the final product. Naphthol **384** was chosen as starting material to provide a two-ring moiety BC. Reduction of that compound with sodium in ammonia followed by hydrolysis gave **385**. The use of the free phenol to protect that ring from reduction is of note. Application of the homoannelation sequence to the ketone gave tricyclic intermediate **(386)**. The remaining aromatic ring was now reduced catalytically. The requisite intermediate **(387)** was isolated following oxidation of the thus obtained alcohol (162). Though the second reduction leads to *cis* hydrogenation, the presence of the carbonyl adjacent to the ring junction allows equilibration to the more stable *trans* compound.

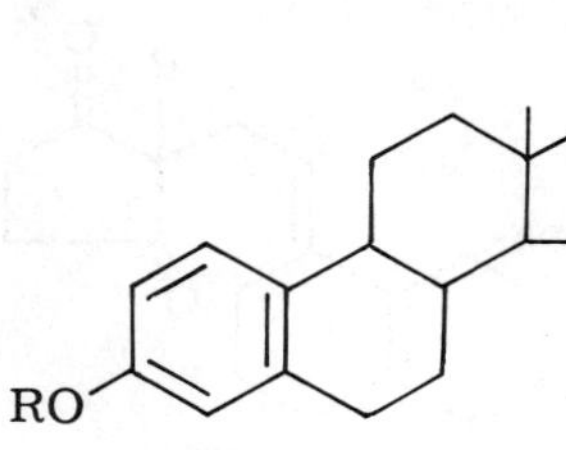

OH

RO

R= glyceryl
(372)

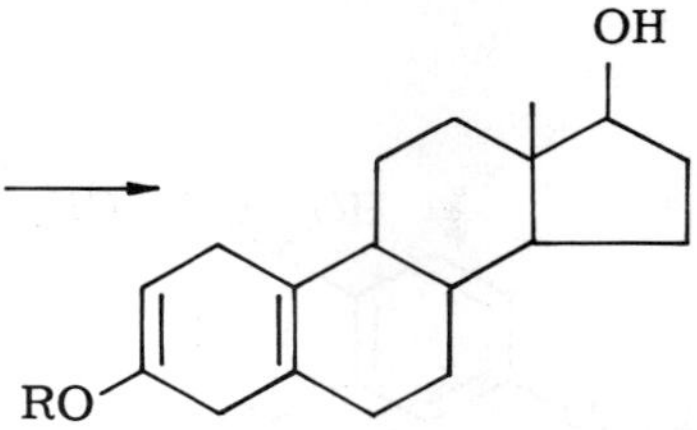

OH

RO

(373)

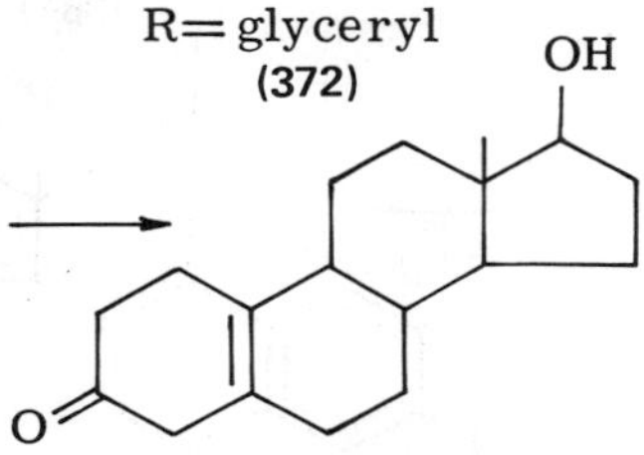

OH

O

(374)

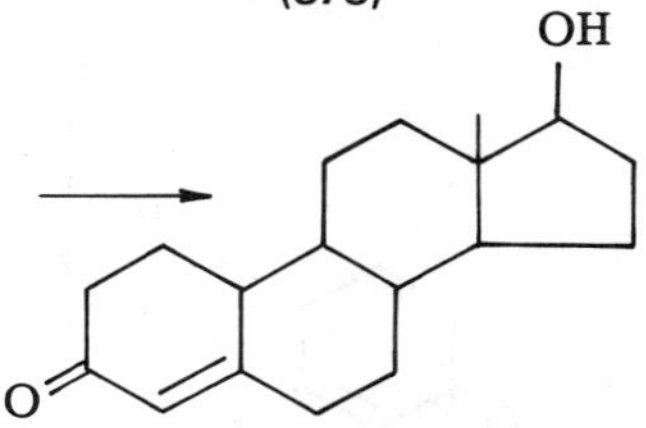

OH

O

(375)

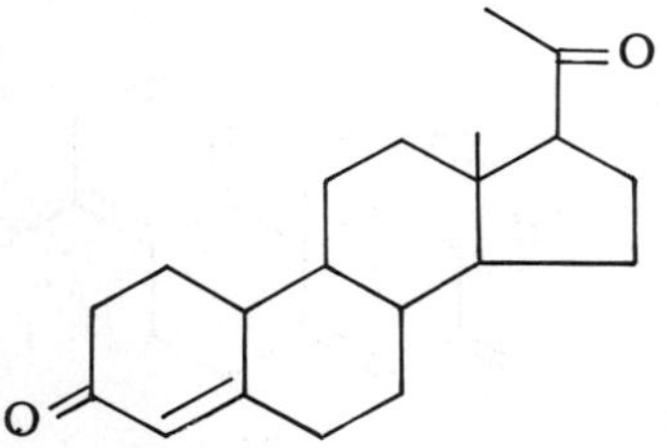

O

O

(376)

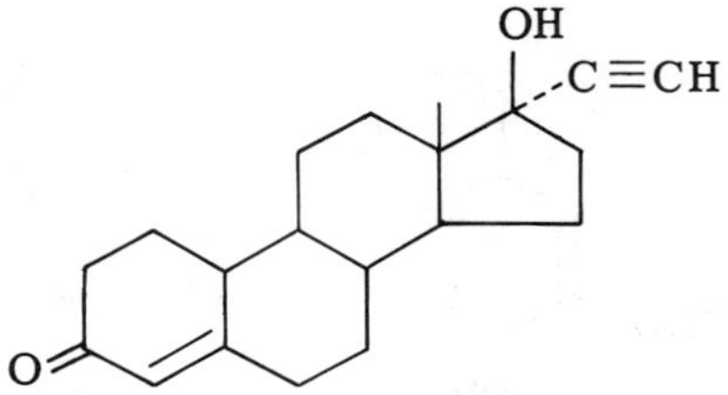

OH
C≡CH

O

(377)

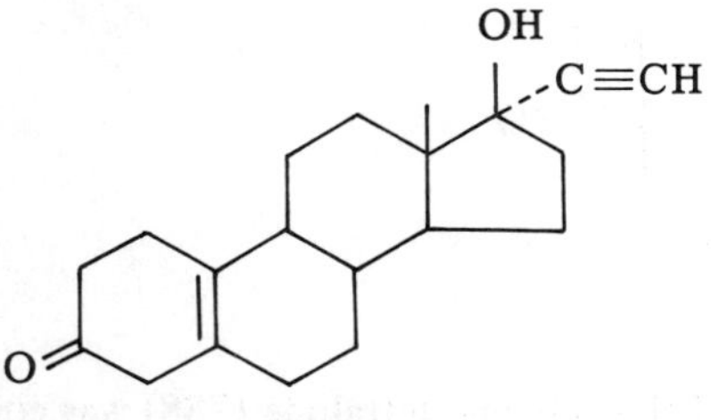

OH
C≡CH

O

(378)

OH

O

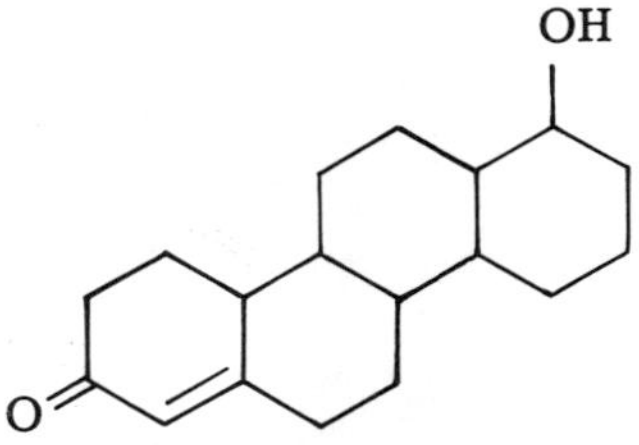

(379)

One of the more widely applied approaches to steroid total synthesis is that which has as its initial aim the construction of a tricyclic intermediate incorporating rings B, C, and D. By deferring construction of the A ring to a later stage, the alkyl substituent at 10 can be varied. The most frequent designs for use of this scheme further resemble each other in carrying the future B ring as an aromatic ring until a late stage of the synthesis.

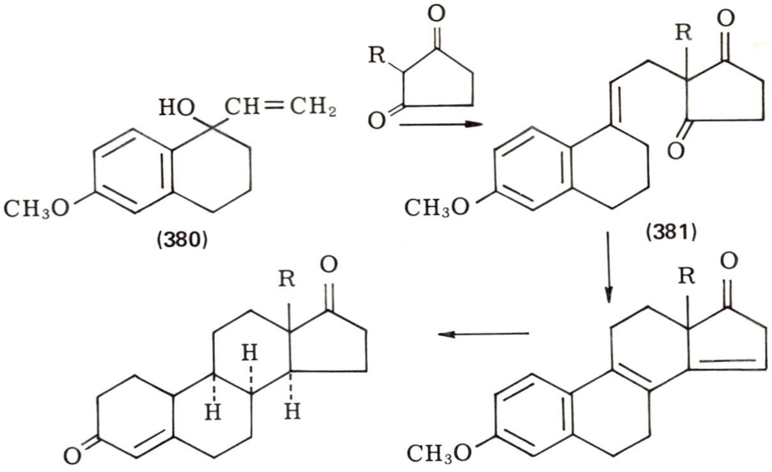

(380) (381) (382) (383)

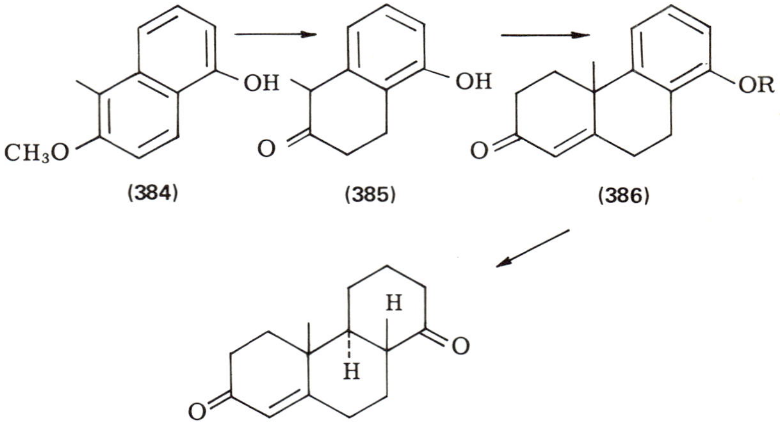

(384) (385) (386) (387)

In one of the early applications of this scheme, tetralone **(388)** was converted by homoannelation to the tricyclic intermediate **(389)** (the D ring was carried as a 6-membered ring for most of the synthesis). The double bond was reduced catalytically to afford the proper C/D ring fusion stereochemistry. Reduction of the alcohol followed by Birch reduction gave the enone **(390)**. The position of the double bond in this compound, interestingly, was used to direct entry of the next group. Thus alkylation of the enolate of this unconjugated ketone went to afford the intermediate **(391)** (8).

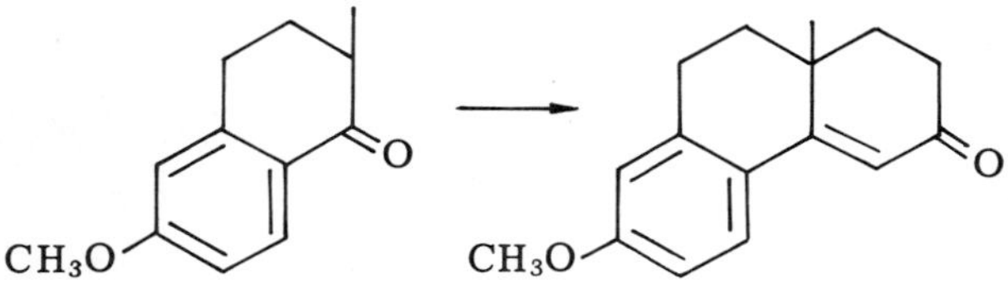

CH_3O O (388) CH_3O O (389)

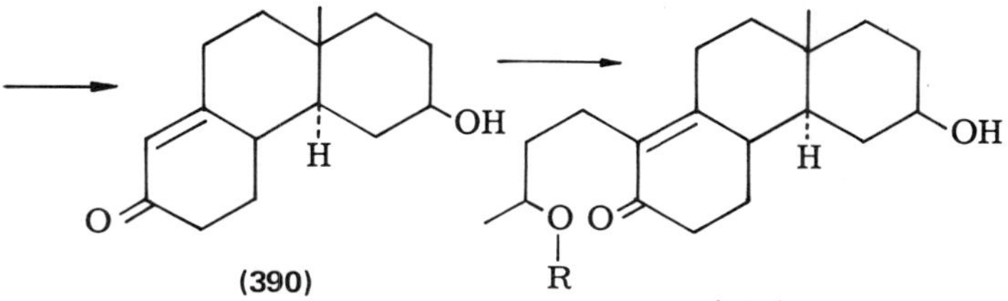

OH H O (390) OH H O O R (391)

A more direct approach to the total synthesis problem used the tricyclic intermediate **(396)**, which possesses the essentially complete BCD portion of the target molecule. By a series of reactions, the tetralone **(392)** was converted to the keto nitrile **(394)**. This gave the tricyclic compound **(395)** when treated under the conditions of the Stobbe condensation. Decarboxylation followed by catalytic reduction and reduction of the ketone gave the tricyclic intermediate **(396)** (163,164). Birch reduction by standard procedures afforded the necessary precursor **(397)** for elaboration of the A ring (165). Starting with the *d*-isomer (resolved at the unsaturated acid stage), the group at Roussel-Uclaf prepared optically active 19-nortestosterone, adrenosterone, and cortisone as well as several other steroid hormones (165,166).

In a modification of this scheme, the ketonitrile **(393)** was alkylated with allyl chloride and the double bond reduced. When subjected to Stobbe condensation, this afforded a tricyclic compound containing an angular propyl group. That intermediate was subjected to the same scheme as above to afford **398**, used then to prepare modified steroids (167).

Reduction of the tricyclic intermediate as the free phenol by means of high-pressure hydrogenation followed by oxidation of the alcohol afforded the saturated compound **(400)** which now possesses the *cis* B/C ring juncture. Elaboration by homoannelation led to isotestosterone (168).

In the total synthesis of the steroid alkaloid conessine, a similar catalytic reduction was used. First, the keto alcohol **(401)** (obtained by reduction of **102**; Section II.B.1.b) was reacted as its tosylate with hydroxylamine. The resulting nitrone **(402)** was reduced to the amine and acetylated. The phenolic methyl group was then removed and the product **(403)** subjected to high-pressure hydrogenation to yield after oxidation **404** (42).

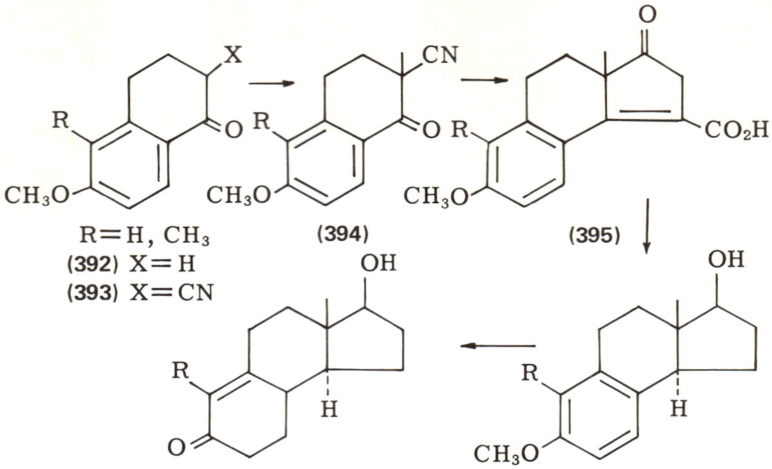

$R=H, CH_3$
(392) $X=H$
(393) $X=CN$

(394) **(395)** **(396)** **(397)**

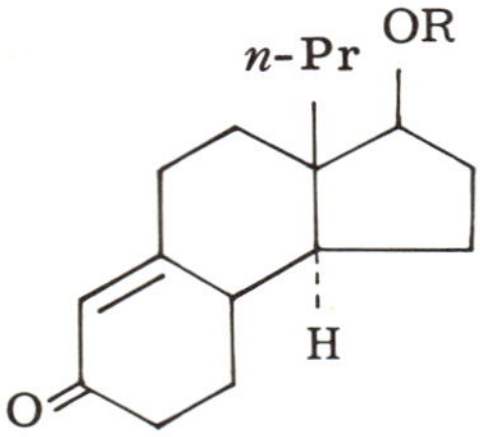

(398)

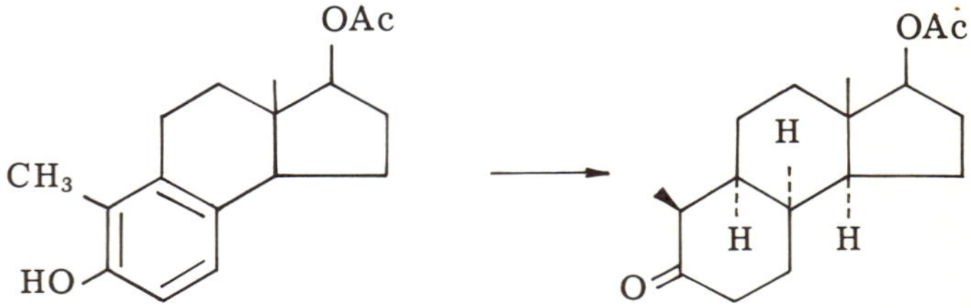

(399) **(400)**

(c) Preparation of 18-Nor-D-homosteroids. In the extensive total synthesis work of the Johnson group the key intermediate in each case was a tetracyclic compound in which the future D ring was present as an anisole moiety. In that work, these compounds were constructed by two successive applications of the homoannelation reaction to the tetralone **(405)** (73). Following some manipulation of the functionality in the A ring appropriate to the synthesis at hand, modification of the anisole was undertaken. First, lithium-ammonia reduction of the styrene bond proceeded to give the desired *trans* B/C junction **(406)**. More vigorous Birch reduction then afforded the mixture of isomeric enol ethers **(407)**

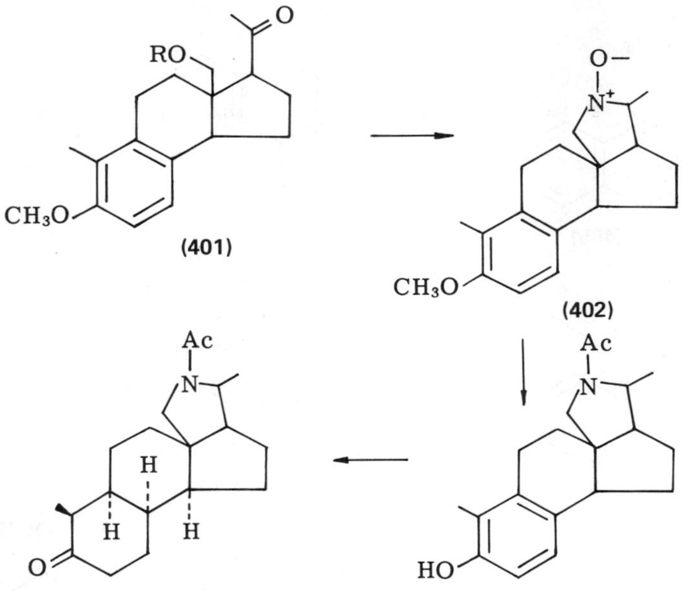

(401) **(402)**

(404) **(403)**

which on hydrolysis gave the two enones (**408a** and **408b**). The former on catalytic reduction gave directly the compound (**409**) used for further transformation. The isomeric enone gave the same compound on reduction in the presence of base. Alternately it proved possible to obtain **409** directly from the mixture of enones by reduction in the presence of base (169,170). The D ring was then modified by the sequence outlined in Section III.A.1.

In an interesting variation on this theme the presence of the aromatic ring was exploited in order to introduce an oxygen function at the future 11 position. Thus treatment of the tetracyclic ketone (**410**) with lead tetraacetate causes that compound to undergo acetoxylation at the benzylic "12" position. That acetoxy group is then readily eliminated to form the styrene (**411**). Treatment of the last intermediate with peracid followed by pyrolysis gives directly the 11-ketone, formed by rearrangement of the epoxide (171).

2. *Reduction of Phenolic Ethers in Tricyclic Terpene Synthesis*

The use of the anisole moiety as a latent cyclohexane in the construction of both triterpenes and triterpenoid alkaloids almost rivals the use of that scheme in steroids. The reasons for this widespread application are much the same as those alluded to above.

A feature common to many of these terpenes is the angular tricyclic system; ring A frequently contains the *gem*-dimethyl group (**411a**). Approaches have been developed in which either ring A or ring C is elaborated from an anisole. The preparation of an intermediate for the total synthesis of α-onocerin

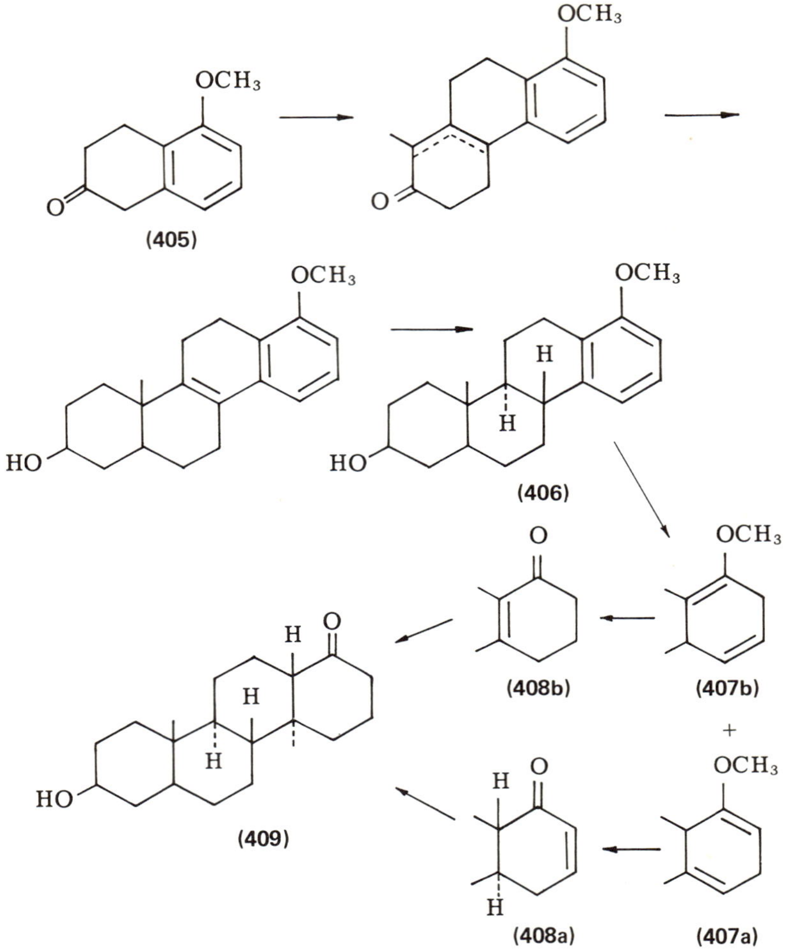

OCH3 OCH3 (405) OCH3 OCH3 H H HO HO (406) O OCH3 H O H (408b) (407b) + H O OCH3 HO H (409) H H (408a) (407a)

illustrates the latter approach. Condensation of 1-methyl-6-methoxy-2-tetralone with methyl vinyl ketone gave the tricyclic nucleus **(412)**. This compound was then subjected to exhaustive methylation. Catalytic reduction went stereospecifically to give **414**. This last on Birch reduction following work-up gave the desired enone (46).

A similar approach was used in the course of the total synthesis of a degradation product of phyllocladene. The tricyclic compound **(416)**, obtained by the same sequence of reactions as **414**, was reduced by means of a Wolff-Kishner reaction. The free phenol was then obtained by demethylation. The aromatic ring in this case was obtained by high-pressure catalytic reduction. Oxidation of the resulting alcohol gave **417**; the carbonyl group adjacent to the ring fusion ensures isomerization to the more stable ring fusion (88).

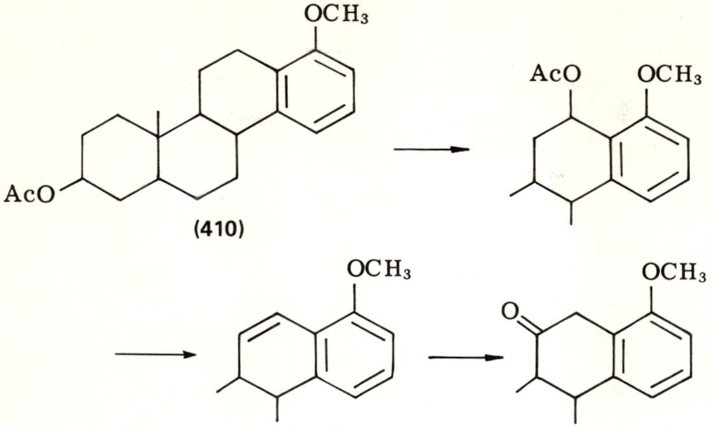

OCH3

AcO

AcO OCH3

(410)

OCH3 OCH3

O

(411)

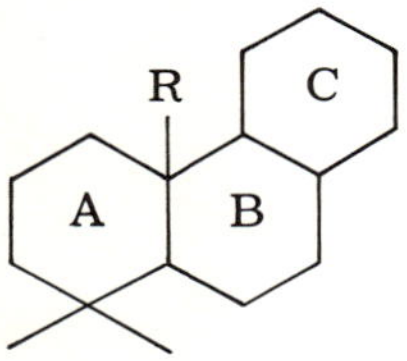

R C

A B

(411a)

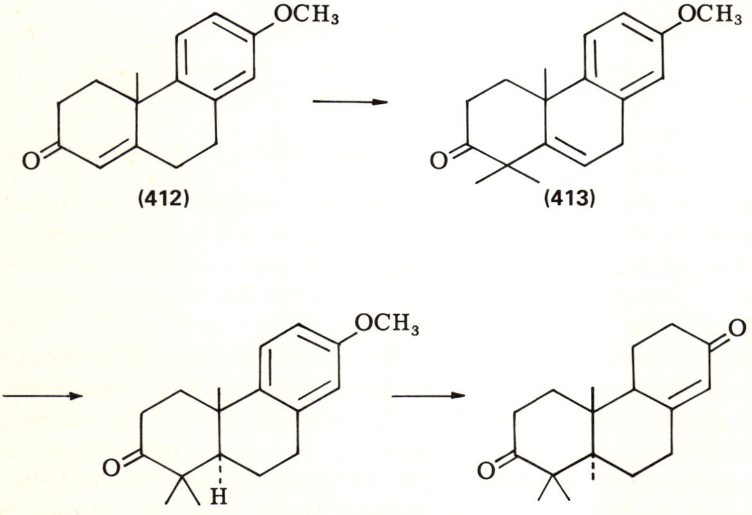

OCH3 OCH3

O O

(412) (413)

OCH3 O

O O

H

(414) (415)

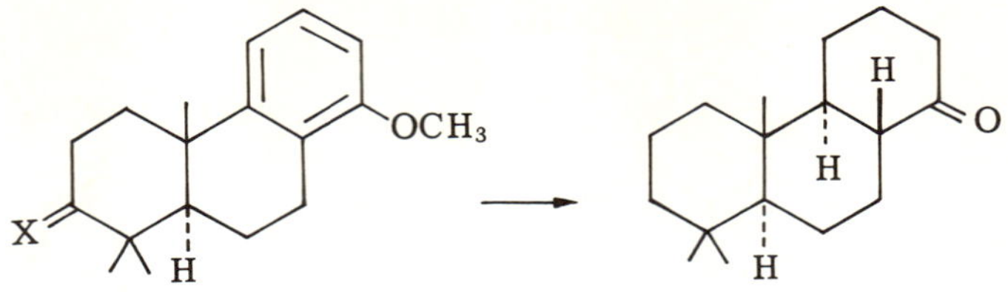

X
OCH3
H
H
H
O
H

(416) X=O
(417) $X=H_2$
(418)

An independent synthesis with the same goal finally went through a similar sequence. The "desketo" compound was obtained in this case by cyclization of the anisole **(419)** by means of polyphosphoric acid. The reduction of the C ring, accomplished in this work by Birch reduction, gave the enone **(421)** (48).

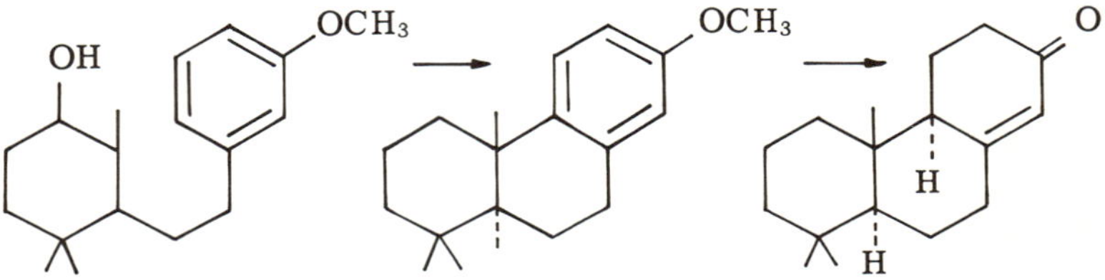

OH
OCH3
OCH3
O
H
H

(419) (420) (421)

The total synthesis of cassaic acid similarly departed from the tricyclic intermediate **(414)**. In this case the ketone was reduced to the alcohol **(422)**. Metal in ammonia reduction afforded the intermediate **(423)** (172).

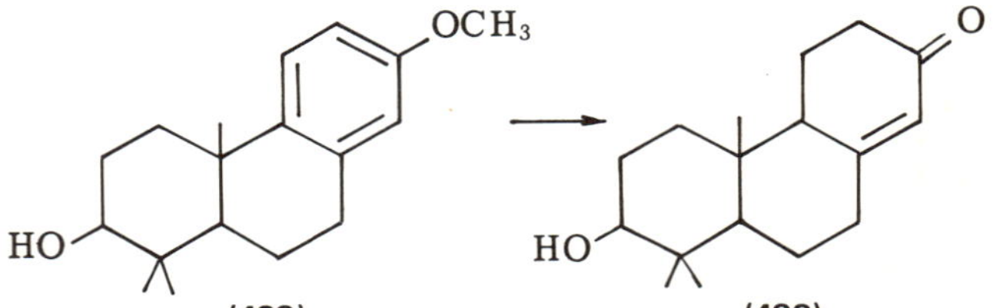

OCH3
HO
O
HO

(422) (423)

The absence of an angular methyl group at the A/B fusion of rimuene permitted the use of an aromatic moiety for the eventual A ring in a total synthesis of this terpene. Manipulation of the functional groups of **424** led to **425**. This compound, on Birch reduction, gave the enone **(426)**, which was converted to **427** by a sequence similar to that used to obtain **414** (173). An analogous scheme starting with 6-methoxy-2-tetralone gave the compound lacking the B/C angular methyl group (174).

This general approach has proven equally applicable to the preparation of more highly functionalized terpenes. In a synthesis of a kaurene derivative, the starting material was again the familiar tetralone. Homoannelation gave the tricyclic ester **(428)** which was then methylated to **429**. Catalytic reduction and

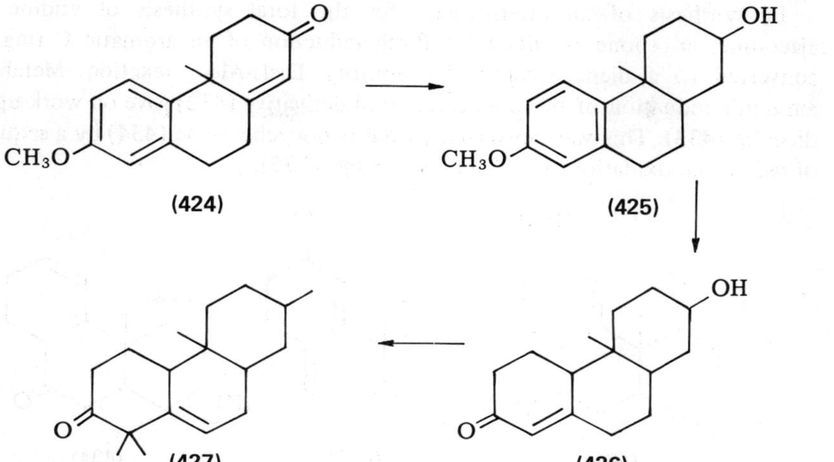

(424) (425) (426) (427)

the removal of the ketone afforded the intermediate **(430)**. Reduction of the aromatic ring was, in this case, achieved by catalytic hydrogenation of the free phenol. Oxidation of the thus obtained alcohol gave **431** (17).

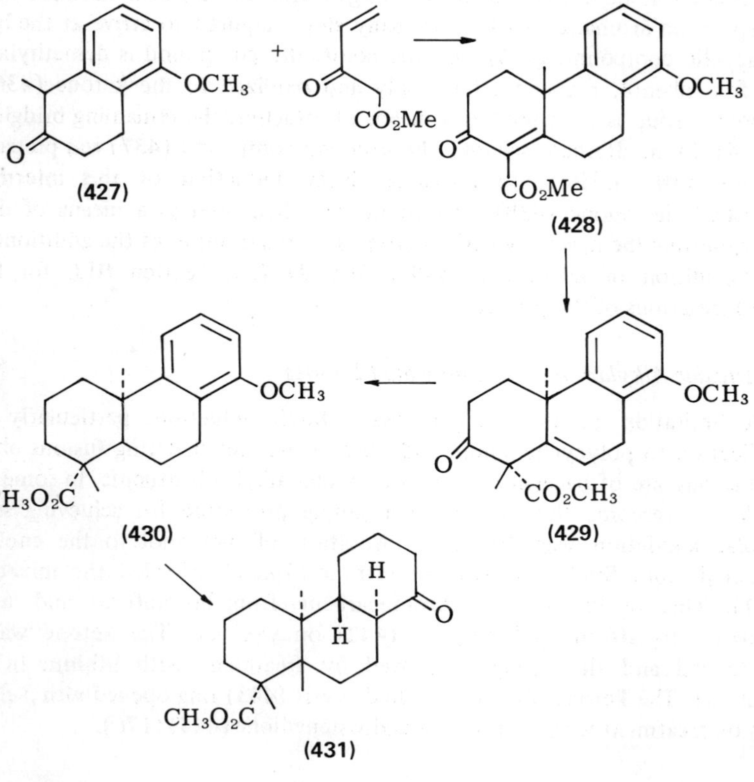

(427) (428) (429) (430) (431)

In synthesis of an intermediate for the total synthesis of atidine and ajaconine, an enone obtained by Birch reduction of an aromatic C ring was converted to a diene receptor for eventual Diels-Alder reaction. Metal and ammonia reduction of the podocarpic acid derivative (**432**) gave on work-up the dienone (**433**). This was converted to the endocyclic diene (**434**) by a sequence of reduction, oxidation, and elimination steps (175).

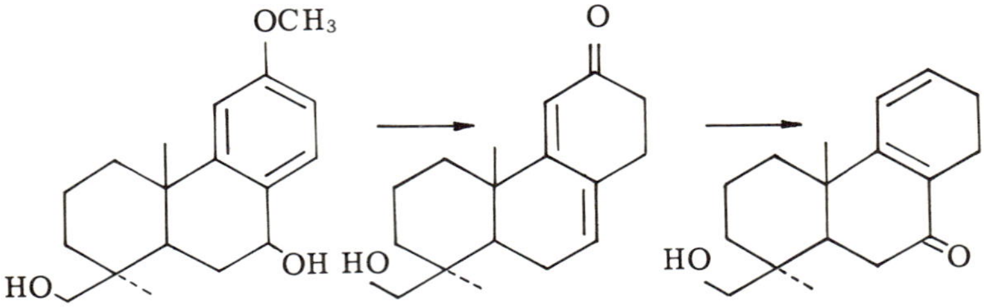

OCH_3 O HO OH HO HO O

(432) **(433)** **(434)**

The latent functionality of the anisole moiety has proven particularly useful in the total synthesis of terpene alkaloids such as garryine and veatchine. The syntheses of the New Brunswick group generally start with a relatively simple two-ring tetralone destined to furnish rings B and C of the final molecule. Ring C is kept as an aromatic ring for the many steps required to arrive at the bridged tetracyclic compound (**435**). At this point, the compound is demethylated to the free phenol, reduced catalytically and oxidized to the ketone (**436**); the carbonyl group is then used as a means of attaching the remaining bridging ring (13,14). In an alternate scheme, the isomeric compound (**437**) was prepared by starting with an isomeric tetralone. Birch reduction of this intermediate furnished the enone (**438**). The enone was then used as a means of directly incorporating the necessary carbon atoms for construction of the additional ring; photoaddition of allene gave **439** (101,102). (See Section III.C for further transformations of the product.)

3. *Angular Alkylation of Unconjugated Enones*

A limitation of the straightforward Birch reduction, particularly in its application to polycyclic compounds, lies in the fact that ring fusions obtained in this way are of necessity devoid of angular alkyl substituents. In some recent work to overcome this deficiency a simple procedure for achieving such an angular alkylation was developed. Addition of methanol to the enol ether obtained from Birch reduction of estrone-17-ketal afforded the mixed ketal (**441**). This readily added dibromocarbene–from bromoform and tertiary butoxide–to afford cyclopropane (**442**) on work-up. The ketone was then reketalized and the halogen removed by treatment with lithium in liquid ammonia. The ketone obtained on hydrolysis (**443**) ring-opened with β-elimination on treatment with acid to give androstenedione (**444**) (176).

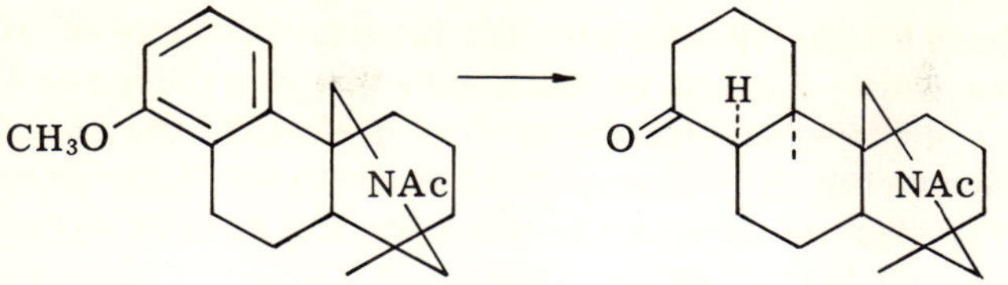

CH_3O NAc NAc O H

(435) (436)

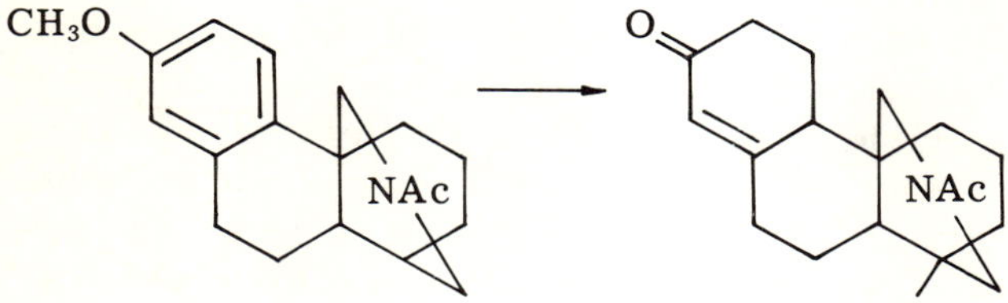

CH_3O O NAc NAc

(437) (438)

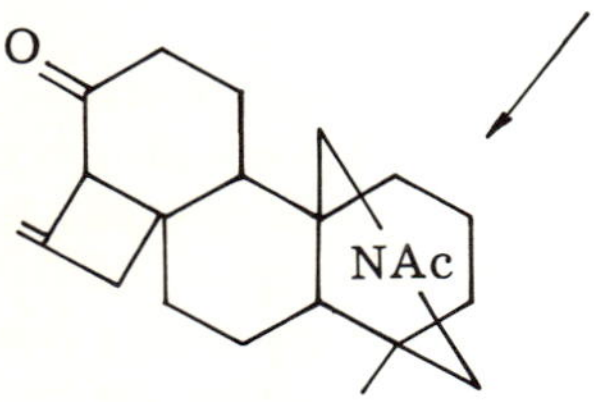

O NAc

(439)

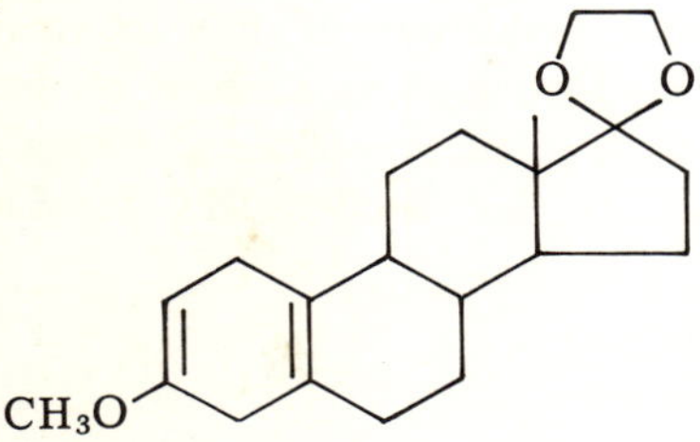

O O O O

CH_3O CH_3O CH_3O

(440) 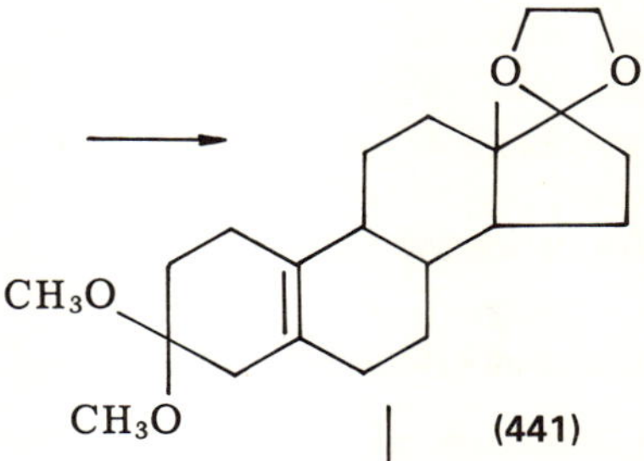 (441)

O O

X_2

O O

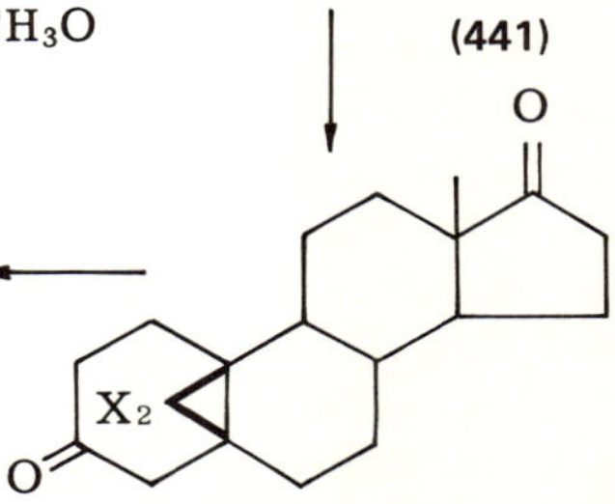

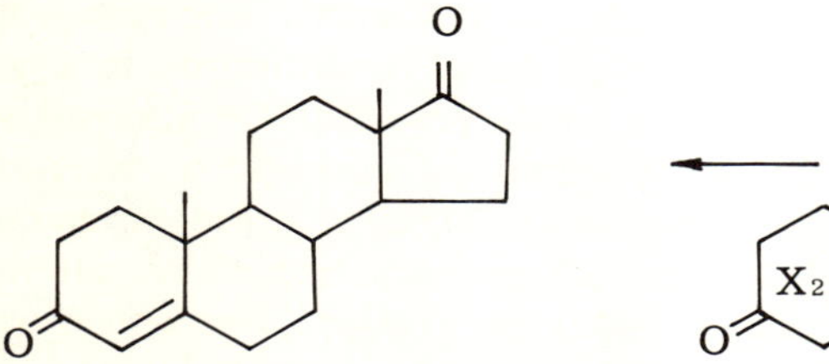

(444)

(442) X=Br
(443) X=H

It was subsequently found that the "carbene" from ethyl dizaoacetate also added to ketals obtained from metal ammonia reductions of anisoles (**445**). Completion of the scheme in this case led to the ketoester (**447**) (177). In that same study it was found difficult to add the reagent to internal olefins from reduction of polycyclic compounds. This shortcoming could, it was subsequently found, be circumvented by arranging for intramolecular delivery of the entering group. In an elegant synthesis for the carbocyclic bridged rings of the terpene alkaloids, first the acids **448** and **452** were obtained by addition of methanol to the Birch reduction products of the corresponding naphthoic acids. These were then converted to the diazoketones by treatment with oxalyl chloride followed by diazomethane. Those products (**449**,**453**) were then decomposed in the presence of copper sulfate to afford the carbene addition products. Treatment with acid led to opening of the cyclopropane ring and formation of the bridged bicyclic ketones **451** and **455**, respectively (178).

CH_3O OCH_3 (**445**) → MeO OMe CO_2Et (**446**) → O EtO_2C (**447**)

COR MeO OMe (**448**) R=OH (**449**) R=CHN_2 → O MeO OMe (**450**) → O O (**451**)

COR MeO OMe (**452**) R=OH (**453**) R=CHN_2 → O MeO OMe (**454**) → O O (**455**)

A somewhat similar reaction sequence has been developed for angular alkylation of the methyl ethers of α-diketones. Thus addition of Simmons reagent to the allyl alcohol (**456**)—obtained on reduction of the ketone—leads to cyclopropane (**457**). This last was converted to the desoxy compound (**458**) by oxidation of the alcohol, followed by Wolff-Kishner reduction. The scheme to the alkylated product (**459**) was completed by treatment with acid (179). A noteworthy feature of this scheme is the control of stereochemistry by the hydroxyl group, a well-known property of the Simmons reagent (180).

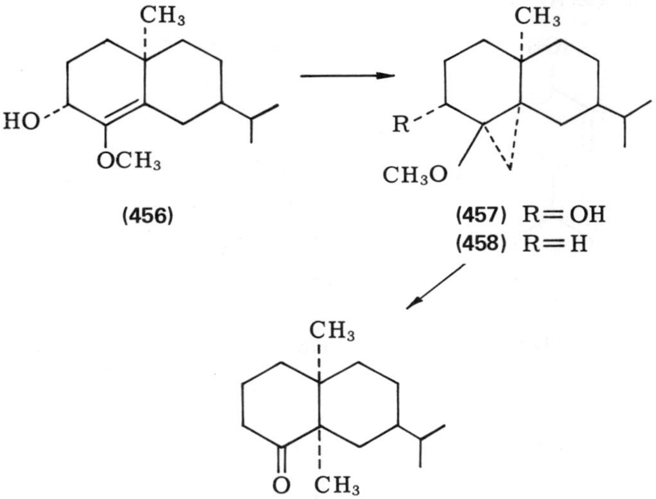

CH_3 HO OCH_3

(456)

CH_3 R CH_3O

(457) R=OH
(458) R=H

CH_3 O CH_3

(459)

B. 4-SPIROCYCLOHEXADIENONES FROM PHENOLS

Spiranes have historically constituted a class of organic compounds for which few methods of any generality were available. The finding that formation of the anion of phenols such as **460** resulted in internal alkylation to **461** thus made these bicyclic compounds readily available (181,182). The versatility of the method was subsequently shown by preparation of **463** and **465** by this general method (183,184).

This reaction was used by Masamune in an elegant method for the construction of the [3.2.1] bicyclooctane system shared by garryine and kaurene. In a model study, naphthol (**466**) was found to be smoothly converted to the bridged dienone (**467**) on treatment with base (185). The initial stages of the total synthesis of those compounds had as goal the tricyclic ketone (**470**). Thus treatment of intermediate (**468**) with base afforded the requisite dienone (**469**), oxygenated in one of the bridges for further manipulation at a later stage. Catalytic reduction gave the desired tricyclic compound (186).

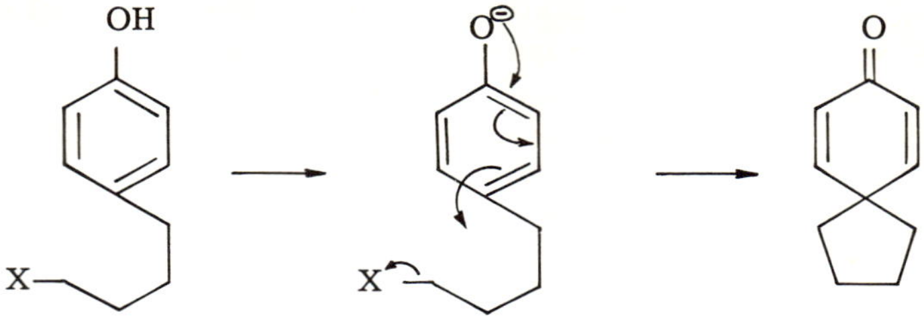

OH

X

(460) X=Br, OBs

O

(461)

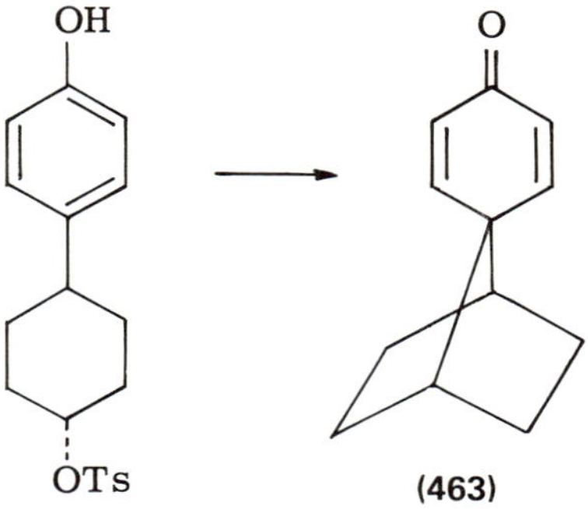

OH

OTs

(462)

O

(463)

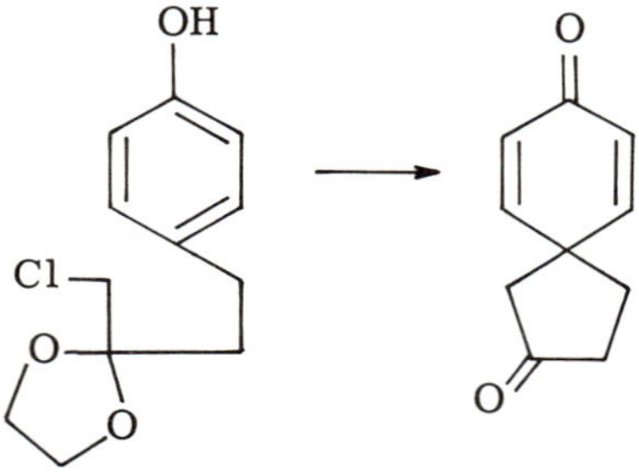

OH

Cl

O

O

(464)

O

O

(465)

R

X

HO

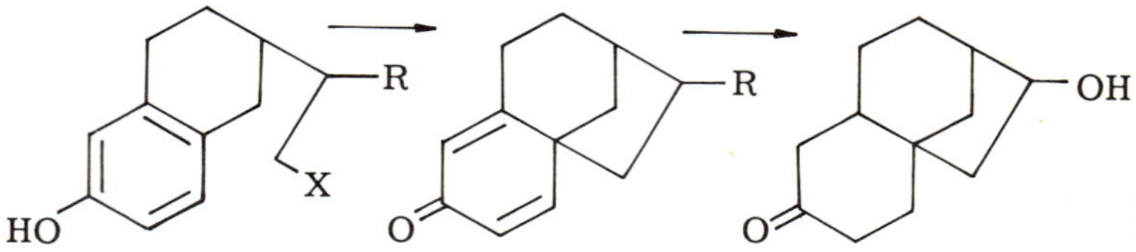

(466) R=H; X=OTs
(468) R=OTHP, X=Br

R

O

(467) R=H
(469) R=OTHP

OH

O

(470)

A careful perusal of the structure of cedrene **(474)** shows that this natural product also incorporates the spirocyclohexane moiety. A pair of independent total syntheses of that compound approached the preparation by preforming that unit. In one of these schemes, the phenol **(471)** was elaborated by a set of straightforward reactions. Treatment with strong base gave the dienone **(472)**. This last was transformed to the diol **(473)** by hydrogenation and then reaction with methylmagnesium bromide. Treatment with acid led to cedrene **(474)** (187).

The other group chose to form their dienone in exactly the opposite sense. Thus these workers prepared the phenol **475**. Treatment with base led to the same cyclohexadienone. The synthesis in this case was completed by formolysis of **476**—obtained from the dienone by reduction, reaction with methylenephosphorane, hydrochlorination-dehydrohalogenation, and finally Grignard reaction (188).

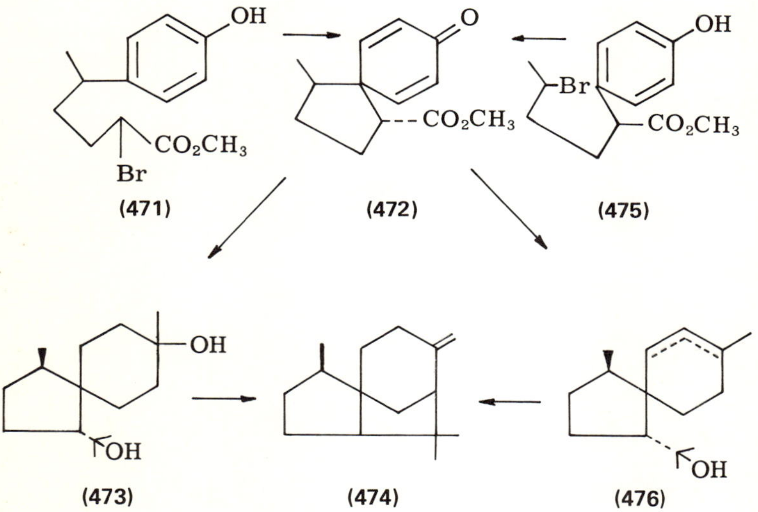

OH, CO_2CH_3, Br **(471)** → O, CO_2CH_3 **(472)** ← OH, Br, CO_2CH_3 **(475)**

OH, OH **(473)** → **(474)** ← OH **(476)**

VI. Electrocyclic Rearrangements in Synthesis

A. *THE CLAISEN REARRANGEMENT*

The aromatic Claisen rearrangement is one of the older reactions known to organic chemistry. Its use in the preparation of aromatic compounds is well documented and relatively straightforward. The aliphatic counterpart of the reaction, though developed only recently, has been put to good use in the synthesis of complex molecules. In part this reaction owes its utility to offering yet a further extension of the latent functionality of conjugated enones, and by that token, the anisole moiety; the starting materials for this reaction are allylic alcohols. Thus we can interrelate these functionalities as shown in eq. (16).

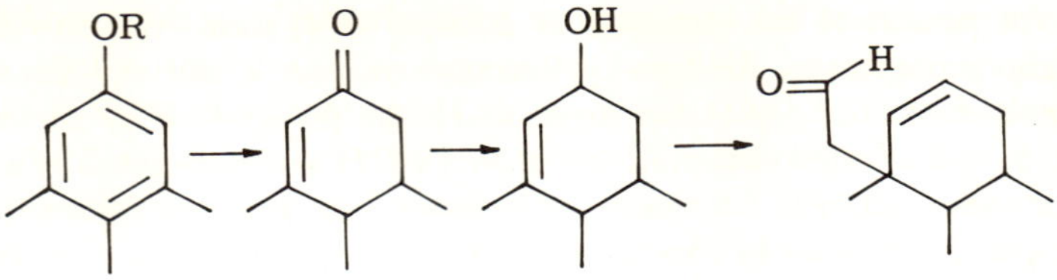

(16)

1. Rearrangement of Allyl Vinyl Ethers

The prototype for this reaction—allyl vinyl ether—will on heating rearrange to pent-4-enal **(478)** (189,190). The finding that vinyl ethers of allylic alcohols can be formed under mild conditions by mercuric sulfate catalyzed exchange with ethyl vinyl ether (191), coupled with a demonstration of the scope of the reaction (192), spurred its application to synthesis. Further work on the reaction showed it to proceed stereoselectively. Thus the ether derived from the β-alcohol **(479)** rearranges to the β-alkylated product **(480)**; though the corresponding α-ether **(481)** affords only a low yield of the aldehyde, this was shown to be the α-isomer **(482)** (193).

(477) (478)

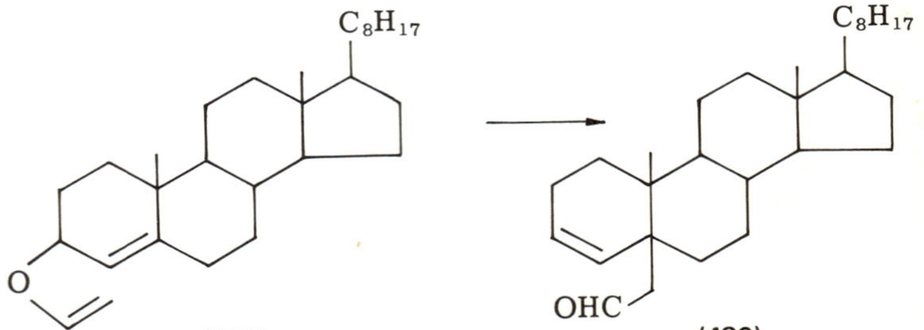

(479) (480)

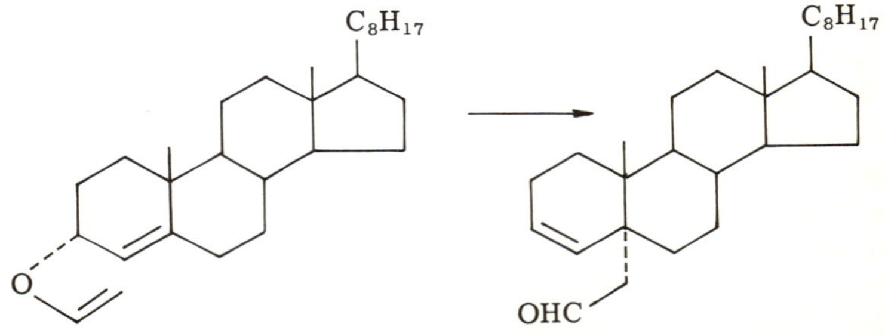

(481) (482)

The rearrangement has been used to good advantage for the introduction of two carbon fragments. Thus in the course of a synthesis aimed at some triterpenes, the enone **(483)** was reduced. The predominant product **(484)** was converted to its vinyl ether. Pyrolysis of this gave the angularly substituted aldehyde **(486)** (48,194). This newly introduced fragment was eventually used in the construction of a bridged ring (see Section II.B.1.c).

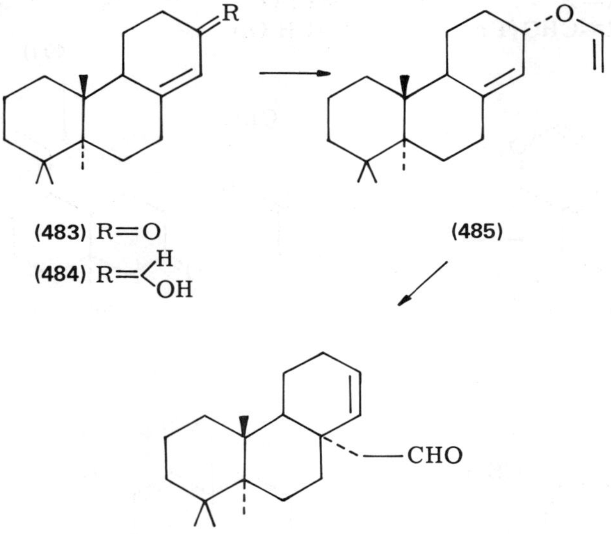

(483) R=O

(484) R=<H, OH

(485)

CHO

(486)

The reaction was used as the key step in a total synthesis of the monomeric precursor to β-onocerin. The initial goal in this synthesis was the allylic alcohol **(490)**. Thus formylation of the bicyclic ketone **(487)** followed by formation of the isopropyl ether gave **488**. Treatment of that enolate with sodium borohydride followed by acid gave the unsaturated aldehyde **(489)**. Reduction followed by exchange with ethyl vinyl ether gave the requisite intermediate. The last was then rearranged to **492** (195,196). In very much the same manner, the vinyl ether **(493)** was prepared from a ketone analogous to **487**. Rearrangement in this case afforded aldehyde **(494)**, prepared as a model for the A/B ring of pentacyclic triterpenes (197).

A variation on this sequence was used to prepare a tricyclic intermediate for the total synthesis of isopimaradiene. Aldehyde **(495)**, obtained by a similar sequence to that used for **489**, was reacted with methylmagnesium bromide to afford the carbinol **(496)**. This last was converted to the vinyl ether and this rearranged to **497**. Appropriate manipulation of the functional groups gave the acetyl-aldehyde **(498)**. Cyclization of that dicarbonyl compound afforded the tricyclic enone **(499)** (198).

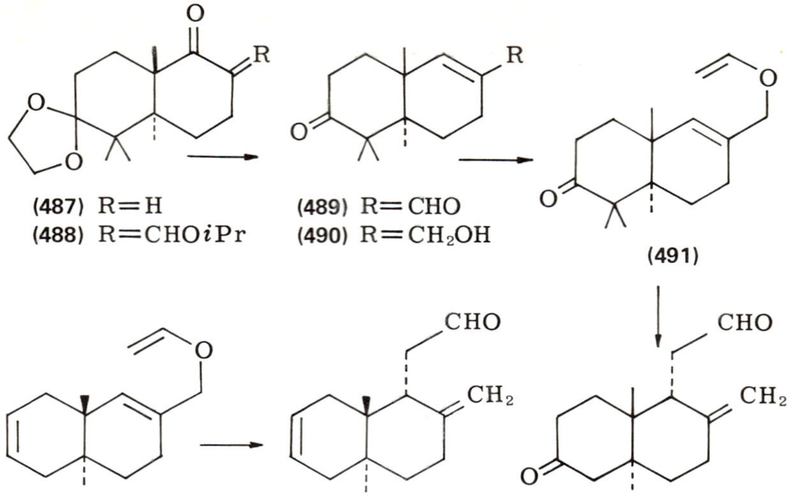

(487) R=H
(488) R=CHO*i*Pr

(489) R=CHO
(490) R=CH_2OH

(491)

(493)

(494)

(492)

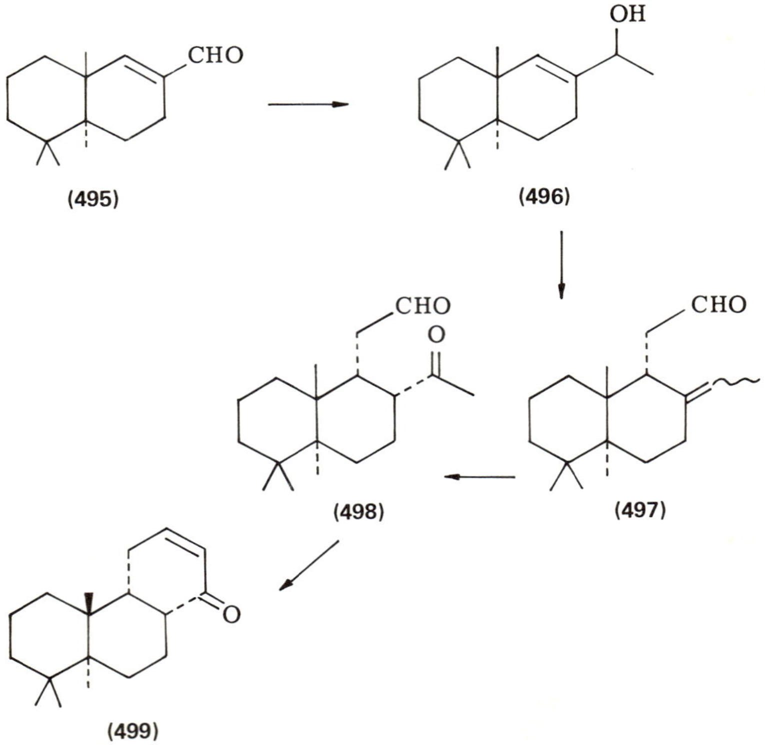

(495)

(496)

(498)

(497)

(499)

The rearrangement was used in a total synthesis of thujopsene to both introduce a quaternary carbon atom and at the same time provide a terminus for eventual intramolecular carbene introduction. Thus pyrolysis of the vinyl ether from the allyl alcohol **(500)** gave aldehyde **501**; this was converted in several steps to the tosylhydrazone of the chain-lengthened aldehyde. The product on treatment with base (to convert the hydrazone to the carbene) followed by irradiation gave a low yield of **503** (199).

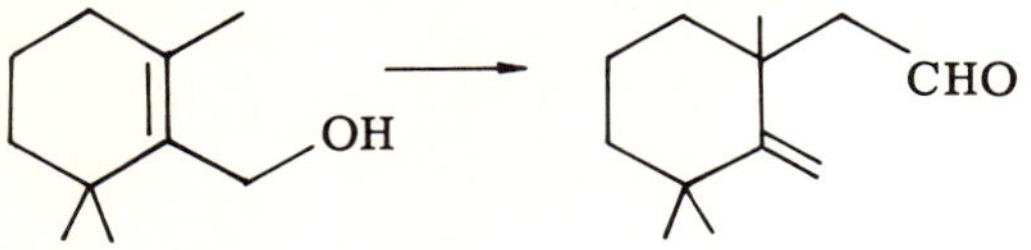

OH CHO

(500) **(501)**

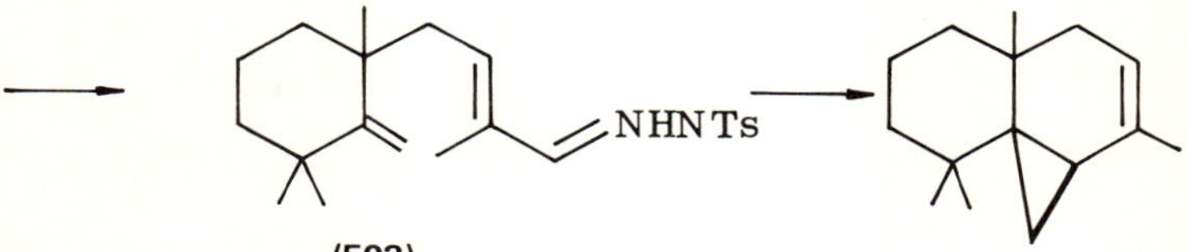

NHNTs

(502)

(503)

The vinyl ether rearrangement was used to good advantage in yet another scheme for the construction of the BCD bridged-ring system of atisine. Allyl alcohol **(504)** (obtained by reduction of the corresponding enone) was in turn converted to the vinyl ether and this pyrolyzed. The resulting aldehyde **(505)** was oxidized and esterified to give **506**. Dieckmann cyclization of that diester followed by work-up gave the bridged-ring compound **(507)** (200).

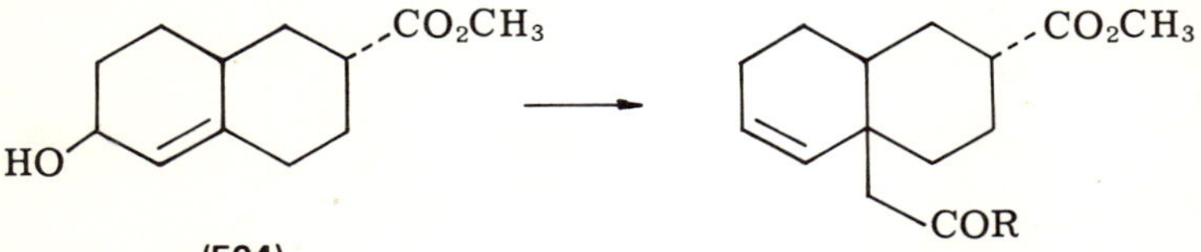

CO_2CH_3 HO CO_2CH_3 COR

(504)

(505) R=H
(506) R=OCH_3

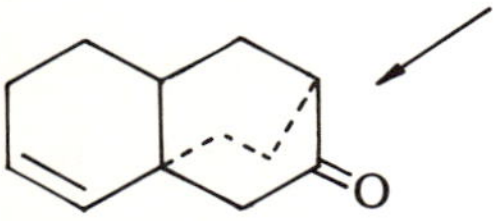

O

(507)

Use of the rearrangement in the preparation of intermediates in a recent total synthesis of ibogamine illustrates exploitation of the reaction to control stereochemistry. Reduction of the enone **(508)** by means of lithium aluminum hydride affords the *cis* compound. Claisen reaction of the corresponding vinyl ether then goes with retention of configuration to give the aldehyde (**510**). The carbonyl group was then reduced by Wolff-Kishner reaction and the remaining oxygen function converted to an amine. Treatment of that product (**511**) gave the aziridine (**512**), which finally became the nonindole moiety of ibogamine (201).

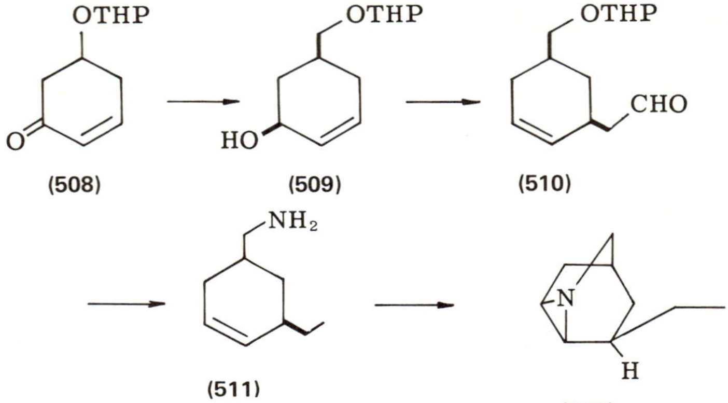

OTHP OTHP OTHP CHO O HO

(**508**) (**509**) (**510**)

NH_2 N H

(**511**) (**512**)

A simple, elegant, and recent total synthesis of bakuchiol methyl ether shows that the usefulness of the method is not restricted to alicyclic compounds. Rearrangement of the vinyl ether of geraniol (**513**) gave the doubly unsaturated aldehyde (**514**). This on reaction with anisylmagnesium bromide followed by careful dehydration (alumina) gave the desired triene (202).

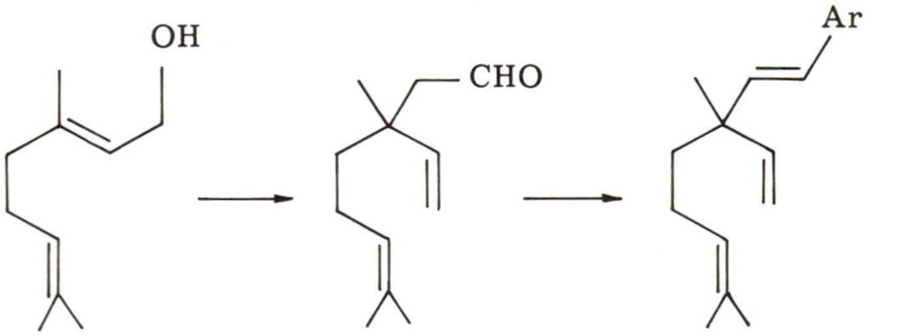

OH CHO Ar

(**513**) (**514**) (**516**)

$Ar = p-CH_3OC_6H_4$

A good deal of lore has accumulated for the exploitation of the relative positions of the double bond and activating group in Diels-Alder adducts. In an effort to circumvent the lack of regiospecificity of some of these reactions and thus make the "adducts" more readily available, Büchi developed an internal version of the Claisen rearrangement. Thus the starting materials were prepared by reaction of both the acrolein dimer (**517**) and methyl vinyl ketone dimer (**518**) with methylenephosphorane. Pyrolysis of the resulting dihydropyrans (**519**,**520**) afforded the products **521** and **522**, which have a double bond and carbonyl disposed as in a Diels-Alder adduct. Reaction of the ester **(523)** (obtained by reaction of **518** with the corresponding phosphorane) under rearrangement conditions gave the *cis* ketoester **(524)**. The isomeric *cis* starting acid gave a mixture of *cis* and *trans* products (203).

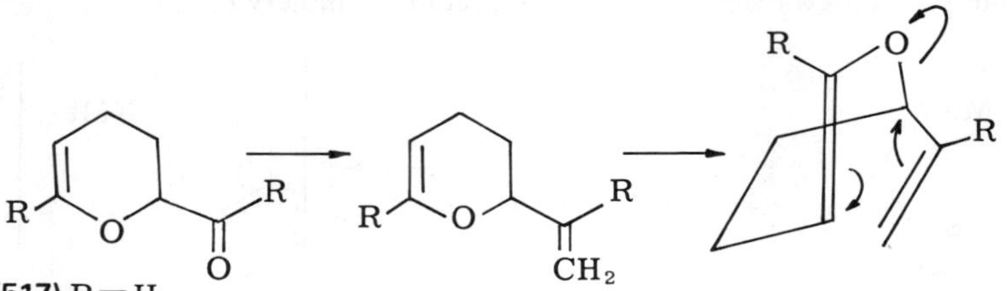

(517) R=H
(518) R=CH_3

(519) R=H
(520) R=CH_3

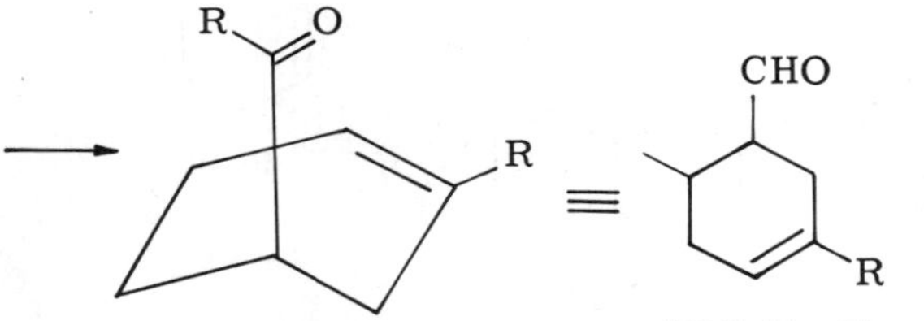

(521) R=H
(522) R=CH_3

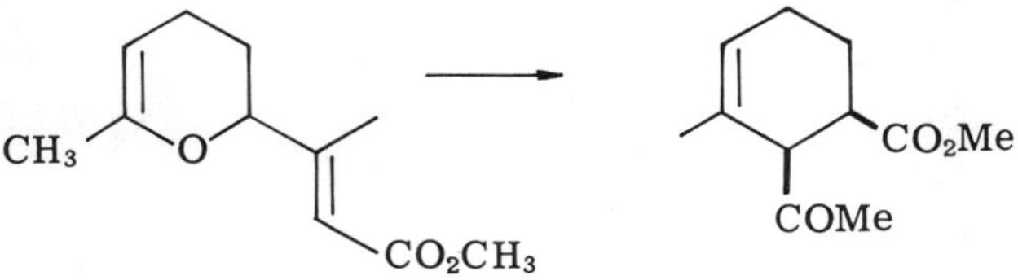

(523)

(524)

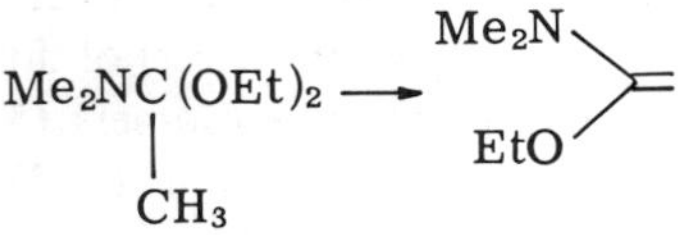

$Me_2NC(OEt)_2$ (with CH_3) →

(525)

2. *Rearrangement of Other Unsaturated Allyl Ethers*

Treatment of dimethylacetamide diethyl acetal with base affords a mixture which consists largely of the unsaturated compound **(525)**. This, when heated with allyl alcohol, leads in good yield to the amide **(526)**. The reaction apparently consists of an alcohol interchange, and then a cyclic rearrangement quite analogous to that of allyl vinyl ethers (204). The reaction on further investigation was found to be quite general. Thus each of the unsaturated cyclohexenols (**527**,**528**) rearranges stereospecifically to the corresponding acetamide (**529**,**530**). This rearrangement then complements that of the vinyl ethers in providing a means of adding an acetamide moiety (205).

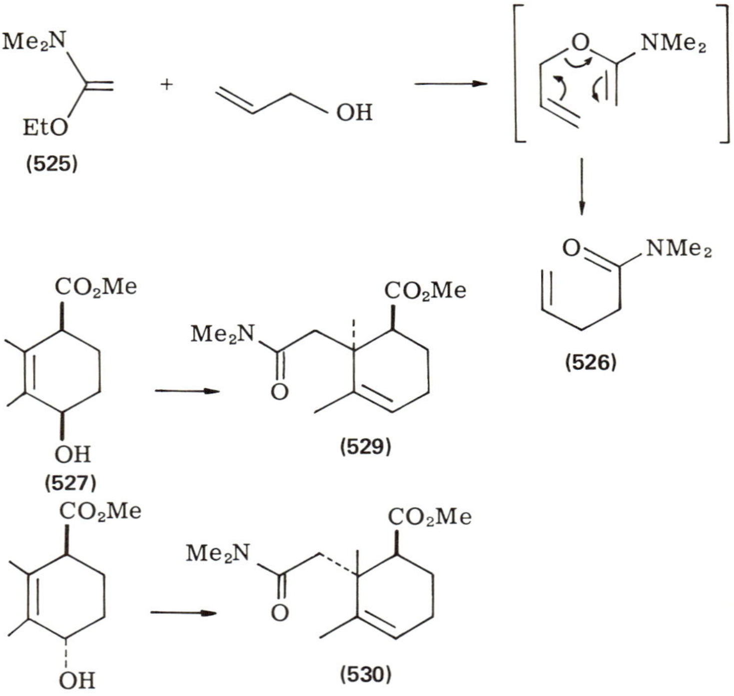

Me_2N EtO (525) + OH → [O NMe_2] → O NMe_2 (526)

CO_2Me OH (527) → Me_2N O CO_2Me (529)

CO_2Me OH (528) → Me_2N O CO_2Me (530)

Not long after this, the reaction was used in a total synthesis of crinine in order to construct the quaternary carbon atom present in that molecule. Thus treatment of the allylic alcohol (**531**) (obtained by reduction of the corresponding enone) with the reagent obtained from the acetal of dimethylacetamide gave the rearranged product as a mixture of stereoisomers (**532**, only one isomer shown). Treatment with base gave the lactam (**533**), which on further manipulation afforded the natural product (206).

The scope of the Claisen-type rearrangement was considerably increased by the recent finding that allyl alcohols undergo a very similar rearrangement with methyl orthoacetate. In this case, too, the first step presumably involves an ester interchange. Loss of methanol leads to the necessary unsaturated system (**534**). The last, on heating, undergoes the observed electrocyclic rearrangement to afford, in this case, an ester (207).

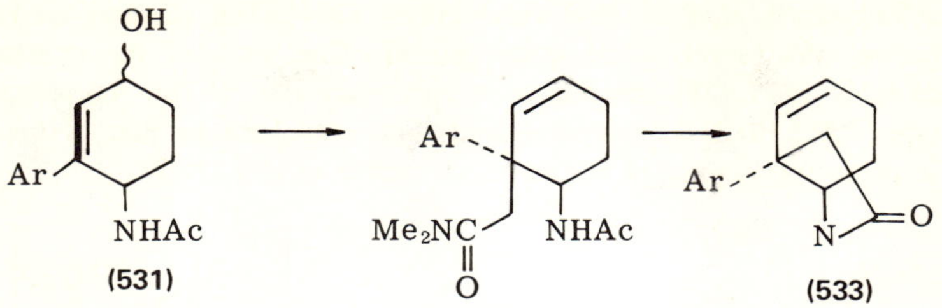

OH

Ar NHAc **(531)** → Ar Me$_2$NC(=O) NHAc **(532)** → Ar N =O **(533)**

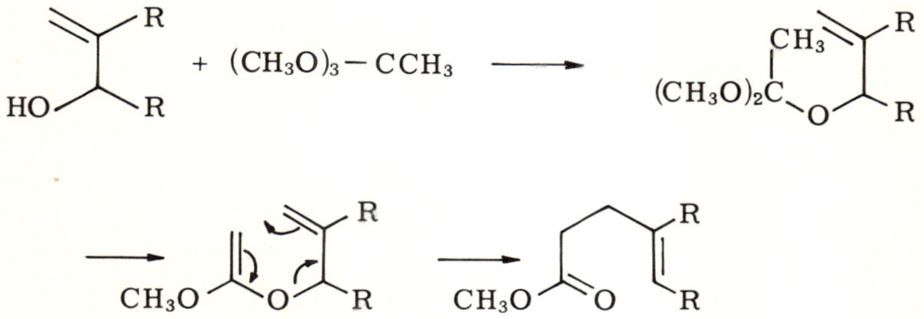

HO R R + $(CH_3O)_3-CCH_3$ → CH_3 $(CH_3O)_2C$ O R R

→ CH_3O O R R **(534)** → CH_3O O R R

Application of this scheme to the allylic alcohol (**535**) gave the product (**536**). Noteworthy features of this reaction are first the incorporation of the acetic acid residue and then the high degree of stereospecificity; the newly formed double bond has the *trans* configuration. Both these features were exploited in a recent total synthesis of squalene. Thus the diester (**538**) was obtained when the bisallylic alcohol (**537**) (obtained by reaction of succindialdehyde with 2-propenyllithium) was heated in the presence of ethyl orthoacetate. The diester thus obtained was reduced to the alcohol and then oxidized to the dialdehyde (**539**). That intermediate was then again treated with isopropenyllithium. The resulting bisallylic alcohol (**540**) was subjected to the vinyl ether rearrangement and the thus obtained dialdehyde (**541**) converted to squalene (**542**) by means of isopropylidene phosphorane (207).

VII. The Use of Heterocyclic Compounds in Synthesis

The author will be the first to admit that the above title is to a degree inappropriate for a section of a chapter; it would be much more fitting as the title of a good sized book. In a chapter such as this, however, we are caught between the Scylla of comprehensiveness and the Charybdis of space limitation. We will thus content ourselves with presenting some examples which illustrate the principle of latency as typified by heterocyclic compounds.

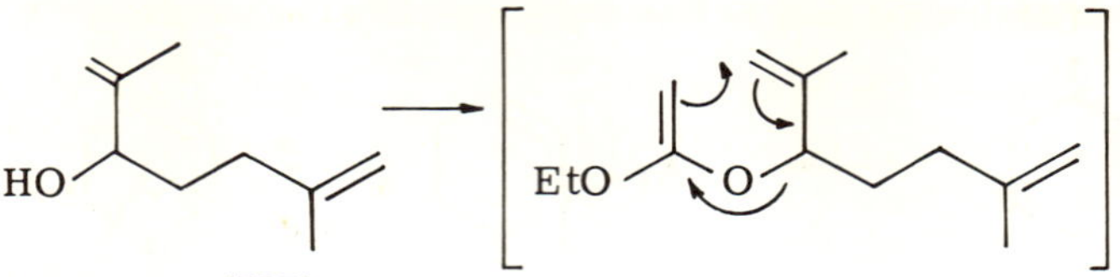

HO

EtO

O

(535)

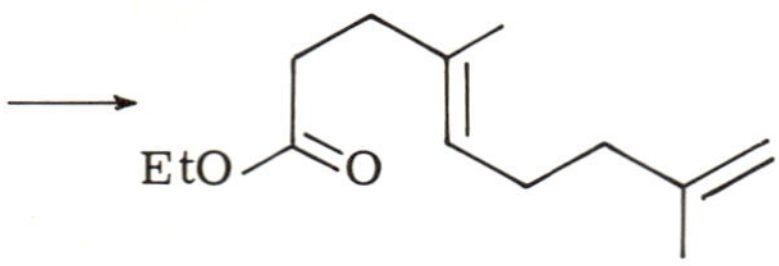

EtO

O

(536)

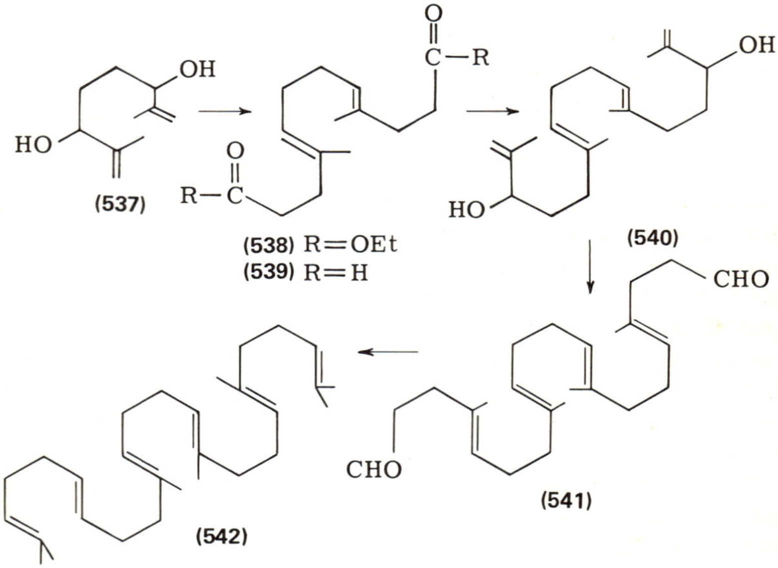

O

C—R

OH

OH

HO

O

(537) R—C

HO

(538) R=OEt
(539) R=H

(540)

CHO

CHO

(541)

(542)

A. ENAMINE REACTIONS WITH RETENTION OF NITROGEN

Enamines, and the closely related imminium salts and enamides, have proven of great value in the formation of new carbon-carbon bonds. In the usual course of events some carbonyl function is activated by formation of the enamine or one of its related functions; following the desired reaction, the product is hydrolyzed to the new carbonyl compound. It has proven possible, however, to use this reaction in such a manner as to take advantage of the enhanced reactivity of enamines while at the same time retaining the nitrogen in the final product.

1. Enamines and Imminium Salts in Forming Quaternary Carbon Atoms

That very principle was invoked in a synthesis aimed at the erythena alkaloids. Cyclization of the amide **(543)** provides the imminium salt **(544)**. This compound at the same time incorporates nitrogen in the proper position for the final compound and also provides the reactive center about which construction of a spiro ring may be started. Thus reaction of the salt with the Grignard reagent from allyl chloride gave **545** (9). In much the same manner, the enamide **(546)** was alkylated to afford the conjugated imminium salt **(547)**. It is of note that addition of a Grignard reagent to this molecule can occur in more than one sense. In fact, addition occurs at the central carbon atom to afford the lyconnotine degradation product **(548)** (208), and similarly, with the allyl reagent, the lycopodium intermediate **(549)** (209).

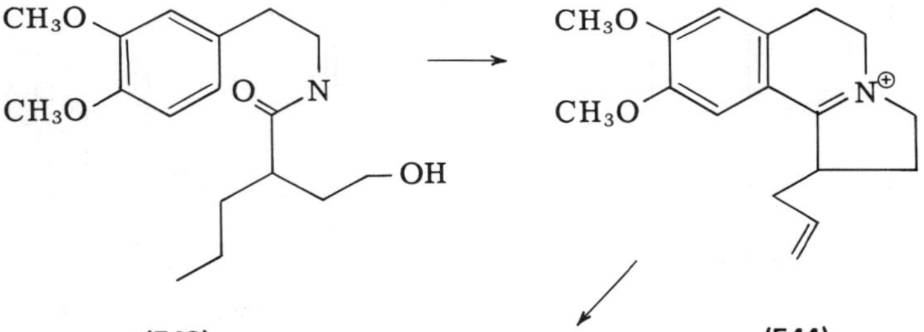

CH_3O CH_3O O N OH → CH_3O CH_3O $N^{\oplus}$

(543) **(544)**

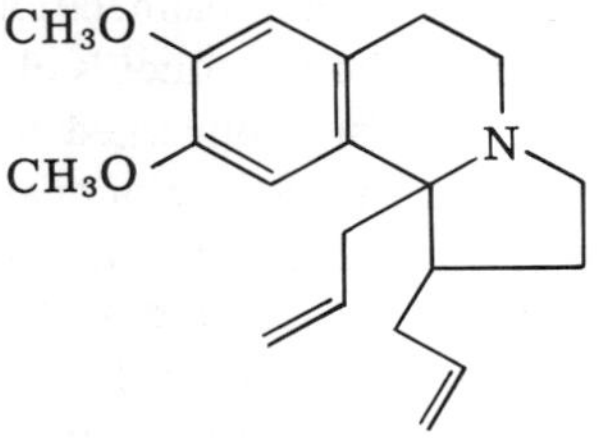

CH_3O CH_3O N

(545)

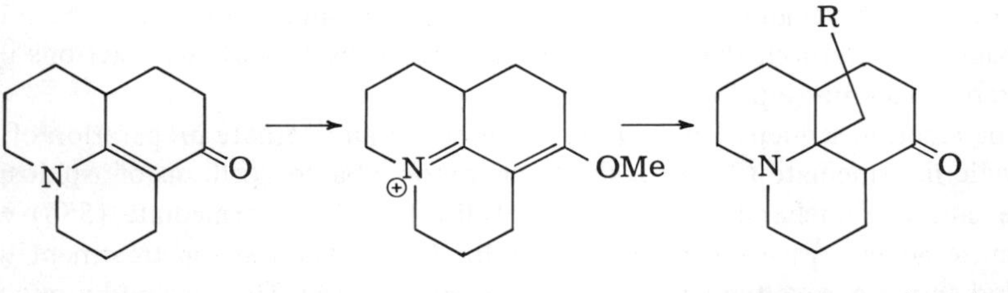

N O → $N^{\oplus}$ OMe → R N O

(546) **(547)** **(548)** $R{=}CH(CH_3)_2$
(549) $R{=}CH{=}CH_2$

Reaction of an enamine as an electrophile was used in an alternate and simple synthesis of the erythina alkaloid skeleton. Treatment of veratrylamine (**550**) with the keto acid (**551**) afforded the "enamine" (**552**). Treatment of this product with polyphosphoric (210) or phosphoric acid (211) led to the Friedel-Crafts cyclization of the enamine to give the spirane (**553**).

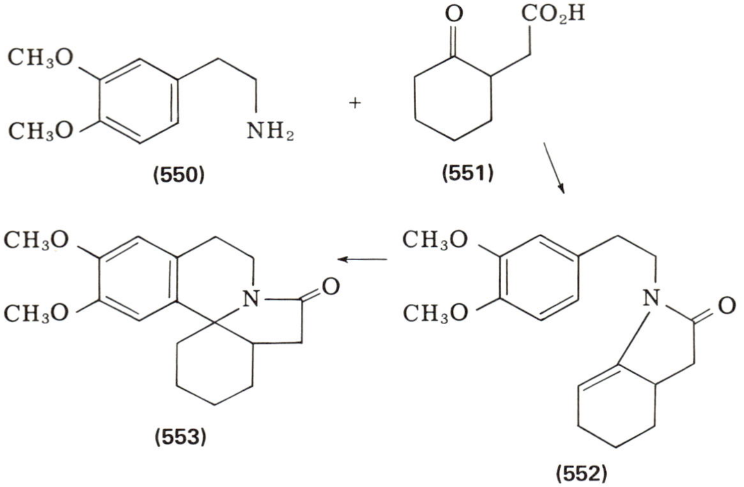

CH_3O CH_3O NH_2 + O CO_2H

(550) (551)

CH_3O CH_3O N O ← CH_3O CH_3O N O

(553) (552)

2. Homoannelation of Enamines

The use of enamines in the Robinson homoannelation procedure is well documented; in all cases, however, the desired product is obtained by hydrolytic removal of the nitrogenous moiety. We are concerned here with those cases where the starting material incorporates an enamine or enamide structure which is to be a part of the synthetic goal. Nitrogen is thus to be retained in the product. In general these procedures call for reaction of the nitrogen compound with a conjugated enone. The reaction can be rationalized as a Michael addition of the enamine, rearrangement of the double bond, and then either acylation (eq. 17) or a modified aldol condensation (eq. 18).

In an alternate homoannelation scheme, an unsaturated acid or ester is reacted with the enamine of a secondary or primary amine. Following the initial Michael condensation, the second bond is formed by any of the reactions used to prepare lactams (eq. 19).

The reaction sequence of eq. (17) was used for an alternate preparation of the tricyclic intermediate (**319**) for the Fischer indole-based synthesis of aspidospermine and quebrachamine. Oxidation of the bicyclic intermediate (**554**) with mercuric acetate gave the required enamine (**555**). The last, on treatment with methyl acrylate, gave the key tricyclic compound (**556**). The carbonyl group was retained in the subsequent manipulations for use in the indole synthesis (212).

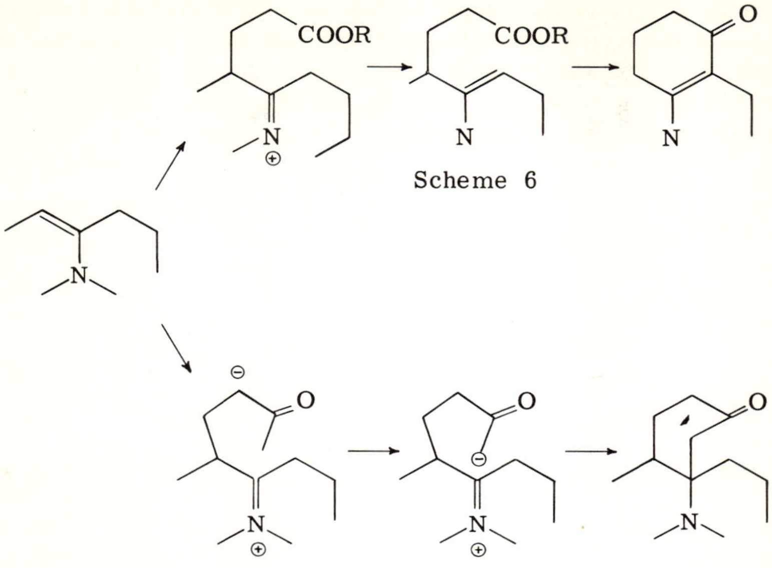

COOR COOR O (17)

N N N

Scheme 6

N

O O O (18)

N N N

Scheme 7

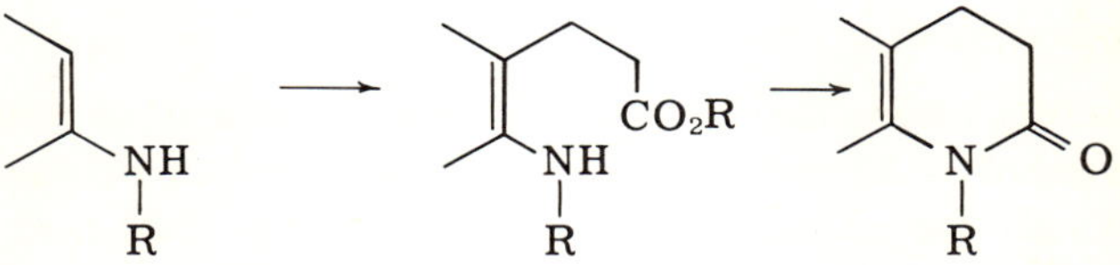

NH R CO$_2$R NH R N O R (19)

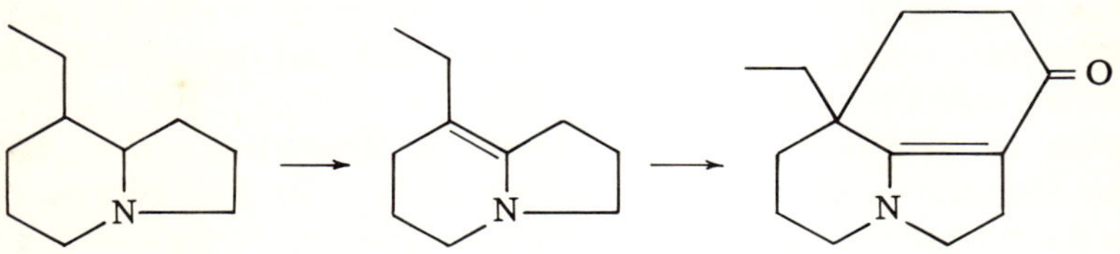

N N N O

(554) (555) (556)

The same general concept was utilized as a two-step sequence in the synthesis of 13-aza-11-keto steroids. Condensation of tetralone **(557)** with the pyrrolidine **(558)** proceeded cleanly to the enamino ester **(559)** (conjugation with the aromatic ring no doubt controls the position of unsaturation). This, on heating, undergoes internal acylation to the steroid-like compound **(560)** (213).

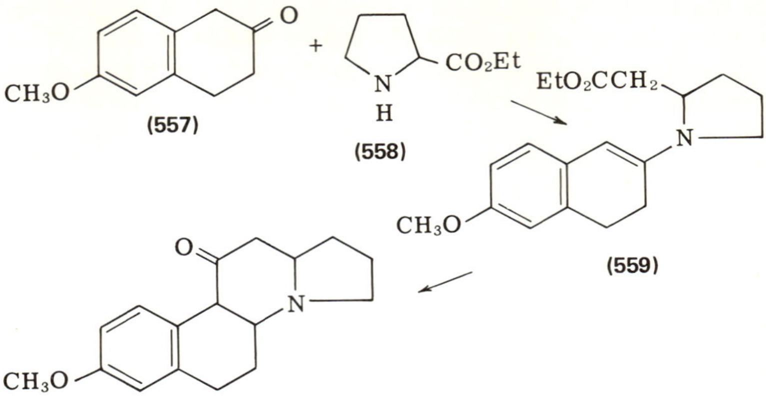

CH_3O (557) + (558) CO_2Et N H → EtO_2CCH_2 N CH_3O (559) → O N CH_3O (560)

Access to an unusual class of indoles was made easier by a variation of this scheme which has as its second step an aldol-type reaction rather than an acylation. Reaction of dihydroresorcinols such as **561** with the dimethyl acetal of an aminoacetaldehyde leads to the enamides **(562)**. These on reaction with acid lead to the indoles **(563)**, probably by way of the free aldehyde (214).

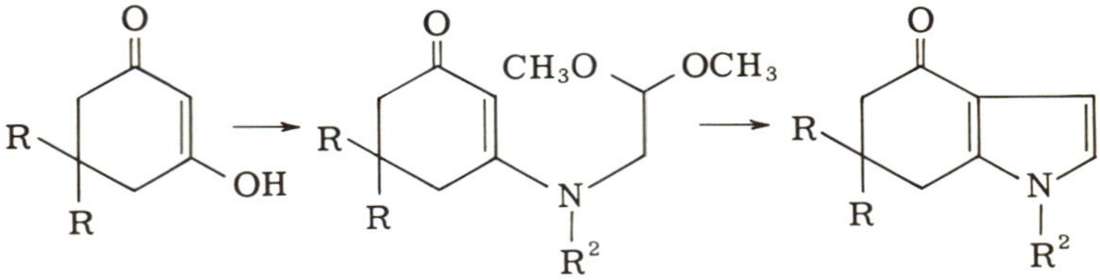

(561) R=H, CH_3 (562) (563) R=H, CH_3, $C_6H_5CH_2$

The total synthesis of mesembrine illustrates the homoannelation by eq. (18). Thus, several groups independently reported that treatment of the pyrroline **(564)** (see Section IV.E for preparation of the starting material) with methyl vinyl ketone afforded *d,l*-mesembrine **(565)** (152,215,216).

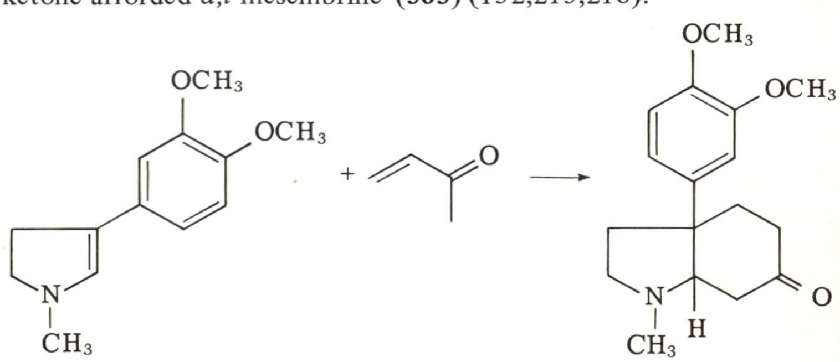

(564) (565)

The scheme which ends in the formation of a lactam was used to good advantage in the preparation of an intermediate for rings A and B of a totally synthetic 4-azasteroid. In order to obtain the oxidation state of an aromatic ring, these workers started their synthesis by condensing ethyl propiolate with the enamide **(566)**. The desired product **(567)** was obtained in a single step, probably by the intermediacy of a Michael adduct (217). In a related reaction the enamide **(568)** was heated with acrylic acid. There was isolated from the last reaction the lycopodium alkaloid intermediate **(569)** (218).

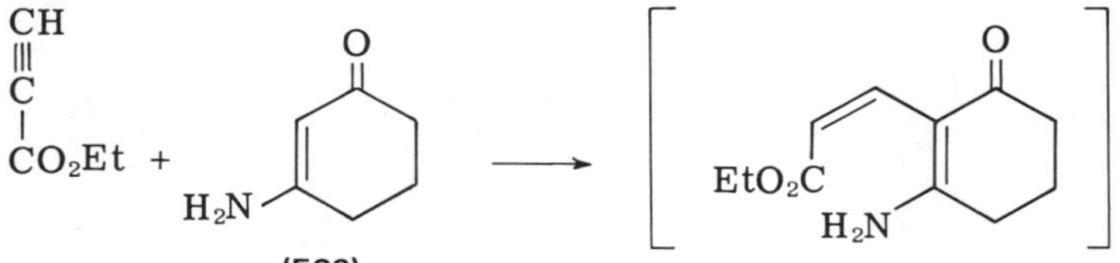

CH≡C–CO_2Et + H_2N **(566)** → [EtO_2C ... H_2N] → **(567)**

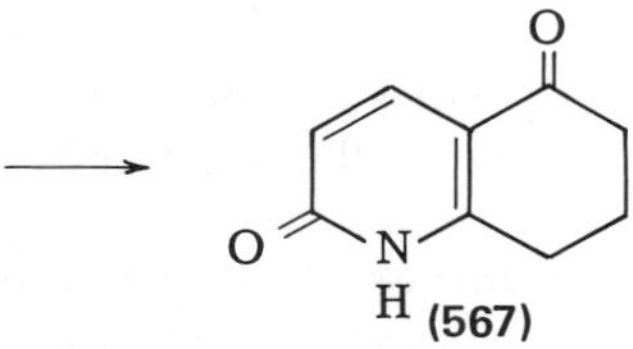

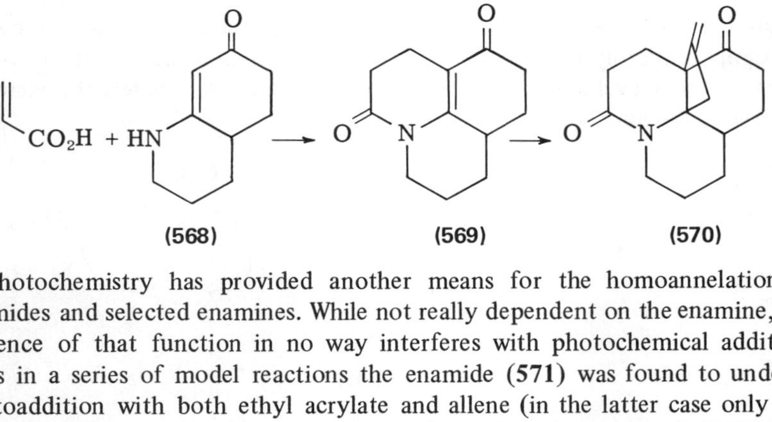

(568) **(569)** **(570)**

Photochemistry has provided another means for the homoannelation of enamides and selected enamines. While not really dependent on the enamine, the presence of that function in no way interferes with photochemical addition. Thus in a series of model reactions the enamide **(571)** was found to undergo photoaddition with both ethyl acrylate and allene (in the latter case only one product is shown, **573**; both isomers are formed in the reaction) (219). The addition of allene to **569** to form the tetracyclic compound **(570)** was a key step towards a synthesis of lycopodium alkaloids (218). Irradiation of the *N*-methylaniline enamines such as **574** was found to result in attack on the aromatic ring and thus entry to some polycyclic indolines such as **575** (220).

3. *3-Acyl-1,4,5,6-tetrahydropyridines as Synthetic Intermediates*

In the usual course of events reduction of a pyridinium salt proceeds rapidly to the hexahydro stage. However, it has recently been found that the reaction

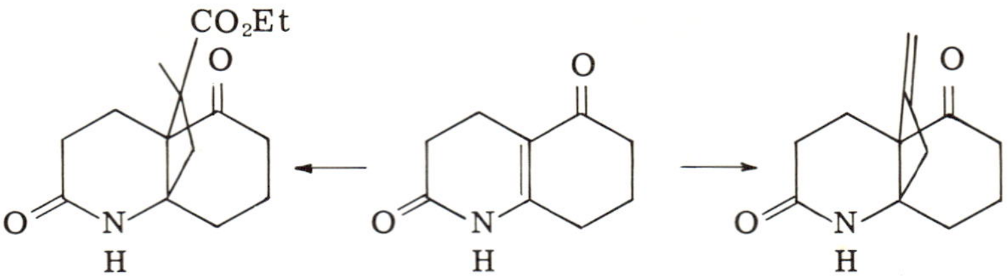

CO_2Et

(572) (571) (573)

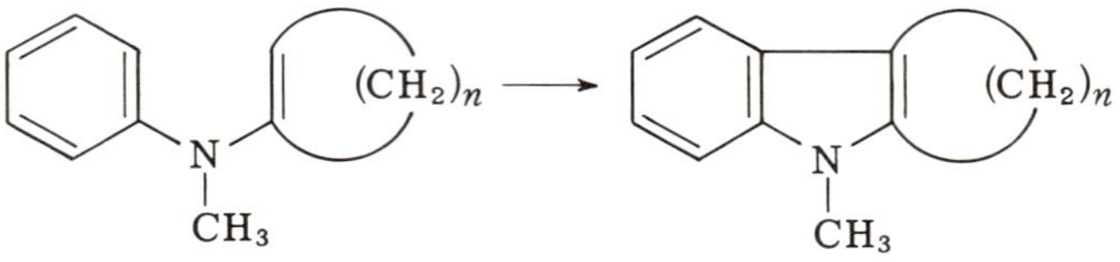

$(CH_2)_n$ $(CH_2)_n$

CH_3 CH_3

(574) (575)

stops at the tetrahydro stage when a carbonyl group is present at the 3 position. This selectivity has been ascribed to the fact that the thus produced enamide is relatively resistant towards further reduction.

The general scheme which has been evolved for use of this partial reduction involves *N*-alkylation of the pyridine with an appropriately substituted halide. Reduction of this function then reveals the latent enamide. This last intermediate is then made to undergo some reaction characteristic of that function; for example, hydrolysis of a side-chain ketal affords an enamide ketone which cyclizes to the bicyclic nitrogen-containing compound. Alternately the enamide cyclizes to an indole 2 position (eq. 20) (221).

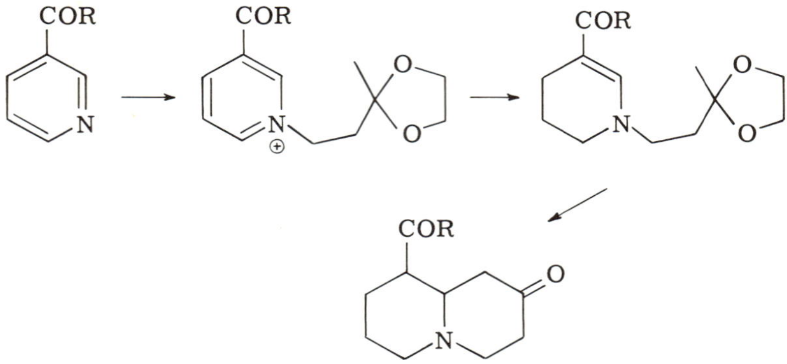

COR COR COR

(20)

COR

In an early application of this scheme, the pyridine **(576)** was condensed with the indole halide **(577)**. Catalytic reduction gave the enamide **(579)**. Treatment of that intermediate with acid led to cyclization to the 2 position of the indole **(580)**. This tetracyclic product was then further modified to afford finally eburnamonine (222). A total synthesis of corynantheidine was accomplished by the same scheme, but starting with the more highly substituted pyridine **581**. Appropriate modification of the side chains of **584** gave the natural product (223). The same general method was employed in the conversion of **585** to the intermediate for corynantheidol (**586**), and conversion of **587** to the epialloyohimbine precursor (**588**) (224).

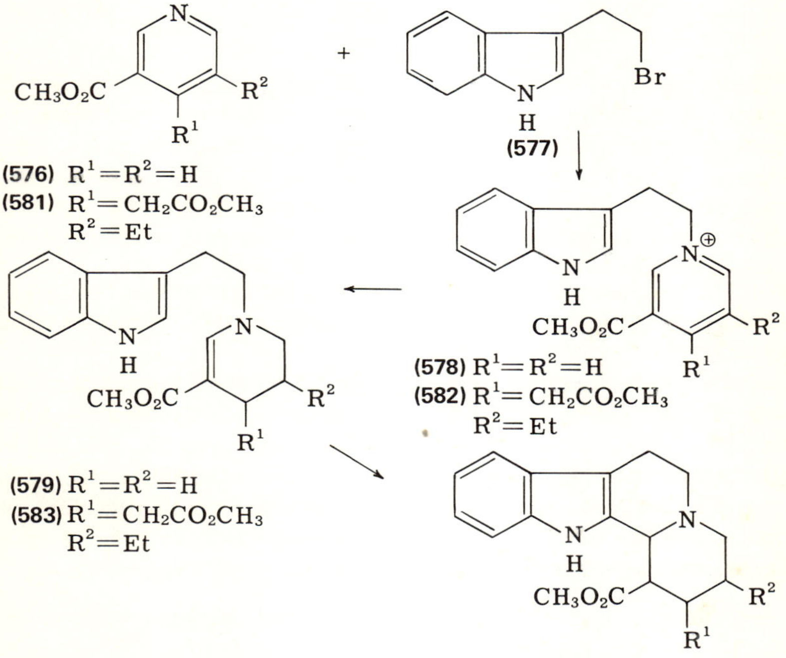

CH_3O_2C R² R¹ + Br N H

(576) $R^1=R^2=H$
(581) $R^1=CH_2CO_2CH_3$
$R^2=Et$

(577)

(578) $R^1=R^2=H$
(582) $R^1=CH_2CO_2CH_3$
$R^2=Et$

(579) $R^1=R^2=H$
(583) $R^1=CH_2CO_2CH_3$
$R^2=Et$

(580) $R^1=R^2=H$
(584) $R^1=CH_2CO_2CH_3$
$R^2=Et$

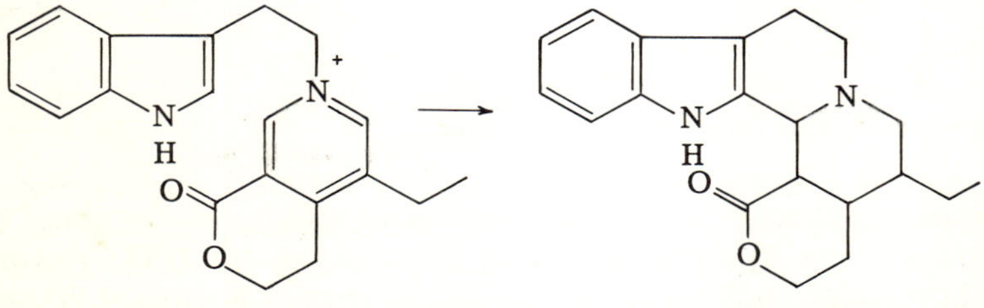

(585) **(586)**

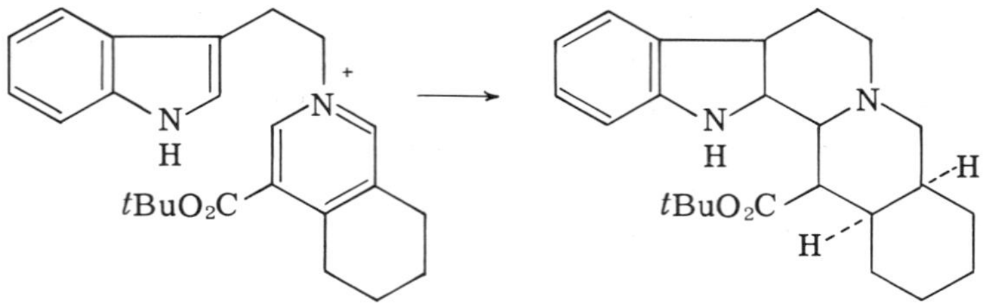

tBuO₂C

(587) (588)

This general reaction scheme has proven particularly suitable for the preparation of bicyclic compounds with nitrogen at the ring fusion. Thus in sequences similar to those above the alkylated acylpyridines, such as **589**, were converted to the enamides and these cyclized to compounds such as **590** (225). It was recently reported that a nitrile group can be used instead of a carbonyl for arresting the reduction. The product from that sequence was used in the total synthesis of lamprolobine (226).

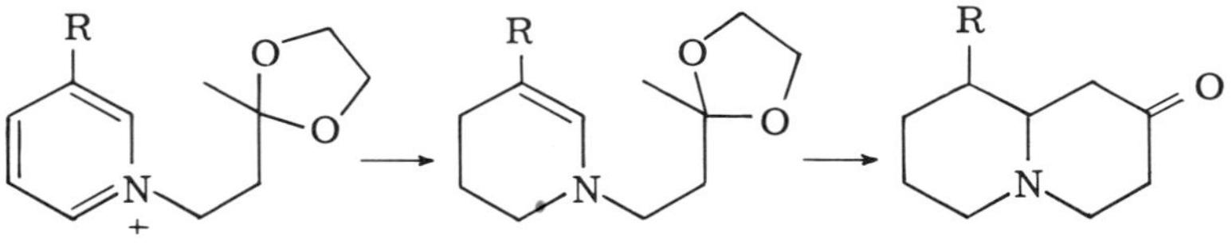

(589) $R=CO_2CH_3$
$R=CN$

(590) $R=CO_2CH_3$
$R=CN$

B. MONOCYCLIC HETEROCYCLES AS LATENT DIFUNCTIONAL MOIETIES

While difunctional fragments are often extremely useful in synthesis, the very possession of two reactive centers makes such molecules difficult to handle. Many schemes have been devised to carry such a grouping through a synthesis in latent form, such as, for example, the olefins discussed at the outset of this chapter. An alternate approach to such a difunctional molecule consists of carrying the moiety as a heterocyclic ring. Many of these are quite stable to a host of reaction conditions; the desired difunctional fragment can be uncovered either by some specific reaction or by modification of the heterocyclic ring in such a manner as to make it labile.

1. Furans as Latent Dicarbonyl Compounds

In more than just a formal sense furans are the cyclic ethers of the enolic form of 1,4-dicarbonyl compounds. The acid-catalyzed ring opening of this heterocyle to a diketone has been long recognized as one of its characteristic reactions. However, that same reaction is often difficult to realize as a synthetic procedure. An elegant way to circumvent this difficulty consists in first adding methanol in an electrolytic step to form the 1,4-dimethoxy furan. This compound, which is the cyclic acetal of the conjugated enedione, can be reduced catalytically to the acetal of the dione; the last compound opens easily to the parent diketone (eq. 21) (227).

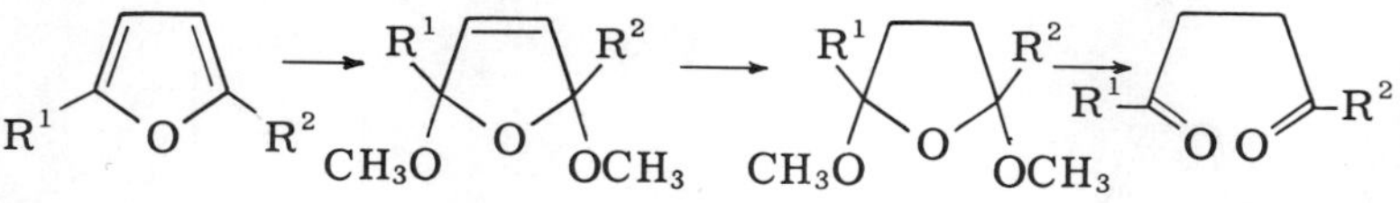

(21)

In the original exploration of this method, 2-acetylfuran (**591**) was converted in a two-step procedure to the intermediate (**592**). This was ring-opened to the trione which then recyclized spontaneously to catechol (227). A closely related scheme afforded a total synthesis of hydrourushiol (**595**). Reaction of the ester (**593**) (obtained by the electrolysis-reduction procedure as applied to methyl furoate) with the Grignard reagent from hexadecane bromide gave, among other products, the ketone (**594**). This was treated with dilute acid to give directly the catechol (**595**) (228).

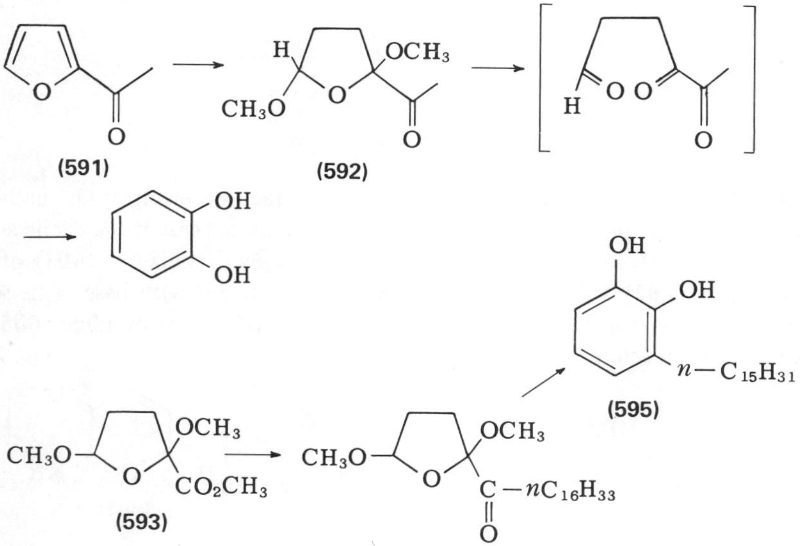

(591) (592)

(595)

(593) (594)

In selected cases the furan can be opened directly to the diketone. This scheme was used in a short elegant synthesis of *cis*-jasmone (**601**), R = Et). Reaction of the lithium salt of 2-methylfuran with the halide (**596**) gave the intermediate (**597**). This was opened to the dione (**600**) by treatment with acid, and this diketone cyclized with base to give jasmone (229). In more recent work by another group, aldehyde (**598**) was used as starting material. Reaction of this with a series of ylides under "salt-free" conditions afforded the *cis* olefins (**597**). These were treated as above to afford jasmone and other rethrolones (230).

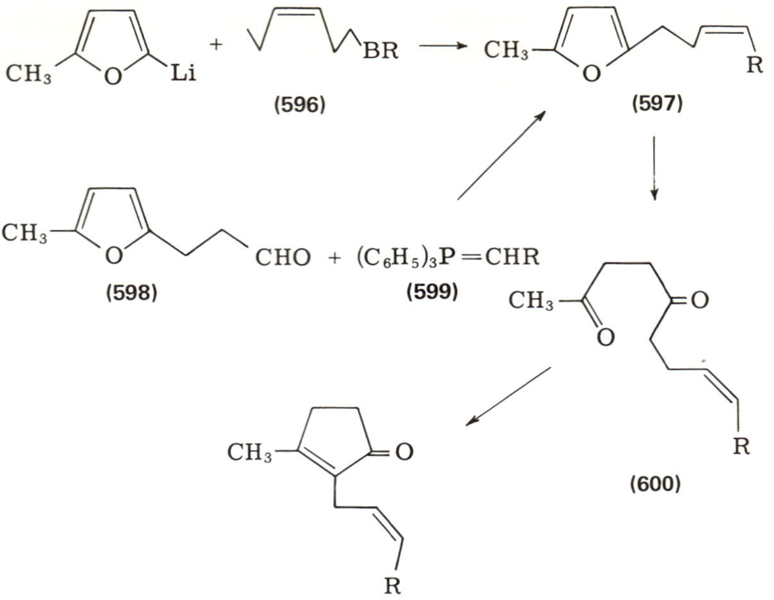

(596)

(597)

(598) (599)

(600)

(601)

$R{=}CH_3,\ C_2H_5,\ CH{=}CH_2$.

It was also found possible to use the product of electrolytic methanol addition without prior reduction; the intermediate in this case is the cyclic acetal of a conjugated enedione. Thus the product of methoxylation **(603)** of the aminofuran **(602)** affords the amine **(604)** on treatment with base. This sensitive compound was not isolated, but converted to 3-hydroxypyridine **(605)** on treatment with acid (231).

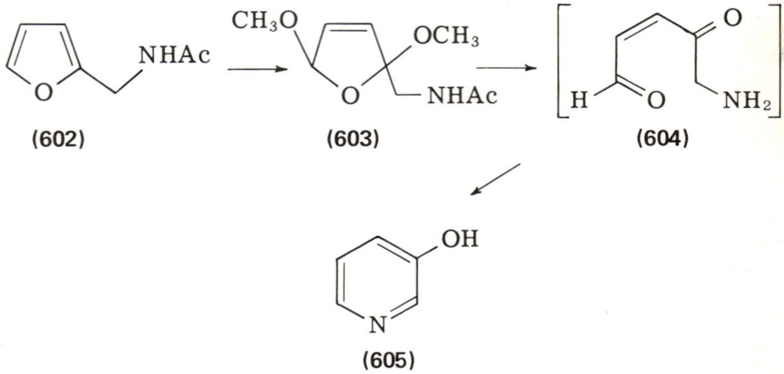

(602) (603) (604)

(605)

This reaction was exploited in an ingenious synthesis of pyridoxin. Diels-Alder reaction of furan with diethyl acetylenedicarboxylate gave, after the loss of acetylene from the intermediate **(606)**, 3,4-dicarbethoxyfuran **(607)**. This was then transformed by a sequence of fairly routine steps to the acetamide (**608**). The latter was methoxylated to **609** and then opened and recyclized as above with concomitant saponification to afford pyridoxin (**610**) (232).

$+ EtO_2CC{\equiv}CCO_2Et \longrightarrow$ [CO_2Et, CO_2Et] (**606**)

EtO_2C CO_2Et (**607**)

$AcOCH_2$ CH_2OAc $NHCOCH_3$ (**608**)

$AcOCH_2$ CH_2OAc CH_3O OCH_3 H $NHCOCH_3$ (**609**)

CH_2OH $HOCH_2$ OH CH_3 N (**610**)

2. *Azlactones in Synthesis*

The accumulation of functionality in the tetracycline antibiotics calls for delicate manipulation in the last stages of the synthesis of these molecules. In particular the labile amino function in the D ring is easily lost. In order to meet this challenge Muxfeldt developed the azlactone as a combined activating and protecting group. In essence the scheme initially takes advantage of the active methylene group in the heterocycle by condensation with an aldehyde. The resulting enone moiety will then undergo annelation with a molecule such as acetone dicarboxylate, probably in a two-step sequence; in the course of the ring formation the azlactone opens to form the corresponding amide (233).

The scheme was first applied to the preparation of a tetracycline derivative. Thus condensation of aldehyde (**612**) with the thioazlactone (**611a**) afforded the derivative (**613**). This compound was then condensed with acetone dicarboxylate to yield the annelated product (**614**). The ketal was hydrolyzed and the last remaining ring formed by a base-catalyzed internal acylation to give **615** (38). In an analogous scheme, the desmethyl aldehyde (**616**) was taken through the same sequence. In the last case, however, the oxygenated azlactone was employed; the

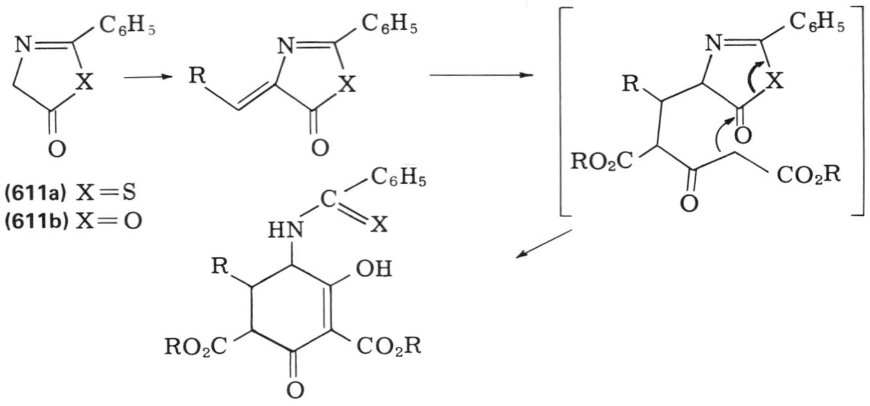

(611a) X=S
(611b) X=O

half amide of acetone dicarboxylate was used in this case to afford more direct access to the final tetracycline (234). In its most elegant application, this scheme was applied to aldehyde **(620)** (see Section II.B.1.a) in the total synthesis of terramycin (39).

The reactivity of the methylene group of an azlactone was used to good advantage in the total synthesis of cyclopenin. Condensation of the azlactone **(622)** with the appropriate aldehyde afforded compound **623.** In this case, the ring was cleaved with sodium methoxide to a mixture of the isomers of **624.** Reduction of the nitro group followed by cyclization of the resulting aminoester gave the desired carbon skeleton (235).

3. *Isoxazoles in Synthesis*

Formation of new rings almost of necessity requires a difunctional molecule. A recurring theme in this work has been the development of methods of carrying such rings through synthesis in some latent form. We have by now seen a spate of schemes which depend on oxidative methods for developing the desired fragment. Under some circumstances a reduction step might be desired to finally unmask the latent function. Such a method is now available by the use of isoxazoles (236).

In its most straightforward form, this scheme starts with the alkylation of a ketone with a 4-chloromethyl isoxazole. The nitrogen-oxygen bond of the heterocycle is then fissioned by hydrogenation. The iminoketone obtained in this way can then be treated with a protic base to form a tricarbonyl compound. This last, on further base treatment, undergoes aldol condensation to form a new ring and loss of the side chain by β-dicarbonyl cleavage (237). Under appropriate conditions, the iminoketone can be induced to react at nitrogen to lead eventually to pyridines (238,239) (eq. 22).

This was used to introduce all but one of the carbon atoms of rings A and B of progesterone in one fell swoop in a recent total synthesis of that molecule. Thus the enolate of dione **(626)** was alkylated with the isoxazole **(627)**;

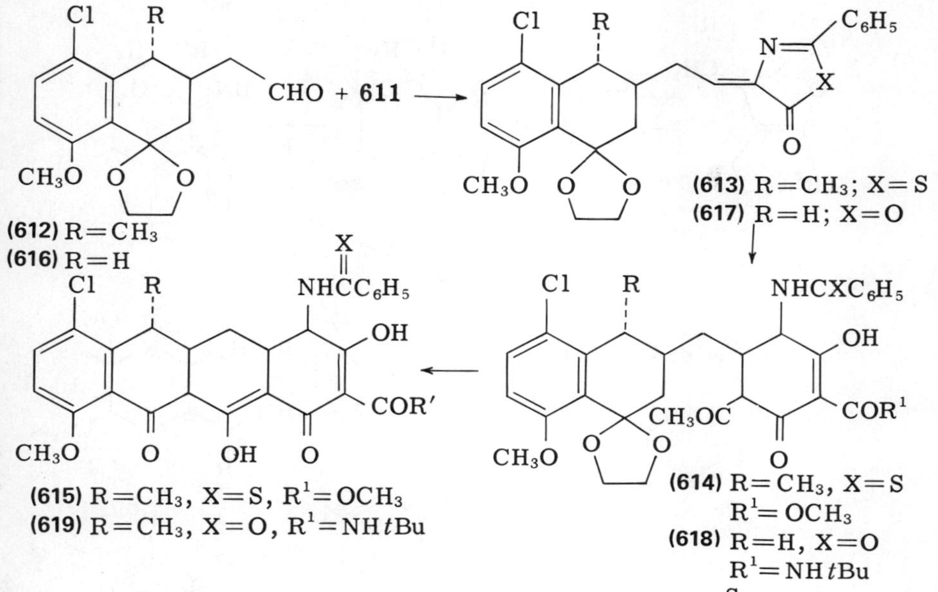

(612) R=CH_3
(616) R=H

(613) R=CH_3; X=S
(617) R=H; X=O

(614) R=CH_3, X=S, R^1=OCH_3
(618) R=H, X=O, R^1=NHtBu

(615) R=CH_3, X=S, R^1=OCH_3
(619) R=CH_3, X=O, R^1=NHtBu

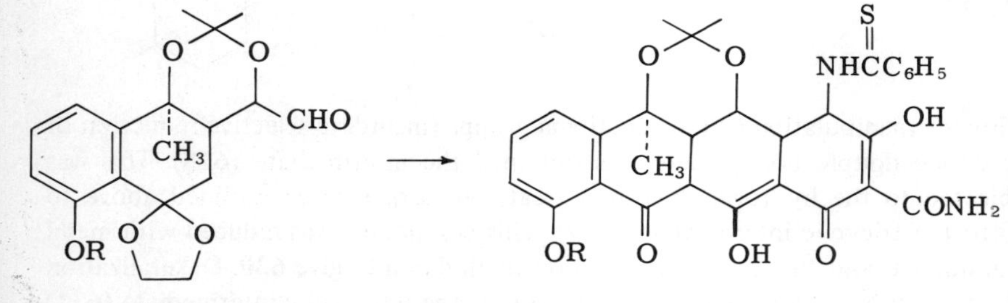

(620)

(621)

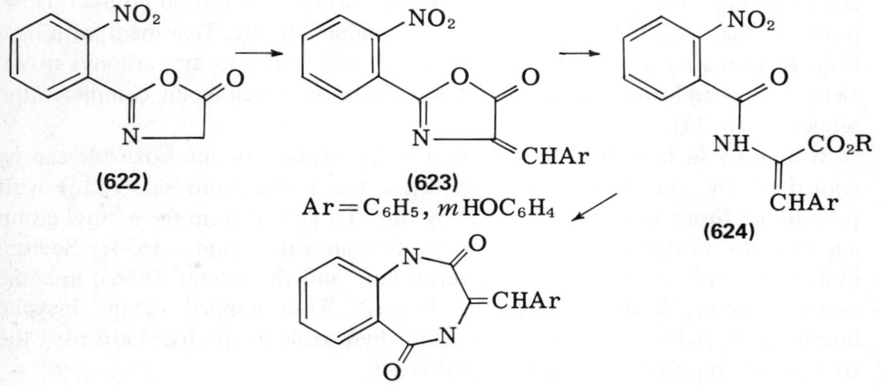

(622)

(623)

Ar=C_6H_5, m HOC_6H_4

(624)

(625)

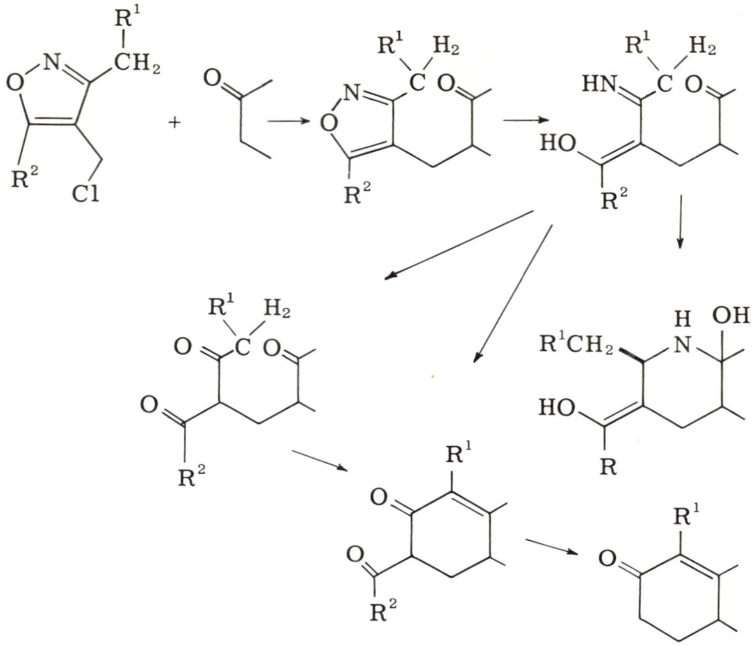

(22)

following manipulation of the functional groups (including selective reduction of the enone double bond) there was obtained the intermediate **(628)**. This was subjected to the hydrogenolysis-base-treatment sequence as outlined above to afford the tricyclic intermediate **(629)**. This compound was reduced with metal in ammonia, and the intermediate anion methylated to give **630.** Deketalization followed by base-catalyzed aldol reaction gave the tetracyclic intermediate **(631)** (49).

In an alternate approach, isoxazoles have been used for the construction of aromatic rings. The sequence starts as before with the alkylation product. Now, however, the isoxazole is converted to the imminium salt. Treatment with base leads to formation of the methylene enamine which adds to the carbonyl group. Dehydration followed by fission of the oxygen-nitrogen bond completes the sequence (eq. 23).

It appears in fact that the direction of cyclization of the isoxazole can be controlled by the choice of base. Thus the isoxazolium salt **(632)** with pyrrolidine forms the enamine **(633)** by loss of a proton from the methyl group adjacent to nitrogen to form finally the aromatic amine **(634)**. Sodium hydroxide leads to the alternate intermediate and the phenol **(636)**; in some cases, however, both products are formed. When applied to the bicyclic intermediate **(637)** this sequence (sodium hydroxide as the base) afforded the tricyclic intermediate to ferruginol **(639)** (240).

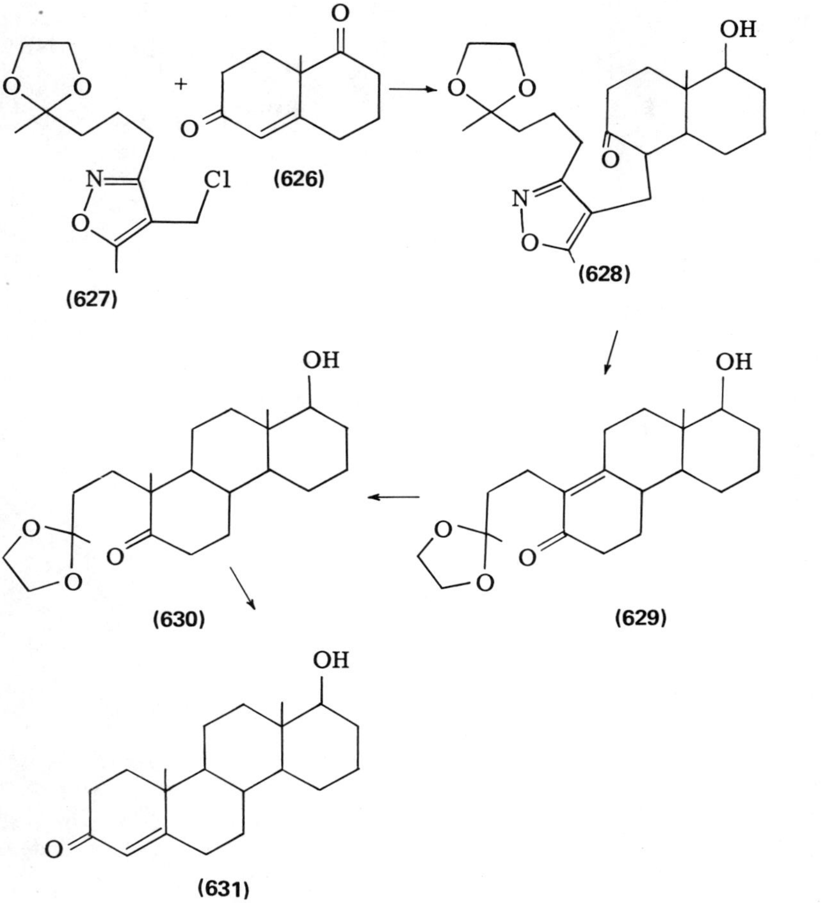

(627) (626) (628)

(630) (629)

(631)

NHR

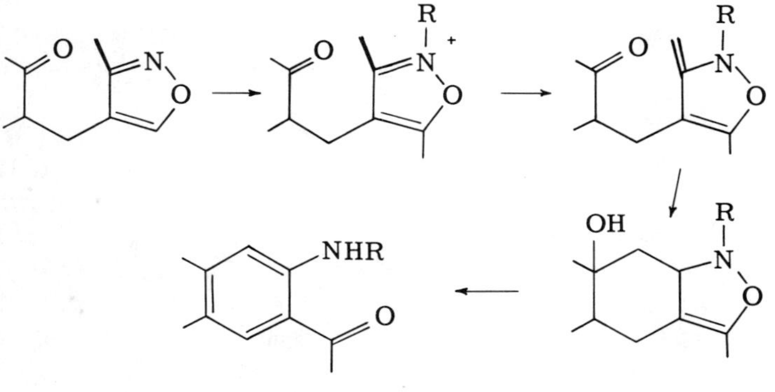

(23)

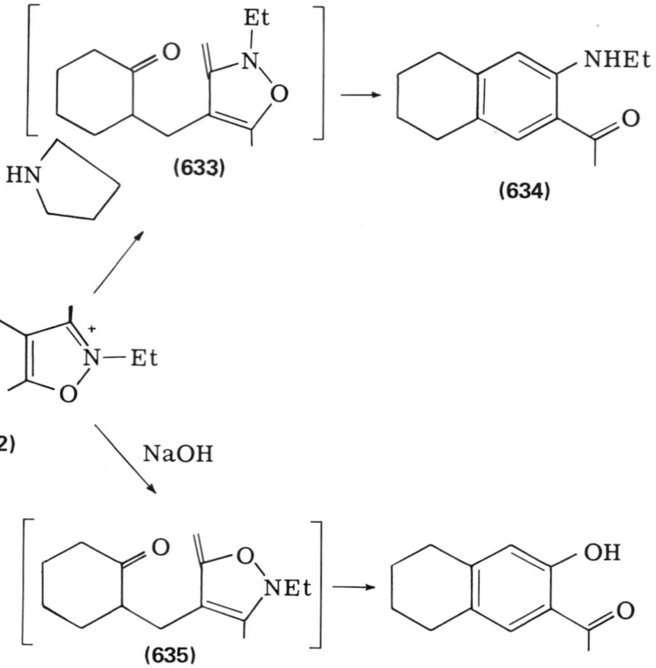

Et

O

N

O

NHEt

O

HN

(633)

(634)

O

N—Et

O

(632)

NaOH

O

O

NEt

OH

O

(635)

(636)

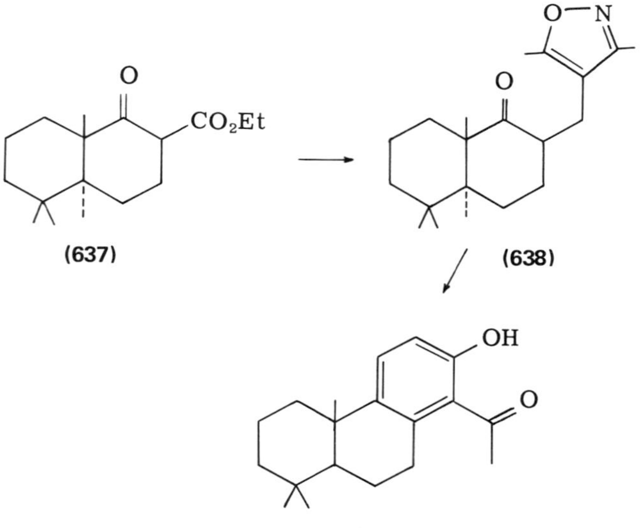

O—N

O

O

CO_2Et

(637)

(638)

OH

O

(639)

ACKNOWLEDGMENTS

First and foremost I wish to express my appreciation to the Upjohn Company for the grant of a sabbatical leave. Though it smack of cliche to say that this chapter would not otherwise have been written, I must say it. It is almost impossible in the press of everyday pressures and responsibilities to sit back to reflect on a field, to say nothing of writing a chapter such as this.

I also thank the Chemistry Department of Princeton University for their hospitality. Departments are, of course, not as yet animate beings. Thus it is with great pleasure that I express my thanks to Dr. Edward C. Taylor of Princeton, not only for making the arrangements for my stay here, but also for some valuable discussions and, when needed, encouragement. Finally it is my pleasure to thank again Miss Carolyn Jarchow of The Upjohn Company for typing the manuscript.

REFERENCES

1. J. F. W. McOmie, in *Advances in Organic Chemistry; Methods and Results,* Vol. 3 (R. A. Raphael, E. C. Taylor, and H. Winberg, eds.), Wiley-Interscience, New York, 1963, pp. 191-294.
2. G. Stork and H. Conroy, *J. Amer. Chem. Soc.,* **73,** 4748 (1951).
3. L. H. Sarett, G. E. Arth, R. M. Lukes, R. E. Beyler, G. I. Poos, W. F. Johns, and J. M. Constantine, *J. Amer. Chem. Soc.,* **74,** 4975 (1952).
4. N. A. Nelson and R. B. Garland, *J. Amer. Chem. Soc.,* **79,** 6313 (1957).
5. F. Giarrusso and R. E. Ireland, *J. Org. Chem.,* **33,** 3560 (1968).
6. R. E. Ireland, P. S. Grand, R. E. Dickerson, J. Bordner, and R. Rydjeshi, *J. Org. Chem.,* **35,** 570 (1970).
7. G. Büchi and H. Wüest, *J. Amer. Chem. Soc.,* **87,** 1589 (1965).
8. G. Stork, H. J. E. Loewenthal, and P. C. Mukharji, *J. Amer. Chem. Soc.,* **78,** 501 (1956).
9. A. J. Manson and K. Wiesner, *Chem. Ind.* (London), **1953,** 641.
10. K. Wiesner, Z. Valenta, A. J. Manson, and F. W. Stonner, *J. Amer. Chem. Soc.,* **77,** 675 (1955).
11. M. Shamma and H. R. Rodriguez, *Tetrahedron Lett.,* **1965,** 4847.
12. R. Pappo, D. S. Allen, Jr., R. U. Lemieux, and W. S. Johnson, *J. Org. Chem.,* **21,** 478 (1956).
13. Z. Valenta, K. Wiesner, and C. M. Wong, *Tetrahedron Lett.,* **1964,** 2437.
14. R. W. Guthrie, W. A. Henry, H. Immer, C. M. Wong, Z. Valenta, and K. Wiesner, *Collect. Czech. Chem. Commun.,* **31,** 602 (1966).
15. K. Wiesner, K. K. Chan, and C. Demerson, *Tetrahedron Lett.,* **1965,** 2893.
16. K. Wiesner and J. Santroch, *Tetrahedron Lett.,* **1966,** 5939.
17. K. Mori and M. Matsui, *Tetrahedron Lett.,* **1966,** 175.
18. R. B. Turner, G. D. Diana, G. E. Fodor, K. Gebert, D. L. Simmons, A. S. Rao, O. Roos, and W. Wirth, *J. Amer. Chem. Soc.,* **88,** 1786 (1966).
19. T. Matsumoto, M. Yanagiya, E. Kawakami, T. Okuno, M. Kakizawa, S. Yasuda, Y. Gama, J. Omi, and M. Matsunaga, *Tetrahedron Lett.,* **1968,** 1127.
20. L. J. Dolby and R. J. Milligan, *J. Amer. Chem. Soc.,* **88,** 4536 (1966).
21. J. E. D. Barton and J. Harley-Mason, *Chem. Commun.,* **1965,** 298.
22. J. Harley-Mason and M. Kaplan, *Chem. Commun.,* **1967,** 915.
23. L. Costedo, J. Harley-Mason, and T. J. Leeney, *Chem. Commun.,* **1968,** 1186.
24. J. P. Kutney, K. K. Chan, A. Failli, J. H. Fromson, G. Gletsos, and V. R. Nelson *J. Amer. Chem. Soc.,* **90,** 3891 (1968).
25. G. Buchi and W. D. MacLeod, *J. Amer. Chem. Soc.,* **84,** 3205 (1962).

25a. W. A. Ayer, W. R. Bowman, G. A. Cooke, and A. C. Soper, *Tetrahedron Lett.*, **1966,** 2021.

26. P. Grafen, H. J. Kabbe, O. Roos, G. D. Diana, T. Li, and R. B. Turner, *J. Amer. Chem. Soc.*, **90,** 6131 (1968).

27. G. Stork, E. E. van Tamelen, L. J. Friedman, and A. W. Burgstahler, *J. Amer. Chem. Soc.*, **73,** 4501 (1951).

28. G. Stork, E. E. van Tamelen, L. J. Friedman, and A. W. Burgstahler, *J. Amer. Chem. Soc.*, **75,** 384 (1953).

29. R. B. Woodward, F. Sondheimer, D. Taub, K. Heusler, and W. M. McLamore, *J. Amer. Chem. Soc.*, **74,** 4223 (1952).

30. L. B. Barkley, M. W. Farrar, W. S. Knowles, and H. Raffelson, *J. Amer. Chem. Soc.*, **75,** 4110 (1953).

31. E. J. Corey and S. Nozoe, *J. Amer. Chem. Soc.*, **85,** 3527 (1963).

32. E. J. Corey and S. Nozoe, *J. Amer. Chem. Soc.*, **87,** 5728 (1965).

33. J. P. Kutney, A. By, T. Inaba, and S. Leong, *Tetrahedron Lett.*, **1965,** 2911.

34. T. Ogawa, K. Mori, M. Matsui, and Y. Sumiki, *Tetrahedron Lett.*, **1967,** 4483.

35. T. Ogawa, K. Mori, and M. Matsui, *Tetrahedron Lett.*, **1968,** 2551.

36. G. Büchi, B. Gubler, R. S. Schneider, and J. Wild, *J. Amer. Chem. Soc.*, **89,** 2776 (1967).

37. G. Büchi and J. A. Carlson, *J. Amer. Chem. Soc.*, **90,** 5336 (1968).

38. H. Muxfeldt, *Angew. Chem.*, **74,** 825 (1962).

39. H. Muxfeldt, G. Hardtmann, F. Kathawala, E. Vedejs, and J. P. Mooberry, *J. Amer. Chem. Soc.*, **90,** 6534 (1968).

40. W. F. Johns, *J. Amer. Chem. Soc.*, **80,** 6456 (1958).

41. G. Stork, H. N. Khastgir, and A. J. Solo, *J. Amer. Chem. Soc.*, **80,** 6457 (1958).

42. G. Stork, S. D. Darling, I. T. Harrison, and P. S. Wharton, *J. Amer. Chem. Soc.*, **84,** 2018 (1962).

43. J. A. Marshall and W. S. Johnson, *J. Amer. Chem. Soc.*, **84,** 1485 (1962).

44. W. S. Johnson, J. F. W. Keana, and W. S. Johnson, *Tetrahedron Lett.*, **1963,** 193.

45. W. S. Johnson, M. F. Semmelhack, M. U. S. Sultanbawa, and L. A. Dolak, *J. Amer. Chem. Soc.*, **90,** 2994 (1968).

46. G. Stork, J. E. Davies, and A. Meisels, *J. Amer. Chem. Soc.*, **81,** 5516 (1959).

47. G. Stork, A. Meisels, and J. E. Davies, *J. Amer. Chem. Soc.*, **85,** 3419 (1963).

48. R. F. Church, R. E. Ireland and J. A. Marshall, *Tetrahedron Lett.*, **17,** 1 (1960).

49. G. Stork and J. E. McMurry, *J. Amer. Chem. Soc.*, **89,** 5464 (1967).

50. E. J. Corey, N. H. Anderson, R. M. Carlson, J. Paust, E. Vedejs, I. Vlattas, and R. E. K. Winter, *J. Amer. Chem. Soc.*, **90,** 3245 (1968).

51. R. Zurflüh, L. L. Dunham, V. L. Spain, and J. B. Siddall, *J. Amer. Chem. Soc.*, **92,** 425 (1970).

52. R. B. Woodward, F. E. Bader, H. Bickel, A. J. Frey, and R. W. Kierstead, *J. Amer. Chem. Soc.*, **78,** 2023 (1956).

53. R. B. Woodward, F. E. Bader, H. Bickel, A. J. Frey, and R. W. Kierstead, *J. Amer. Chem. Soc.*, **78,** 2657 (1956).

54. R. B. Woodward, F. E. Bader, H. Bickel, A. J. Frey, and R. W. Kierstead, *Tetrahedron*, **2,** 1 (1958).

55. E. E. van Tamelen, M. Shamma, A. W. Burgstahler, J. Wolinsky, R. Tamm and P. E. Aldrich, *J. Amer. Chem. Soc.*, **80,** 5006 (1958).

56. E. E. van Tamelen, L. J. Dolby, and R. G. Lawton, *Tetrahedron Lett.*, **19,** 30 (1960).

57. E. E. van Tamelen, J. P. Yardley, and M. Miyano, *Tetrahedron Lett.*, **1963,** 1011.

58. S. Masamune, S. K. Ang, C. Egli, N. Nakatsuka, S. K. Sarkar, and Y. Yasunari, *J. Amer. Chem. Soc.*, **89,** 2506 (1967).

59. J. A. Findlay, W. A. Henry, T. C. Jain, Z. Valenta, K. Wiesner, and C. M. Wong, *Tetrahedron Lett.*, **1962,** 869.

60. R. B. Woodward, M. P. Cava, W. D. Ollis, A. Hunger, H. U. Daeniker, and K. Schenker, *J. Amer. Chem. Soc.*, **76**, 4749 (1954).
61. R. B. Woodward, M. P. Cava, W. D. Ollis, A. Hunger, H. U. Daeniker, and K. Schenker, *Tetrahedron,* **19**, 247 (1963).
62. G. Stork, R. A. Kretchmer, and R. H. Schlessinger, *J. Amer. Chem. Soc.*, **90**, 1647 (1968).
63. K. Wiesner and A. Philipp, *Tetrahedron Lett.*, **1966**, 1467.
64. K. Wiesner, A. Philipp, and Pak-tsun Ho, *Tetrahedron Lett.*, **1968**, 1209.
65. W. Nagata, S. Hirai, K. Kawata, and T. Okumura, *J. Amer. Chem. Soc.*, **89**, 5046 (1967).
66. W. Nagata, S. Hirai, T. Okumura, and K. Kawata, *J. Amer. Chem. Soc.*, **90**, 1650 (1968).
67. For good general discussion of this scheme for total synthesis see W. S. Johnson *J. Amer. Chem. Soc.*, **78**, 6278 (1956).
68. W. S. Johnson, *J. Amer. Chem. Soc.*, **66**, 215 (1944).
69. M. W. Goldberg and S. Studer, *Helv. Chim. Acta,* **24**, 478 (1941).
70. W. S. Johnson, D. K. Banerjee, W. P. Schneider, and C. D. Gutsche, *J. Amer. Chem. Soc.*, **72**, 1426 (1950).
71. W. S. Johnson, D. K. Banerjee, W. P. Schneider, C. D. Gutsche, W. E. Shelberg, and L. J. Chinn, *J. Amer. Chem. Soc.*, **74**, 2832 (1952).
72. W. S. Johnson, I. A. David, H. C. Dehm, R. J. Highet, E. W. Warnhoff, W. D. Wood, and E. T. Jones, *J. Amer. Chem. Soc.*, **80**, 661 (1958).
73. W. S. Johnson, B. Bannister, B. M. Bloom, A. D. Kemp, R. Pappo, E. R. Rogier, and J. Szmuszkovicz, *J. Amer. Chem. Soc.*, **75**, 2275 (1953).
74. W. S. Johnson, B. Bannister, and R. Pappo, *J. Amer. Chem. Soc.*, **78**, 6331 (1956).
75. W. S. Johnson, R. Pappo, and A. D. Kemp, *J. Amer. Chem. Soc.*, **76**, 3353 (1954).
76. W. S. Johnson, R. Pappo, and W. F. Johns, *J. Amer. Chem. Soc.*, **78**, 6339 (1956).
77. W. S. Johnson, B. Bannister, R. Pappo, and J. E. Pike, *J. Amer. Chem. Soc.*, **77**, 817 (1955).
78. W. S. Johnson, B. Bannister, R. Pappo, and J. E. Pike, *J. Amer. Chem. Soc.*, **78**, 6354 (1956).
79. W. S. Johnson and D. S. Allen, Jr., *J. Amer. Chem. Soc.*, **79**, 1261 (1957).
80. W. S. Johnson, D. S. Allen, R. R. Hindersinn, G. N. Sausen, and R. Pappo, *J. Amer. Chem. Soc.*, **84**, 2181 (1962).
81. W. Parker, R. Ramage, and R. A. Raphael, *J. Chem. Soc.*, **1962**, 1558.
82. W. S. Johnson, D. G. Martin, R. Pappo, S. D. Darling, and R. A. Clement, *Proc. Chem. Soc.*, **1957**, 58.
83. W. S. Johnson, J. C. Collins, R. Pappo, and M. R. Rubin, *J. Amer. Chem. Soc.*, **80**, 2585 (1958).
84. W. S. Johnson, J. C. Collins, Jr., R. Pappo, M. R. Rubin, P. J. Kropp, W. F. Johns, J. E. Pike, and W. Bartman, *J. Amer. Chem. Soc.*, **85**, 1409 (1963).
85. T. Sakan, A. Fujino, F. Murai, A. Suzui, and Y. Butsugan, *Bull. Chem. Soc. Japan,* **33**, 1737 (1960).
86. S. A. Achmad and G. W. K. Cavill, *Proc. Chem. Soc.*, **1963**, 166.
87. L. R. C. Barclay, C. E. Milligan, and N. D. Hall, *Can. J. Chem.*, **40**, 1664 (1962).
88. R. B. Turner and P. E. Shaw, *Tetrahedron Lett.*, **18**, 24 (1960).
89. R. E. Ireland and R. C. Kierstead, *J. Org. Chem.*, **27**, 703 (1962).
90. R. E. Ireland and R. C. Kierstead, *J. Org. Chem.*, **31**, 2543 (1966).
91. E. E. van Tamelen and M. Shamma, *J. Amer. Chem. Soc.*, **76**, 950 (1954).
92. E. E. van Tamelen, M. Shamma, and P. Aldrich, *J. Amer. Chem. Soc.*, **78**, 4628 (1956).
93. G. Stork and R. K. Hill, *J. Amer. Chem. Soc.*, **76**, 949 (1954).
94. G. Stork and R. K. Hill, *J. Amer. Chem. Soc.*, **79**, 495 (1957).

95. E. E. van Tamelen, P. E. Aldrich, and T. J. Katz, *Chem. Ind.* (London), **1956,** 793.
96. E. E. van Tamelen, P. E. Aldrich, and J. B. Hester, Jr., *J. Amer. Chem. Soc.,* **79,** 4817 (1957).
97. E. J. Corey, N. M. Weinshenker, T. K. Schaaf, and W. Huber, *J. Amer. Chem. Soc.,* **91,** 5675 (1969).
98. E. J. Corey and R. Noyori, *Tetrahedron Lett.,* **1970,** 311.
99. E. J. Corey, Z. Arnold, and J. Hutton, *Tetrahedron Lett.,* **1970,** 307.
100. F. Nakatsubo, Y. Kishi, and T. Goto, *Tetrahedron Lett.,* **1970,** 381.
101. R. W. Guthrie, A. Philipp, Z. Valenta, and K. Wiesner, *Tetrahedron Lett.,* **1965,** 2945.
102. R. W. Guthrie, Z. Valenta, and K. Wiesner, *Tetrahedron Lett.,* **1966,** 4645.
103. H. Dugas, M. E. Hazenberg, Z. Valenta, and K. Wiesner, *Tetrahedron Lett.,* **1967,** 4931.
104. K. Wiesner, V. Musil, and K. J. Wiesner, *Tetrahedron Lett.,* **1968,** 5643.
105. C. A. Grob and P. W. Schiess, *Angew. Chem. Int. Ed.,* **6,** 1 (1967).
106. A. Eschenmoser and A. Frey, *Helv. Chim. Acta,* **35,** 1660 (1952).
107. R. B. Clayton and H. B. Henbest, *Chem. Ind.* (London), **1953,** 1315.
108. R. B. Clayton, H. B. Henbest, and M. Smith, *J. Chem. Soc.,* **1957,** 1982.
109. P. S. Wharton, *J. Org. Chem.,* **26,** 4781 (1961).
110. P. S. Wharton, G. A. Hiegel, and R. V. Coombs, *J. Org. Chem.,* **28,** 3217 (1963).
111. P. S. Wharton and G. A. Hiegel, *J. Org. Chem.,* **30,** 3254 (1965).
112. E. J. Corey, R. B. Mitra, and H. Uda, *J. Amer. Chem. Soc.,* **85,** 362 (1963).
113. E. J. Corey, R. B. Mitra, and H. Uda, *J. Amer. Chem. Soc.,* **86,** 485 (1964).
114. W. Nagata, T. Sugasawa, M. Narisada, T. Wakabayashi, and Y. Hayase, *J. Amer. Chem. Soc.,* **85,** 2343 (1963).
115. J. A. Marshall and G. L. Bundy, *Tetrahedron Lett.,* **1966,** 3359.
116. J. A. Marshall and C. J. V. Scanio, *J. Org. Chem.,* **30,** 3019 (1965).
117. J. A. Marshall, C. J. V. Scanio, and W. J. Iburg, *J. Org. Chem.,* **32,** 3750 (1967).
118. C. A. Grob, H. R. Kiefer, H. Lutz, and H. Wilkens, *Tetrahedron Lett.,* **1964,** 2901.
119. J. A. Marshall and G. L. Bundy, *J. Amer. Chem. Soc.,* **88,** 4291 (1966).
120. J. A. Marshall and G. L. Bundy, *Chem. Commun.,* **1967,** 854.
121. H. J. Liu, Z. Valenta, J. S. Wilson, and T. T. J. Yu, *Can. J. Chem.,* **47,** 509 (1969).
122. P. Wieland, G. Anner, and K. Miescher, *Helv. Chim. Acta,* **36,** 1803 (1953).
123. P. Gaffen, H. J. Kabbe, O. Roos, G. D. Diana, T. Li, and R. B. Turner, *J. Amer. Chem. Soc.,* **90,** 6131 (1968).
124. J. A. Marshall, N. H. Andersen, and P. C. Johnson, *J. Amer. Chem. Soc.,* **89,** 2748 (1967).
125. J. A. Marshall and N. H. Andersen, *Tetrahedron Lett.,* **1967,** 1219.
126. K. Biemann and G. Spiteller, *Tetrahedron Lett.,* **1961,** 299.
127. G. Stork and J. E. Dolfini, *J. Amer. Chem. Soc.,* **85,** 2872 (1963).
128. J. P. Kutney, N. Abdurahman, P. LeQuesne, E. Piers, and I. Vlattas, *J. Amer. Chem. Soc.,* **88,** 3656 (1966).
129. J. Harley-Mason, Atta-Ur-Rahman, and J. A. Beisler, *Chem. Commun.,* **1966,** 743.
130. J. P. Kutney, W. J. Cretney, P. L. LeQuesne, B. McKague, and E. Piers, *J. Amer. Chem. Soc.,* **88,** 4756 (1966).
131. J. Harley-Mason and Atta-Ur-Rahman, *Chem. Commun.,* **1967,** 208.
132. J. Harley-Mason and Atta-Ur-Rahman, *Chem. Commun.,* **1967,** 1048.
133. B. A. Dadson, J. Harley-Mason, and G. H. Foster, *Chem. Commun.,* **1968,** 1233.
134. S. Julia, M. Julia, and M. Davis, *C. R. Acad. Sci., Paris,* **248,** 2489 (1959).
135. S. Julia, M. Julia, and M. Davis, *Bull. Soc. Chim. Fr.,* **1960,** 169.
136. K. Tanabe, R. Takasaki, K. Sakai, R. Hayashi, and Y. Morisawa, *Chem. Pharm. Bull.* (Tokyo), **10,** 1126 (1962).
137. J. Tadanier, *J. Org. Chem.,* **28,** 1744 (1963).
138. R. M. Moriarty and T. D. D'Silva, *J. Org. Chem.,* **28,** 2445 (1963).
139. P. B. Solman, *J. Org. Chem.,* **28,** 3559 (1963).

140. G. Just and C. Simonovitch, *Tetrahedron Lett.*, **1967**, 2093.
141. W. P. Schneider, U. Axen, F. H. Lincoln, J. E. Pike, and J. L. Thompson, *J. Amer. Chem. Soc.*, **90**, 5895 (1968).
142. U. Axen, F. H. Lincoln, and J. L. Thompson, *Chem. Commun.*, **1969**, 303.
143. W. P. Schneider, *Chem. Commun.*, **1969**, 304.
144. J. Schreiber, W. Leimgruber, M. Pesaro, P. Schudel, and A. Eschenmoser, *Angew. Chem.*, **71**, 637 (1959).
145. J. Schreiber, W. Leimgruber, M. Pesaro, P. Schudel, T. Threlfall, and A. Eschenmoser, *Helv. Chim. Acta*, **44**, 540 (1961).
146. E. J. Corey, M. Ohno, P. A. Vatakencherry, and R. B. Mitra, *J. Amer. Chem. Soc.*, **83**, 1251 (1961).
147. E. J. Corey, M. Ohno, R. B. Mitra, and P. A. Vatakencherry, *J. Amer. Chem. Soc.*, **86**, 478 (1964).
148. E. J. Corey and S. Nozoe, *J. Amer. Chem. Soc.*, **86**, 1652 (1964).
149. E. J. Corey and S. Nozoe, *J. Amer. Chem. Soc.*, **87**, 5733 (1965).
150. G. Stork and H. T. Cheung, *J. Amer. Chem. Soc.*, **87**, 3783 (1965).
151. R. V. Stevens, M. C. Ellis, and M. P. Wentland, *J. Amer. Chem. Soc.*, **90**, 5576 (1968).
152. S. L. Keeley, Jr., and F. C. Tahk, *J. Amer. Chem. Soc.*, **90**, 5584 (1968).
153. A. J. Birch, *J. Chem. Soc.*, **1944**, 430; see Chapter 1 of this Volume for the most recent review of this reaction.
154. A. J. Birch and S. M. Mukherji, *Nature*, **163**, 766 (1949).
155. L. Miramontes, G. Rosenkranz, and C. Djerassi, *J. Amer. Chem. Soc.*, **73**, 3540 (1951).
156. C. Djerassi, L. Miramontes, G. Rosenkrantz, and J. Sondheimer, *J. Amer. Chem. Soc.*, **76**, 4092 (1954).
157. F. B. Colton, U.S. Pat. 2,691,028 (1954).
158. W. S. Johnson, H. C. Dehm, and L. J. Chinn, *J. Org. Chem.*, **19**, 670 (1954).
159. S. N. Ananchenko and I. V. Torgov, *Tetrahedron Lett.*, **1963**, 1553.
160. H. Smith, G. A. Hughes, G. H. Douglas, G. R. Wendt, G. C. Buzby, R. A. Edgren, J. Fisher, T. Foell, B. Gadsby, D. Hartley, D. Herbst, A. B. A. Jansen, K. Ledig, B. J. McLoughlin, J. McMenamin, T. W. Pattison, P. C. Phillips, R. Rees, J. Siddall, J. Siuda, L. L. Smith, T. Tokolics, and D. H. P. Watson, *J. Chem. Soc.*, **1964**, 4472.
161. For a fuller discussion of steroid total synthesis, see P. D. Klimstra and F. B. Colton in *Contraception: The Chemical Control of Fertility* (D. Lednicer, ed.), Dekker, New York, 1969, p. 69.
162. J. W. Cornforth and R. Robinson, *J. Chem. Soc.*, **1946**, 676.
163. W. S. Johnson, J. W. Petersen, and C. D. Gutsche, *J. Amer. Chem. Soc.*, **69**, 2942 (1947).
164. D. K. Banerjee, S. Chatterjee, C. N. Pillai, and M. V. Bhatt, *J. Amer. Chem. Soc.*, **78**, 3769 (1956).
165. L. Velluz, G. Nomine, J. Mathieu, E. Toromanoff, D. Bertin, J. Tessier, and A. Pierdet, *C. R. Acad. Sci., Paris*, **250**, 1084 (1960).
166. L. Velluz, G. Nomine, and J. Mathieu, *Angew. Chem.*, **72**, 725 (1960).
167. L. Velluz, G. Nomine, R. Bucourt, A. Pierdet, and Ph. Dufay, *Tetrahedron Lett.*, **1961**, 127.
168. D. K. Banerjee, V. Paul, and S. K. Balasubramanian, *Tetrahedron Lett.*, **12**, 23 (1960).
169. W. S. Johnson, E. R. Rogier, and J. Ackerman, *J. Amer. Chem. Soc.*, **78**, 6322 (1956).
170. W. S. Johnson, W. A. Vredenburgh, and J. E. Pike, *J. Amer. Chem. Soc.*, **82**, 3409 (1960).
171. W. S. Johnson, A. D. Kemp, R. Pappo, J. Ackerman, and W. F. Johns, *J. Amer. Chem. Soc.*, **78**, 6312 (1956).
172. R. B. Turner, O. Buchardt, E. Herzog, R. B. Morin, A. Riebel, and J. M. Sanders, *J. Amer. Chem. Soc.*, **88**, 1766 (1966).
173. R. E. Ireland and L. N. Mander, *J. Org. Chem.*, **32**, 689 (1967).

174. R. E. Ireland and L. N. Mander, *Tetrahedron Lett.*, **1964**, 3453.
175. L. H. Zalkow, B. Kumar, D. H. Miles, J. Nabors, and N. Schnautz, *Tetrahedron Lett.*, **1968**, 1965.
176. A. J. Birch, J. M. Brown, and G. S. R. Subba Rao, *J. Chem. Soc.*, **1964**, 3309.
177. R. D. Stipanovic and R. B. Turner, *J. Org. Chem.*, **33**, 3261 (1968).
178. D. J. Beames and L. N. Mander, *Chem. Commun.*, **1969**, 498.
179. E. Wenkert and D. A. Berges, *J. Amer. Chem. Soc.*, **89**, 2507 (1967).
180. W. G. Dauben and G. H. Berezin, *J. Amer. Chem. Soc.*, **85**, 468 (1963).
181. S. Winstein and R. Baird, *J. Amer. Chem. Soc.*, **79**, 756 (1957).
182. A. S. Dreiding, *Holv. Chim. Acta*, **40**, 1812 (1957).
183. R. Barner, A. S. Dreiding, and H. Schmid, *Chem. Ind.* (London), **1958**, 1437.
184. S. Dorling and J. Harley-Mason, *Chem. Ind.* (London), **1959**, 1551.
185. S. Masamune, *J. Amer. Chem. Soc.*, **83**, 1009 (1961).
186. S. Masamune, *J. Amer. Chem. Soc.*, **86**, 288 (1964).
187. E. J. Corey, N. N. Girota, and C. T. Mathew, *J. Amer. Chem. Soc.*, **91**, 1557 (1969).
188. T. G. Crandall and R. G. Lawton, *J. Amer. Chem. Soc.*, **91**, 2127 (1969).
189. C. D. Hurd and M. A. Pollack, *J. Amer. Chem. Soc.*, **60**, 1905 (1938).
190. C. D. Hurd and M. A. Pollack, *J. Org. Chem.*, **3**, 550 (1939).
191. R. L. Adelman, *J. Amer. Chem. Soc.*, **77**, 1669 (1955).
192. A. W. Burgstahler, *J. Amer. Chem. Soc.*, **82**, 4681 (1960).
193. A. W. Burgstahler and I. C. Nordin, *J. Amer. Chem. Soc.*, **83**, 198 (1961).
194. R. F. Church, R. E. Ireland, and J. A. Marshall, *J. Org. Chem.*, **31**, 2526 (1966).
195. R. F. Church, R. E. Ireland, and J. A. Marshall, *Tetrahedron Lett.*, **1961**, 34.
196. R. F. Church, R. E. Ireland, and J. A. Marshall, *J. Org. Chem.*, **27**, 1118 (1962).
197. R. E. Ireland and J. A. Marshall, *J. Org. Chem.*, **27**, 1620 (1962).
198. R. F. Church and R. E. Ireland, *J. Org. Chem.*, **28**, 17 (1963).
199. G. Büchi and J. D. White, *J. Amer. Chem. Soc.*, **86**, 2884 (1964).
200. R. A. Finnegan and P. L. Bachman, *J. Org. Chem.*, **30**, 4145 (1965).
201. W. Nagata, S. Hirai, T. Okumura, and K. Kawata, *J. Amer. Chem. Soc.*, **90**, 1650 (1968).
202. N. P. Damodaran and S. Dev, *Tetrahedron Lett.*, **1967**, 2897.
203. G. Büchi and J. E. Powell, Jr., *J. Amer. Chem. Soc.*, **89**, 4559 (1967).
204. H. Meerwein, W. Florian, N. Schon, and G. Stopp, *Justus Liebigs Ann. Chem.*, **641**, 1 (1961).
205. A. E. Wick, D. Felix, K. Steen, and A. Eschenmoser, *Helv. Chim. Acta*, **47**, 2425 (1964).
206. H. Muxfeldt, R. S. Schneider, and J. B. Mooberry, *J. Amer. Chem. Soc.*, **88**, 3670 (1966).
207. W. S. Johnson, L. Werthemann, W. R. Bartlett, T. J. Brocksom, T-t. Li, D. J. Faulkner, and M. R. Petersen, *J. Amer. Chem. Soc.*, **92**, 741 (1970).
208. Z. Valenta, P. Deslongchamps, R. A. Ellison, and K. Wiesner, *J. Amer. Chem. Soc.*, **86**, 2533 (1964).
209. H. Dugas, R. A. Ellison, Z. Valenta, K. Wiesner, and C. M. Wong, *Tetrahedron Lett.*, **1965**, 1279.
210. A. Mondon, *Chem. Ber.*, **92**, 1472 (1959).
211. A. Mondon and K. F. Hansen, *Tetrahedron Lett.*, **14**, 5 (1960).
212. M. E. Kuehne and C. Bayha, *Tetrahedron Lett.*, **1966**, 1311.
213. U. K. Pandit, K. deJonge, G. J. Koomen, and H. O. Huisman, *Tetrahedron Lett.*, **1967**, 3529.
214. J. M. Bobbitt and C. P. Dutta, *Chem. Commun.*, **1968**, 1429.
215. R. V. Stevens and M. P. Wentland, *J. Amer. Chem. Soc.*, **90**, 5580 (1968).
216. T. J. Curphey and H. L. Kim, *Tetrahedron Lett.*, **1968**, 1441.

217. M. A. T. Sluyter, U. K. Pandit, W. N. Speckamp, and H. O. Huisman, *Tetrahedron Lett.*, **1966**, 87.
218. K. Wiesner, I. Jirkovsky, M. Fishman, and C. A. J. Williams, *Tetrahedron Lett.*, **1967**, 1523.
219. E. H. W. Böhme, Z. Valenta, and K. Wiesner, *Tetrahedron Lett.*, **1965**, 2441.
220. O. L. Chapman and E. L. Eian, *J. Amer. Chem. Soc.*, **90**, 5329 (1968).
221. For a recent review, see E. Wenkert, *Accounts Chem. Res.*, **1**, 78 (1968).
222. E. Wenkert and B. Wickberg, *J. Amer. Chem. Soc.*, **87**, 1580 (1965).
223. E. Wenkert, K. G. Dave, R. G. Lewis, and P. W. Sprague, *J. Amer. Chem. Soc.*, **89**, 6741 (1967).
224. E. Wenkert, K. G. Dave, and F. Haglid, *J. Amer. Chem. Soc.*, **87**, 5461 (1965).
225. E. Wenkert, K. G. Dave, and R. V. Stevens, *J. Amer. Chem. Soc.*, **90**, 6177 (1968).
226. E. Wenkert and A. R. Jeffcoat, *J. Org. Chem.*, **35**, 515 (1970).
227. J. T. Nielsen, N. Elming, and N. Clauson-Kaas, *Acta Chem. Scand.*, **9**, 9 (1955).
228. W. R. Boehme, *J. Amer. Chem. Soc.*, **82**, 498 (1960).
229. G. Büchi and H. Wüest, *J. Org. Chem.*, **31**, 977 (1966).
230. L. Crombie, P. Hemesley, and G. Pattenden, *J. Chem. Soc., C*, **1969**, 1024.
231. N. Clauson-Kaas, N. Elming, and Z. Tyle, *Acta Chem. Scand.*, **9**, 1 (1955).
232. N. Elming and N. Clauson-Kaas, *Acta Chem. Scand.*, **9**, 23 (1955).
233. H. Muxfeldt, J. Behling, G. Grethe, and W. Rogalski, *J. Amer. Chem. Soc.*, **89**, 4991 (1967).
234. H. Muxfeldt and W. Rogalski, *J. Amer. Chem. Soc.*, **87**, 933 (1965).
235. J. D. White, W. E. Haefliger, and M. J. Dimsdale, *Tetrahedron*, **26**, 233 (1970).
236. G. Stork, *Pure Appl. Chem.*, **9**, 131 (1964).
237. G. Stork and J. E. McMurry, *J. Amer. Chem. Soc.*, **89**, 5463 (1967).
238. M. Ohashi, H. Kamachi, H. Kakisawa, and G. Stork, *J. Amer. Chem. Soc.*, **89**, 5460 (1967).
239. G. Stork, J. Danishefsky, and M. Ohashi, *J. Amer. Chem. Soc.*, **89**, 5459 (1967).
240. M. Ohashi, T. Maruishi, and H. Kakisawa, *Tetrahedron Lett.*, **1968**, 719.

THE STRUCTURE OF RYANODINE

By KAREL WIESNER, *Natural Products Research Center, University of New Brunswick, Fredericton, Canada*

CONTENTS

I. Introduction

The insecticidal principle ryanodine was isolated from the flacourtiaceous plant *Ryania speciosa* Vahl by Folkers and his collaborators (1). The powdered stems of this plant are being used in New Brunswick for the control of the European corn borer, and we became interested in this problem chiefly because of the ready availability of the plant material, which was donated to us in unlimited quantities by S. B. Penick and Company, New York.

Ryanodine, $C_{25}H_{35}O_9N$, suffers a very easy dehydration either on mild acid treatment or on sublimation. The resulting compound, anhydroryanodine $C_{25}H_{33}O_8N$, lends itself much better to degradation studies than ryanodine itself. Because of this fact and since the two compounds were believed to be in a relatively simple relationship, we decided to first elucidate the structure of anhydroryanodine. In the year 1962 the absolute stereo structure **28a** was assigned (2) to this product and the first phase of our structural studies was complete.

The study of ryanodine was initiated at the University of New Brunswick in 1951. At that time we had established that ryanodine was an ester of pyrrole-α-carboxylic acid and of ryanodol, $C_{20}H_{32}O_8$.

According to analysis, active hydrogen determination, and the absence of carbonyl groups in the ir spectrum, ryanodol appeared to be a pentacyclic compound with seven or six hydroxyls and one or two ether bridges. On treatment with acid, ryanodol was found to suffer a dehydration analogous to the dehydration of ryanodine and to yield anhydroryanodol, $C_{20}H_{30}O_7$. This dehydration was connected with the appearance of a carbonyl group in the ir spectrum at 1721 cm^{-1}, which was later shown to be due to a δ-lactone group. Anhydroryanodol consumed 1 mole of periodate and yielded anhydro-oxoryanodol, $C_{20}H_{28}O_7$. This compound was found to be extremely labile to alkali and disintegrated on basic treatment into a number of acids, one of which was identified as isobutyric acid. At this stage these preliminary findings were reported (3), and the problem was temporarily abandoned.

When the study of ryanodine was resumed in 1960 the chances of achieving a rapid solution of the problem were substantially better: we had gained experience in the study of polyfunctional polycyclic compounds which gave us confidence in the ultimate resolution of ryanodine chemistry. A careful study of various ryanodol derivatives by nmr spectroscopy revealed five carbon-methyl groups, of which two were, of course, part of the isopropyl known to be present (see fragmentation, Scheme 1).

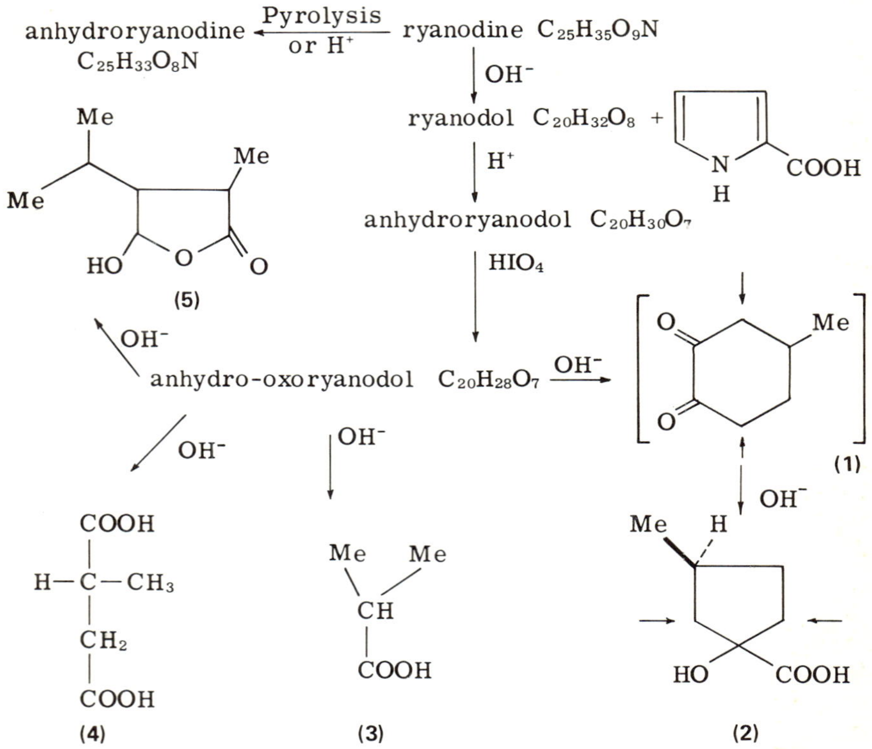

anhydroryanodine $C_{25}H_{33}O_8N$ ← (Pyrolysis or H^+) ryanodine $C_{25}H_{35}O_9N$

ryanodine → (OH^-) ryanodol $C_{20}H_{32}O_8$ + pyrrole-2-carboxylic acid (N–H, COOH)

ryanodol → (H^+) anhydroryanodol $C_{20}H_{30}O_7$

anhydroryanodol → (HIO_4) anhydro-oxoryanodol $C_{20}H_{28}O_7$

anhydro-oxoryanodol → (OH^-) (5) [lactone: Me, Me, Me, HO, O, O]

anhydro-oxoryanodol → (OH^-) (1) [cyclohexanedione: O, O, Me] → (OH^-) (2) [cyclopentane: Me, H, HO, COOH]

anhydro-oxoryanodol → (OH^-) (4) COOH–C(H)(CH₃)–CH₂–COOH

anhydro-oxoryanodol → (OH^-) (3) (Me)(Me)CH–COOH

Scheme 1

II. The Structure of Anhydroryanodine

Anhydroryanodol revealed in the nmr spectrum a fully substituted double bond which carried one of the five methyl groups. The formation of an orthoacetate diacetate which possessed no hydroxyl and the presence of the aforementioned δ-lactone accounted for all the oxygen.

We next turned to the identification of the fragmentation products obtained by basic treatment of oxoanhydroryanodol. The fragmentation products were separated by partition chromatography on silicic acid with water as the stationary phase and chloroform with a varying percentage of butanol as the moving phase. The structures of the individual compounds were elucidated by very simple methods (4) which do not require a detailed description (2-5).

It was possible to demonstrate that compound **2** originated from a benzilic acid rearrangement of the intermediate cyclohexandione **1**.

In compound **2** the two methylenic groups marked by arrows are nonexchangeable for deuterium in an alkaline deuterated medium. In the intermediate **1**, on the other hand, such an exchange reaction is possible. We have therefore performed the fragmentation reaction in heavy water and obtained the acid **2** fully deuterated in the two marked methylenic groups—a decisive proof for the origin of compound **2** from a 6-membered ring.

By lithium aluminium hydride reduction of the methyl ester of **2** a glycol was obtained which on periodate cleavage yielded (–)-3-methyl-cyclopentanone of known absolute configuration. This finding later became the basis of the absolute configuration assignment of anhydroryanodine.

The yields of all compounds in the fragmentation were such that no overlap of any of the products except **3** and **5** was possible and, consequently, all carbon of anhydroryanodol was accounted for. At this stage we came to the conclusion that while the fragmentation scheme probably holds the key to the complete structure of anhydroryanodol, it would require an extremely lucky idea to deduce the structure from these data alone. In fact, we had speculated on the basis of a biogenetic hypothesis derived from the fragmentation scheme and arrived at structures closely related to the right one but never completely correct even in skeleton.

Consequently, we decided that we needed an "idea experiment," for instance a dehydrogenation result, to help us orientate the fragments with respect to each other. After some unsuccessful dehydrogenation studies, we turned to the treatment of anhydroryanodol with refluxing hydroiodic acid and red phosphorus. This reaction gave a mixture of products from which the γ-lactone, $C_{20}H_{26}O_2$, **6** was isolated in good yield. A clue to the structure of **6** was found in an uncharacterized amorphous fraction from the same experiment which gave on selenium dehydrogenation a hydrocarbon $C_{19}H_{22}$ formulated as **7**. Compound **7** was clearly a fluorene by its uv spectrum and showed in the nmr spectrum four aromatic methyls ($\tau = 7.40$, 7.60, 7.68, and 7.77) and two benzylic methyls [singlet (6H) at $\tau = 8.63$]. The nmr spectrum also showed four aromatic hydrogens at $\tau = 2.4$-3.0. In view of the common origin of **6** and **7**, one methyl group was placed on ring C and three on ring A.

Since in compound **6** the absence of a substituent in the position marked by an arrow was proved (by degradation to compound **11**), it was necessary to distribute the three methyls on ring A in **7** as shown, to account for the relationship of **6** and **7** by a relatively simple Wagner rearrangement (5).

The nmr spectrum of **7** and the ir spectrum of **6** considered together enabled us also to place the aromatic methyl in ring C. The ir of **6** was in clear agreement

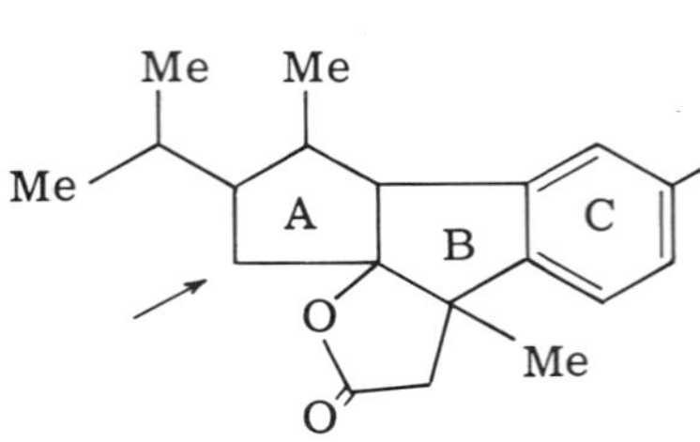

(6)

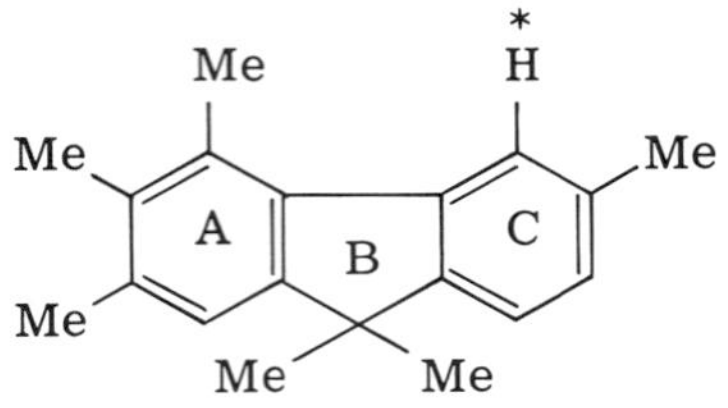

(7)

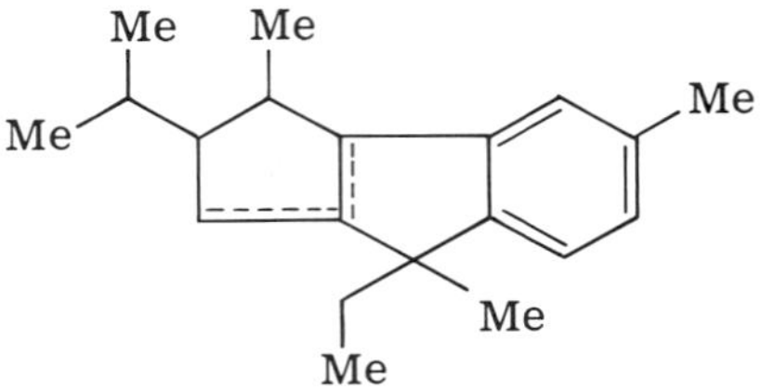

(8) and (9)

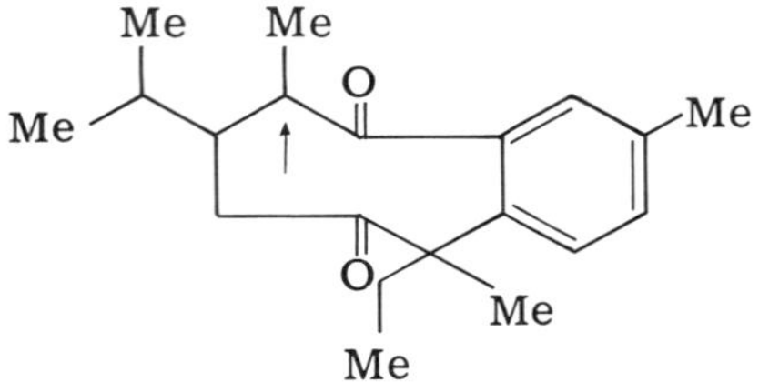

(10)

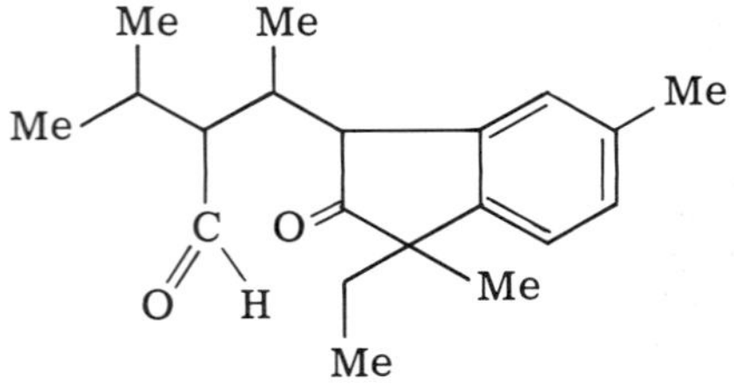

(11)

with the presence of a 1,2,4-substituted benzene ring. Since in compound **7** the hydrogen coming to resonance at the lowest field must be the one marked by the asterisk (unshielding effect of the aromatic ring A) and since the lowest peak in the nmr spectrum of **7** is a singlet, we were able to select the 1,2,4 substitution shown and discard the second possible alternative.

The structural studies performed on **6** are too lengthy to be recounted in detail (6). Suffice it to say that they included the conversion of **6** into the two unsaturated isomers **8** and **9**, and the two seco derivatives **10** and **11** by standard methods. A preparative Kuhn-Roth oxidation of **6** gave a mixture of acetic and isobutyric acid. These data would still permit an exchange in the position of the methyl and isopropyl groups in ring A. The alternative chosen was proved in the following manner. Bromination of the diketone **10** yielded a monobromo derivative in which the bromine atom must have substituted a position α to a keto group, and occupied by a methyl. One methyl doublet in the nmr spectrum of the diketone **10** ($\tau = 9.29, 9.40$) has shifted downfield ($\tau = 8.06$) and become a singlet in the spectrum of the monobromo derivative. This is only possible in the arrangement of the methyl and isopropyl groups portrayed in **10**, and the bromination must have occurred at the position marked by the arrow.

The entire nmr spectrum of compound **6** is in excellent agreement with the structure, and the assignments of the various peaks are shown in Scheme 2.

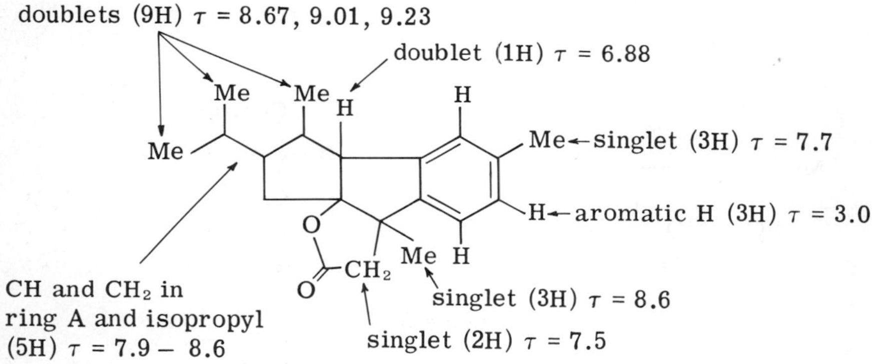

Scheme 2

Of course, there was no assurance at this stage that in the formation of **6** extensive rearrangements did not occur. However, we were pleased to see that the structure of compound **6** may be simply dissected to yield without overlap the skeletons of the fragments obtained in the basic degradation of oxoanhydroryanodol. This is illustrated in Scheme 3.

Studies of functional groups and their environment in anhydroryanodol which we performed at this point increased the areas of known skeletal identity between anhydroryanodol and the lactone **6** to such an extent that we finally accepted the identity of the entire skeleton of the two compounds as extremely probable. A particularly useful derivative in this respect was anhydroryanodol orthoacetate diacetate **12**. This compound was hydrolyzed by mild treatment with alkali to an orthoacetate monoacetate which in turn was oxidized to the fully substituted cyclopentenone **13**.

The ir frequency of the δ-lactone group is high due to strain caused by the orthoacetate formation. Anhydroryanodol itself is a typical δ-lactone (ir max 1720 cm^{-1}).

The assignment of the lactone hydroxyl as tertiary and the two acetyl groups as secondary follows from the fact that neither anhydroryanodol nor the

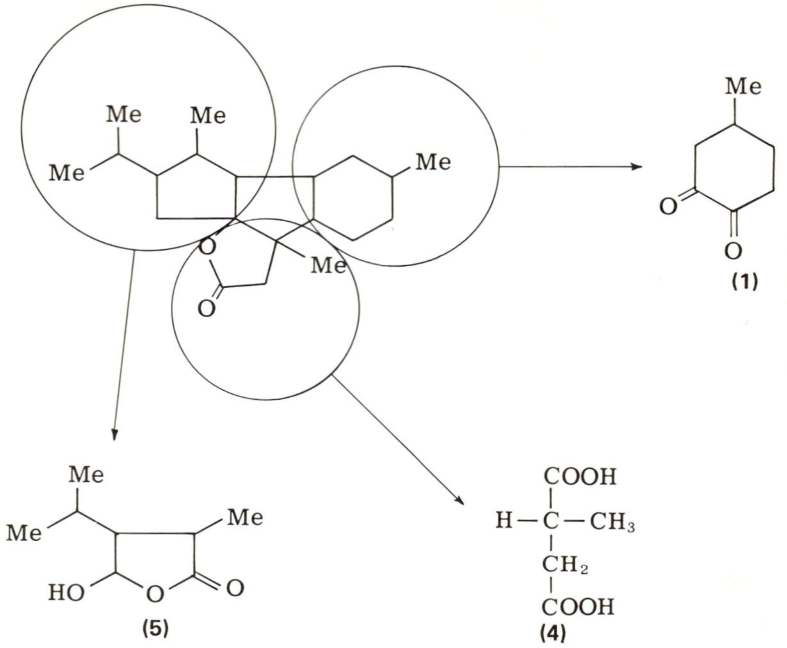

(1)

(5)

(4)

Scheme 3

orthoacetate **12** with both acetyl groups removed by hydrolysis shows any hydrogen unshielded by a lactonic hydroxyl in the nmr spectrum.

It may now be easily seen that the presence of the elements **14**, **15**, and **16** is proved in compound **13**. The presence of the isopropyl group in the α position of the cyclopentenone is not only to be expected on the basis of the structures of lactone **6** and the methyl isopropyl γ-lactol **5**, but follows from the fact that the α position is not occupied by hydrogen or methyl (nmr) and is not a ring junction (violation of the Bredt rule). Consequently, the isopropyl group is the only group present in anhydroryanodol eligible to occupy this position. The proof of the presence of the element **15** in anhydroryanodol has been discussed already in connection with the fragmentation reaction. Finally, the acetic acid side-chain **16** which forms the δ-lactone is attached (nmr) to a quaternary carbon. By exclusion, it becomes clear that this structural element must be the source of methyl succinic acid in the fragmentation reaction. Consequently, we must place a methyl group as indicated in **16**. The objection that the methyl group in methyl succinic acid may not be derived from a methyl of anhydroryanodol, but from a methylenic group by breaking of a bond, is easily met by the previously mentioned fragmentation experiment run in heavy water. The methyl succinic acid isolated from this experiment contained no deuterium in the methyl group. If the methyl group was derived from a fragmentation, the

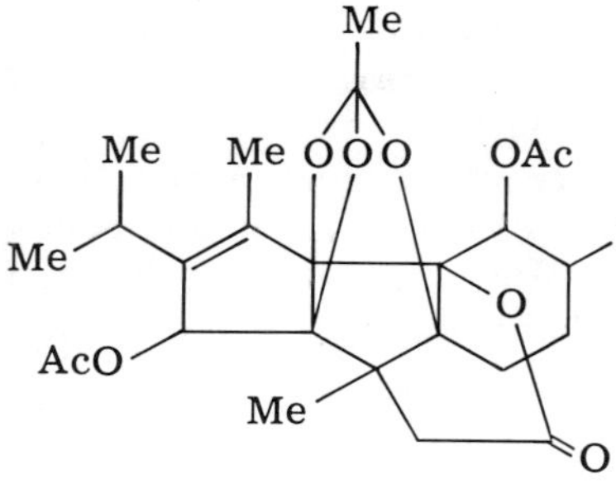

(12)

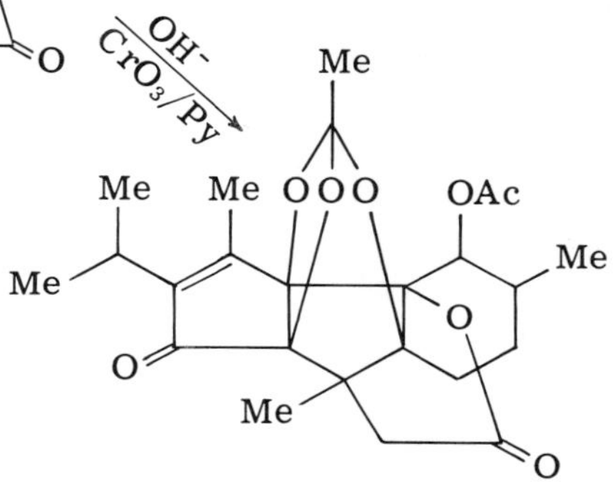

OH^-

CrO_3/Py

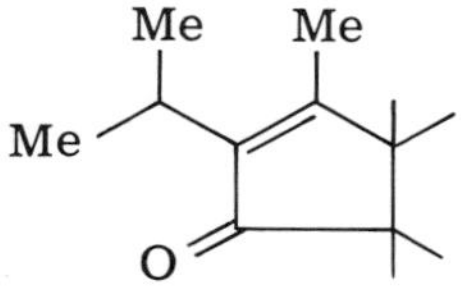

(14)

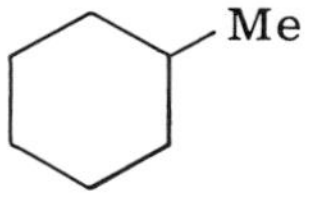

(13)

(15)

(16)

Cyclopentenone

uv: λ_{max} 237.5 mμ, log ϵ = 4

ir: 1710 cm^{-1}

nmr. No vinylic H

β methyl group 7.94 ppm

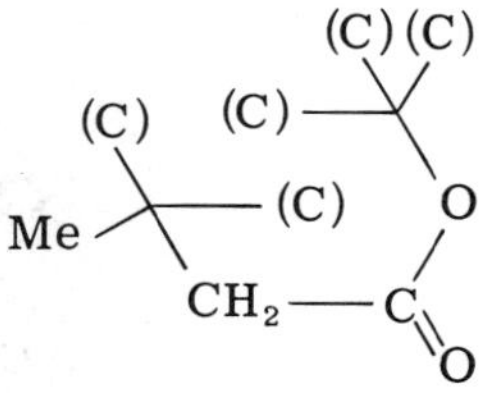

δ-Lactone $>C-CH_2-C(=O)-O$

ir: 1768 cm^{-1}

nmr. Singlet 2H 7.46 ppm

$H-C-O-Ac$

ir: 1738 cm^{-1}

nmr. Doublet 1H 4.28 ppm

Singlet 3H 7.83 ppm

intermediate in this reaction would have picked up at least one deuterone from the medium.

The structural elements **14**, **15**, and **16** may now be assembled into very few complete skeletal structures, one of which is identical with the skeleton of lactone **6**. It may be shown by a careful analysis of all of them that only this last one can explain all the chemistry and especially provide us with a simple and unexceptionally plausible explanation of the fragmentation reaction.

If anhydroryanodol is **17** we may represent anhydro-oxoryanodol as **19** with a masked α,β-unsaturated keto group. This masking explains both the absence of a uv chromophore and a sharp uptake of only one mole of periodate by **17**. The base-catalyzed fragmentation may then be rationalized by the simple and unexceptional Schemes 4 and 5 which do not require any comment.

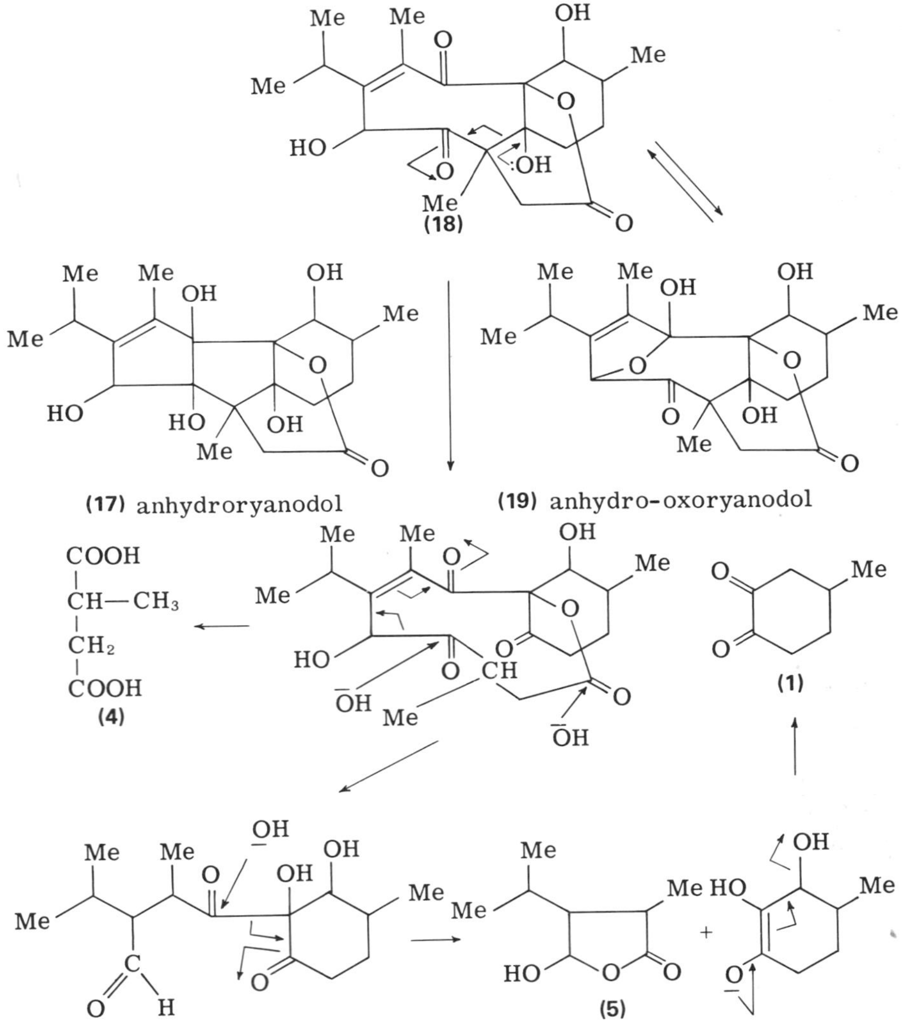

(18)

(17) anhydroryanodol

(19) anhydro-oxoryanodol

(4)

(1)

(5)

Scheme 4

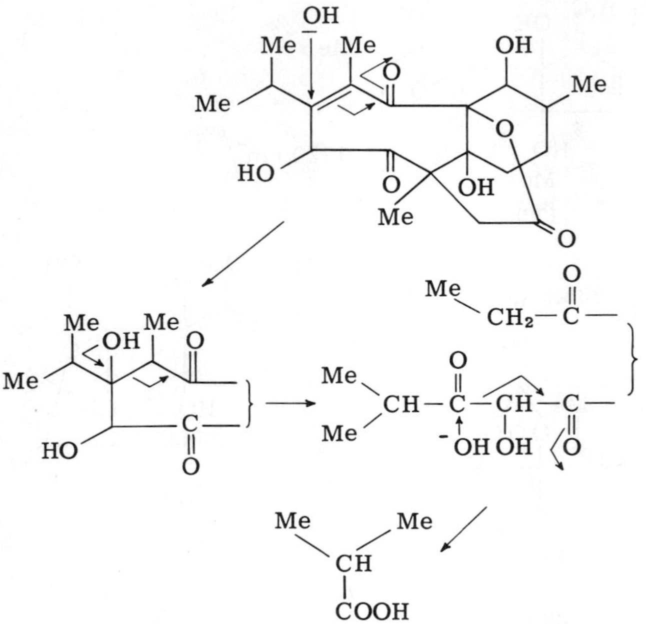

(3)

Scheme 5

For the further corroboration of the anhydroryanodol structure **17**, the bisanhydro series is of importance. Bisanhydroryanodol **20** is obtained by treatment of ryanodol with 20% sulfuric acid at 90°C. It yields 1 mole of formaldehyde on ozonolysis and consumes 1 mole of periodate.

Hydrogenation of bisanhydroryanodol yields the dihydro derivative **21**. Both **20** and **21** may be converted to the corresponding orthoacetates monoacetates. The identity of the skeletal structure of anhydro- and bisanhydroryanodol is proved by the facile conversion of anhydroryanodol orthoacetate monoacetate to the orthoacetate monoacetate of **20** by thionyl chloride in pyridine. If the orthoacetate monoacetate of **21** represented by **22** is reduced by lithium aluminium hydride, the triol **23** is obtained. Cleavage of this compound by 1 mole of lead tetraacetate yields the tertiary aldehyde **25** [nmr singlet (1H) τ = 0.53] by an aldol condensation of the primary cleavage product **24**. This sequence proves clearly that the former lactone hydroxyl, the ring C hydroxyl, and the ring C methyl group are located on adjacent carbons.

With the structure of anhydroryanodol determined, the only remaining structural problem was the location of the pyrrole-α-carboxylic residue in anhydroryanodine. It was possible to solve this question very simply in the following manner. Anhydroryanodine was converted into the orthoacetate diacetate **26**. This was at first puzzling, since anhydroryanodine must possess

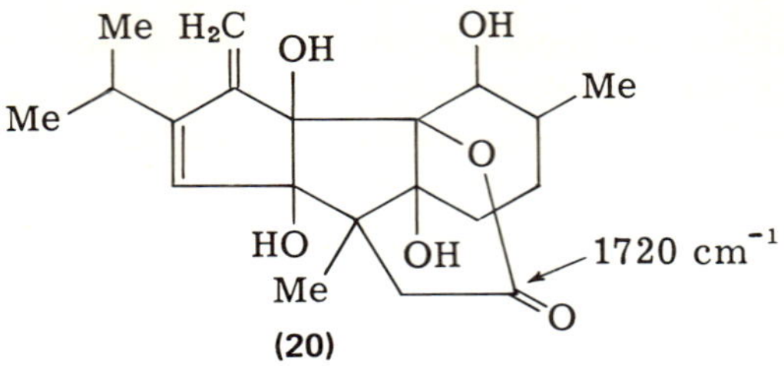

(20)

uv λ_{max} 235.5 (log ϵ = 4.17)

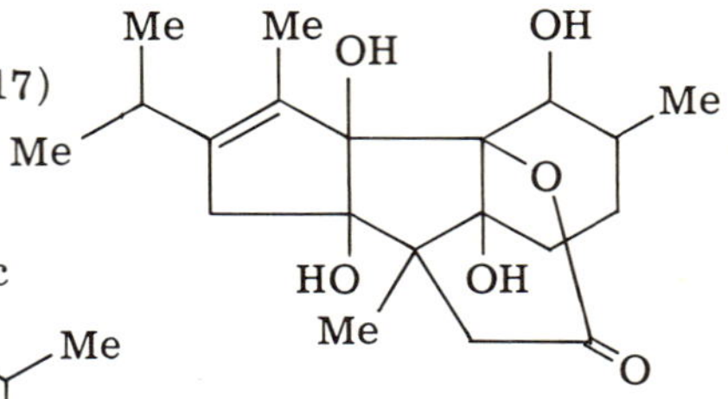

1720 cm^{-1}

(21)

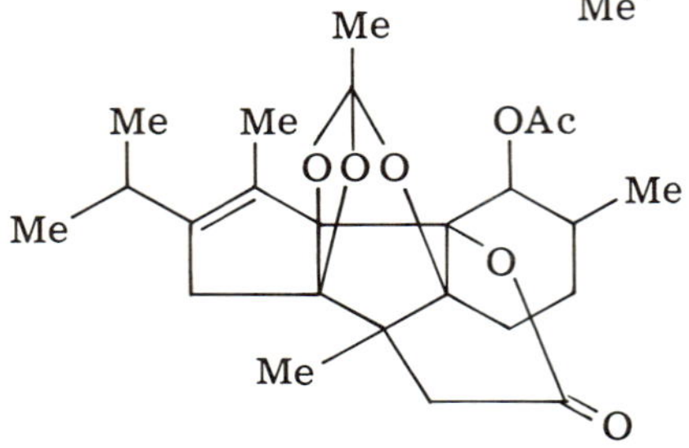

(22)

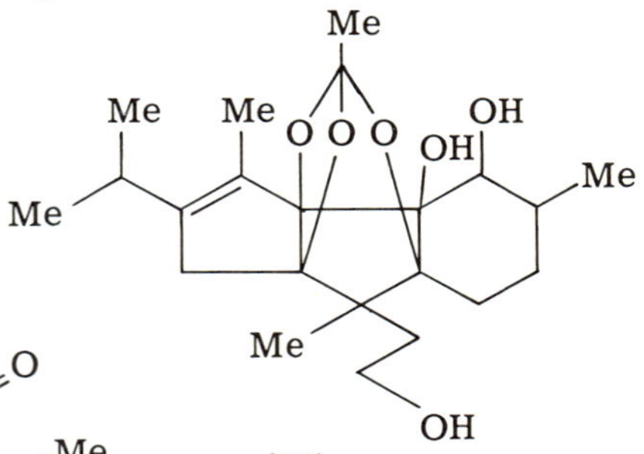

(23)

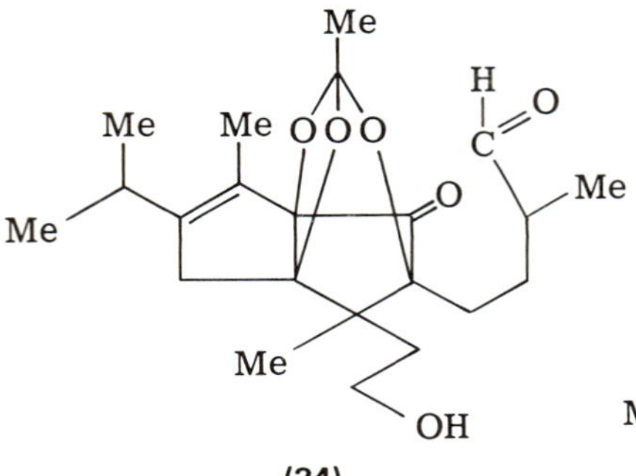

(24)

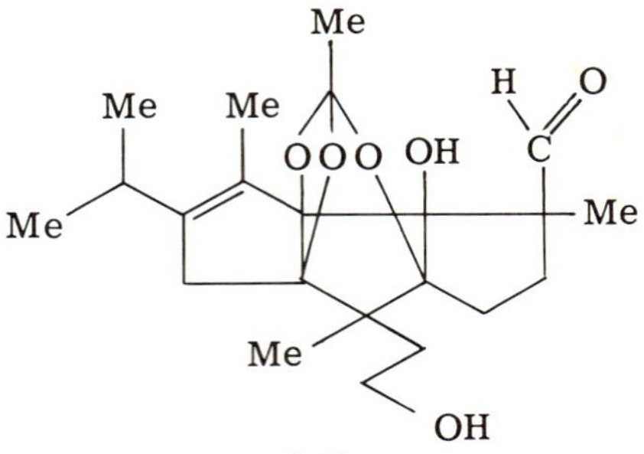

(25)

one less free hydroxyl than anhydroryanodol (i.e., four hydroxyls). However, it was soon realized from the changed uv absorption that one acetyl group has entered the pyrrole nucleus. The same chromophore is obtained if methyl pyrrole-*a*-carboxylate is subjected to the identical acetylation procedure.

We next tried to hydrolyze selectively the allylic ester (as we had done before in the case of anhydroryanodol orthoacetate diacetate) to determine the identity of the acyl group esterfied to the allylic hydroxyl. This turned out to be impossible, since the new acetylpyrrole-α-carboxylic ester group was sufficiently acidic to dissociate completely to an anion in dilute alkali and thus render the ester group immune to hydrolysis. Compound **26** was therefore methylated with diazomethane and converted into the nonacidic *N*-methyl derivative. This compound was not characterized, but it hydrolyzed smoothly to anhydroryanodol orthoacetate monacetate **27** which has been previously shown to possess a free allylic hydroxyl in ring A (by oxidation to the cyclopentenone **13**).

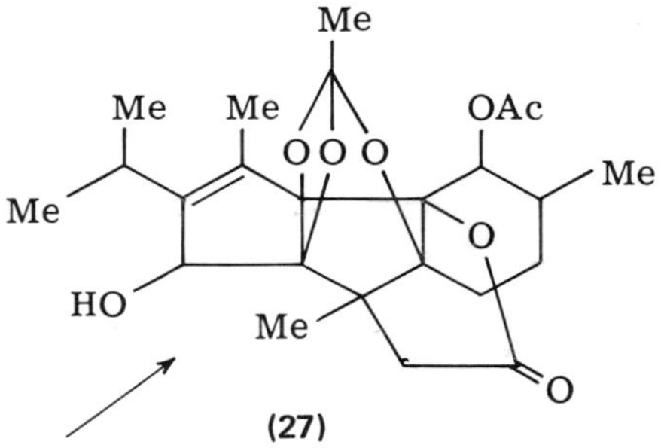

(27)

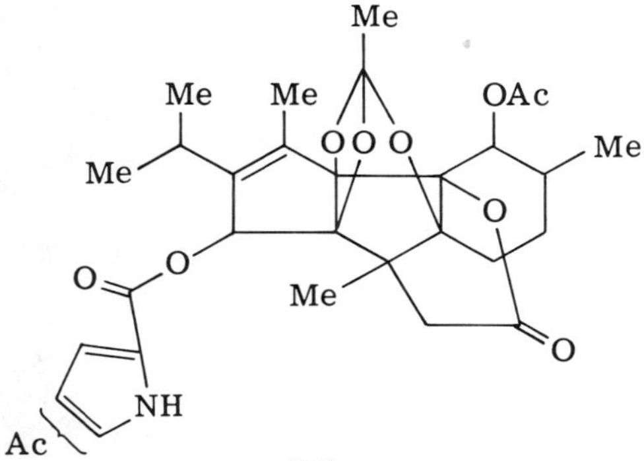

(26)

uv: λ_{max} 275 mμ (ϵ = 17,250)
224 mμ (ϵ = 27,000)
uv: In 0.01*M* alkali
λ_{max} 310 mμ (ϵ = 19,200)
251 mμ (ϵ = 18,800)

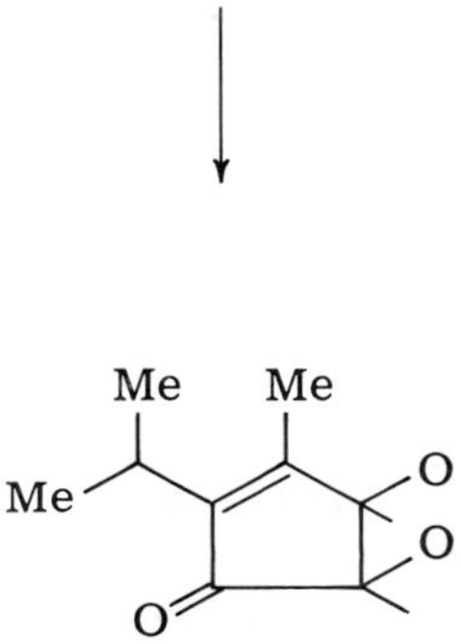

(13)

III. The Absolute Stereochemistry of Anhydroryanodine

The stereochemistry and absolute configuration of anhydroryanodine turned out to be a relatively simple problem. It was possible to show (2) without any difficulties and with the use of already available experimental data that anhydroryanodine must be represented by the complete stereo structure **28.**

The argument is best started with the absolute configuration of the ring C methyl group. This configuration is known, since this particular asymmetric carbon was isolated in the form of the methylcyclopentanone **30** obtained from oxoanhydroryanodol via the cyclohexandione **29**.

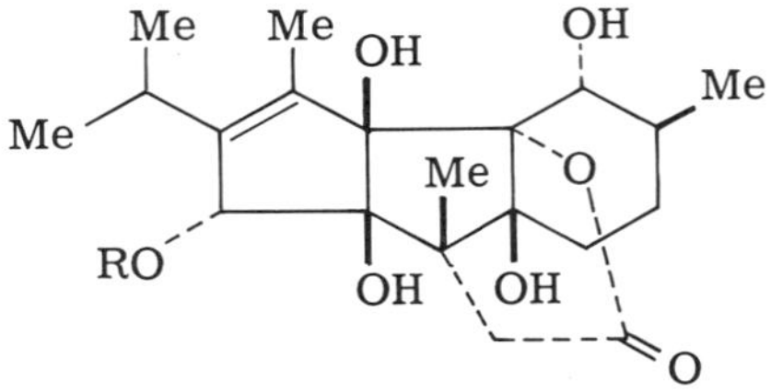

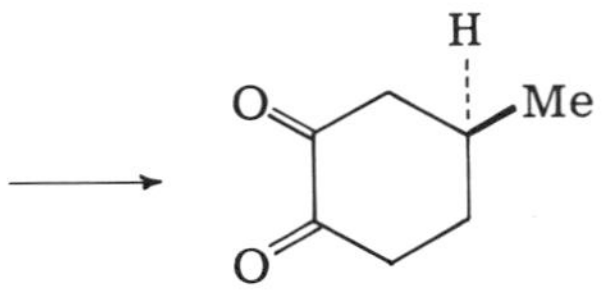

(29)

(28a) R = pyrrole-2-carbonyl-O—

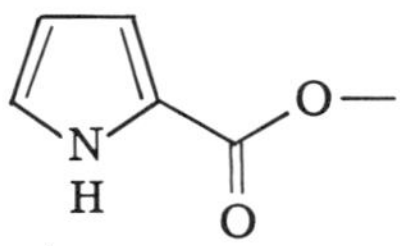

(28b) (R = H)

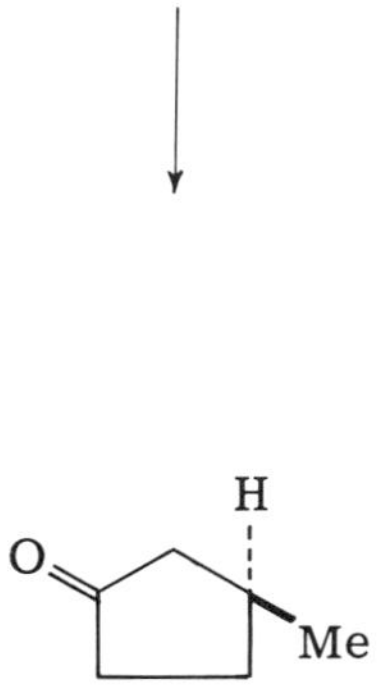

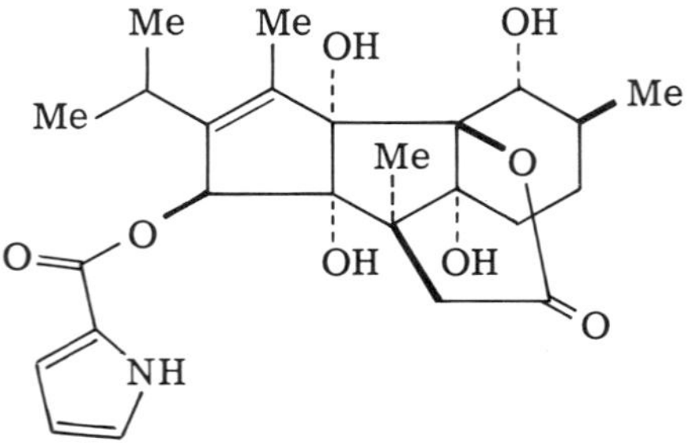

(31)

(30)

The ring C hydroxyl must be *trans* to the methyl group since, in many nmr spectra of various anhydroryanodine derivatives, the hydrogen unshielded by this hydroxyl is found as a sharp doublet with a separation of 8.5-10.5 Hz. Such a splitting shows clearly that the hydrogen in question and its neighbor must be *trans* diaxial. The configuration of the remaining asymmetric centers follows automatically from the limitations imposed on the system by the lactone group and by the formation of the orthoacetate derivatives.

The only exception is the configuration of the allylic hydroxyl in ring A. This is represented as *trans* to the free α-diol group to explain the uptake of only 1 mole of periodate by anhydroryanodol **28b**. The ditertiary *cis* diol reacts faster, and further cleavage is prevented by the hemiketal formation between one keto group and the allylic hydroxyl.

This assignment is now corroborated by some nmr spectroscopic observations. All anhydroryanodol derivatives possessing in some form the allylic hydroxyl show in the nmr spectrum the methylenic group adjacent to the lactone carbonyl as a quadruplet.

On the other hand, all derivatives where the allylic hydroxyl is removed and replaced by a double bond or a methylene group or a keto group, show the methylene adjacent to the lactone carbonyl as a singlet.

Such an influence of the allylic hydroxyl on the unequal shielding of the two hydrogens in question is only possible if the configuration is as postulated in **28**. The only other stereochemical alternative for anhydroryanodine is portrayed in structure **31**. It suffers from the disadvantage that it would be necessary to assume the boat form as the preferential conformation of ring C, to explain the observed splitting displayed by the hydrogen unshielded by the ring C hydroxyl.

IV. The Structure of Ryanodine

After the elucidation of the absolute stereo structure of anhydroryanodine **28a** and anhydroryanodol **28b**, the problem which remained to be solved was clearly the nature of the dehydration reaction leading to these two derivatives and the definition of the structures of ryanodol and ryanodine. Surprisingly, this seemingly minor task turned out to be much more difficult and interesting than expected. The correct structure of ryanodine **44** was finally rigorously deduced and disclosed by me in a lecture at Harvard University on November 1, 1966 (cf. Ref. 7). After this announcement a sample of the *p*-bromobenzyl derivative of ryanodol was submitted to Maria Przybylska at the National Research Council Laboratories, Ottawa, for X-ray crystallographic examination. In November 1967 Srivastava and Przybylska communicated to the author the solution of this problem in terms of structure **51** (cf. Ref. 8). We now discuss the deduction of the ryanodine structure in somewhat greater detail, since it was mainly chemical reasoning rather than various spectral measurements which made the final structural assignment possible.

Ryanodol is a fully saturated compound with five carbon-methyl groups. Two of these appear in the nmr spectrum as singlets, the remaining ones as doublets. Careful deuterium-exchange studies performed on ryanodol gave a value of six or seven active hydrogens. Since ryanodol shows no carbonyl absorption in the ir spectrum, six or seven oxygens must be hydroxyls and the remaining one or two must be ether bridges.

According to this functional-group analysis and the molecular formula $C_{20}H_{32}O_8$, there are then two possibilities for the skeletal structure of ryanodol.

1. Ryanodol contains three carbon rings, six hydroxyls, and two ether bridges.

2. Ryanodol contains four carbon rings, seven hydroxyls, and one ether bridge.

It will be seen in the sequel that an automatic adoption of possibility *1* on the basis of the established structures **28** for the anhydro compounds leads to an impasse, and that it is possibility *2* which in fact is correct.

One of the hydroxy groups of ryanodol is distinctly acidic. As a result of this fact ryanodol can be extracted much more rapidly from an aqueous solution of bicarbonate than from an equally concentrated solution of sodium hydroxide. J. M. Los (Amsterdam) has kindly determined the pK_a of ryanodol and found a value of 11.4 ± 0.2. The most likely group to be responsible for a dissociation constant of this order of magnitude is a hemiacetal or hemiketal hydroxyl.

An important key to the structure of ryanodol was provided by the characterization of the beautifully crystalline trisecoryanodol ($C_{20}H_{26}O_8$) which is formed on oxidation of ryanodol with 3 moles of periodate (9). It is interesting that we have been able to assign rigorously to this compound the structure **32** merely by elucidating completely its functionality.

In the ir spectrum trisecoryanodol **32** shows a single sharp hydroxyl band at 3595 cm^{-1}, a broad very strong carbonyl band between 1780 and 1740 cm^{-1} assigned to the enol lactone, δ-lactone, and formate, and a sharp band at 1695 cm^{-1} assigned to the enol lactone double bond.

However, it is the nmr spectrum of **32** which is particularly revealing. It shows a singlet (lH), $\tau = 1.93$, for the formate hydrogen, and a doublet (lH), $\tau = 6.12$, which is assigned to the hydrogen (marked by arrow in **32**) unshielded by the ether oxygen and adjacent to the ring C secondary methyl group. The signal of this hydrogen is characterized by its splitting (10 Hz), and is present in all ryanodol and anhydroryanodol derivatives. Its presence in the nmr spectrum of **32** shows clearly that this particular region of the molecule remained undisturbed by the periodate oxidation. It is important to note that the above-mentioned doublet is the only hydrogen unshielded by oxygen in the nmr spectrum of **32**. The δ-lactone, which is well known to us from the anhydro derivatives, is revealed in trisecoryanodol by a quadruplet (2H) centered at $\tau = 7.2$ and assigned to the methylene group adjacent to the lactone carbonyl. The methyl group signals are also in excellent agreement with **32**. A singlet (3H), $\tau = 8.18$, is assigned to the vinylic methyl group, while the tertiary methyl unshielded by the enol lactone carbonyl appears as a singlet (3H) at $\tau = 8.6$. Finally, three doublets (9H), $\tau = 8.91$-9.19, are assigned to the remaining three secondary methyl groups. Inspection of **32** reveals that it actually contains a masked β-diketone, since it can yield **37** as an intermediate of a mild hydrolysis. In agreement with this, trisecoryanodol was found to generate quantitatively isobutyric acid with dilute alkali. The same experiment also verified the assignment of the 1.93-ppm peak in the nmr spectrum of trisecoryanodol to the formate ester, since formic acid was simultaneously produced.

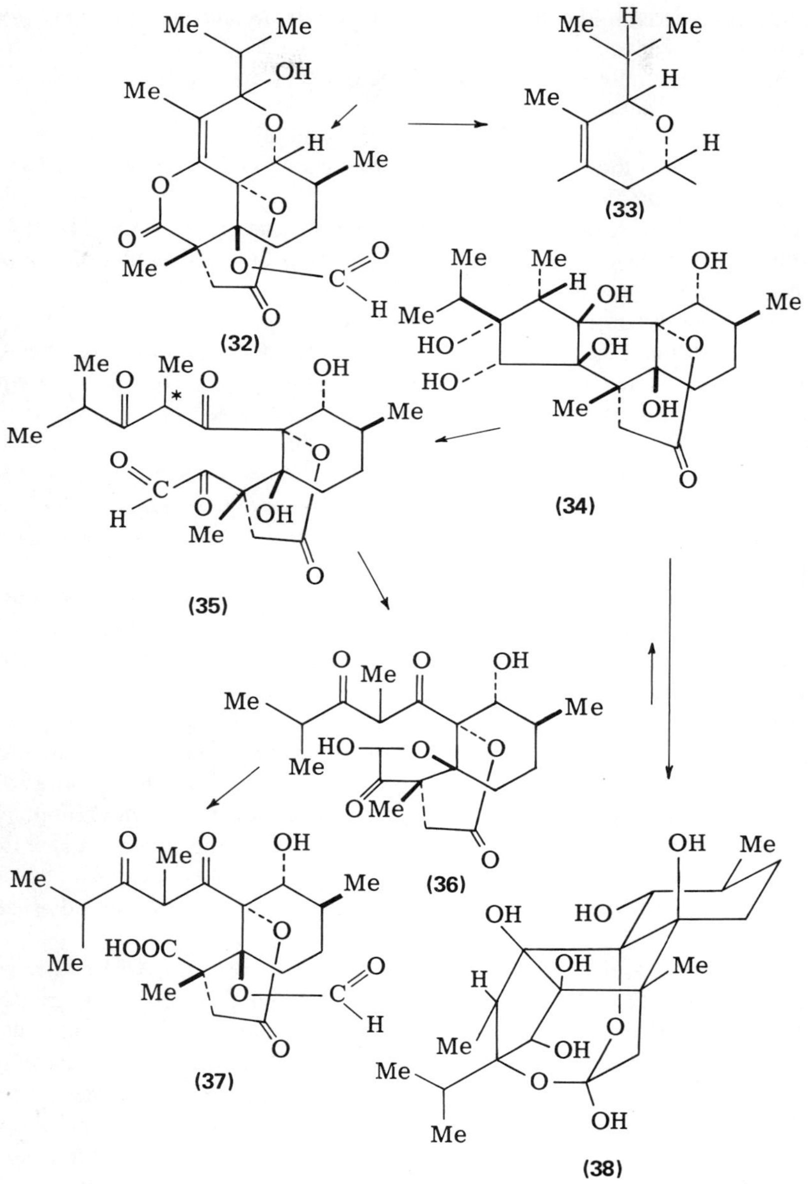

Finally, the presence of an allylic hemiketal in **32** was rigorously proved as follows. Hydrogenation of **32** with platinum oxide in methanol yielded a desoxy compound, $C_{20}H_{26}O_7$, represented by the partial structure **33**. The ir spectrum of **33** is essentially the same as that of **32**, but it *contains no hydroxyl peak.* It is clear that the only hydroxyl of **32** was hydrogenolytically replaced by a hydrogen. Now if the hydrogenolytically replaced hydroxyl was indeed a hemiketal hydroxyl, the nmr spectrum of **33** must show a new signal for a hydrogen unshielded by oxygen. This is in fact observed. The nmr spectra of **32** and **33** are essentially the same except for the presence, in the spectrum of **33**, of a narrow doublet (lH) at $\tau = 5.9$.

Thus practically all the features of trisecoryanodol **32** are proved, and if we give due consideration to the known structure of anhydroryanodol **28b**, the entire structure of **32** is secure.

With the structure of anhydroryanodol **28b** and trisecoryanodol **32** rigorously established, the stage is set for the final deduction of the structure of ryanodol and ryanodine.

Any structure of ryanodol worthy of consideration must explain the formation of **28b** and **32** in a plausible manner. The simplest (but as we shall see not correct) solution of the problem which satisfies this requirement is shown in **34**. This structure portrays the formation of **28b** as a simple dehydration of a tertiary alcohol. Also the representation of the genesis of **32** is facile. Periodate cleavage of the two *cis* glycols in **34** generates the triketoaldehyde **35**, which in its hemiacetal form **36** consumes a third mole of periodate and yields **37**. This last compound may give the structure **32** merely by enol lactonization and hemiketal formation.

The most serious objection against representing ryanodol as **34** is the absence of carbonyl absorption in the ir spectrum of this compound. This objection could conceivably be met by assuming that compound **34** exists predominantly as the hemilactal tautomer **38**.

The steric compulsion for the oxo-cyclo tautomeric equilibrium **34** ⇆ **38** is very favorable, but only one such case (10) is known in the chemical literature.

The compound in question is tetrodotoxine and it is significantly more acidic than ryanodol. An even stronger objection against formula **38** for ryanodol was its inability to explain the so-called "isoseries" of ryanodol derivatives (*vide infra*).

It was therefore necessary to continue the search for an alternative formulation.

The rationalization of the formation of trisecoryanodol **32** by the series **35** → **36** → **37** → **32** seemed too good to be due to coincidence. Consequently it was imperative to ascertain if there exists a formula isomeric with **34** ⇆ **38** which can yield **35** by two glycol cleavages, and at the same time be capable of plausibly explaining the anhydro reaction. *A systematic consideration of the problem has revealed that no such structure can be constructed by manipulating the functional groups in the carbon skeleton of* **34** *or any alternative tricarbocyclic system.* Thus the only possibility which remained to be investigated was the above-mentioned possibility *2,* in other words, a ryanodol structure with four carbon rings, seven hydroxyls, and one ether bridge. Such a structure cannot yield the diseco intermediate **35** merely by two glycol cleavages, but it is necessary to postulate an additional carbon-carbon bond cleavage by a retro aldol reaction. *As soon as we reach this conclusion the* *problem is solved.* If the three keto groups and the aldehyde group of **35** have been generated by cleavages of two glycols, then the bond severed by the reverse aldol reaction must have originally connected the carbon marked by the asterisk in **35** and the carbon of the lactone carbonyl.

Thus we arrive at structure **39** for the diseco precursor of **35** and structure **40** for ryanodol.

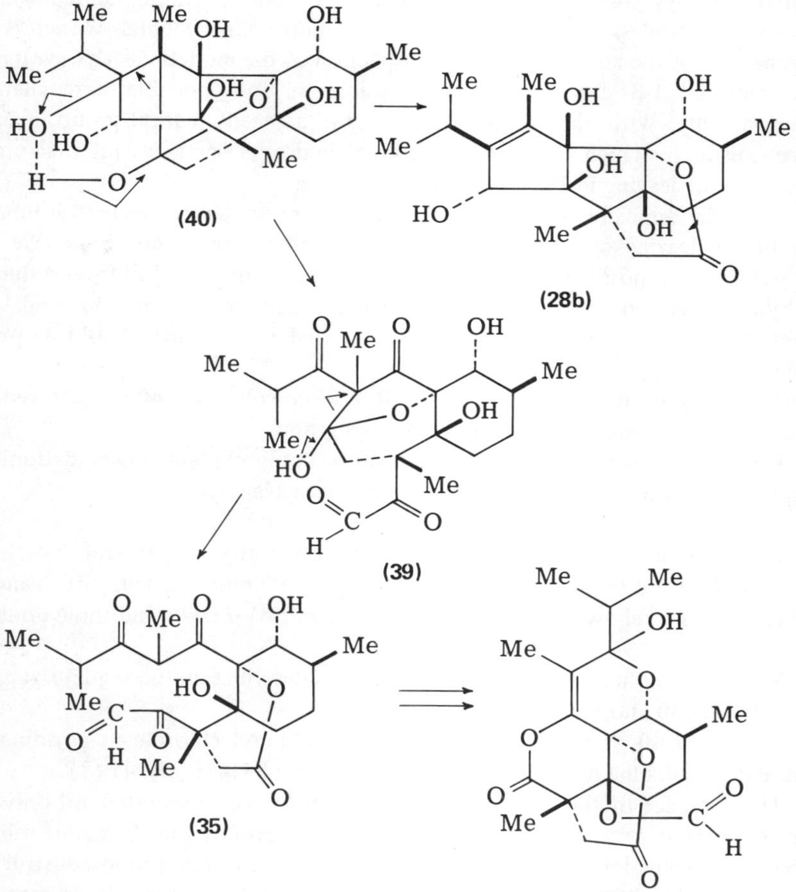

(40)

(28b)

(39)

(35)

(32)

We see now that the genesis of trisecoryanodol **32** from the ryanodol structure **40** is almost as simple as before.

Periodate cleavage of two α-diol groups in **40** yields **39**. A reverse aldol reaction (marked by arrows in **39**) transforms **39** to the intermediate **35**. Finally, the conversion of **35** to trisecoryanodol is the same as shown earlier.

The next question to be asked is whether the ryanodol formula **40** can explain the formation of anhydroryanodol **28b**. The answer is clearly affirmative. A 1-3 diol cleavage portrayed by the arrows in **40** seems extremely favorable. It simultaneously generates the lactone carbonyl, the tetrasubstituted double bond, and relieves steric strain.

One detail of **40** which remains to be settled is the configuration of the carbon carrying the isopropyl and hydroxyl groups. All the remaining stereochemistry of **40** follows automatically from the rigorously known absolute stereochemistry of **28**. The only clue for this remaining stereochemical assignment (unless one would do some more experimental work) is the mechanism of the anhydro reaction. Inspection of the models clearly reveals that the required 1-3 diol cleavage is plausible in both possible stereochemical arrangements. With the isopropyl *cis* to the adjacent methyl group, a cyclic mechanism, in which the acidic hemiketal hydroxyl serves as internal proton donor to the leaving tertiary hydroxyl, can operate.

In the opposite stereochemical arrangement we have excellent antiplanarity of the leaving hydroxyl and the fragmented carbon-carbon bond. We have chosen the first possibility as the more likely one on the tenuous ground that the anhydro reaction proceeds also on high vacuum sublimation and that, consequently, a cyclic mechanism with internal protonation should be preferable.

The X-ray results of Przybylska (cf. Ref. 8) corroborated **40** and showed that the isopropyl is *trans* to the adjacent methyl group.

While at this stage of our discussion **40** has only explained degradations also explainable by **38**, it already shows two distinct advantages:

1. It is in better agreement with the observed acidity of ryanodol.

2. It is in better agreement with the observed nmr spectrum of ryanodol, which seems to show the five methyl groups as *two singlets* and three doublets.

We shall see that the two degradation series described in the sequel rigorously corroborate **40** and eliminate **38**.

Ryanodol **40** yielded, on treatment with thionyl chloride in pyridine and subsequent alkaline hydrolysis, isoanhydroryanodol, $C_{20}H_{30}O_7$, **41** (7).

The presence of the isopropylidene group in **41** is corroborated not only by a singlet (6H) at $\tau = 8.23$ in the nmr spectrum, but also by the characterization of acetone obtained on ozonolysis. In agreement with the proposed structure, compound **41** shows no absorption in the carbonyl region of the ir spectrum. Compound **41** consumed rapidly 1 mole of periodate and gave the oily monoseco derivative **42**. This material shows an ir spectrum (carbonyl maxima at 1751 and 1682 cm^{-1}) in agreement with a ketone in a 5-membered ring, and another conjugated ketone in a larger ring. The uv spectrum of **42** (strong end absorption with shoulder at 265 mμ) is in agreement with a twisted α,β-unsaturated ketone chromophore. Nmr spectroscopy and ozonolysis again corroborated the presence of the isopropylidene group. Clearly the seco derivative **42** has been formed by an α-diol cleavage, followed by an isomerization of the resulting α-ketol into its conjugated isomer.

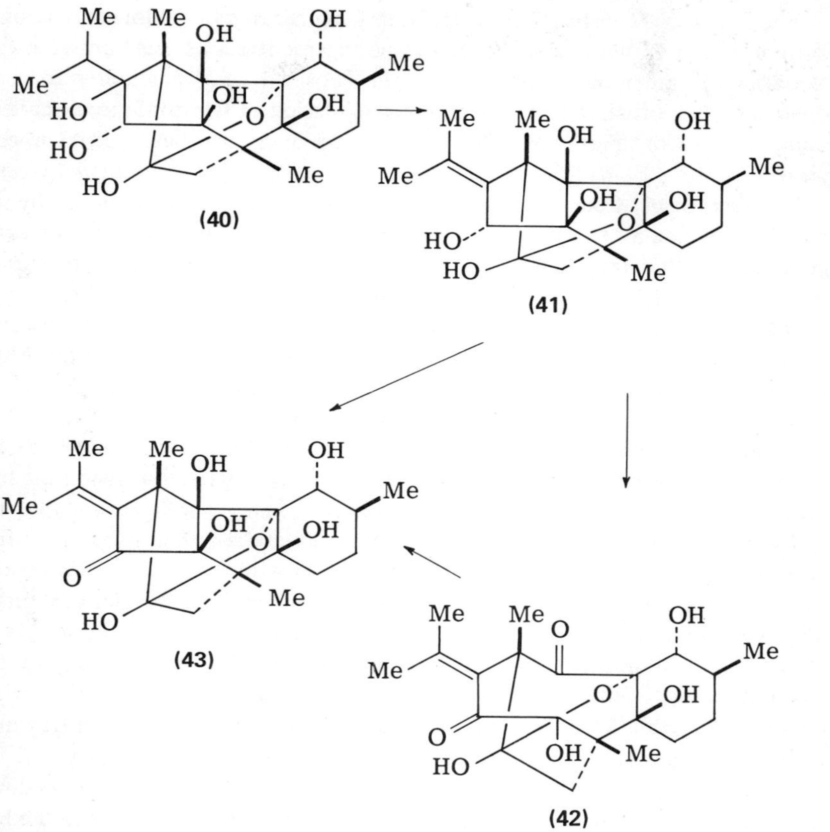

When **42** is treated with a dilute sodium carbonate solution a new crystalline compound ($C_{20}H_{28}O_7$) is obtained in good yield. The structure **43** must be assigned to this material. Ir maxima at 1708 and 1620 cm^{-1} and the uv spectrum (λ_{max} = 270 mμ, log ϵ = 3.9) are in agreement with the proposed structure. This is further corroborated by the nmr spectrum of **43** which shows two singlets, each containing three hydrogens, at τ = 7.75 and 7.97. These must be assigned to the two methyls of the isopropylindene group, strongly unshielded, since they are located in the β-position of an α,β-unsaturated ketone.

If the proposed relative position of the isopropylidene group and the ketone in **43** is correct, then the hydrogens of the isopropylidene methyl groups must be vinylogously activated by the keto group and exchangeable for deuterium. This was indeed observed, since the peaks in the nmr spectrum of **43** assigned to the two methyls in question disappeared when **43** was heated with sodium methoxide in deuterated methanol.

The reaction **42** → **43** may thus be simply explained as an aldol condensation. If we examine **43** we notice that it can be regarded as a ketone formed by the oxidation of the allylic hydroxyl in **41**. Thus a simple possibility to verify rigorously the entire scheme by a direct oxidation of **41** to **43** presents itself. Compound **41** on treatment with platinum and oxygen in fact gave **43** in good yield.

The series of transformations just discussed is not explainable by the ryanodol formula **38**. It constitutes a direct corroboration of the entire periphery of the ring, which carries the isopropyl group, and thus a direct corroboration of the ryanodol structure **40**.

Since we know the complete structure of anhydroryanodine **28a**, and since there does not appear to be any reason why the pyrrole-α-carboxylic acid residue should have changed its site of attachment during the anhydro reaction, we can write for ryanodine the complete structure **44**.

Finally, I mention briefly a reaction sequence which seemed baffling for a few years after it was discovered, but became readily understandable on the basis of the correct complete structure of ryanodine.

Ryanodine consumes rapidly 1 mole of periodate and yields the beautifully crystalline seco derivative **45**. This is in agreement with the site of attachment of the pyrrole-α-carboxylic residue, since **44** has only one free α-diol group. It is interesting that **45** may be reconverted to ryanodine **44** in a yield of about 10% by a catalytic hydrogenation with platinum oxide in glacial acetic acid (11). This reaction, which puzzled us when we discovered it in 1951, is clearly a pinacol-type reduction of the macrocyclic diketone. Compound **45** on treatment with mild alkali gave the diene dione **49** (12) in surprisingly good yield. This remarkable outcome is now easy to understand. A dealdolization (marked with arrows in **45**) and aldol condensation of the resulting enolate ion with the keto group marked by the asterisk in **45** produces the intermediate **46**. Compound **46** contains already the entire system of **49** and may be converted to **49** via the intermediates **47** and **48** by standard base-catalyzed reactions which do not require any comment.

In conclusion, I would like to discuss the possible biogenesis of ryanodine. It is clear from formula **50** that the skeleton of ryanodine may be readily derived by a proper coiling of the standard diterpeniod precursor geranyl geraniol. On the other hand, such a cyclization cannot possibly produce all the rings and substituents of the ryanodine molecule and consequently some of these must be due to secondary processes. As a preliminary to a detailed study of the biogenesis of ryanodine, a systematic search for a more simple ryanodine precursor in various *Ryania* species would seem to be indicated.

The final formula **51** shows the structure of the *p*-bromobenzyl derivative of ryanodol, prepared in my laboratory by Lizzie Jay and determined by X-ray crystallographic methods by Srivastava and Przybylska (8). To our surprise the *p*-bromobenzylation took place not on the acidic hemiketal but on a tertiary hydroxyl.

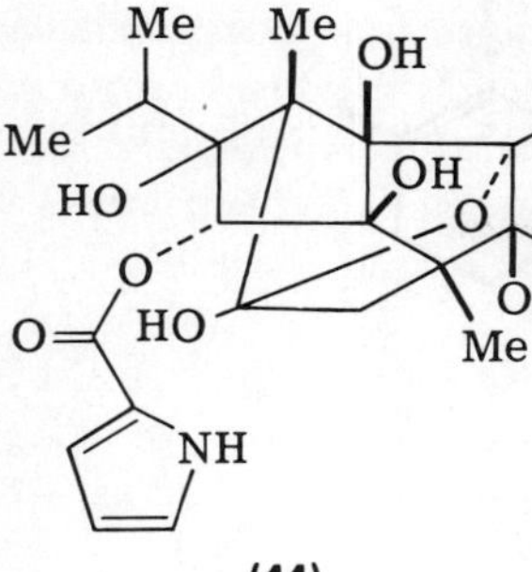

(44)

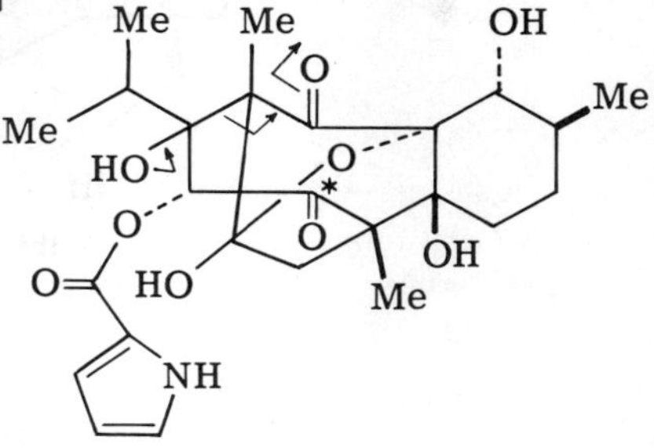

(45)

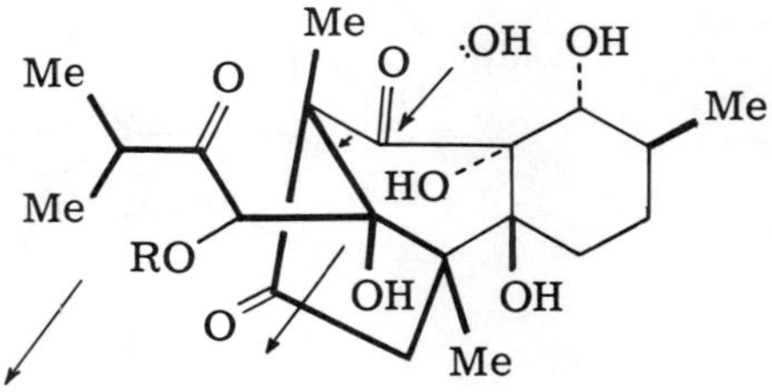

(46)

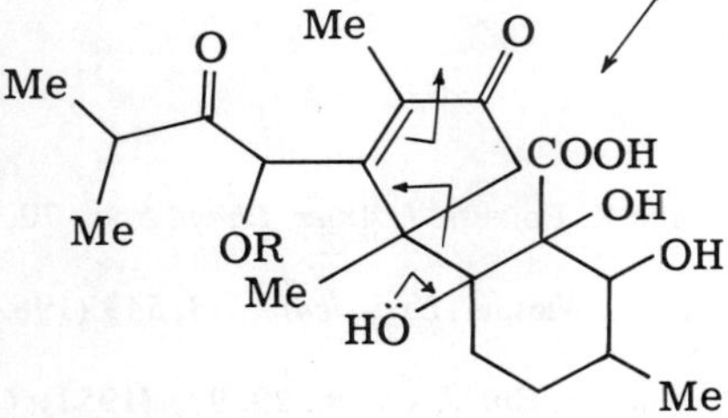

(47)

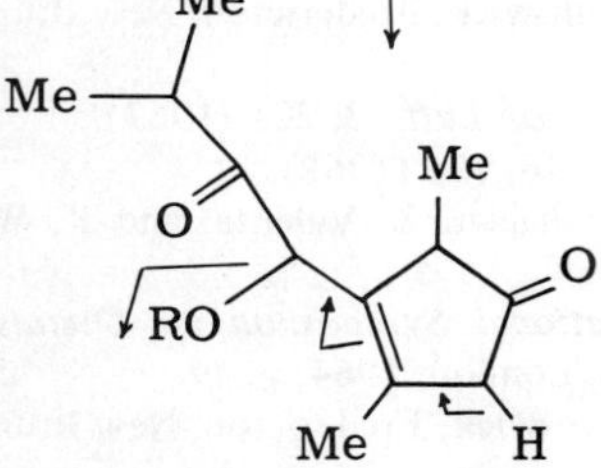

(48)

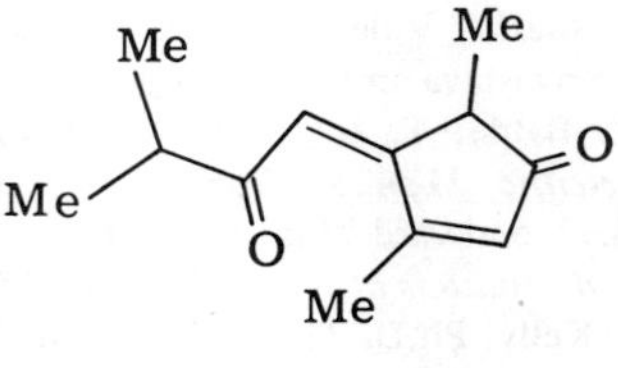

(49)

(50)

(51)

REFERENCES

1. E. F. Rogers, F. R. Koniuszy, J. Shavel, and K. Folkers, *J. Amer. Chem. Soc.*, **70**, 3086 (1948).
2. D. R. Babin, T. P. Forrest, Z. Valenta, and K. Wiesner, *Experientia*, **18**, 549 (1962); cf. K. Wiesner, *Pure Appl. Chem.*, **1963**, 258.
3. R. B. Kelly, D. J. Whittingham, and K. Wiesner, *Can. J. Chem.*, **29**, 905 (1951); *Chem. Ind.* (London), **1952**, 857.
4. D. R. Babin, J. A. Findlay, T. P. Forrest, F. Fried, M. Götz, Z. Valenta, and K. Wiesner, *Tetrahedron Lett.*, **15**, 31 (1960).
5. Z. Valenta and K. Wiesner, *Experientia*, **18**, 111 (1962).
6. T. P. Forrest, Ph.D. Thesis, University of New Brunswick, Fredericton, New Brunswick, 1962.
7. K. Wiesner, Z. Valenta, and J. A. Findlay, *Tetrahedron Lett.*, **3**, 221 (1967).
8. S. N. Srivastava and M. Przybylska, *Can. J. Chem.*, **46**, 795 (1968).
9. D. R. Babin, T. Bogri, J. A. Findlay, H. Reinshagen, Z. Valenta, and K. Wiesner, *Experientia*, **21**, 425 (1965).
10. R. B. Woodward, Special lecture, *3rd International Symposium on Chemistry of Natural Products at Kyoto (Japan)*, Butterworths, London, 1964, p. 49.
11. R. B. Kelly, Ph.D. Thesis, University of New Brunswick, Fredericton, New Brunswick, 1953.
12. F. Fried, Ph.D. Thesis, University of New Brunswick, Fredericton, New Brunswick, 1962.

THE APPLICATION OF PROTON MAGNETIC RESONANCE SPECTROSCOPY TO STRUCTURE IDENTIFICATION IN POLYCYCLIC AROMATIC MOLECULES

By K. D. BARTLE and D. W. JONES, *School of Chemistry, The University, Bradford, Yorkshire, BD71DP England*

CONTENTS

I. Introduction

A. NUCLEAR MAGNETIC RESONANCE SPECTROSCOPY

Many atomic nuclei have a nonzero spin number I and, therefore, possess a small magnetic moment. In the presence of a magnetic field, H_0, such magnets

Manuscript received November 1969.

can assume $2I + 1$ orientations, each associated with a quantized energy level (1,2). Thus nuclei of ^{1}H or ^{13}C, which have $I = 1/2$, can assume two possible energy levels, and resonance absorption of energy can occur between them by the application of a radiofrequency, ν_0, given by the resonant condition

$$2\pi\nu_0 = \gamma H_0 \tag{1}$$

Here the magnetogyric ratio, γ, is a characteristic of the nuclear species; for protons, $\gamma = 2.675 \times 10^{-4}\ \text{sec}^{-1}\ \text{Oe}^{-1}$ so that $\nu_0 = 60$ MHz for a laboratory field H_0 of 14,100 Oe. For the abundant isotopes ^{12}C and ^{16}O, on the other hand, $I = 0$ and no nuclear magnetic resonance spectrum can be obtained.

Since the first experiments on nuclear magnetic resonance (nmr) spectroscopy in 1945 and the discovery of the chemical shift four years later, the application of the technique to chemistry has grown to such an extent that it now plays a part in something like one in every five or six chemical papers published. High-resolution nmr effects were first detected with ^{19}F and ^{31}P nuclei, and the increasing availability of multinuclei spectrometers is encouraging the already appreciable use of a wide range of magnetic nuclei in inorganic and organic chemistry. For ^{1}H nuclei, the small electron density means that the high-resolution effects are much finer, but the wealth of structural information that can be extracted from these (3,4) is such that the great bulk of nmr studies in organic chemistry involve resonance of ^{1}H. This has only become feasible because of the complementary development and fairly widespread availability of high-resolution magnets and spectrometers. For the highest magnetic fields in use for high-resolution nmr spectroscopy (5), very few 220 MHz spectrometers are in existence. Already, however, studies of carcinogenic and other polycyclics (6,7) point to the valuable simplifications that can be effected so as to allow full spectral analyses impracticable at lower field strengths.

The early application of proton magnetic resonance to polycyclic hydrocarbons has been extensive, although this has arisen as much from the spectroscopic and theoretical interest of the compounds (8) as from the need to establish molecular structure and stereochemistry. In particular, the notion of magnetic shielding by ring currents in aromatics (see Section II.B) has provided a fruitful, if at times controversial, spur to more accurate pmr measurements in polycyclics and to corresponding calculations of pmr parameters. The unique ability of pmr to sense some aspects of molecular symmetry (see, for example, Section II.A) has proved especially valuable in enabling positions of substitution to be ascertained in polynuclear aromatic derivatives. The ease with which orientation of radicals introduced may be established by this nondestructive technique from quite small quantities of material often contrasts sharply with more time-consuming (and sometimes more ambiguous) chemical methods, whether these involve synthesis, degradation, or conversion to compounds of known structure.

It is a commonplace that any physical method of structure elucidation should be applied in conjunction both with other physical methods and with chemical methods. Equally, an analytical technique based on any one branch of physics must, in principle, be unique in its ability to yield certain information; pmr spectroscopy is distinctive in that it is based on a minute *magnetic* probe. In this chapter, the complementary value of ultraviolet (uv), visible, and infrared (ir) spectroscopy and of mass spectroscopy will be cited in a number of cases. However, our aim is naturally to emphasize situations where pmr was either used exclusively or provided especially valuable information for structure determinations among derivatives of naphthalene, biphenyl, biphenylene, acenaphthene, acenaphthylene, fluorene, anthracene, phenanthrene, triphenylene, triptycene, naphthacene, chrysene, pyrene, perylene, violanthrene, dihydropleiadene, and dibenzo[*g,p*]chrysene. Some cases in which aromatic rings are separated from each other by heterorings are included but, in general, neither large rings displaying aromaticity nor heterocyclics are covered. Broadly, the examples in Section II are arranged to illustrate a progressive increase in complexity in pmr interpretation from the simplest considerations of symmetry in Section II.A (often enough, these are quite elegant, since most pmr interpretations are less empirical than many ir correlations) to the full spectral analyses of Section II.C and the more advanced experimental techniques of Section II.D. Examples in Section III are drawn from the extensive applications of pmr made in the fields of natural products, coal tar and petroleum, and cavity substituents, while in Section IV attention is directed to conformational aspects. Obviously any individual pmr problem in polynuclear chemistry may involve several of these topics and its ultimate inclusion in one section rather than another will usually be in order to emphasize one aspect.

B. CHEMICAL SHIFTS

In liquids and solutions, as well as in gases, rapid random molecular tumbling motion averages the relatively large direct magnetic dipolar fields responsible for the breadth of lines in solid samples. Thus, given a spectrometer of sufficient resolution, sensitivity, and stability, it may be possible to observe fine structural features of the pmr spectrum arising from interactions with nonzero time average between the resonating nucleus and (*1*) electrons and (*2*) other nuclei *via* the electrons. The secondary magnetic field produced by orbital motion of electrons induced by the applied field is not averaged to zero by field fluctuations originating in the molecular motion of the liquid or solution, and the local magnetic field at the nucleus A is given by

$$H^{A}_{local} = H_0 \, (1 - \sigma^{A}) \qquad (2)$$

where H_0 is the laboratory (external) magnetic field and σ^A is the screening constant appropriate to the chemical environment of the screened nucleus. Evidently, σ^A is some measure of the way in which the environment of the nucleus in question differs from that of a completely isolated or "bare" nucleus.

The screening is, in fact, anisotropic and should be represented by a tensor σ^A; in a liquid, one observes σ^A as the mean of the three principal components of σ^A. Nuclear screening causes resonance to occur at different values of the applied field for chemically distinct nuclei; differences in resonant field strength for nuclei of the same kind with different environments are called chemical shifts. The term high-resolution pmr spectroscopy is generally applied to nuclear resonance measurements under conditions of resolution such that chemical shifts of nonequivalent protons in the molecule can be distinguished, and the fine structure of resonance signals due to spin-spin interaction (see Section I.C) may be observed.

The magnitude of the chemical shift must be taken with reference to some standard, for which the "bare" nucleus mentioned above is obviously not feasible. In practice, the reference standard may be either external, for example, contained in a glass capillary, or, more usually, internal, in which case any correction for bulk susceptibility is eliminated. The magnetically isotropic molecule tetramethylsilane (TMS) which gives a single line at relatively high field, is the most commonly used internal reference (9) for the magnetic resonance of protons. Since chemical shifts are proportional to the magnitude of H_0, the chemical-shift difference between two nuclei H-1 and H-2 becomes

$$\delta H_0 = (\sigma_1 - \sigma_2)\, H_0 \tag{3}$$

where σ_1 and σ_2 are the screening constants for nuclei 1 and 2, and δ is useful for reporting chemical-shift differences because spectrometers operate at different resonance frequencies. Resonance separations are most conveniently measured in frequency units (Hz). If a signal is shifted n Hz downfield from TMS, then

$$\delta = 10^6 n \nu^{-1} \tag{4}$$

where ν is the oscillator frequency, and δ is in ppm. The arbitrary value of 10 is often ascribed to the TMS line and the shifts are then designated τ values given by

$$\tau = 10 - 10^6 n \nu^{-1} \tag{5}$$

While the majority of shifts in organic compounds fall within the τ range 0-10, some shifts for strongly shielded protons will have $\tau > 10$, and shifts for very acid protons may have negative τ values. We shall not adopt this system.

The suitability of TMS as a reference has received much critical attention (10-13). Ideally, exact determinations of δ require measurements at infinite dilution in the gas phase (12) whereas, for the sake of convenience, most measurements are made in the liquid phase. It must be borne in mind that the shielding actually experienced by TMS in a given solution will always be slightly different from that experienced by any other solute present, whatever solvent is employed. This is particularly true of polycyclic molecules containing some sterically hindered protons, which are less accessible to the shielding or deshielding influences of the solvent than either unhindered protons in the same molecule or those of the reference. For example, while protons of most polycyclic molecules are respectively shielded and deshielded (14,15) by C_6D_6 and $CDCl_3$ relative to an "inert" solvent such as cyclohexane (16), H-1 of both benzo[*c*]- and dibenzo[*c,g*]phenanthrenes shows effects of opposite sign (16). Similar effects have been observed for benzenes substituted with bulky alkyl groups (17). On the other hand, as long as the significance of chemical-shift differences is not overestimated, TMS remains a reasonable reference for nonpolar solutes in nonpolar solvents (13). In view of the interaction between TMS and aromatic solvents, Laszlo *et al.* (12) have recommended that neopentane (also isotropic, volatile, and with singlet resonance), and possibly cyclohexane or di-*t*-butylmethane, are to be preferred to TMS for precise work involving aromatic solvents. The alternative of an external reference eliminates these interactions, but magnetic susceptibilities (χ) determined to within 0.25% are required (12) if chemical shifts are to be measured with an accuracy of 0.01 ppm. Pmr methods of determining χ, recently reviewed (18), are incapable of such precision.

The chemical shift measured for an individual hydrogen atom, δ^A, in a molecule provides a single shielding parameter, σ^A, which is a measure of the electric currents induced at that atom. Direct calculation of σ^A (19) is impossible for all but the smallest molecules (20). *Faute de mieux,* the screening is divided (21,22) into separate atomic contributions: (*1*) circulations in the atom itself, σ^{AA}, both diamagnetic and paramagnetic; and (*2*) interactions across space due to circulations on other atoms: (*a*) $\sigma^{A,\text{ring}}$, nonlocalized, in other words, associated with ring currents in π systems; and (*b*) σ^{AB}, localized, in other words, associated with other atoms or bonds.

Discussions of overcrowded molecules also take into account (23) what are termed proximity effects; in general these may be expected to influence both (*1*), through intramolecular dispersion forces, and (*2*), by distortion of normal bond angles and distances.

The importance of each of these terms in determining σ^A in polynuclear aromatic molecules, their derivatives, and heterocyclic analogs has been reviewed (24-26) but, of these contributions, $\sigma^{A,\text{ring}}$ is of particular interest in aromatic hydrocarbons. Despite controversy about the origin of chemical shifts (27,28) and magnetic susceptibilities in polynuclears, so-called π-electron ring currents induced by the applied magnetic field (29) are generally accepted as the dominant contributions. Pauling's semiclassical theory (30) was first used to account for magnetic susceptibilities and their anisotropies in benzenoid hydrocarbons, and the influence of an external field in Hückel molecular-orbital (MO) theory was discussed by London (31). The magnetic field promotes ring fluxes which may be regarded as either specific to a given ring (32) or as line currents in the component bonds (33). The two methods commonly used for calculating ring-current effects give similar results; in one, due to McWeeny (34), ring-current shifts can be obtained without first deriving the current itself; the secondary field is determined directly by means of a test dipole. Ring currents calculated by the Pople method give chemical shifts indirectly by combination with distance-dependence tables, for example, those of Johnson and Bovey (35), subsequently modified and extended (36,37), which allow for the distribution in space of the π-cloud. The tables comprise a list of calculated proton shieldings from a unit ring current tabulated in terms of the cylindrical coordinates of the proton position with respect to the ring center. References in the literature to the Johnson and Bovey tables are legion, but caution is necessary in their application; for example, Jonathan *et al.* found (38) that chemical shifts for a range of polynuclear hydrocarbons calculated in this way were considerably higher than the experimental values. It seems probable that only a part of the total magnetic anisotropy of benzene can be ascribed to the ring current, and that the remainder is a result of contributions from electrons in σ bonds and from localized π electrons (39,40); Dailey (41) showed that much better agreement was obtained if calculated shifts in polynuclears were multiplied by a factor 0.60, equivalent to taking a new calculated MO value for the ring-current contributions to the anisotropy of benzene as $-30 \times 10^{-12}\ m^3\ mol^{-1}$.

For most organic chemists, chemical-shift calculations are probably less important than a knowledge of the range of values for protons of different structural environments in polynuclears, and of the influence of substituents. Several years ago, Martin (42) classified aromatic protons on the basis of their immediate environments and, on the evidence chiefly of first-order spectral analyses, he has listed (15,42) some of their chemical shifts (Table I). Since that time, more complete analyses have been published for many spectra of unsubstituted hydrocarbons (7,24,38,43-49). Martin's classification (42) may be illustrated for 4-phenyldibenzo[*b,l*]chrysene (**1**):

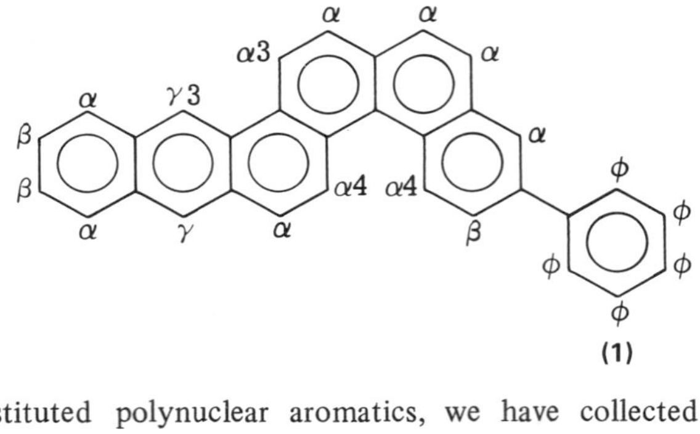

(1)

For substituted polynuclear aromatics, we have collected chemical-shift effects for a number of systems and in Tables II to X list values of substituent shielding values (SSV)–cf. substituent chemical shift (50)–for different groups.

TABLE I

Classification (42) of Proton Types in Polynuclear Aromatic Hydrocarbons with Approximate Chemical Shift, δ ppm in $CDCl_3$

Type	Designation	Symbol	Signal multiplicity	δ
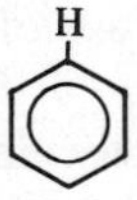 H	Nonangular [or peninsular (23)] H on isolated ring	H-ϕ	Singlet	7.27
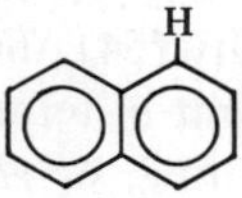 H	H of α-naphthalene type	H-α	Multiplet	7.78
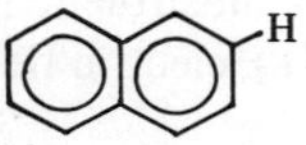 H	H of β-naphthalene type	H-β	Multiplet	7.42
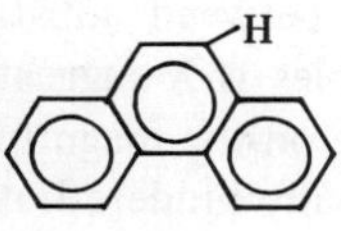 H	H of *meso*-phenanthrene type	H-α β	Singlet	7.72
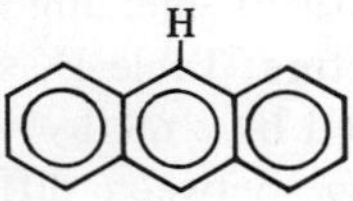 H	H of *meso*-anthracene type	H-γ	Singlet	8.38
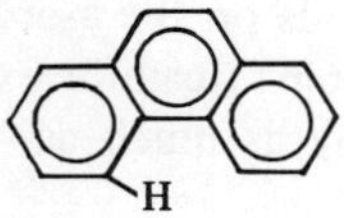 H	Angular [or bay (23)] angular H of phenanthrene type	H-α 3	Multiplet	8.67
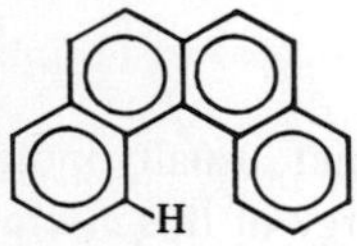 H	Angular H of ("fjord")benzo[*c*]-phenanthrene type ("fjord")	H-α 4	Multiplet	9.15
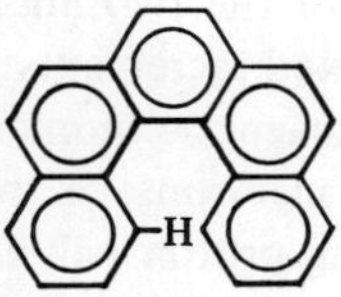 H	Angular H of dibenzo[*c*,*g*]-phenanthrene type	H-α 5	Multiplet	8.52
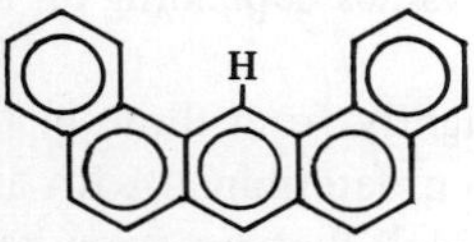 H	Double angular phenanthrene H of *meso*-anthracene type[a]	H-γ 3,3	Singlet	9.97[b]

[a]"Double bay".

[b]In CCl_4(69).

The SSV is defined as the shielding in parts per million (i.e., positive value corresponds to shift to high field) measured with respect to the same proton in the unsubstituted hydrocarbon in the same solvent at the same concentration. Of the several compilations of SSV in benzene available (1,2), that of Ballantine and Pillinger (51) seems to embrace most examples. Recent analogous work in the olefinic series (52,53) yielded SSV for 36 groups. The use of additivity schemes in the benzene series has been reviewed (54).

The SSV figures contained in Tables II-X are, of course, wholly empirical and are intended to help locate substituent orientation. While the prediction of chemical shifts in aromatic systems is a very remote possibility (54), the effects of substituents have at least been rationalized to some extent in terms of the migration of charge from the substituent to the aromatic ring, or *vice versa* (55-57). However, there may be large contributions from substituent magnetic anisotropies (58,59), local van der Waals interactions (60,61), electric fields (62), and changed ring currents (63,64).

It is hardly surprising, therefore, that the variation between substituent effects in different polynuclear systems is as wide as Tables II-X suggest. The overall environment of a group can exert a strong influence on the magnitude of its SSV: thus when the nitro group in naphthalene is sterically hindered, its *peri* SSV of −0.79 (Table IV) is reduced to 0.025 (65). Substituent shielding values may be sensitive to more subtle details of molecular structure; Table IX shows how the chemical shift of a "bay" or $\alpha 3$ proton confronted by a methyl group experiences effects of different magnitude and *even sign* between different systems (23). The *ortho* SSV of halogens apparently depends on the π-order of the intervening bond (24), but even this subtle effect (66,67) cannot explain why H-6 of 5-bromobenzo[*a*]biphenylene is deshielded by as much as −1.02 ppm compared with the parent hydrocarbon (68).

C. *SPIN-SPIN COUPLINGS*

In liquids, nuclear spins within one molecule may interact, usually indirectly along the bonds rather than directly across space. The energy of this interaction is proportional to the scalar product of the spin vectors of the two nuclei. In simple terms, coupling may be considered to arise from fixed increments in the total magnetic field produced at a given nucleus by the magnetic moments of neighboring nuclei and transmitted by the bonding electrons; a proton undergoing resonance absorption with a single neighboring proton will experience an effective magnetic field which has two possible values depending on the spin orientation of the second nucleus.

Coupling constants in the substituted rings differ slightly from those in the parent hydrocarbon. However, so few of them have been determined with any precision [e.g., naphthalenes (78,95); phenanthrenes (24)] that no rules have been derived to enable small changes in J to be useful in structure determination in derivatives. Brügel (103) has analyzed the spectra of numerous di- and trisubstituted naphthalenes but the additivity relationships for J, well known in the monocyclic series (74,79,104), have not been evaluated. The most important generalization about coupling constants in polycyclic nuclear systems is that the usual ranges of aromatic J values are $J_{ortho} = 6\text{-}10$ Hz; $J_{meta} = 1\text{-}3$ Hz; and $J_{para} = 0\text{-}1$ Hz, while interring couplings are absent or less than 1 Hz. Thus the relative disposition of two protons can usually be deduced by inspection if J between them can be measured. The point should be stressed, however, that spacings between lines in a pmr spectrum do not always correspond to a coupling constant. Furthermore, J_{meta} may be as small as 0.6 or 0.7 Hz in some cases, for example, acenaphthylene (95,105), fluoranthene, and benzofluoranthenes (48,106).

The relative signs of couplings in polynuclears are taken as the same (positive), by analogy with the monocyclic series (107,108), but few have been determined. In benzo[k]fluoranthrene, J_{12}, J_{13} and J_{23} all have the same sign (48).

Jonathan *et al.* observed (38) that, in six alternant polynuclear hydrocarbons, J_{ortho} was related to the Hückel MO π-bond order of the intervening carbon-carbon bond, $P_{nn'}$, by

$$J_{ortho} = 12.7P_{nn'} - 1.1 \tag{6}$$

TABLE II[a]

SSV of Various Groups in Benzene Determined in CCl_3 or CCl_4 Solution[b]

Substituent	Ortho		Meta		Para	
OH	0.45	(78)	0.10	(69)	0.40	(45)
O-alkyl	0.45	(449)	0.10	(340)	0.40	(184)
OCOR	0.20	(60)	−0.10	(52)	0.20	(41)
NH_2	0.55	(3)	0.15	(4)	0.55	(1)
CH_2—	0.10	(80)	0.10	(104)	0.10	(43)
CH<	0	(45)	0	(69)	0	(15)
CH=CH—R	−0.10	(30)	0	(86)	−0.10	(30)
CHO	−0.65	(16)	−0.25	(11)	−0.10	(1)
COR	−0.70	(54)	−0.25	(129)	−0.10	(10)
COOH (R)	−0.80	(6)	−0.25	(7)	−0.20	(2)
Cl	−0.10	(16)	0	(44)	0	(2)
Br	−0.10	(6)	0	(5)	0	(2)
NO_2	−0.85	(6)	−0.10	(4)	−0.55	(1)

[a]From Ref. 51.

[b]5-10% w/v; additivity of substituent effects is assumed. Figures in parentheses indicate the number of examples used to arrive at each value.

TABLE III

SSV of Halogens and Cyano Group in Some Aromatic Hydrocarbons

			Substituent, X				
			Fluorine	Chlorine	Bromine	Iodine	Cyano
X-benzene	*ortho*		+0.28[a] +0.31[b]	−0.03[n] −0.02[b]	−0.18[n] −0.22[b]	−0.39[n] −0.40[b]	−0.35[c] −0.36[n] 
	meta		+0.02[a] +0.02[b]	+0.02[n] +0.03[b]	+0.08[n] +0.09[b]	+0.21[n] +0.25[b]	−0.13[c] −0.18[n]
	para		+0.18[a] +0.22[b]	+0.09[n] +0.12[b]	+0.04[n] +0.03[b]	+0.03[b]	−0.28[n]
					X position[m] 1 2		
X-naphthalene[h]		ν_1			– −0.26 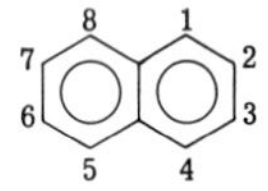		
		ν_2			−0.37 –		
		ν_3			+0.11 −0.24		
		ν_4			−0.01 +0.12		
		ν_5			−0.04		
		ν_6			−0.10		
		ν_7			−0.17		
		ν_8			−0.50		
3-X-phenanthrene[k]		ν_1			−0.10 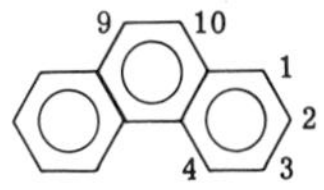		
		ν_2			−0.17		
		ν_4			−0.09		−0.36
		$\nu_{9,10}$			0		
9-X-phenanthrene		ν_8	−0.31[d]	-0.52_5[d]	−0.47[d] −0.58[e]	−0.37[d]	-0.45_5[d,f] −0.40[e]
		ν_{10}	$+0.38_5$[d]	−0.04[d]	−0.29[d] −0.45[e]	-0.73_5[d]	−0.50[d] −0.40[e] 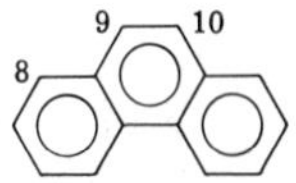

1-X-pyrene[g]	ν_{10}	−0.35			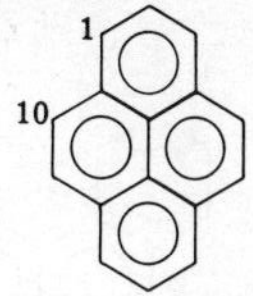
	ν_4	−0.57[h]	−0.55	−0.53	
5-X-benzo[*a*]biphenylene	ν_6	−1.02	−1.02	−0.52	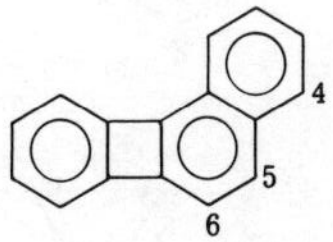
	ν_2	−0.19			
1-X-azulene[j]	ν_3	+0.01			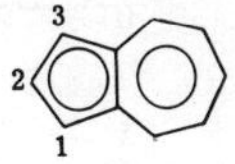

[a]In CCl_4 (70).
[b]In cyclohexane (71); analyses from deuterated derivatives.
[c]In cyclohexane (72); values based on additivity in disubstituted compounds.
[d]In CS_2 (73).
[e]In $CDCl_3$ (68).
[f]−0.42 in 9-cyano-10-methyl derivative[d].
[g]In $CDCl_3$ (75).
[h]−0.70 in 5,6-dibromo derivative.
[j]In tetrahydrofuran (76).
[k]In $CDCl_3$ (77).
[l]In CCl_4 (78).
[m]2,6-Dibromo derivative.
[n]Ref. 74.

TABLE IV

SSV of Nitro Group in Some Aromatic Hydrocarbons

Benzene	*ortho*	*meta*	*para*							
	−0.92[a]	−0.25[a]	−0.39[a]							
	−0.95[b]	−0.21[b]	−0.33[b]							
	−0.95[c]	−0.26[c]	−0.38[c]							
Naphthalene[c]	ν_1	ν_2	ν_3	ν_4	ν_5	ν_6	ν_7	ν_8		
1	–	−0.83	−0.18	−0.33	−0.17	−0.20	−0.35	−0.79[d,e]		
2	−0.93	–	−0.78	−0.12	−0.15	−0.22	−0.17	−0.20		
Phenanthrene[h]	ν_1	ν_2	ν_3	ν_4	ν_5	ν_6	ν_7	ν_8	ν_9	ν_{10}
1	–	−0.60		-0.26_5	0	−0.15			−0.21	−0.58
2	-0.92_5	–	−0.78	-0.11_5	-0.03_5	−0.11	-0.14_5	-0.08_5	-0.11_5	-0.11_5
3	-0.12_5	-0.82_5	–	−0.89	-0.07_5	−0.13	-0.12_5	-0.06_5	−0.22	-0.07_5
9	−0.11	−0.13	−0.17	-0.11_5	−0.07	-0.12_5	-0.16_5	−0.61	–	-0.74_5
Triphenylene[f]	ν_1	ν_2	ν_3	ν_4	ν_{12}					
1	–	<−0.27		<−0.25	+0.48					
2	−0.67	–	−0.42							

	ν_2	ν_{12}					
Pyrene[g]							
1	-0.63	-0.67					

	ν_2	ν_3	ν_4	ν_5	ν_6	ν_7	ν_8
Azulene							
1	-0.25^{i}	$+0.31^{i}$	-0.21^{i}				-1.15^{i}
	$-0.76^{i,k}$		$-0.31^{i,k}$		$-0.39^{i,k}$	$-0.66^{i,k}$	
	-0.65^{j}	-0.10^{j}					

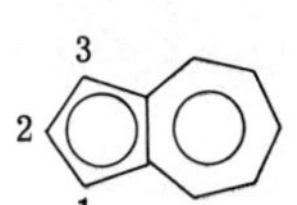

[a] In CCl_4 (70).

[b] In cyclohexane (71); analyses from deuterated derivatives.

[c] In CCl_4 (65).

[d] -0.025 if NO_2 sterically hindered[c].

[e] -0.61 in 1,4-dinitro derivative (80); in cyclohexane.

[f] In $CDCl_3$ (81).

[g] In $CDCl_3$ (75).

[h] In $CDCl_3$ (24).

[i] Solvent not specified (82).

[j] In THF (76).

[k] Guiazulene series (82).

TABLE V

SSV of Hydroxyl, Methoxyl and Acetoxyl Groups in Some Aromatic Hydrocarbons

		Substituent, X		
		Hydroxyl[f]	Acetoxyl	Methoxyl
X-benzene	*ortho*	+0.53[a] +0.56[k]	+0.21[j] +0.25[k]	+0.50[a] +0.43[b] +0.48[k]
	meta	+0.17[a] +0.12[k]	+0.02[j] +0.03[k]	+0.12[a] +0.04[b] +0.09[k]
	para	+0.45[a] +0.45[k]	+0.13[k]	+0.45[a] +0.37[b] +0.44[k]

		X-position		X-position	
		1	2	1	2
X-naphthalene	ν_1		+0.60[e]		+0.78[c] +0.85[d]
	ν_2	+0.71[e]		+0.74[c] +0.82[d] +0.68[e]	
	ν_8	−0.34[e]		−0.42[c] −0.42[d] −0.49[e]	
9-X-anthracene[c,g]	ν_1			−0.30	
	ν_2			+0.02	

		X-position						
		1	2	3	4	5	6	
2-X-triphenylene[i]	ν_1		+0.22					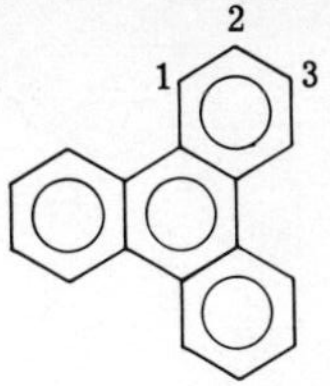
	ν_3		+0.23					
X-benzo[c]-phenanthrene	ν_1	–	+0.60	+0.13	+0.42			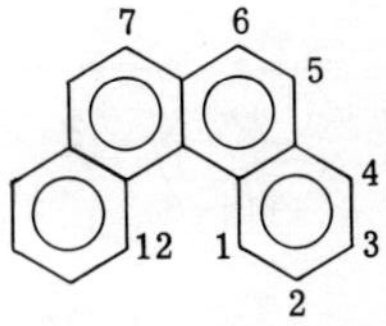
	ν_2	+0.45	–	+0.33				
	ν_3		+0.37	–	+0.67			
	ν_4	+0.30		+0.63	–	−0.58		
	ν_5				−0.62	–	+0.62	
	ν_6					+0.65	–	
	ν_7						−0.63	
	ν_{12}	>+1.00						

[a] In CCl_4 (79).

[b] In cyclohexane (71); analyses from deuterated derivatives.

[c] In CCl_4(83); this reference also contains data for methylthio derivatives of polynuclear aromatic hydrocarbons.

[d] In CCl_4 (84).

[e] In $CDCl_3$ (84).

[f] Pmr spectral data have been reported (85) for 2-hydroxyazulene; but this compound shows keto/enol tautomerism.

[g] 9,10-Dimethoxy derivative.

[h] In $CDCl_3$ (86).

[i] In $CDCl_3$ (81).

[j] Solvent not specified (88).

[k] In CCl_4 (74).

TABLE VI

SSV of Carbonyl Groups in Some Aromatic Hydrocarbons

X-benzene

	Substituent X: CHO	COCH$_3$	COPh	COOR	COOH
ortho	-0.56[a] -0.58[j]	-0.48[b] -0.62[h]	-0.74[c] -0.71[h]		
meta	-0.22[a] -0.21[j]	-0.14[h]	-0.10[c] -0.11[h]		
para	-0.29[a] -0.28[j]	-0.21[h]	-0.20[c] -0.21[h]		

X-naphthalene

	CHO, X-Position 1	CHO, X-Position 2	COCH$_3$, X-Position 1	COCH$_3$, X-Position 2
ν_1	–	-0.48[p,q]	–	-0.50[r,q]
ν_2			-0.64[p,q]	–
ν_3			-0.50[p]	
ν_8	-1.25[p]		-0.87[p]	

X-phenanthrene

	CHO, X-Position 2	CHO 3	CHO 9	COCH$_3$, X-Position 2	COCH$_3$ 3	COCH$_3$ 9	COOR, X-Position 2	COOR 3	COOR 4	COOR 9	COOH, 9-Position[m]
ν_1	-0.45[j]		-0.20_5[k]	-0.58[k] -0.57[j]	-0.09_5[k]	-0.02[k]	-0.67[j]			-0.04_5[k]	-0.09[k]
ν_2	–	-0.37[j]	-0.10[k]	–	-0.56[k] -0.47[j]	0[k]	–	-0.57[j]		-0.01[k]	-0.01_5[k]
ν_3	-0.40[j]	–	-0.17[k]	-0.54_5[k] -0.50[j]	–	-0.06[k]	-0.58[j]	–		-0.06_5[k]	
ν_4		-0.40[j]	-0.04_5[k]	-0.04_5[k]	-0.61_5[k] -0.53[j]	$+0.04_5$[k]		-0.63[j]	–	$+0.01_5$[k]	$+0.11$[k]
ν_5			-0.04_5[k]	-0.04[k]	-0.13[k]	$+0.02_5$[k]			$+0.49_5$[k]	-0.01[k]	-0.10_5[k]
ν_6			-0.08_5[k]	-0.07[k]	-0.12[k]	0[k]			$+0.09_5$[k]	-0.02[k]	-0.04_5[k]
ν_7			-0.13_5[k]	-0.10[k]	-0.11_5[k]	-0.03_5[k]			-0.02_5[k]	-0.06_5[k]	-0.09[k]
ν_8			-1.53_5[k] -1.33[j]	-0.05[k]	-0.07_5[k]	-0.95_5[k] -0.82[j]				-1.19_5[k] -1.05[j]	-1.12_5[k]
ν_9			–	-0.10_5[k]	-0.11[k]	–			-0.05[k]	–	–
ν_{10}			-0.51[k] -0.38[j]	-0.10_5[k]	-0.11[k]	-0.43[k] -0.42[j]			-0.05[k]	-0.77[k] -0.72[j]	-0.82_5[k]

				2-Position	2-Position	1-Position	
X-triphenylene[d]	ν_1			−0.53	−0.47	–	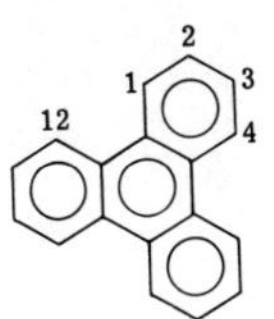
	ν_2			–	–	−0.3	
	ν_3			−0.38	−0.38		
	ν_4					−0.2	
	ν_{12}					+0.50	
		1-Position		1-Position	1-Position	1-Position	
X-pyrene[o]	ν_2					−0.63	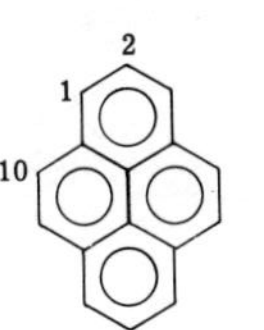
	ν_{10}	−1.13		−1.02	−1.19[g]	−1.17	
			X-Position 1	X-Position 2		2-Position	
X-biphenylene[e]	ν_1		–	−0.40		−0.43	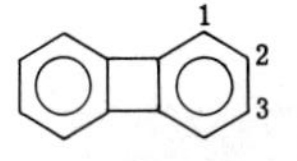
	ν_2		−0.42	–		–	
	ν_3			−0.67		−0.80	
						5-Position	
X-benzo[a]biphenylene[e]	ν_4					−1.38	
	ν_6					−0.98	

TABLE VI (continued)

		1-Position	1-Position	
X-azulene	ν_2	-0.33^n	-0.45^n -0.37^f	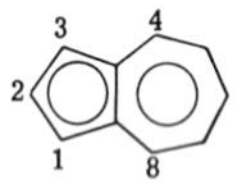
	ν_3	$+0.04^n$	$+0.06^n$	
	ν_4		-0.20^f	
	ν_8		-1.69^f	

[a]In cyclohexane (71); analyses from deuterated derivatives.
[b]In $CDCl_3$ (87).
[c]In cyclohexane (72); values based on additivity in disubstituted compounds.
[d]In $CDCl_3$ (81).
[e]In $CDCl_3$ (68).
[f]In $CDCl_3$ (75).
[g]COC_6H_5COOH.
[h]In CCl_4 (74).
[j]In $CDCl_3$ (89).
[k]In CS_2 (24).
[m]3-Ethyl derivative.
[n]In THF (76).
[o]Solvent not specified (90).
[p]In $CDCl_3$ (87).
[q]Value obtained by substracting ν_1 in naphthalene (15).
[r]In CCl_4 (84).

TABLE VII

SSV of Nitrogen-containing Groups in Some Aromatic Hydrocarbons

(a) Amino group	*ortho*	*meta*	*para*
Benzene	+0.81[a]	+0.26[a]	+0.65[a]
	+0.76[b]	+0.20[b]	+0.63[b]
	+0.75[i]	+0.25[i]	+0.65[i]

	Position	ν_{10}
Phenanthrene[e]	9	+0.81

		ν_2
Pyrene[d]	1	+0.53

		ν_1	ν_2	ν_3	ν_4	ν_{12}
Triphenylene[f]	1	–	+0.63		+0.47	−0.63
	2	+0.73	–	+0.65		

TABLE VII (continued)

		ν_1	ν_2	ν_3	ν_4	ν_5	ν_6	ν_7	ν_{12}
Benzo[*c*]phenanthrene[g]	1	–	+0.65		+0.30				+1.00 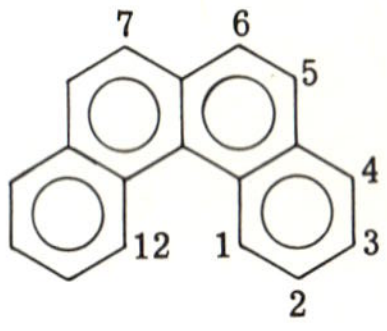
	2	+0.80	–	+0.60					
	3	+0.15	+0.47	–	+0.83				
	4	+0.55	+0.15	+0.70	–				
	5					–	+0.75		
	6					+0.63	–		

(b) Dimethylamino and acetylamino groups

		Substituent X	
		Dimethylamino	Acetylamino
X-benzene	*ortho*	+0.68[a] +0.60[b] +0.66[i]	−0.22[h] 
	meta	+0.18[a] +0.10[b] +0.18[i]	+0.04[h]
	para	+0.67[a] +0.62[b] +0.67[i]	

X-naphthalene[c]		X-position	
		1	2
	ν_1	–	+1.01
	ν_2	+0.54	–
	ν_3		+0.46
	ν_8	+0.34	

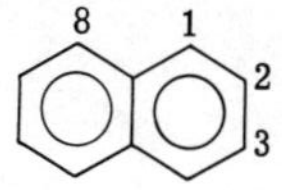

1-X-pyrene[d]		X-position
		1
	ν_2	+0.22

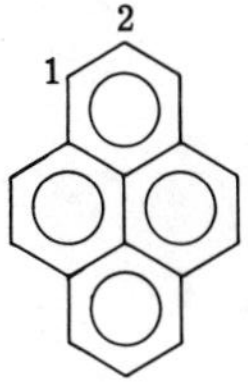

X-benzo[c]phenanthrene[g]		X-position				
		2	3	4	5	6
	ν_1	−0.38		+0.12		
	ν_2	–				
	ν_3		–			
	ν_4		−0.50	–	−0.40	
	ν_5			−0.42	–	−0.40
	ν_6				−0.48	–
	ν_7					−0.43

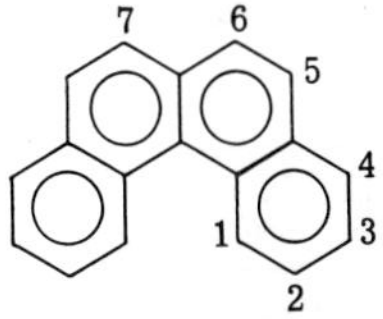

[a]In CCl_4 (79).
[b]In cyclohexane (71); analyses from deuterated derivative.
[c]In CCl_4 (83).
[d]In $CDCl_3$ (75).
[e]In CS_2 (91).
[f]In $CDCl_3$ (81).
[g]In $CDCl_3$ (86).
[h]In a polar solvent (88).
[i]In CCl_4 (74).

TABLE VIII

SSV of the Methyl Group in Some Aromatic Hydrocarbons

	Position					
		ortho	*meta*	*para*		
Benzene		+0.186[a]	+0.074[a]	+0.155[a]		
		+0.17[b]	0[b]	+0.17[b]		
		+0.201[m]	+0.124[m]	+0.220[m]		
	Position	ν_1	ν_2	ν_3	ν_4	ν_8
Naphthalene	1[c]	–	+0.29	+0.24	+0.21	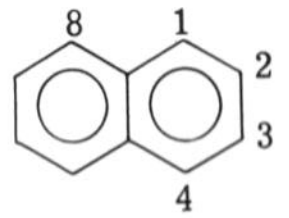
	1[d]	–	+0.33	+0.28	+0.23	
	1[e]	–				−0.14
	2[f]	+0.40	–	+0.28	+0.16	
		ν_1	ν_2	ν_3	ν_4	ν_{10}
Phenanthrene[g]	1	–	$+0.13_5$	$+0.10_5$	$+0.08_5$	−0.24 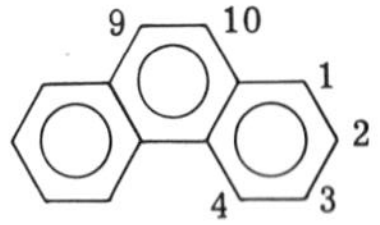
	2	$+0.22_5$	–	$+0.19_5$	+0.14	+0.07
	3	+0.10	$+0.16_5$	–	$+0.22_5$	$+0.03_5$
	4	+0.15	$+0.11_5$	+0.17	–	$+0.03_5$
	9					$+0.16_5$
		ν_1	ν_2	ν_3	ν_4	
Triphenylene[h]	1[i]	–	+0.17	+0.17	+0.12	
			±0.05	±0.05	±0.08	
	2[j]	+0.22	–	+0.23	+0.14	

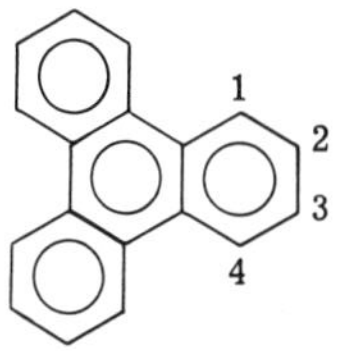

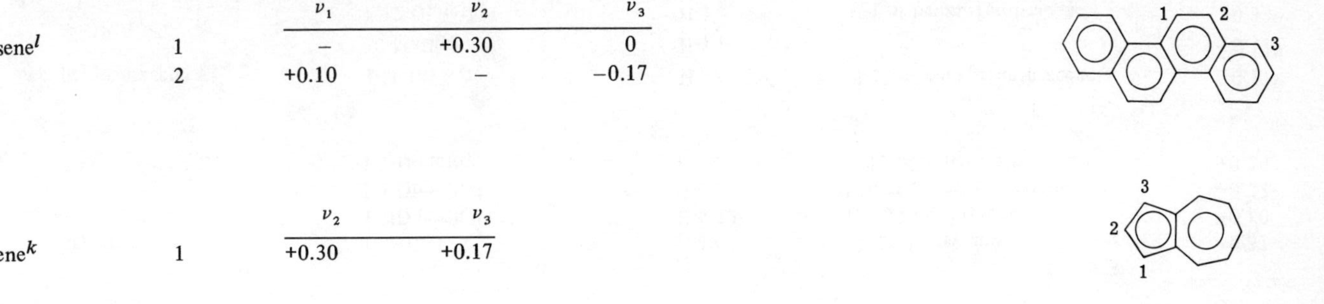

		ν_1	ν_2	ν_3	
Chrysene[l]	1	–	+0.30	0	
	2	+0.10	–	−0.17	

		ν_2	ν_3	
Azulene[k]	1	+0.30	+0.17	

[a]Neat liquid (92).
[b]In CCl_4 (93).
[c]In CS_2 (94).
[d]In CCl_4 (95); 1,8-dimethyl derivative.
[e]In cyclohexane (80); 1,4-dimethyl derivative.
[f]2,7-Dimethyl derivative (94).
[g]In CCl_4 (24).
[h]In CS_2 (96).
[i]1,6,11-Trimethyl derivative.
[j]2,7-Dimethyl derivative.
[k]See Ref. 97.
[l]See Ref. 98.
[m]In CCl_4 (99).

TABLE IX

SSV of Methyl Group in Bay (23) Situations in Polynuclear Hydrocarbons

Series	Derivative	Opposed proton	Standard for comparison	SSV
Phenanthrene[k] 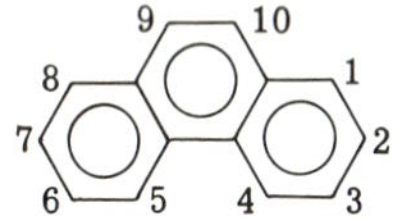	4-Methyl	H-5	H-4 of phenanthrene	-0.26[a,b]
	2,5-Dimethyl	H-4	H-4 of 2-methylphenanthrene	-0.28[c]
	2,3,5-Trimethyl	H-4	H-4 of 2,3-dimethylphenanthrene	-0.16[c]
Triphenylene 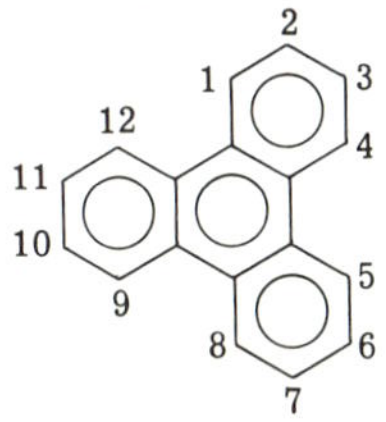	1-Methyl	H-12	H-5, H-8, H-9	±0.05[d]
	1,3-Dimethyl	H-12	H-5, H-8, H-9	0[e]
	1,6,11-Trimethyl	H-12	H-5	±0.05[e]
	1,3,6,11-Tetramethyl	H-12	H-5	±0.05[e]
Chrysene[f] 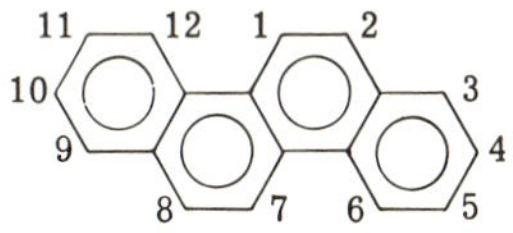	1-Methyl	H-12	H-12 of chrysene	-0.32
	1,7-Dimethyl	H-6,12	H-6,12 of chrysene	$+0.10$
	1,8-Dimethyl	H-12	H-6 of 2-methylchrysene	-0.22
	1,2-Dimethyl	H-12	H-12 of 2-methylchrysene	$+0.20$
Benz[*a*]anthracene 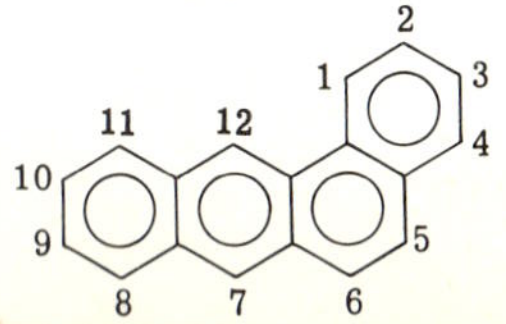	1-Methyl	H-12	H-12 of benz[*a*]anthracene	-0.17[g]
	12-Methyl	H-1	H-1 of benz[*a*]anthracene	$+0.17$[h]
	7,12-Dimethyl	H-1	H-1 of benz[*a*]anthracene	$+0.33$[e,i]
	8,12-Dimethyl	H-1	H-1 of benz[*a*]anthracene	$+0.22$[h]

Perylene 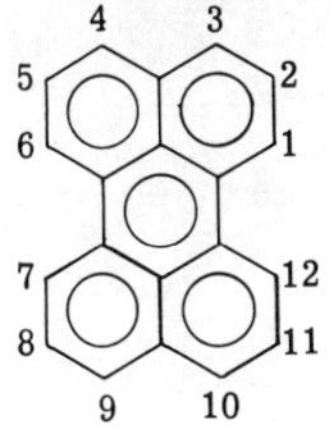	1-Methyl	H-12	H-6, H-7	-0.16^{j}
Benzo[*c*]phenanthrene 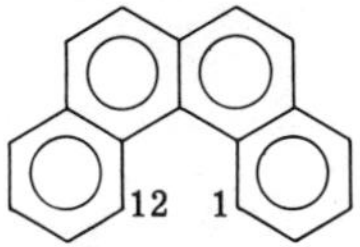	1-Methyl[m]	H-12	H-12 of benzo[*c*]phenanthrene	$+1^{l}$

[a]In CCl_4 (24).
[b]In $CDCl_3$ (89).
[c]In CCl_4 (100).
[d]In CS_2 (91).
[e]In CS_2 (23).
[f]In $CDCl_3$ (98).
[g]In CCl_4 (69).
[h]In $CDCl_3$ (47).
[i]In CS_2 (7).
[j]Solvent not specified (101).
[k]Chemical shift of H-5 in 1,2,3,4,-tetramethylphenanthrene suggests (102) a shielding relative to the corresponding proton in phenanthrene.
[l]In $CDCl_3$ (96).
[m]"Fjord" (α4) situation.

TABLE X

SSV of 9-Alkyl Groups in Phenanthrene[a,b]

Derivative	ν_8 (peri)	ν_{10} (*ortho*)
9-Methyl	-0.15_5	$+0.16_5$
9-Ethyl	-0.22	$+0.16$
9-*n*-Propyl	-0.21_5	$+0.17_5$
9-Isopropyl	-0.28	$+0.10$
9-Isobutyl	-0.21	$+0.18_5$
9-*t*-Butyl[c]	-0.68	-0.06
9-*t*-Pentyl	-0.64	$+0.03$
9-Cyclopentyl	-0.32	$+0.10_5$

[a]In CCl_4 and CS_2 (24).

[b]Spectra of 9-*n*-dodecyl and 9-ethylphenanthrene (100) suggest (24) similar δ values for H-8 in these compounds.

[c]The *t*-butyl group has the following SSV in benzene: *ortho*, −0.02; *meta*, −0.08; *para*, −0.21 (99).

Equation (6) has been extended to 42 couplings in 15 alternants (109), and deviations (6) arising from steric effects examined. Different values for the constants in eq. (6) have been proposed (110) and the relation has been extended to 5-membered rings (110) and to both nitrogen and oxygen heterocycles (111). In the nonalternant hydrocarbons fluoranthene, benzo[*k*]-fluoranthene, and benzo[*g,h,i*]fluoranthene, coupling constants follow a similar relation (48), while couplings in substituted polynuclears involving fluorine (73) and thallium (112) also depend on bond order. Applications of eq. (6) include provision of evidence for bond fixation in biphenylene (113), a decision between alternative solutions of the spectra of certain alkyl naphthalenes (95), and the prediction of the type of spectrum to be expected in cyclophanes (114).

The existence of spin-spin couplings across four or more bonds (115) has been established with the aid of double-irradiation experiments in nine parent polycyclic hydrocarbons and a large number of derivatives (116). The majority of these couplings are between ring and methyl protons, or between protons on different rings. The relation between these couplings, tabulated with many others from the literature and MO bond orders, has been reviewed (117). *Ortho* and *para* ring methyl (benzylic) couplings have negative sign (118) and larger magnitudes than the positive $J^{meta}_{benzylic}$; such couplings could find application in

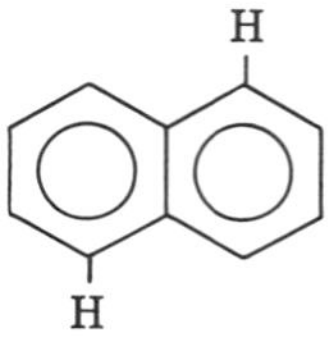

H

H

epi

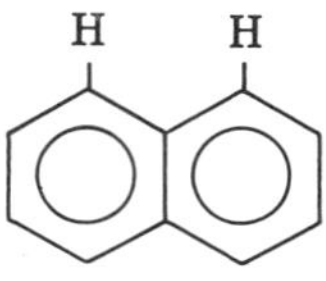

H H

peri

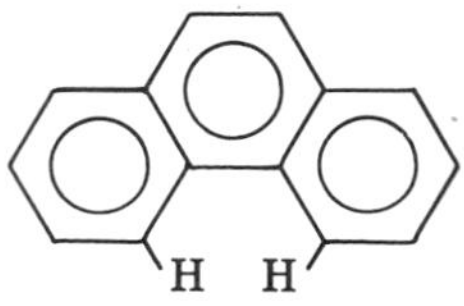

H H

bay

structure determination by pmr. Inter-ring couplings, which tend to be larger in derivatives than in parent hydrocarbons, follow the sequence $J^{epi} > J^{peri} > J^{bay}$ and may also be diagnostic in the spectra of unknown structures.

D. *ANALYSIS OF PROTON MAGNETIC RESONANCE SPECTRA OF POLYNUCLEARS*

Proton magnetic resonance (pmr) spectra may be roughly classified (1) into two categories: those involving chemical shifts which are large in comparison with the spin-spin coupling constants, J, and those involving smaller shifts of the same order as the J's. The classification is imprecise as to the limit of J, which varies with the type of system, but the following notation is adopted (119). Symbols A, B, . . . , are used to denote nonequivalent nuclei where J_{AB}/δ_{AB} is relatively large; commonly A refers to the least shielded nucleus, B to the next least shielded, and so on. M, N, . . . , (or K, L, . . . , or P, R, . . . ,) then refer to another set with signals close neither to those of **A**, **B**, . . . , nor to those of the set X, Y, ... Equivalent nuclei are described by the same symbol, for example, A_2 or B_3, and nuclei with the same chemical shift but different couplings to other nuclei [i.e., chemically equivalent but magnetically nonequivalent (120)] are differentiated by primes. Thus AA′BB′X would imply that the two A nuclei differ from each other magnetically but not chemically; they are coupled to the two nonequivalent B nuclei to an extent comparable with the chemical-shift difference between A and B; and all four AA′BB′ nuclei are coupled to X, which has a shift far removed from those of A, A′, B, and B′.

The complexity of the proton spin system in the spectrum of a given molecule depends on the experimental conditions; clearly, increase of the magnetic field of the spectrometer diminishes the ratio of J to δ (5). Furthermore, in polycyclics the small inter-ring or benzylic couplings may be erased by double resonance, treated as a first-order perturbation, or (most frequently and regrettably) neglected. Under such circumstances, the pmr spectra of alternant polynuclear hydrocarbons and their derivatives can vary in complexity from the 2-spin AX system with trivial solution to the AA′BB′C of biphenyls. Methods of analyzing every variant of pmr spectra have been extensively tabulated and reviewed (1,121-123), but it seems worthwhile to outline approaches to polynuclear hydrocarbons and derivatives.

With the above simplification, condensed benzenoid hydrocarbons yield spin systems involving a maximum of four nuclei. The original analyses of **AA′BB′** spectra from compounds with D_{2h} (naphthalene, anthracene) and D_{3h} (triphenylene) symmetries were made with the assumption of negligible *para* coupling since it was not thought possible to determine ($J_{AA'}$ + $J_{BB'}$) directly (Ref. 38 and others therein). Later work used the rule due to Dischler and Englert (124), and to which attention was drawn by Abraham (125), which allows ($J_{AA'} + J_{BB'}$) to be found by a direct full analysis of such spectra (48,96,113,126). Computer analysis has been made (78) of the AA′BB′ of naphthalene.

Although the 4-spin spectra of polynuclears with lower symmetry are strictly **ABCD**, they often approximate rather closely to **ABMX** (1a). For example, in phenanthrene, H-4 is shifted from H-2 and H-3, which form a group at

M
A
B
X

(1a)

higher field, with the shift of H-1 intermediate (24,127). The frequencies and intensities resulting from a factorization of the ABCD secular equations (128) as ABMX with $J_{AM} = J_{BX}$ and $J_{BM} = J_{AX}$ were presented by Batterham, Tsai, and Ziffer (127) along with nomograms (Figure 1) which show the changes in appearance of the AB, M, and X groups with $(\nu_A - \nu_B)$ (47,127) for typical values of J. These nomograms have been used to extract chemical shifts from a number of rather complex spectra [phenanthrenes (127); benz[*a*]anthracenes, (47); dibenzothiophen and other condensed sulfur heterocycles (129); carcinogenic polynuclear hydrocarbons (7); and pentacyclic hydrocarbons (46)], since they are particularly useful for recognizing the M and X groups corresponding to a given **AB** of **ABMX**.

The ABMX system has been treated rigorously (i.e., without symmetry restrictions on coupling constants) by Lee and Sutcliffe (130) for 1,1,2-trifluoro-1-bromo-2-chloroethane, and for polynuclear systems by Bartle (131), who has described the ABMX treatment of the spectra of substituted phenanthrenes (24), benzo[*c*]phenanthrene (44), and fluorene (49). In each case the preliminary ABMX parameters were refined by computing as ABCD. More recently,

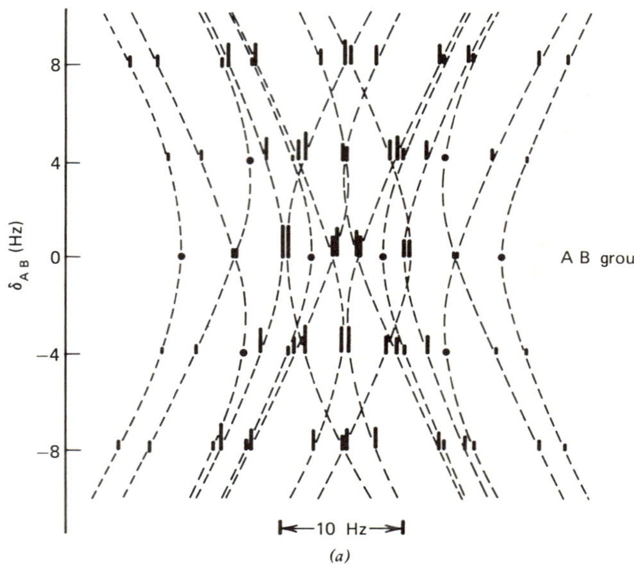

(a)

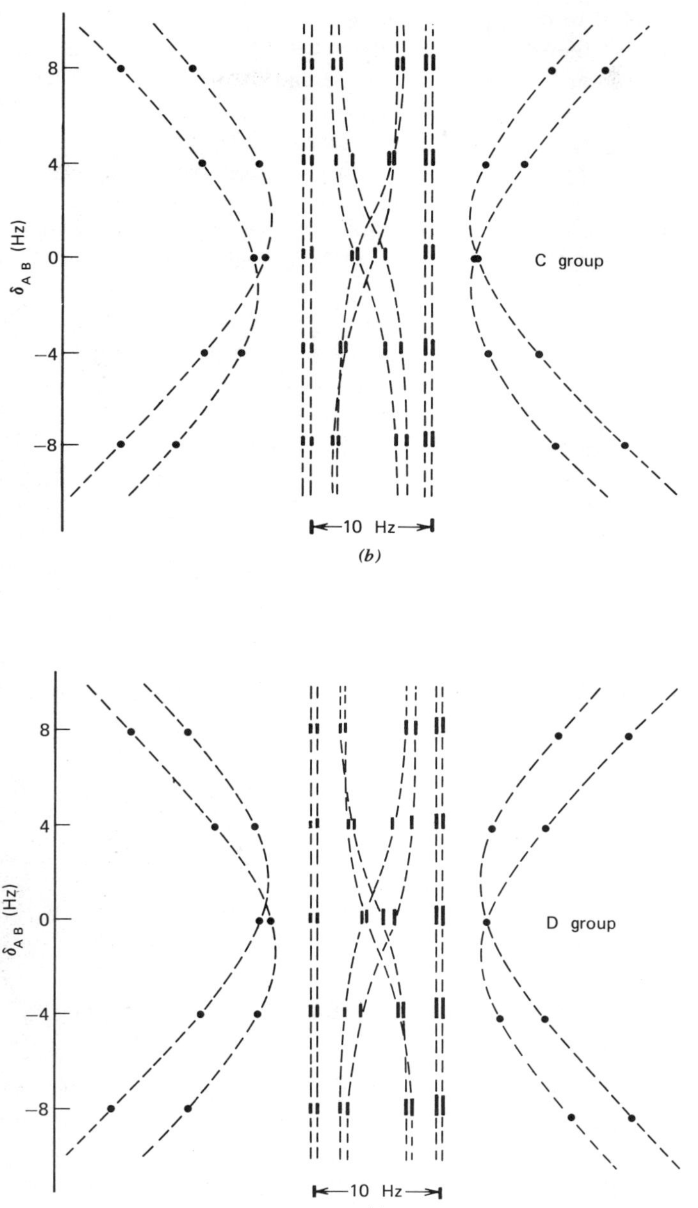

Figure 1 Nomograms (after 127) showing appearance of (*a*) AB, (*b*) C, and (*c*) D regions of ABCD/ABMX spectra of polynuclear aromatic systems for various values of chemical shift difference, δ_{AB}.

least-square iterative computer methods have been applied to 4-spin polynuclear systems [phenanthrene (45); benz[*a*]pyrene (6); and to the 5-spin spectra of substituted azulenes (76), biphenyl, and pentadeuterobiphenyl (132)].

Three-spin systems are rather common in the polynuclear series. Some of these, for example, pyrene (38) and azulene (133,134), are AB_2, while the AMX solution may be applicable in certain cases, for example, triphenylenes (96) and perylenes (135). A number of computer treatments of ABC spectra have appeared [fluoranthene and benzofluoranthenes (48,106); acenaphthene and acenaphthylene (95,106)] but the **ABX** approximation is frequently applied, either alone, or followed by refinement as **ABC**. The situations illustrated by **2a** and **2b** often arise. For situation **2a** [e.g., 2- and 3-alkylphenanthrenes and phenanthrene-9,10-quinones (24,73,136); 2- and 3-alkylnaphthalenes (94)], the ABX approximation is fairly closely followed, since the couplings of A and B to **X** are weak. In **2b**, **R** may be part of another condensed ring, as in perylene (38) and its derivatives (135), but, in those cases studied, has most usually been methyl or another group. Care must be taken that a correct ABX procedure is followed (2,8,137,138), and that attempts (87) are not made to extract J_{AX} and J_{BX} from the **X** region, since no spacings correspond even approximately to these parameters. If the group R exerts similar *ortho* and *meta* SSV, the condition may be approached where $|\delta_A - \delta_B| \simeq 1/2\ (J_{AX} - J_{BX})$ so that (95) one AB subspectrum appears to be degenerate, and fewer than the usual eight lines are detected in the AB region. An example of such a spectrum is 1,5-dimethylnaphthalene (95,139); many other examples have recently appeared (95) along with a discussion of this effect.

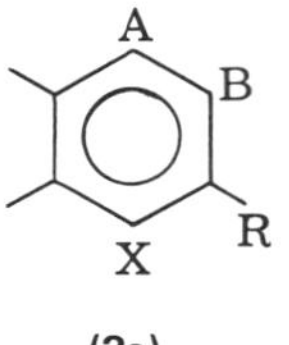

A B R X

(2a)

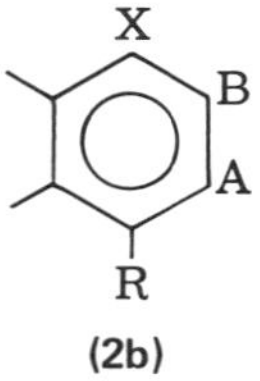

X B A R

(2b)

II. Proton Magnetic Resonance Methods of Structure Determination in Polycylic Molecules

Information for establishment of the structure of an organic molecule may be derived from pmr spectroscopy undertaken at different levels of refinement. In the simplest cases, mere inspection of a spectrum may suffice; problems needing successively more detailed interpretations of pmr spectra are discussed in Sections A, B, and C.

A. SYMMETRY CONSIDERATIONS

Ring closure of **3** may involve electrophilic attack at either C-8 or C-2 in the naphthalene nucleus (140): reduction then leads to one of the hydrocarbons

4a or **4b**. The observation of two CH_2 signals at $\delta = 4.38$ and 4.6 ppm, each with intensity appropriate to two protons, proves that 7,12-dihydro-8,9-benzo-pleiadene **4b** is the product. In **4a** the identical CH_2 environments would lead to a singlet for this structure, while a mixture of **4a** and **4b** would show either three signals, or (if the signals from the two molecules happened to coincide) two peaks of unequal intensity.

2 8 CH_2COOH (3) —Polyphosphoric acid→ CH_2 $C{=}O$ or CH_2 $C{=}O$ —Zn in fused $ZnCl_2/NaCl$→ CH_2 CH_2 (4a) CH_2 CH_2 (4b)

Equally simple considerations reveal the symmetry of substitution in 2,7-dimethyltriphenylene, the product of reaction between the organolithium derivative of *o*-fluorobromobenzene and 4,4′-dimethyl-2-fluoro-2′-iodobiphenyl, thought to occur *via* an aryne mechanism (141). The pmr spectrum of the hydrocarbon (96) shows a single methyl line corresponding to six protons. If this approach is to be used correctly, care must be taken that magnetic equivalence criteria are rigorously applied before a decision is made between structures. Structure **5** was chosen (142) as the structure of the dihydro intermediate isolated during the dehydrogenation (143,144) of 7-phenylbenz[*a*]anthracene to dibenzo[*a,e*]fluoranthene, on the grounds that the pmr signal of the methylene hydrogens was split. Other "plausible" structures such as **6** contain "isolated" methylene groups which were expected to appear as singlets (142). Now H_a and H_b in **6** are equivalent only if H_c and the 6-membered rings lie in the same plane; otherwise an AB spectrum is forecast. Admittedly, δ_a and δ_b might be sufficiently similar to produce a (singlet) collapsed AB spectrum, but this is unlikely. A better test would be the form of the methine signal (A of ABC in **5**, singlet in **6**).

In a synthesis of dibenz[*a,h*]fluorene, Rahman and Tombesi (145) reported 100 MHz spectra of an intermediate diketone, the symmetry of which quite

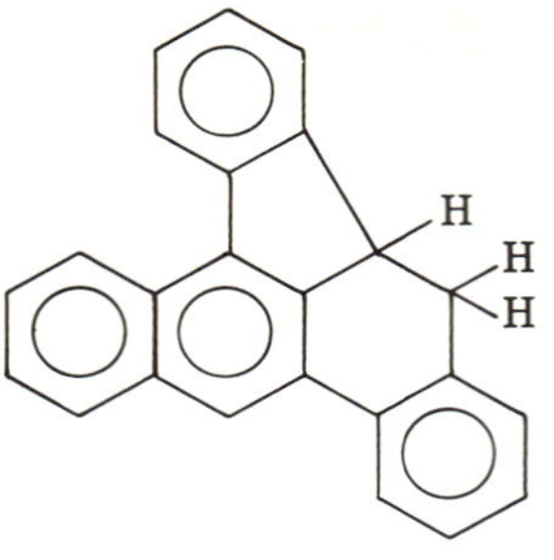

(5)

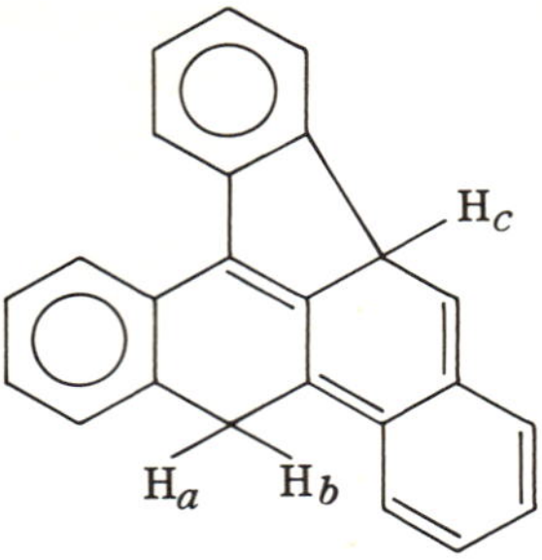

(6)

excluded structures **7a** and **7b**; both have a mirror plane perpendicular to the molecular plane. Two *ortho* doublets, centered on 7.8 and 7.3 ppm ($J = 8.0$ Hz) and two *para* singlets at 8.2 and 7.4 ppm ($J = 0$) were observed, corresponding to structure **7c**.

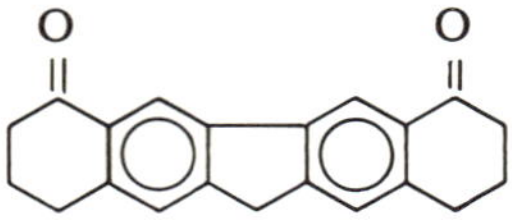

(7a)

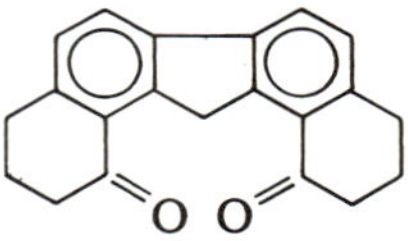

(7b)

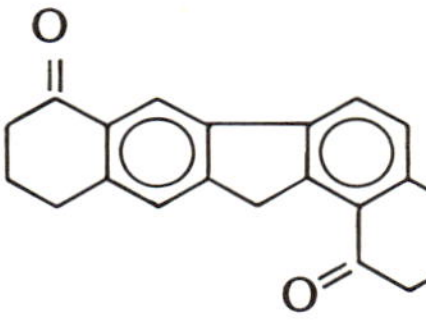

(7c)

As an example of magnetic equivalence in a polynuclear molecule with higher symmetry than C_2, 2,3,6,7,10,11-hexamethoxytriphenylene **8** was isolated in low yield when veratrole was acetylated in the presence of aluminum chloride at –278°K (146). The observation of only two signals in the pmr spectrum, with intensity ratio 3:1, at $\delta = 4.14$ and 7.19 ppm, respectively, demonstrated the D_3 symmetry. A similar pattern proved the structure of the corresponding hexamethyltriphenylene (147).

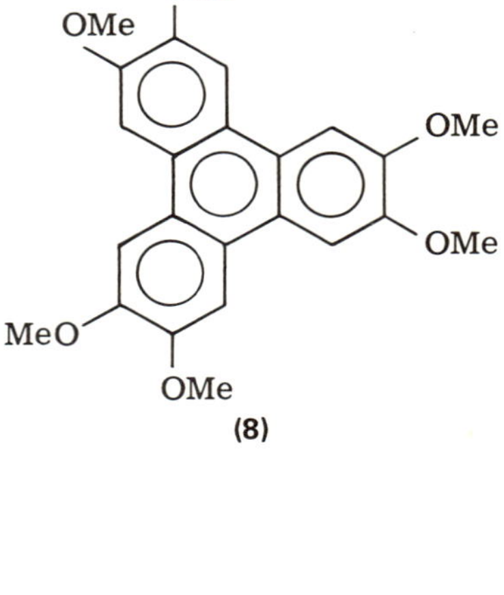

(8)

A tetramethylnaphthalene (148) and a tetramethylanthracene (149), both isolated from high-boiling petroleum distillates, were shown to have D_{2h} symmetry by pmr. Each gave singlet methyl resonances (12 protons) and the former gave a singlet for the aromatic hydrogens. Chemical shifts (*vide infra*) (149) and comparison with known samples (148,149) then enabled 2,3,6,7-tetramethylnaphthalene (**9**) (rather than the 1,4,5,8 isomer) and 1,4,5,8-tetramethylanthracene (**10**) (and not 2,3,6,7) to be identified.

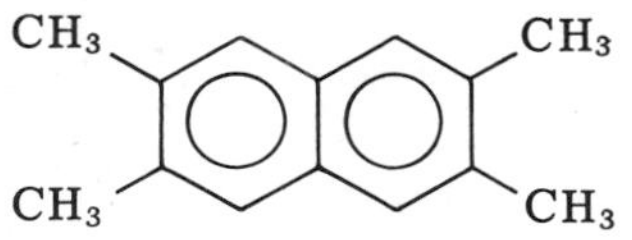

(9)

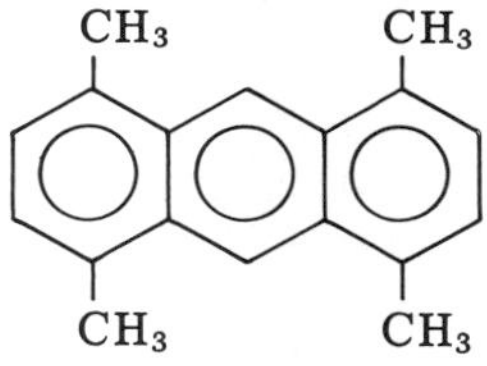

(10)

B. *CHEMICAL SHIFTS*

If symmetry considerations alone are inadequate to assign a structure, the chemical shifts of the protons of substituents may supply sufficient information. Yew *et al.* (150) measured 48 methyl shifts in 24 mono-, di-, tri-, tetra-, penta-, and hexamethylnaphthalenes. They found that a given CH_3 shift $\delta_{\alpha,\beta}$ could be expressed by the equation:

$$\delta_{\alpha,\beta} = \delta^0_{\alpha,\beta} + \Delta\delta_1 + \Delta\delta_2 + \Delta\delta_3 + \Delta\delta_4 + \Delta\delta_5 \qquad (7)$$

where $\delta^0_\alpha = 2.64$ ppm (for α-methyl derivative); $\delta_\beta{}^0 = 2.48$ ppm (for β-methyl) derivatives); $\Delta\delta_1 = 0.10$ ppm (shielding) contribution to the shift of any methyl group for an adjacent methyl substituent; $\Delta\delta_2 = 0.05$ ppm (shielding) contribution to the shift of any methyl group for each nonadjacent methyl substituent in the same ring; $\Delta\delta_3 = 0.01$ ppm (shielding) contribution to the shift of any methyl for each additional methyl on the opposite ring (excepting 1,8 and 4,5 substitution); $\Delta\delta_4 = 0.20$ ppm (deshielding) contribution to the shift of a methyl on C-1 if there is a methyl on C-8 or *vice versa*; $\Delta\delta_5$ = 0.06 or 0.13 ppm (deshielding) for, respectively, 1,2,3-tri- or 1,2,3,4-tetramethyl substitution (effects ascribed to steric interaction and distortion of the ring). Values of $\Delta\delta_n$ were determined by a least-square process, and eq. (7) then gave calculated shifts for CH_3 of methylnaphthalenes which differed from experiment by a maximum of 0.04 ppm. Yew *et al.* were able to apply these correlations to the recognition of polymethylnaphthalenes in petroleum (151); their methods were also of particular value in following the migration of methyl groups after the intramolecular cyclization of allyldimethylbenzylacetic acids in the preparation of tetramethylnaphthalenes (152,153).

Rather similar, though less systematic, considerations allowed assignment of structures to methylnaphthalenes obtained during dehydrogenation studies on terpenoids (154), and to a dimethylchrysene (155) obtained (156) by degradation of a steroid rearrangement product. Gore *et al.* (157) found δ_{CH_3} of considerable diagnostic value for the identification of substituted 2,3-dimethyl naphthalenes. Substitution at the 1 position causes separation of the methyl proton resonances by 0.17-0.33 ppm and at the 5 position by not more than 0.08 ppm. 6-Substituted compounds exhibit an unresolved CH_3 signal. Carruthers *et al.* (149) suggested that the deduction of the orientation of methyl substituents in the anthracene nucleus by pmr was preferable to the uncertain results of the unsupported synthesis by Elbs pyrolysis (158), or by a reaction involving heating in the presence of aluminum chloride. The CH_3 resonances of methylanthracenes can readily be distinguished from each other since they follow (Table XI) the same sequence of shifts as the aromatic protons of anthracenes, namely, *meso*, $> \alpha > \beta$ (38). A general correlation between δ_{CH_3} and the corresponding unsubstituted δ_H was first noted by MacLean and Mackor (159) for seven compounds, and more recently has been extended to include a number of other methyl polynuclears (160) (Figure 2*a*). As Figure 2*b* shows for many ethyl polynuclears, and as was suggested earlier (24) for ethylphenanthrenes, CH_2 and CH_3 shifts also depend on position and could be used to find the orientation of ethyl groups.

TABLE XI

Chemical Shifts[a] of Methyl Protons in Polymethylanthracenes (149)

Isomer	Methyl group		
	α	β	*meso*
1-Methyl	2.70	–	–
2-Methyl	–	2.52	–
9-Methyl	–	–	2.89
1,4-Dimethyl	2.69	–	–
9,10-Dimethyl	–	–	3.03
1,3,6-Trimethyl	2.70	2.45	–
2,3,6-Trimethyl	–	2.52(3H),2.45(6H)	–
1,2,3,4-Tetramethyl	2.66	2.37	–
1,3,5,7-Tetramethyl	2.70	2.44	–
1,3,6,7-Tetramethyl	2.70	2.40	–
1,3,6,8-Tetramethyl	2.72	2.45	–
1,4,5,8-Tetramethyl	2.72	–	–

[a] δ in ppm, for 5 to 10% solutions in $CDCl_3$.

Koptyug *et al.* (161) used calculated CH_3 chemical shifts, rather than comparison with known compounds, to identify a heptamethylfluorene obtained, together with decamethylbiphenyl, by the reaction of 4,4′-dihydroxybiphenyl with methanol on alumina (162). For a 5% solution in $CDCl_3$, methyl

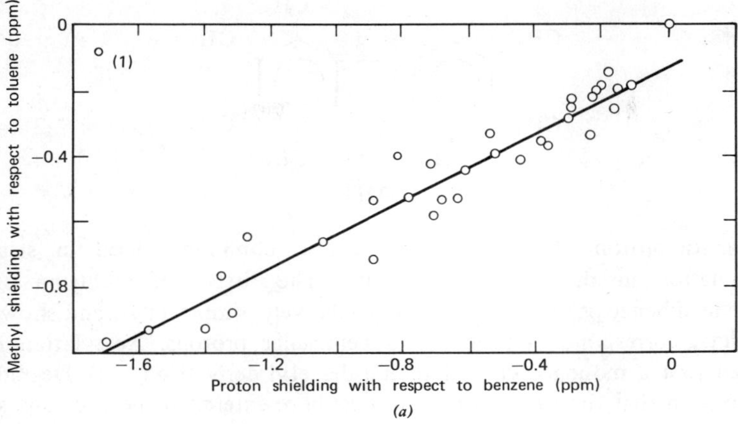

(a)

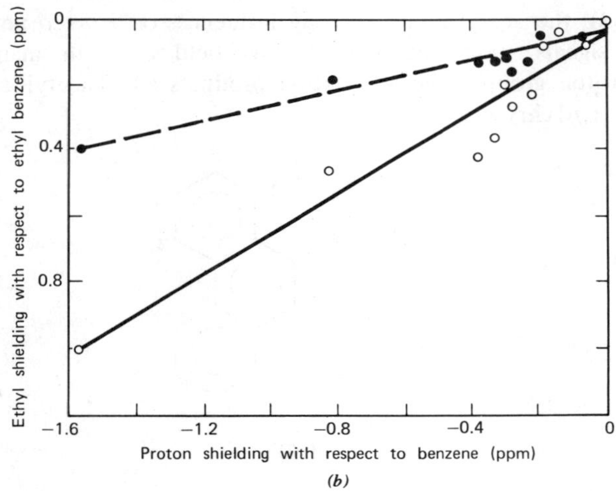

(b)

Figure 2 Graphs of proton shielding against (*a*) methyl (after 160) and (*b*) ethyl shielding in polynuclear hydrocarbons. For (*a*) the regression line $\delta_{CH_3} = 0.51\ \delta_H - 0.13$ is calculated with the divergent point (1) 1-methylbenzo[*c*]phenanthrene, excluded. In (*b*), the (full) regression line, $\delta_{CH_2} = 0.607\ \delta_H - 0.049$, corresponds to the open circles for CH_2 of ethyl; the dashed line, $\delta_{CH_3} = 0.226\ \delta_H - 0.043$, corresponds to the closed circles for CH_3 of ethyl.

signals are found at $\delta = 2.33$ (18 protons) and $\delta = 2.69$ ppm (3 protons). By assuming that the fluorene molecule is flat, Koptyug *et al.* (161) calculated that the 4 and 5 methyls will be shifted by about 0.4 ppm to low field relative to the signals of the other CH_3 groups in the fluorene system. Since only one methyl signal is so shifted, the "vacant" position must be 5, so that the compound is 1,2,3,4,6,7,8-heptamethylfluorene (**11**).

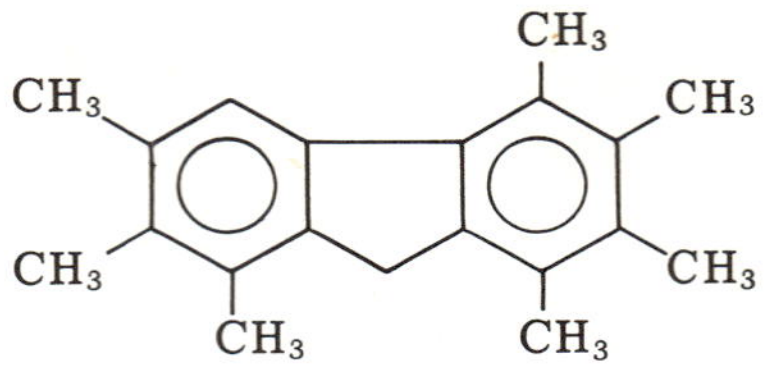

(11)

Aromatic-proton chemical shifts are also commonly used in structure determination in the polynuclear series. The first-order spectrum of the hexacycle dibenzo[*g,p*]chrysene (**12**) is relatively simple (15), and shows two multiplets, corresponding to bay and peninsular protons. Acetylation of **12** afforded first a monoacetyl- and then a diacetyl derivative (163). Degradation (164) proved that the first substituent must have entered at the 1- or 3-position, and that the second had entered ring C; there are, therefore, four possibilities for the disubstituted derivatives—1,9; 1,11; 3,9; and 3,11. Shifts in the pmr spectra of the acetyl derivatives showed that substitution had occurred at a peninsular position (163); the acetyl group strongly influences only *ortho* protons (Table VI), while signals were detected to low field of both main bay- and peninsular-proton groups. The acetylation products are 3-acetyl- and 3,11-diacetyldibenzo[*g,p*]chrysenes.

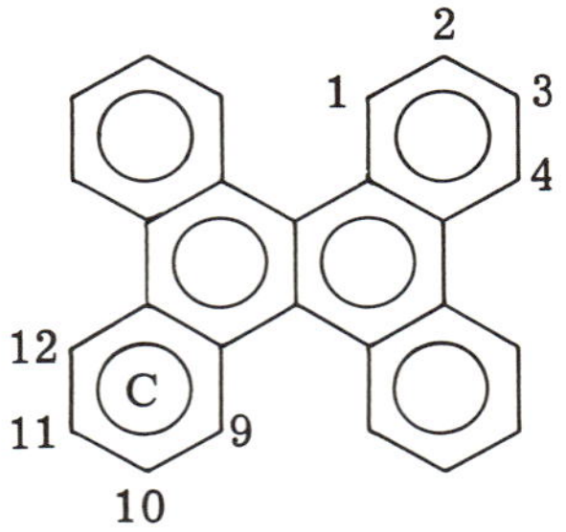

(12)

The low-field shift of the bay proton resonances of phenanthrene relative to H-1,3 and H-9 allowed Wood and Mallory (165) to distinguish the photoconversion products of 3-trifluoromethylstilbene. The solid isomer isolated from the approximately 1:1 mixture was identified as 2-trifluoromethylphenanthrene (**13**), since its pmr spectrum exhibits a low-field multiplet at $\delta = 8.50$-8.82 ppm with area corresponding to two protons—the H-4 and H-5 signals. The second, liquid, 4-isomer (**14**) exhibited only a 1-proton multiplet (centered at 8.66 ppm) to low field of the main aromatic group. Chemical shifts of aromatic protons were also used by Blackburn *et al.* (77) to confirm the products of irradiation of 4,4′-dibromo- and 4,4′-dicyanostilbenes as 3,6-dibromo- and 3,6-dicyanophenanthrenes.

CF_3 $\xrightarrow[I_2,\ air]{h\nu}$ CF_3 + CF_3

(13)

(14)

No survey of structural studies in polycyclic chemistry would be complete without mention of the elegant syntheses of the helical aromatic molecules hexa-, hepta-, octa-, and nonahelicenes **(15-18)** and [13]helicene in high yield by photo-induced cyclization (166-168) following the Wood and Mallory (165) method for phenanthrenes. Pmr spectra were recorded and approximate assignments made both for intermediates and for products ranging from phenanthrene and benzo[*c*]phenanthrene up to nonahelicene, with special attention to the unambiguously synthesized hexahelicene. Not only was it possible thereby to establish the structures of the desired intermediates and to confirm the identity of the product helicenes in the more ambiguous syntheses, but the very presence in a fraction of signals in the region of 8.3 ppm downfield from TMS was accepted as sufficient evidence for discarding the fraction. Furthermore, the absence of signals at fields lower than 8.00 ppm in the products excludes the isomeric structures which could, in theory, result from the photocyclizations. The observation of resonances at fields as high as about 6.2 ppm is strong evidence for structures in which protons are held above aromatic rings, and thus experience ring-current *shieldings.* The pmr spectrum of another unusual molecule (169), pyracylene **(19)**, emphasizes its closer affinity with the [12]annulene system rather than with naphthalene.

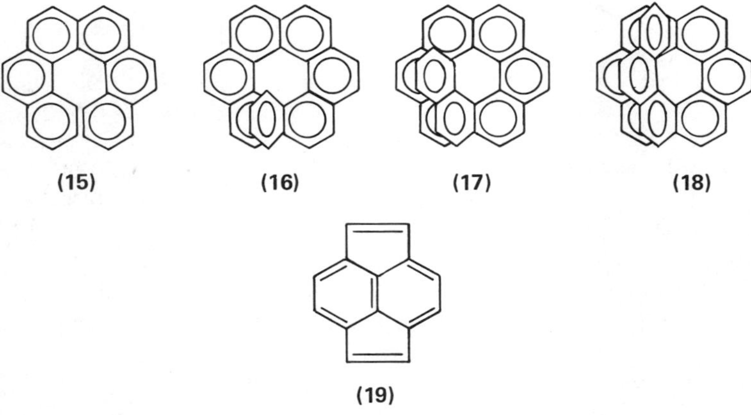

(15) **(16)** **(17)** **(18)**

(19)

As was pointed out for the helicenes, the comparison of chemical shifts for a range of related compounds can be a fruitful source of structural information. Martin and co-workers (87) also used this approach in their unambiguous determination of the structure of 34 cyclic ketones formed by cyclization of acids AR–$(CH_2)_n$–COOH ($n \geqslant 2$) by inspection of pmr spectra. They found that cyclic C=O groups shield *ortho* ring protons from –19 to –47 Hz at 60 MHz and *peri* protons from –32 to –39 Hz; *meta* and *para* protons are shifted much less by a C=O group. The chemical shifts of 450 aromatic phenolic compounds have been surveyed (51) to facilitate prediction of the orientation in unknown polysubstituted phenols, while Reed (170) has deduced empirical rules for predicting the shift influences of OH, CH_3, and Cl groups.

Pmr also provides evidence for the structures of compounds derived from the product (**21**), of the intramolecular cyclization (171) of **20** under mild Friedel-Crafts conditions (172). For example, a two-proton singlet ($\delta = 4.47$ ppm), cf $\delta = 4.59$ in 7*H*-benzanthracene itself (**23**), along with detection of ethoxy signals, confirms structure **22** for the compound obtained (171) by merely boiling **21** with ethanol. Campbell and Scott (173) have reported a similar remarkably easy ether formation. A singlet 2-proton resonance at $\delta = 4.49$ was diagnostic (171) of CH_2 in the hydroxy tautomer (**24**) formed by heating **21** *in vacuo.*

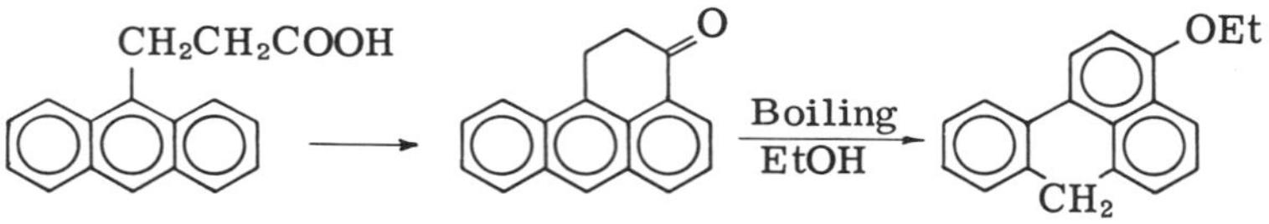

CH_2CH_2COOH (20) → (21) $\xrightarrow[\text{EtOH}]{\text{Boiling}}$ (22) [OEt, CH_2]

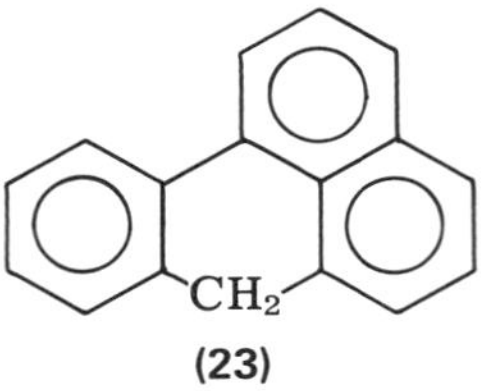

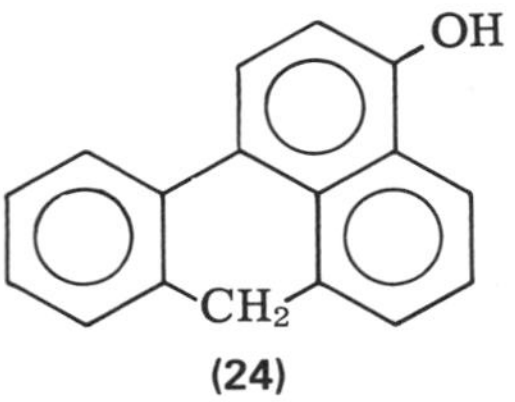

(23) [CH_2] (24) [OH, CH_2]

The dicarboxylic acid analog of **20** can also be cyclized (174) to the perylene derivative (**25**) whose pmr spectrum (171) confirms its structure. Another example of the use of a chemical shift to help establish the structure of a condensation product of an anthracene derivative is the bianthryl (**26**), obtained from the reduction of 2,6-dimethoxyanthraquinone (175). This shows a singlet for the *meso* protons at $\delta = 8.41$ ppm, compared with 8.48 ppm for the corresponding protons of anthracene (38,176).

Even when the gross structure has been defined by a single-crystal X-ray analysis, pmr may help to resolve a detailed stereochemical point (177) such as

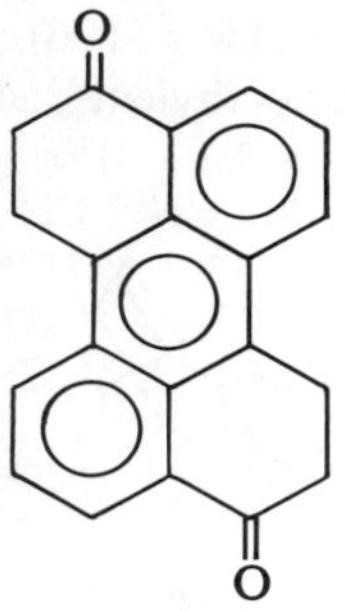

(25)

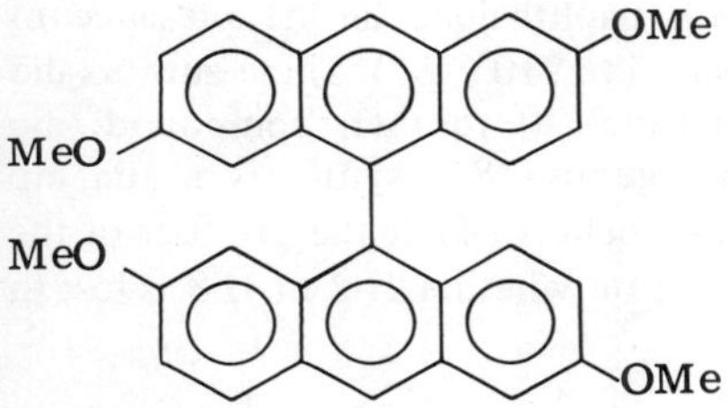

(26)

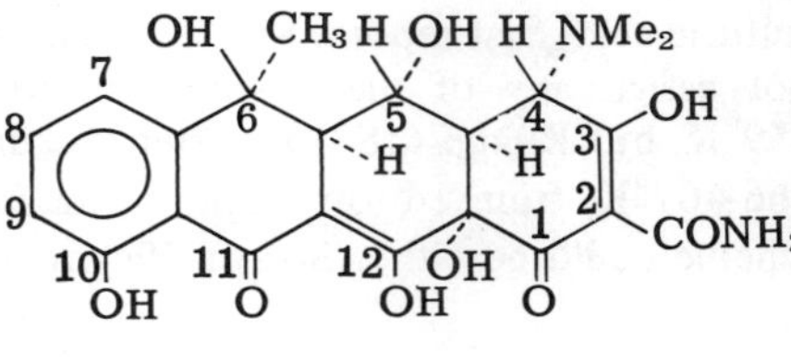

(27)

the *trans* arrangement of hydroxyl groups on carbons 5 and 6 in terramycin (**27**). A neat use of chemical-shift arguments is represented by the identification of the products of the diene reaction between naphthacene and acrylonitrile (178), which gave two isomers, **28a** and **28b**. Now for 5,12-dihydro-5,12-ethanonaphthacene (**28c**), the H-6,11 singlet is found at 7.62 ppm. Correspondingly, in the spectrum of one of the acrylonitrile adducts, there is a 2-proton singlet at 7.65 ppm; this compound is evidently **28a**.

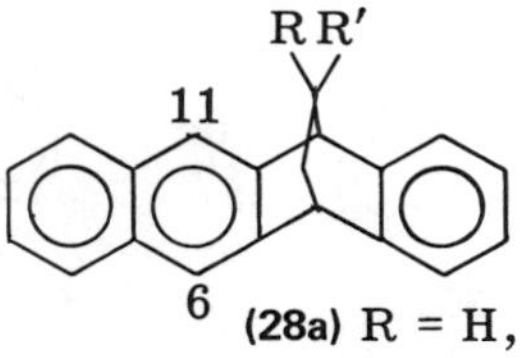

(**28a**) R = H, R′ = CN
(**28b**) R = CN, R′ = H
(**28c**) R = R′ = H

For the other product, however, a 1-proton singlet is found at $\delta = 7.70$ (H–6) and another at $\delta = 7.84$–the resonance of H-11 deshielded by the magnetic anisotropy of the **CN** group. A calculation of this anisotropy shift, using distances measured from Dreiding models and a value of $\Delta\chi_{CN} = 16.5 \times 10^{-12}\ m^3\ mol^{-1}$ (179) gave 0.136 ppm, in almost exact agreement with the experimental $\delta_{11} - \delta_6 = 0.14$ ppm. Other rather similar values of $\Delta\chi_{CN}$ have been reported by

Whipple (180) and Smith and Shoulders (176). The latter authors satisfactorily accounted for δ_{AB} in the tetracyanoethylene adduct (**29**) of anthracene.

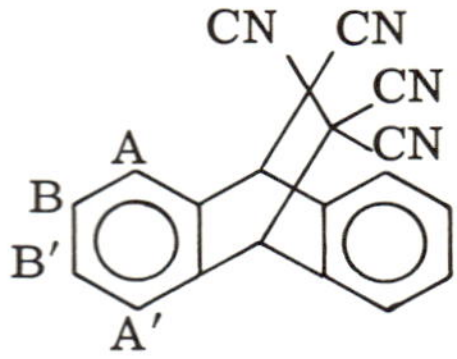

(29)

On the basis of chemical shifts, Martin *et al.* (181) identified the product of the condensation of thiosalicylic acid with naphthalene in the presence of sulfuric acid, first reported by Davis and Smiles in 1910 (182). These authors did not assign any of the possible structures (**30a**)-(**d**) to their compound, mp 459°K, but Knapp (183) reported the unambiguous (181) synthesis of **30a**, mp 466-467°K, from 1-bromonaphthalene in low yield. In fact, the product of the scheme could be either **30a** or **30c**, depending on whether H-2 or H-8 is lost in

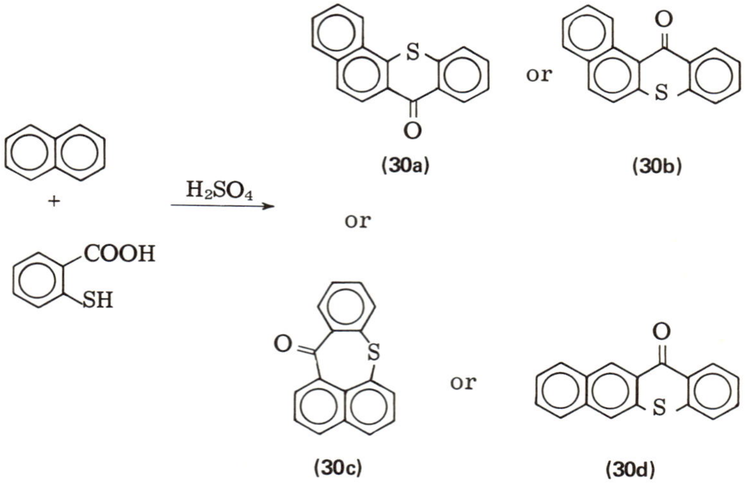

H_2SO_4

(30a) (30b) (30c) (30d)

the dehydration step; when Knapp carried out the reaction with 1-bromo-2-methoxynaphthalene, he obtained the methoxy derivative of **30c**. Martin *et al.* (181) resolved this problem by repeating Davis and Smiles' procedure and recording the 60 MHz pmr spectrum of their product (mp 469-470°K). A comparison with the spectra of related thiochromones and thioxanthene-9-ones then allowed structures **30b-d** to be rejected; the 7*H*-benzo[*c*]thioxanthen-7-one (**30a**) was compatible with the observed spectrum.

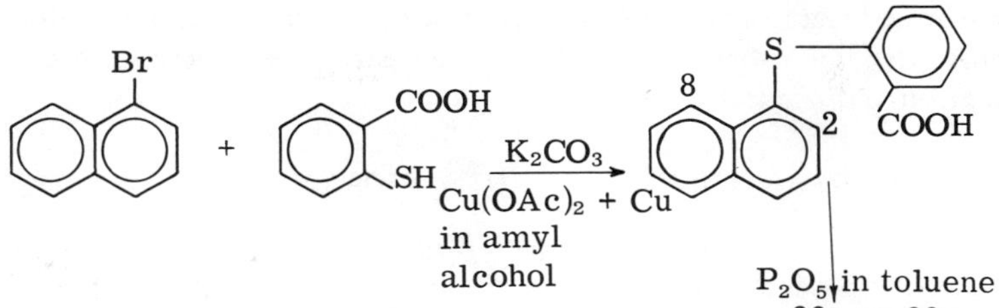

Br COOH SH K_2CO_3 $Cu(OAc)_2$ + Cu in amyl alcohol 8 S COOH 2 P_2O_5 in toluene **30a** or **30c**

Recently, Fieser and Sachs reported the solution by pmr of another long-studied problem (184). Wichelhaus (185) first described the oxidation of naphthalene-1,2-quinone to a compound $C_{20}H_{10}O_5$, later shown by Hooker and Fieser (186) to have structure **31a** or **31b**.

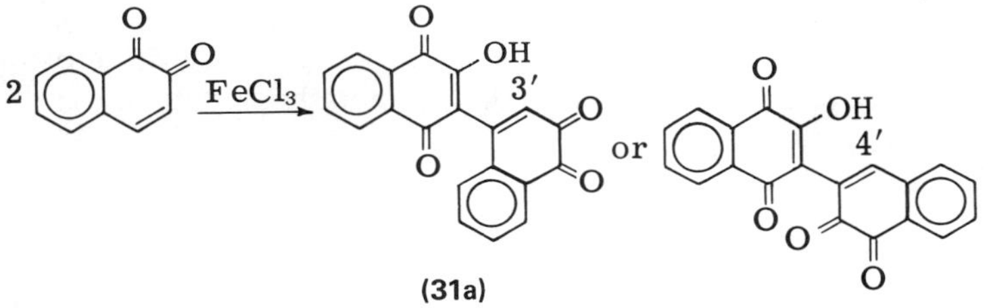

2 $FeCl_3$ OH 3′ or OH 4′

(31a) **(31b)**

Since zinc-dust distillation of the oxidation product gave 2,2′-binaphthyl (186), it was concluded that the structure is **31b**. This evidence became invalid when Copeland *et al.* (187) found that three isomeric binaphthyls undergo rapid interconversion at 763°K and that the sequence of stabilities is 2,2′ > 1,2′ > 1,1′. Furthermore, zinc-dust distillation of the oxidation product (188) prepared by another route, of naphthalene-1,2-quinone gave a product which did not depress the melting point of authentic 1,2′-binaphthyl; this evidence suggests structure **31a**. Fieser and Sachs (184) determined the pmr spectrum of the oxidation product (in dioxane solution at 323°K) in which a sharp 1-proton singlet assigned to H-3′ of **31a** or H-4′ of **31b** is found at 6.30 ppm downfield from the hexamethylsiloxane reference, in addition to multiplets in the aromatic region. In the spectrum of the model compound **(32)**, H-3 resonated at $\delta = 5.83$ ppm but the H-4 signal in **33** was masked by the aromatic protons

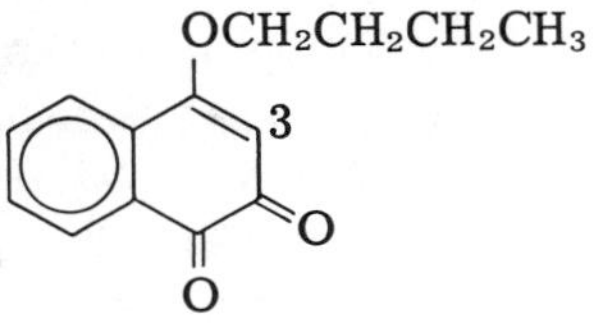

$OCH_2CH_2CH_2CH_3$ 3 4 CH_3 CH_3

(32) 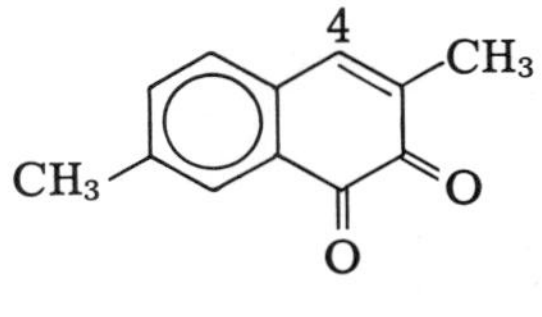 **(33)**

at *much lower* field. The $\delta = 6.30$ singlet is thus assigned to a β proton of an *o*-quinone ring, in other words, H-3′ of the structure **31a** of the oxidation product of naphthalene-1,2-quinone.

C. COMPLETE INTERPRETATION OF PROTON MAGNETIC RESONANCE SPECTRUM

Consideration of the entire pmr spectrum in both aliphatic and aromatic regions, taking couplings, integrals, and chemical shifts into account, is of course a more powerful source of significant structural information.

In the synthesis of some stable, severely overcrowded hydrocarbons, quite simple pmr measurements have enabled the rearrangements to be established (189-191). For example, the pmr spectrum (189) of 1,2-*bis*diphenylmethylene indane (**34**), obtained by the interaction of diphenylketen with 2-diphenylmethylene-1-indanone, has two methylene protons forming an AB quartet ($J = 21$ Hz) at 3.2 and 1.8 ppm; phenyl group A causes their environments to differ. A high-field aromatic doublet at 6.4 ppm, shown also by the 3,3-dimethyl derivative, is attributed to the 7 proton shielded by ring D.

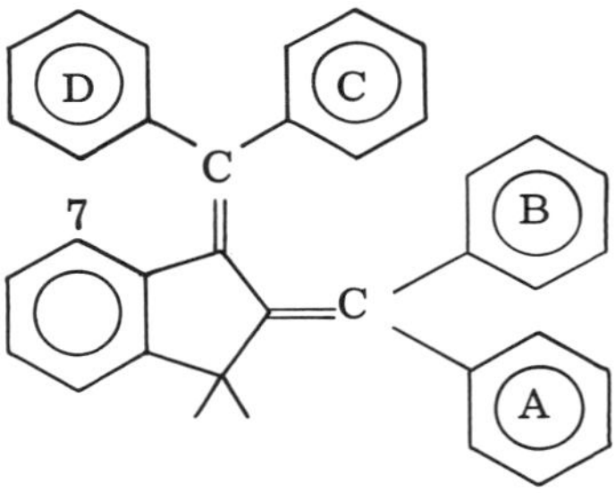

(34)

The reaction of 3-chloro-1,2-naphthoquinone with dimethylbutadiene yields not only a 1:1 adduct, of which Ansell and Murray (192) confirmed the structure by pmr, but also a "further reaction product"; to this, Fieser and Dunn (193) assigned the structure **35**, that is, a 1:2 adduct less the elements of HCl. In contradiction of **35**, Ansell and Murray (192) found in the spectrum 6 aromatic protons (7.15-7.85 ppm multiplet), a hydroxy proton (lost on

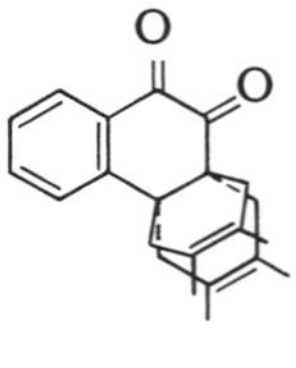

(35) 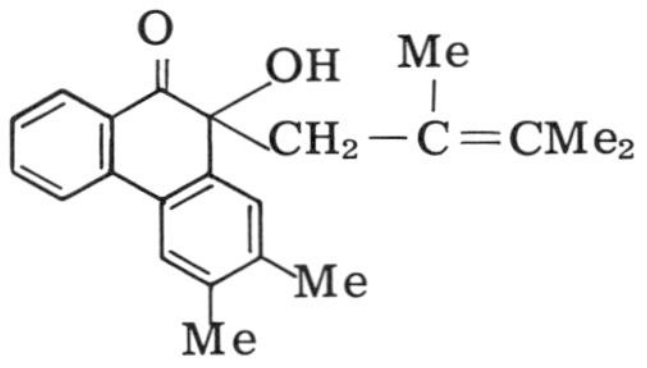 (36) 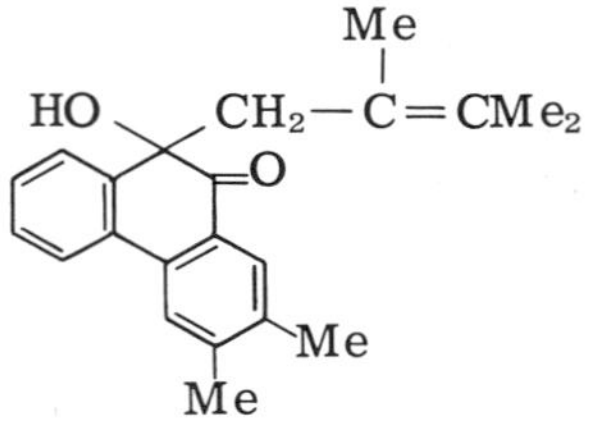 (37)

deuteration), a methylene AB quartet (2.34 and 2.75 ppm, $|J| = 14$ Hz), two aromatic methyl groups (2.30 and 2.37 ppm), two vinyl methyl groups at 1.50 ppm, and a further vinyl methyl group at 1.28 ppm. It was concluded that the structure is **36** or **37**, or a mixture of the two.

Electrolytic reduction in alkaline solutions of aromatic aldehydes and ketones is a useful route to the corresponding pinacols (194) which has been applied to many benzenoid and heterocyclic carbonyl compounds. Reduction of 1-acetylnaphthalene at a mercury cathode (195) gives the substituted tetrahydro-1,4-methano-3-benzoxepin **(38)** instead of the expected pinacol. In the pmr spectrum of **38**, 22 protons are evident, of which 11 are aromatic and 6 are from two uncoupled methyl groups (one, acetyl, at $\delta = 2.29$ ppm and one at 1.47

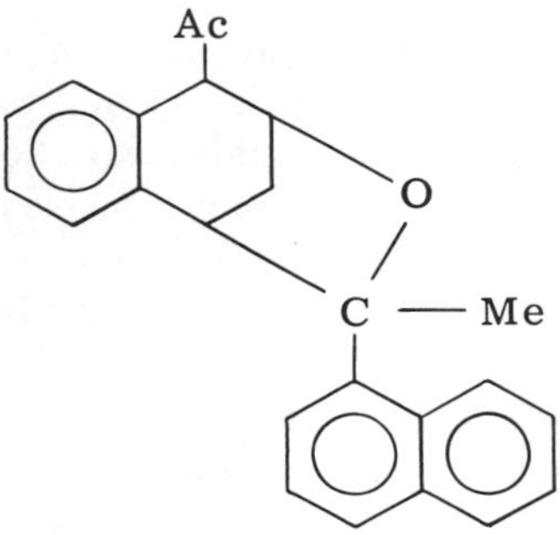

Ac

O

C — Me

(38)

ppm); the shifts and coupling constants of the five aliphatic protons ($H_{A\text{-}E}$) were confirmed by double resonance. After the material had been recrystallized from dioxane/D_2O in presence of a trace of NaOD, the spectrum revealed neither H_C nor an acetyl methyl resonance, and the H_A quartet had collapsed to a doublet. If follows that H_C must be attached to the carbon bearing the acetyl group and be adjacent to H_A. The signal from H_D appears as a multiplet owing to coupling to H_A, H_B, and H_E, while the H_D and H_E signals have the correct positions and coupling constant (J_{DE} = 12 Hz) for nonequivalent methylene protons. The H_D must be flanked by H_A and H_B in order to account for J_{AD} (6.0 Hz) and J_{BD} (3.5 Hz). H_E is coupled detectably only to H_D. Thus the arrangement shown by **38a** accounts for the observed pmr spectrum of **38**.

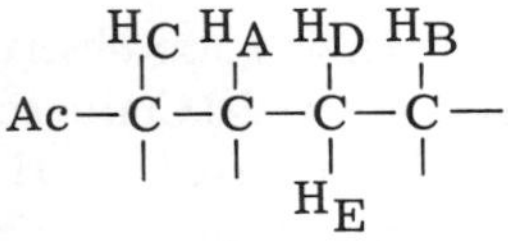

$$\begin{array}{ccccccccc} & & H_C & & H_A & & H_D & & H_B \\ & & | & & | & & | & & | \\ Ac & - & C & - & C & - & C & - & C - \\ & & | & & | & & | & & | \\ & & & & & & H_E & & \end{array}$$

(38a)

When this information is combined with the need to accommodate **39**, of which the structure was shown both by pmr and synthesis, as the product of acid treatment of **38**, the structure is established. Bavin *et al.* (136) were able to

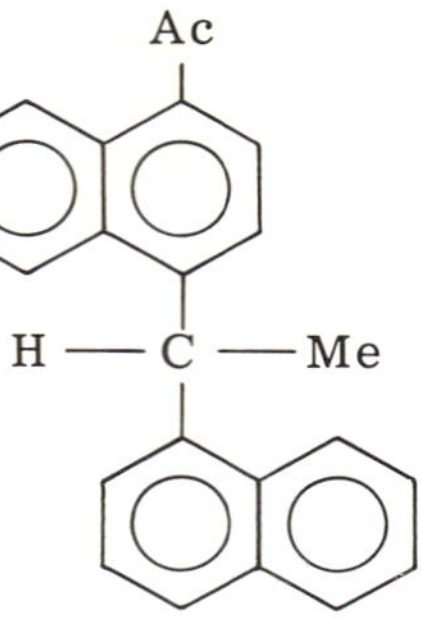

(39)

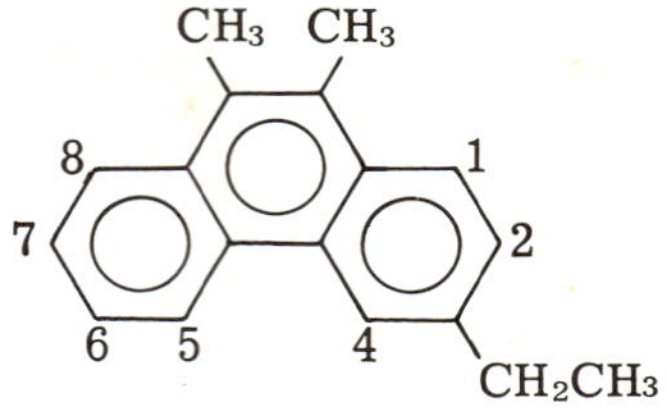

(40)

identify an ethyl-substituted 9,10-dimethylphenanthrene (**40**) from a study of its 60 MHz pmr spectrum. The unknown compound was obtained by reduction of an acetyl-9-10-dimethylphenanthrene; this acetylation was investigated in the expectation that this system would be more amenable than phenanthrene (196) to an examination of the postulated (197) interconversion of 2- and 3-acetylphenanthrenes. Reduction of the only acetyl derivative isolated gave an ethyl derivative; in its pmr spectrum, lines of the H-5-H-8 ABMX pattern can be picked out. The broad singlet 0.2 ppm to high field of H-5 has intensity equivalent to one proton and is in the position predicted for H-4, with small *meta* and *para* couplings. H-8 is superposed on two slightly broadened lines separated by 8.4 Hz—presumably from H-1 *ortho*-coupled to another proton with $J = 8.4$ Hz. In confirmation, there are at high field two doublets, with centers 8.4 Hz apart, which are assigned to H-2 with *ortho* and *meta* splittings. The conclusion that the ethyl group is in the 3 position is confirmed by the ring spectrum of the original acetyl compound. The ABMX pattern is again present, from H-5 to H-8, but now the broad H-4 resonance is shifted to low field of H-5 by the characteristic action of the $COCH_3$ group. The corresponding observation of broad singlets for H-1 of 2-substituted naphthalenes allowed Girdler, Gore, and Hoskins (198) to identify acetylation products of 2-bromo- and 2-acetamidonaphthalenes. By even simpler methods, Bil and Bernstein (199) proved that the dinitro compound they isolated from the products of nitration of 2,6-dimethylnaphthalene was the 1,5 isomer (**41**). A 6-proton singlet for the equivalent methyl groups was found at $\delta = 2.55$ ppm, while in the aromatic region there was a 4-proton AB system, with $J_{AB} = 8.5\text{-}9.0$ Hz, typical of *ortho* coupling. A 4-proton AB system, with $J = 9$ Hz, was also observed (200) in the spectrum of the diacetyl derivative prepared by Friedel-Crafts acetylation of 2,6-dimethoxyanthracene; a 2-proton singlet was observed for H-9,10 and the structure can only be **42**. Acetylation products of other derivatives of 2,6-dihydroxyanthracene were also amenable to study by pmr (201,202), as were the halogenation products of 2,6-dihydroxyanthracene itself (203).

A consideration of intensities in the pmr spectrum (204) led to confirmation of **43** as the structure of the product of reduction by KI solution of the

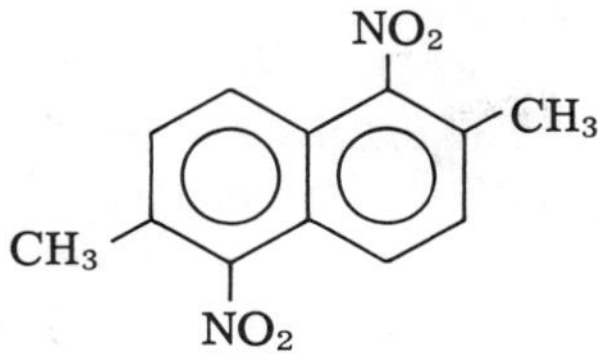

(41) 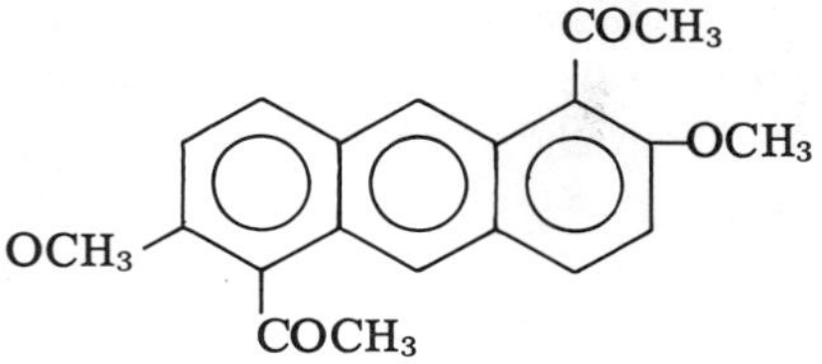 **(42)**

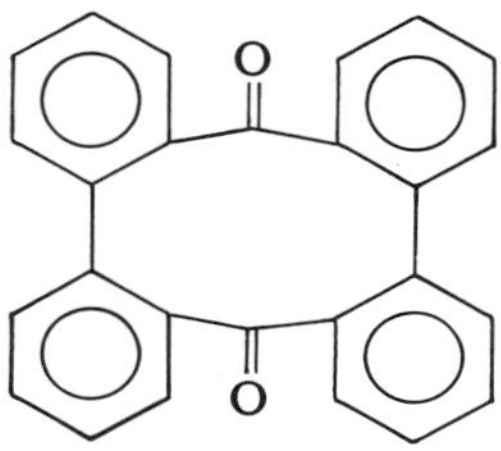

(43)

compound formed from dibenzo[*g,p*]chrysene (**12**) and ozone at 195°K. Three aromatic multiplets were observed with intensity ratio 1:2:1 from an ABMX spectrum. Only attack at the "central" bond in **12** could lead to a compound with such a pattern.

Spectra of unsymmetrical alkylnaphthalenes can be interpreted (151) as the superposition of two spectra corresponding to different rings. Thus, for 2,3,6-trimethylnaphthalene, the aromatic spectrum consists of an ABX pattern similar to that of 2,6- and 2,7-dimethylnaphthalenes and a strong singlet from H-1 and H-4, which have resonance positions similar to one another and to the 1,4 hydrogens of 2,3-dimethylnaphthalene. By this method of analysis, the aromatic-proton pattern for any tri- or tetramethyl naphthalene can be predicted and used for identification.

Since coupling constants are not much affected by substitution, Yew and Mair (151) expected naphthalenes with alkyl substituents other than methyl groups to give similar aromatic spectra. In this way, 2-methyl-6-isopropylnaphthalene, 2-methyl-6-*n*-propylnaphthalene, and 1-ethyl-3,7-dimethylnaphthalene from petroleum were identified by comparing their pmr spectra in the aromatic region with those of 2,6-dimethylnaphthalene and 1,3,7-trimethylnaphthalene. The effects of alkyl groups on *ortho* chemical shifts are not uniform however; measurements on 9-alkylphenanthrenes (24) and alkylbenzenes (205) reveal that aromatic shifts depend on the number of carbon atoms substituted on the α carbon of the alkyl group. The high-field *ortho* shift decreases with increasing bulkiness of the alkyl substituent, and becomes about zero for *t*-butyl.

Substitution of a bromine atom on a benzene ring shifts a *meta* proton to higher field (72,206-120 and Table III); Kim and Anderson (139) confirmed that this was also the case in the naphthalene series by recording the spectra of

1-bromo-4-methylnaphthalene (**44**), which shows the effect, and 1-bromo-3,4-dimethylnaphthalene (**45**), which, of course, does not. The same conclusion may be drawn from the chemical shifts in 1-bromonaphthalene and a number of dibromonaphthalenes which Paviot and Hoarau (78) determined by iterative analysis.

Since the spectrum of a bromination product (211) of 1,5-dimethylnaphthalene contained an 8 Hz doublet, centered as high as $\delta = 6.63$ ppm and with components broadened by unresolved long-range benzylic coupling, the compound must be 4-bromo-1,5 dimethylnaphthalene (**46**).

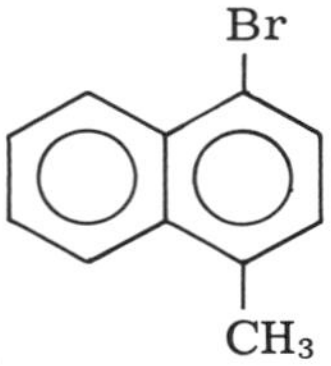

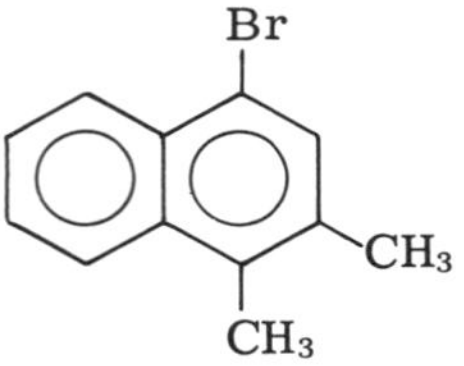

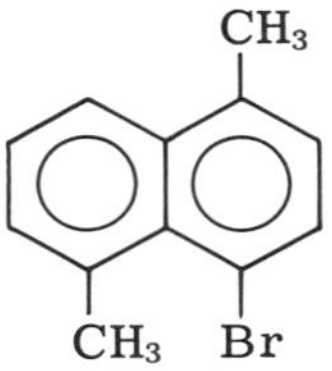

(**44**) (**45**) (**46**)

Wells and Alcorn (212) have published a list of chemical-shift effects produced by a nitro group in naphthalenes. These increments, later revised by Wells (65) (Table IV), which depend directly on the position of the substituent, have enabled Richer and co-workers to locate substituent positions in derivatives of 1,7-dimethoxy- (213), 2,7-dimethoxy- (214,217), 2,6-dimethyl- (215), 2,6-di-*t*-butyl- (216), and 2,7-di-*t*-butylnaphthalenes (217). Coupling patterns were also helpful in the identification of halogen derivatives.

Chang *et al.* (218) used pmr methods in the identification of the nitration products of 2,3-dimethylnaphthalene. Despite the claim that a halogen atom has virtually no effect on the position of pmr resonances, pmr spectra have been as useful with chloro and bromo as with nitro derivatives of diethoxy- and dimethoxynaphthalenes (219). Among other dicyclic aromatic structures confirmed by studies of the complete pmr spectrum are 3,5,6-tribromoacenaphthylene (220), 4,5-disubstituted 1,2-dibromoacenaphthylenes (221), the reaction product of acenaphthylene with dinitrogen tetroxide (222), further chlorinated and brominated chloro- and bromoacenaphthenes (223), and a number of C-1- and C-2-methylated acenaphthenes (224).

Pmr spectroscopy has proved particularly valuable in identifying the products of the alkylation (225,226) of polynuclear aromatic hydrocarbons by alkyllithiums. In the first report of such a reaction, Dixon and Fishman (225) proved from integral area ratios that mono- and di-*t*-butyl naphthalenes resulted from the reaction between naphthalene and *t*-BuLi in decalin; in the suspected *t*-butylnaphthalene, the $(CH_3)_3$ singlet at $\delta = 1.4$ has area 1.3 compared with 1.0 for the aromatic multiplet between $\delta = 6.85$ and 8.30. The calculated ratio is also 1.3:1.0. Similarly, the three *t*-Bu singlets at $\delta = 1.33$, 1.56, and 1.88 ppm in the spectrum of the mixture of di-*t*-butyl isomers have total area exactly three times that of the aromatic multiplet.

By similar arguments, Zieger and Rosenkranz (227) showed that a product of the reaction between perylene (**47**) and *n*-butyllithium was a monosubstituted *n*-butylperylene; furthermore, the α-methylene resonance is shifted to lower field than the similarly located protons in 2- and 3-*n*-hexylperylenes and 2-ethylperylenes. Pmr also suggests that ethyllithium yields 1-ethylperylene from perylene (228). Both the CH_2 and CH_3 resonances have greater δ values than their counterparts in the 2- and 3-isomers, while only three aromatic protons resonate in the bay chemical-shift region. Methylation of perylene with methyllithium (101) gives a methyldihydroperylene which dehydrates to 1-methylperylene.

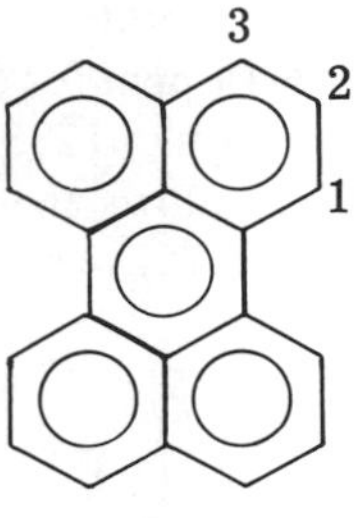

3
2
1

(47)

In another growth area of polynuclear chemistry, synthesis by aryne reactions (229,230), pmr has been widely applied. For example, the structure of 9-phenyl-9,10-dihydrophenanthrene (**48**), resulting from the reaction of benzyne with styrene (231), is proved by spectroscopy, in particular pmr. On the other hand, pmr indicated that benzyne reacts across the 9,10 positions of 9-vinylanthracene, to give 9-vinyltriptycene (**49**), rather than with the olefinic bond (232). In the pmr spectrum of **49**, H-2 and H-3 resonances [A and M of AMX

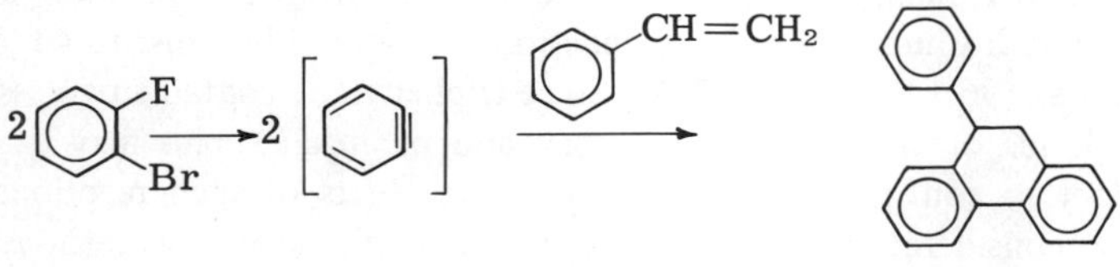

2 F Br → 2 [ ]   CH=CH₂ →

(48)

CH=CH₂ →

H(1) C=C H(2) H(3)

H(10)

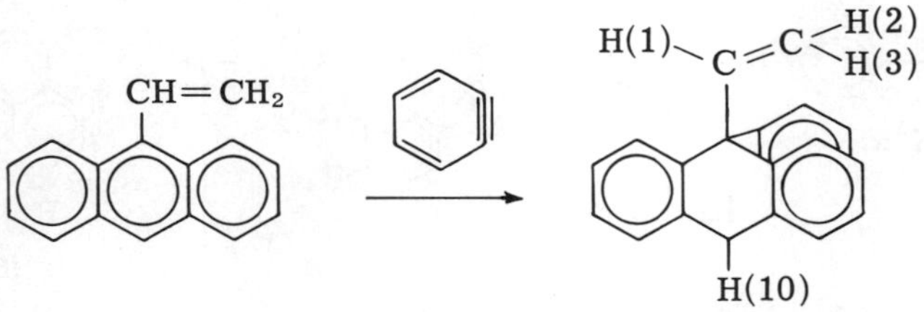

(49)

with J_{12} = 11.5 Hz (*cis*), J_{13} = 11.5 Hz (*trans*), and J_{23} = 2 Hz (*gem*)] as well as the H-10 singlet are found in the region δ = 5.0 to 6.5 ppm. The H-1 signals are obscured by the 12-proton aromatic multiplet. Klanderman and Perkins determined the composition of the product mixtures from the nitration of triptycene (**50a**) with nitric acid in acetic anhydride from the areas of the bridgehead proton absorptions (233). 1-Nitrotriptycene (**50b**) has a peak for one bridgehead proton near δ = 6.6 ppm (234) and a small peak was detected at δ = 6.55 ($CDCl_3$) and 6.51 ppm [$(CD_3)_2SO$] whose area suggested (233) only 2.3% α-nitration. The main products are 2-nitrotriptycene (**50c**), 2,6-dinitro- (**50d**), and 2,7-dinitrotriptycenes (**50e**), and the analyses for these were carried out by integration of the bridgehead proton areas at δ = 5.66 (**50a**), 5.95 and 5.90 for **50c**, and 6.22, 6.18, and 6.14 ppm for **50d** and **50e**. Structure **50a** itself has received a detailed pmr analysis (235) involving double resonance; decoupling of the methine protons allows the aromatic protons to assume an AA′BB′ system. An earlier analysis (176) was reported without spin-decoupling.

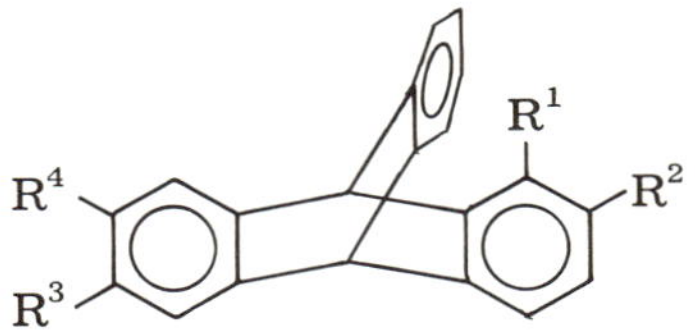

(**50a**) $R^1 = R^2 = R^3 = R^4 = H$
(**50b**) $R^1 = NO_2$; $R^2 = R^3 = R^4 = H$
(**50c**) $R^2 = NO_2$; $R^1 = R^3 = R^4 = H$
(**50d**) $R^2 = R^3 = NO_2$; $R^1 = R^4 = H$
(**50e**) $R^2 = R^4 = NO_2$; $R^1 = R^3 = H$

Substituted triphenylenes (**51**) may be prepared from 2-halogeno-2′-biphenyl Grignard and lithium reagents by reaction with suitably substituted benzyne intermediates (96,141,147,236,237). Since triphenylene contains two essentially different kinds of aromatic protons, bay and peninsular, pmr may be used to good effect to confirm the identity of the products of such reactions. While symmetry considerations may suffice in favorable cases (see, for example,

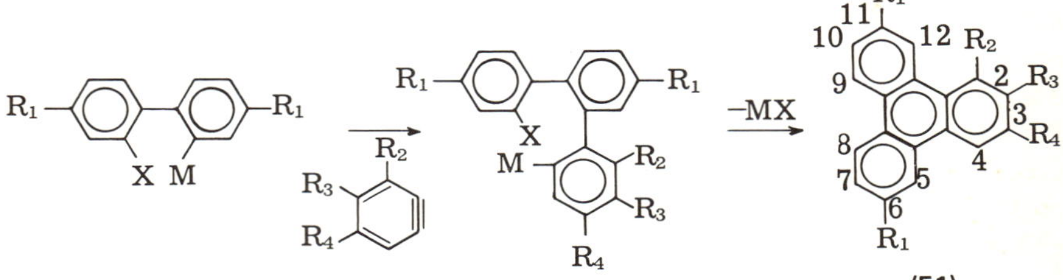

(**51**)

Section II.A), study of the whole spectrum is more usually required; thus if $R_1 = R_3 = H_1$ and $R_2 = R_4 = CH_3$, 1,3-dimethyltriphenylene is expected (237) for (**51**). In confirmation, the spectrum shows a multiplet, centered at δ = 8. 48 ppm, of intensity four protons, assigned fo H-5, H-8, H-9, and H-12; slightly upfield (δ = 8.20) is a broad singlet from H-4. A 4-proton (peninsular) multiplet is found at δ = 7.48 ppm and the H-2 singlet at δ = 7.22. The two methyl signals are in typically bay and peninsular methyl shift regions (96) with δ = 3.00 (1-CH_3) and 2.50 (3-CH_3) ppm.

A recent application of pmr to provide conclusive proof of structure is a study of the addition of ethyl diazocetate to anthracene and phenanthrene (238). Previously, the reaction has been interpreted (239) as an addition to the most reactive bond in the molecule, and formation of the esters **52**,**53** has been cited in support of this suggestion (240-242). Since no proof of the structure of **52** had been reported, and phenanthrene adducts other than **53** had been found (243,244), Graham *et al.* (238) reexamined both reactions. The pmr spectrum of the methyl ester of **52** confirms the structure and establishes that it is the *trans* isomer; signals are found at δ = 0.99 ppm (intensity equivalent to 1 proton, with triplet structure, $J = 4$ Hz, assigned to CHCOOMe); between δ = 2.3 and 2.7 ppm (1H, multiplet, H-2); δ = 3.12 ppm (1H, quartet, $J = 4$ and 9 Hz, H-1); δ = 2.63 ppm (3H, singlet, —$COOCH_3$); δ = 6.1 to 6.6 ppm (2H, multiplet, H-3 and H-4),

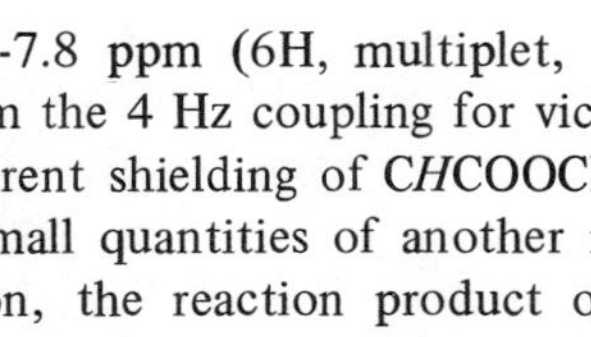

(52)

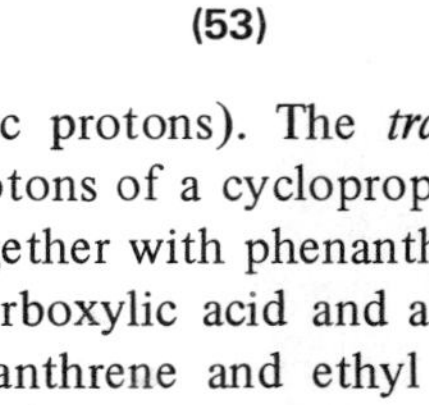

(53)

and δ = 7.2-7.8 ppm (6H, multiplet, aromatic protons). The *trans* geometry follows from the 4 Hz coupling for vicinal protons of a cyclopropane ring, and the ring-current shielding of C*H*$COOCH_3$. Together with phenanthrene-9-acetic acid, and small quantities of another monocarboxylic acid and a dicarboxylic acid fraction, the reaction product of phenanthrene and ethyl diazoacetate yielded three cyclopropane derivatives: **53** (acid), **54**, and **55**. Pmr of **53** and the methyl ester of **54** confirmed that they are stereoisomers: each spectrum showed eight aromatic protons, no olefinic signals, and methyl singlets at δ = 3.63 and 3.26, respectively. The remaining protons form an AX_2 system in **53** with $J = 4$ Hz, but an AB_2 in **54**, with $J_{AB} = 8.66$ Hz. In the *trans* ester (**53**), the C*H*COOMe signal at δ = 1.00 ppm was shielded relative to the corresponding proton in **54** (δ = 2.25 ppm); the COOCH_3 resonances showed the opposite effect since those protons of the *cis* ester (**54**) were shielded by the aromatic rings. These observations are fully in accord with models of these compounds.

The pmr spectrum of the ester of **55** is also consistent with the structure, and with the *trans* configuration.

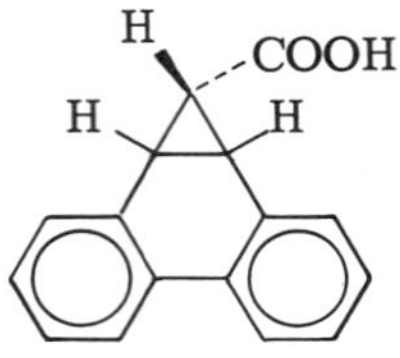

(54)

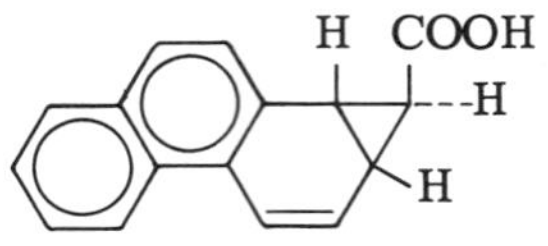

(55)

A final example in this section concerns the reactions of the p-Me-$C_6H_4SO_2^{\cdot}$ radical (**56**) generated (245) by thermolysis of toluene-*p*-sulfonyl iodide in presence of copper powder. Pmr was of considerable diagnostic help (246) in proving that such radicals add to the *meso* positions of both anthracene and naphthacene. For example, with anthracene, the product was 9,10-dihydro-9,10-ditoluene-*p*-sulfonylanthracene (**57**); its pmr spectrum showed singlets from two methyl groups at $\delta = 2.25$, and another at 4.73 ppm (2H *meso* hydrogens), together with 16 aromatic protons in bands between $\delta = 6.28$ and 7.55 ppm. With 9-methylanthracene, **56** gave 9-anthryltoluene-*p*-sulfonylmethane (**58**), identified (246) from its pmr spectrum: CH_3 at 2.27 ppm, and CH_2 at $\delta = 5.8$, as well as a *meso* hydrogen signal at $\delta = 8.40$ and 12 aromatic protons between 6.9 and 7.1 ppm.

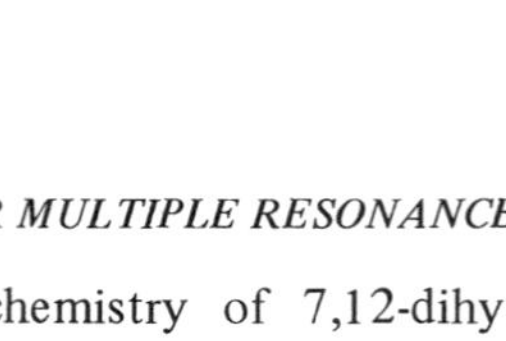

(57)

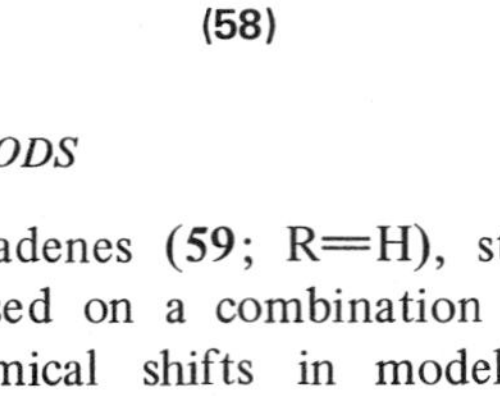

(58)

D. NUCLEAR MULTIPLE RESONANCE METHODS

In the chemistry of 7,12-dihydropleiadenes (**59**; R═H), structural and stereochemical assignments have been based on a combination of symmetry arguments (see Section II.A) and chemical shifts in model compounds (247,248). Neither method, however, could determine whether the chlorine in **59** (R═Cl), which exists as a stable conformer (247), is axial or equatorial.

Spin decoupling was used to clarify this point in the following way: the C-12 protons give rise to an AB quartet centered at 4.88 ppm with $\delta_{AB} = 1.60$ ppm. The lowest-field pair of this system has half-height width $\Delta = 2.8$ Hz but it narrows to $\Delta = 1.9$ Hz when the aryl frequency is irradiated, and collapses to a

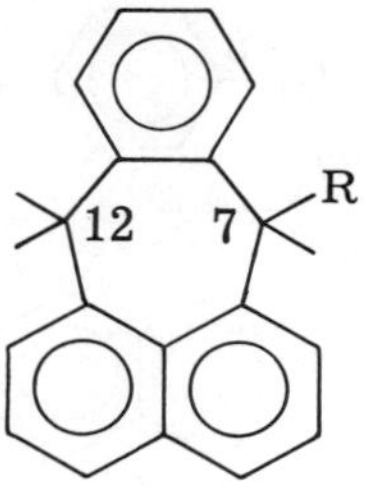

(59)

$\Delta = 1.9$ Hz singlet when the proton is decoupled simultaneously from aryl and high-field AB protons in a triple-irradiation experiment. Since benzylic coupling is favored when the side-chain carbon-hydrogen bond is perpendicular to the plane of the aromatic system (115,249,250) the low-field C-12 proton is evidently axial, a conclusion consistent with chemical-shift arguments (247,248). With Δ for H-7 also 1.9 Hz, like that of H-12_{equ} and the "residual" Δ of H-12_{ax}, H-7 must be equatorial, and the chlorine atom in **59** (R═Cl) is therefore axial. Lansbury (251) has also used intramolecular nuclear Overhauser effects (NOE), whose applications have been reviewed recently (252), in assigning structures to geometric isomers resulting from carbonyl addition of ethyllithium to 7-(12H)-pleiadenone followed by dehydration. Chemical shifts in **60-62** are listed in Table XII, with the arbitrary assumption that methyl and methine signals at lower field are *syn* to the naphthalene ring. While internally consistent,

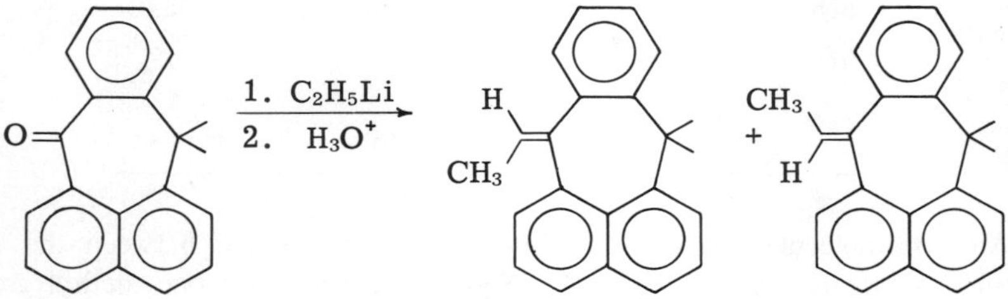

(60a) **(60b)**

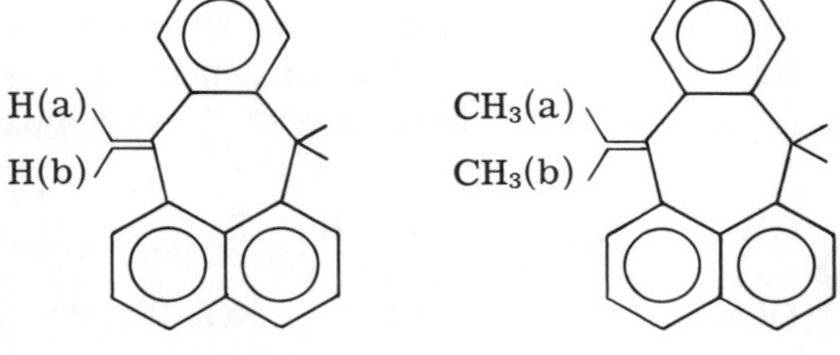

(61) **(62)**

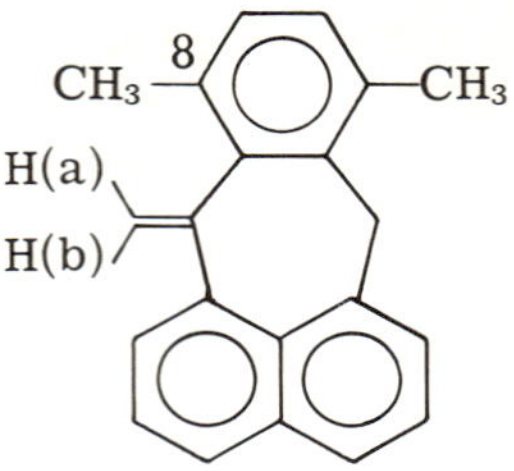

(63)

these assignments might be reversed; double-resonance experiments on **63** could remove the ambiguity. A model suggested that the vinyl proton *syn* to the methyl at C-8 should be close enough to allow an NOE experiment and, indeed, irradiation of the low-field methyl signal in **63** produced a 20% signal-intensity increase in the upfield vinyl proton, which must, therefore, be *syn* to phenyl. These observations fully confirm the assumptions in Table XII and allow **60a** and **b** to be identified.

TABLE XII

Chemical Shifts (ppm downfield from TMS) in 7-Ethylidene-, 7-methylene-, and 7-isopropylidene-7,12-dihydropleiadenes

Compound		δ_{CH_3}	δ_H
60a		1.93	5.80
60b		1.70	6.05
61	a	–	5.28
	b	–	5.63
62	a	1.73	–
	b	1.90	–

An intramolecular NOE also allowed a decision (253) between the two possible structures, **64** and **65**, of the product, which has one methyl group fewer, of irradiation of 2,3-dimethoxy-5-nitronaphthalene in an alkaline medium. Saturation of the methoxy signal led to 21% enhancement of the H-4 resonance, while H-1 showed no change. The compound is, therefore, 2-hydroxy-3-methoxy-5-nitronaphthalene (**64**). In related work (254), NOE experiments not only allowed identification of 2-methoxy-3-hydroxy-6-nitronaphthalene (**66**) as the product of irradiation of 2,3-dimethoxy-6-nitronaph-

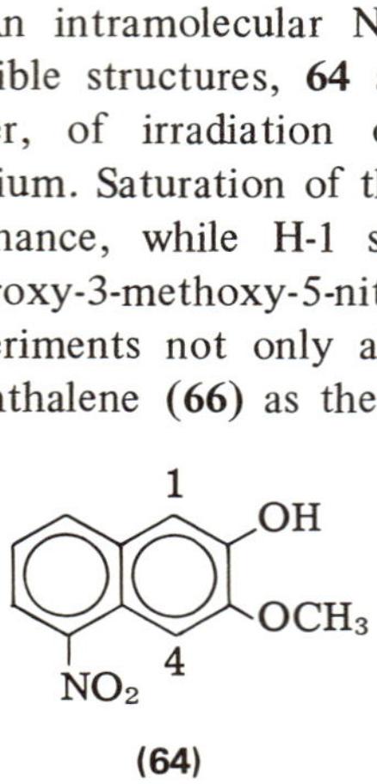

(64) 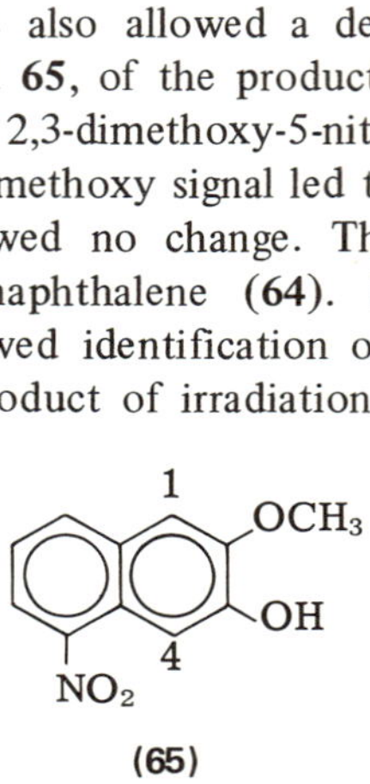 **(65)** 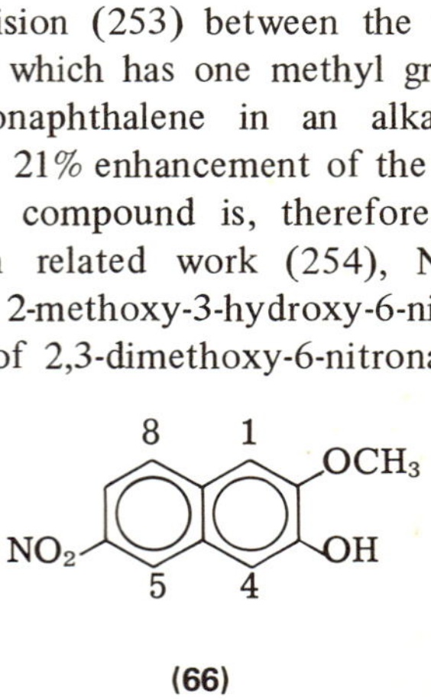 **(66)**

thalene, but also confirmed assignments of the aromatic signals. Saturation of the H-4 peak enhanced the H-5 signal, and *vice versa*, while similar results were obtained for H-1 and H-8.

Single-resonance pmr spectroscopy still left some doubt between structures **67a** and **b** for the *o*-quinone methide trimer formed when 4-methoxy-2,6-dimethyl phenol is oxidized by silver oxide in THF (255). Multiple-resonance (double, triple, and spin-tickling) experiments not only enabled the overlapped regions of ring IV CH_2 protons and the ring III/IV bridgehead proton to be clarified in positive support of structure **67a**, but they also demonstrated long-range couplings.

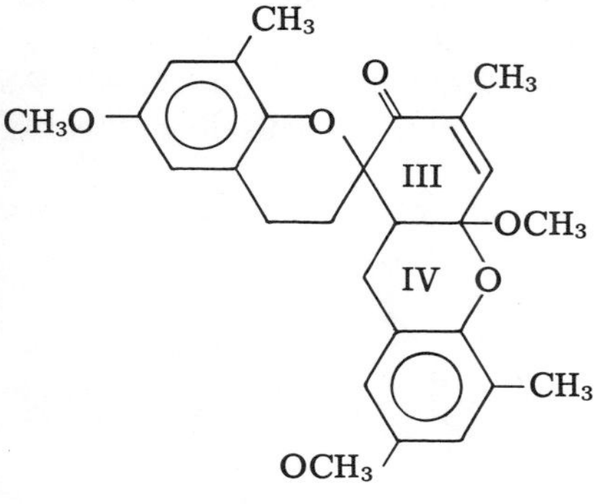

(67a) 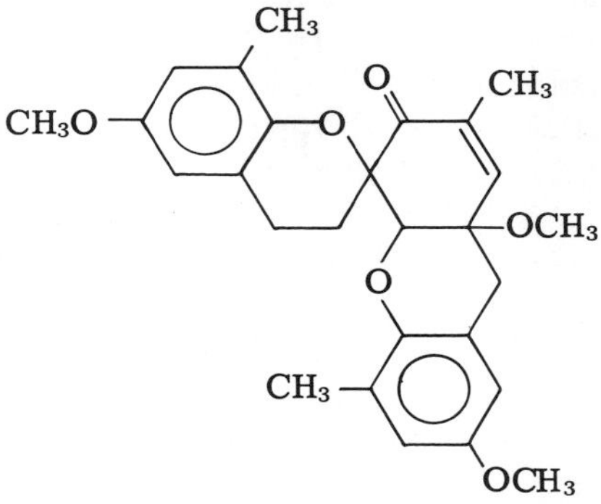 **(67b)**

While, as discussed in Section II.B, naphthacene adds acrylonitrile across *para* positions, the pmr spectrum of the product (256,257) with naphthalene confirms 1,2-*ortho* addition to yield the cyclobutane (**68**). Multiplets at $\delta = 7.14$ (4H), 3.54 (3H), and 2.65 (2H) were assigned to aromatic, methine, and methylene protons, respectively; each vinyl proton gave a 5 Hz doublet. Spin-decoupling strongly suggests that the nitrile group is located on C-8, since the additional splitting of one vinyl doublet is removed when the methylene frequency (2.65 ppm) is irradiated. Since negligible vinyl coupling is expected to H-8, the CH_2 group is most likely to be in the 7 position.

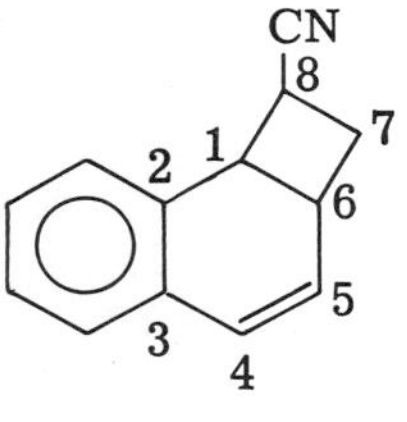

(68)

Mattingley and Zweig (258) irradiated dilute aqueous methanolic 2-methoxynaphthalene with uv light to yield 5,6,11,12-tetrahydro-5,12,6,11-[1,2,3,4,]-butanetetrayldibenzo[*a,e*]cyclooctene-13,16-dione (**69**) or (**70**). The single-resonance pmr spectrum was consistent with either of the novel dibenzopentacyclic diketone structures (**69** and **70**): in addition to an 8-proton aromatic multiplet at 6.9 ppm, there were three aliphatic multiplets of equal intensity, centered on 3.1 (H_a), 3.6 (H_x), and 4.4 (H_m) ppm, and corresponding to a calculated XMAA'M'X' system with $J_{MX} = J_{M'X'} = 8.0$, $J_{AX} = J_{A'X'} = -1.5$, $J_{AM} = J_{A'M'} = J_{AA'} = 10.6$, $J_{A'X} = J_{AX'} = J_{M'X} = J_{MX'} = J_{A'M} = J_{AM'} = J_{MM'} = 0$ Hz. Differentiation between **69** and **70** in favor of **69** was effected by double-resonance examination of the reduction product—**71** or **72**. In structure **71**, H_h would be coupled to H_a' and H_x but not to H_m' whereas in **72** it would be coupled to H_m and H_x' but not H_a. In the spectrum of the reduction product in deuterated methanol, the change to a quartet of just the 4,4 ppm octet identifies this as H_h. Irradiation of this region causes no change to H_m (at 3.8 ppm) in accordance with **71** but not **72**; further, the collapse of multiplets at 2.5 (H_x) and 3.4 (H_a) ppm to give a similar pattern to that of the diketone precursor helps to confirm the proton assignments.

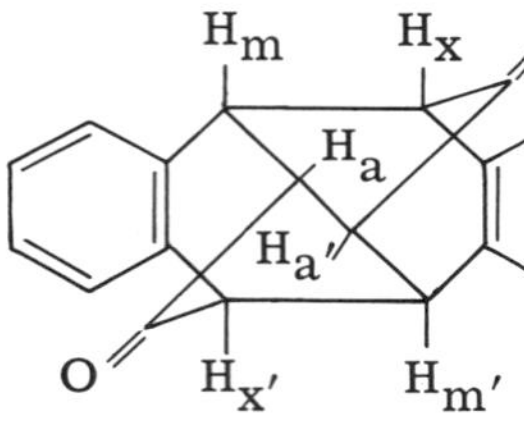

(69) Reduction with LiAlH (71)

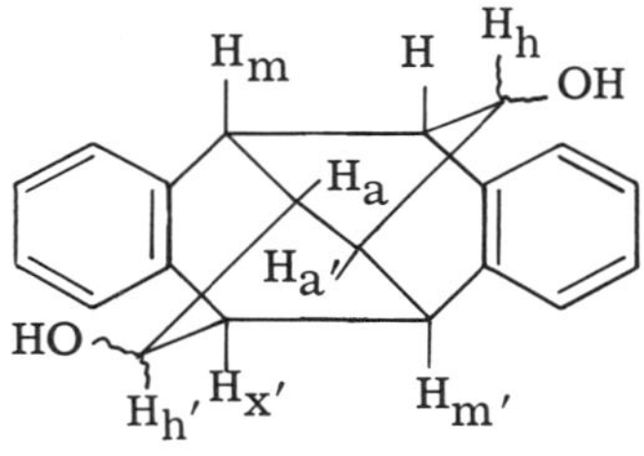

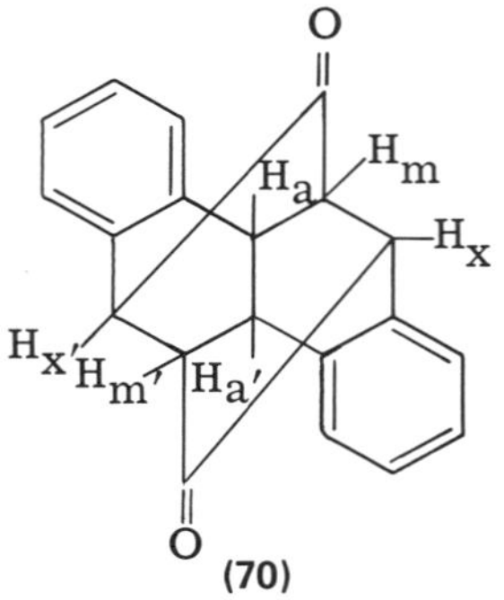

(70) 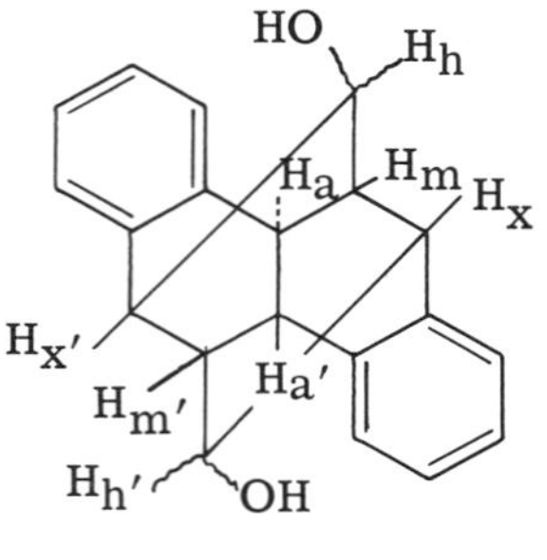 (72)

Many of the techniques outlined in Sections II.A-D are incorporated in the very comprehensive application by Berg, Jakobsen, and Johansen (259) of nmr spectroscopy to the orientation of alkyl substituents in pyrene (**73**) and phenanthrene following Friedel-Crafts reactions. When **73** reacts with $AlCl_3$ and *t*-butyl (or *t*-amyl) chloride, the resulting dialkylpyrene has been shown to be 2,7 rather than the reported (260) 1,6. The di-*t*-butylpyrene has a very simple pmr spectrum in $CDCl_3$ solution, with nine protons at 1.57, two at 7.98, and two at 8.17 ppm, which revealed no spin-spin splitting despite extensive variation of solvent. This lack of coupling is sufficient to determine the orientation as 2,7 (**74**); the 7.98 ppm singlet was ascribed to $\alpha\beta$ positions, and the 8.17 singlet to α protons because of the low-field *ortho* shift expected from *t*-butyl groups. There are two doublet $^{13}C{-}^{1}H$ satellites above and below the aromatic ^{1}H spectrum. One doublet with 1.7 Hz coupling is typical of *meta* coupling between two α protons, thus confirming the 2,7 orientation; the other coupling (8.9 Hz) is marginally lower than had previously been encountered (75) for *ortho* coupling between two $\alpha\beta$ positions in pyrene. (The $^{13}C{-}^{1}H$ coupling constants were estimated as 156 (α), 154 (β), and 159 Hz ($\alpha\beta$). Yet further confirmation of the assignment of the α protons came from an NOE experiment; irradiation at the *t*-butyl frequency left the upfield aromatic signal unchanged but caused a 20% reduction in the intensity of the downfield signal.

Secondary alkylation, on the other hand, went in the expected 1, 3, 6, and 8 positions so that there was no trace of a 2,7-diisopropyl derivative. The aromatic spectrum of tetraisopropylpyrene showed singlets in intensity ratio 2:1. That at 8.33 ppm was assigned to the $\alpha\beta$ position and that at 7.97 ppm to the β position; the difference is relatively large because the alkyl substituents shift *ortho* protons to higher field and *peri* protons to lower field. If substitution had been 4,5,9,10, which would be consistent with the observed aliphatic spectrum (a 4H heptet at 4.09 ppm and a 24H doublet at 1.53 ppm, both with $J = 6.9$ Hz), then the aromatic spectrum would have been AB_2.

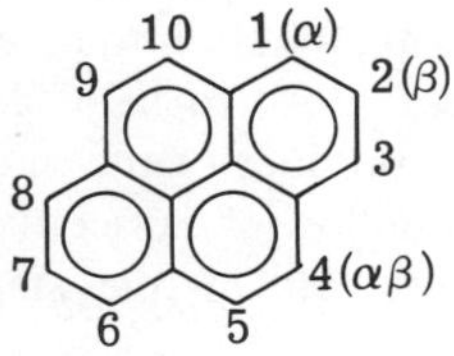

10 1(α) 2(β) 3 4($\alpha\beta$) 5 6 7 8 9

(73)

(74)

H(3) H(4) H(1)

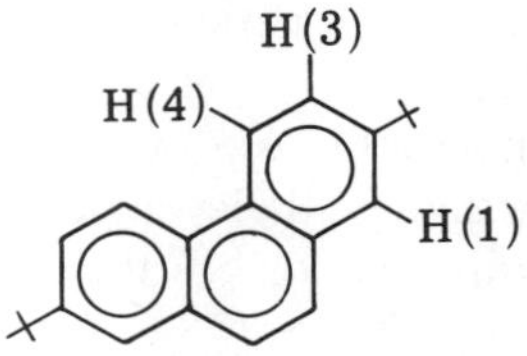

(75)

For phenanthrene, Berg *et al.* (259) identified the *tert*-butylation product as 2,7- (**75**) rather than 3,9-di-*t*-butylphenanthrene. They analyzed the aromatic spectrum of **75** as ABX (with a superimposed H-9 singlet) and refined it as ABC. Partial decoupling experiments on **75** and on di-*t*-butylphenanthrenequinone (which also received an ABX/ABC spectral analysis) demonstrated that all the *ortho* coupling constants have the same sign, presumably positive.

III. Some Special Applications of Proton Magnetic Resonance to Structure Elucidation in Polynuclears

A. *STRUCTURAL STUDIES OF POLYNUCLEAR NATURAL PRODUCTS AND RELATED COMPOUNDS*

Drastic dehydrogenation to yield polynuclear hydrocarbons and their derivatives has long been recognized as an important approach to the elucidation of the skeletal structure (261) of a wide variety of natural products. The steroids, for example, which include sterols, bile acids, certain hormones, cardiac glycosides, and some alkaloids, all contain the perhydrocyclopentenophenanthrene skeleton (**76**); they may yield methylcyclopentenophenanthrenes, of which Diel's hydrocarbon (**77**) is often an important constituent, as well as mono-, di-, and trimethylphenanthrenes substituted in any position other than 4 and 5. Furthermore, alkyl phenanthrenes often result from di-and higher

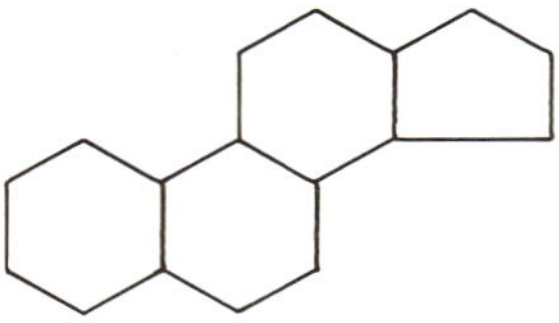

CH_3

(76) 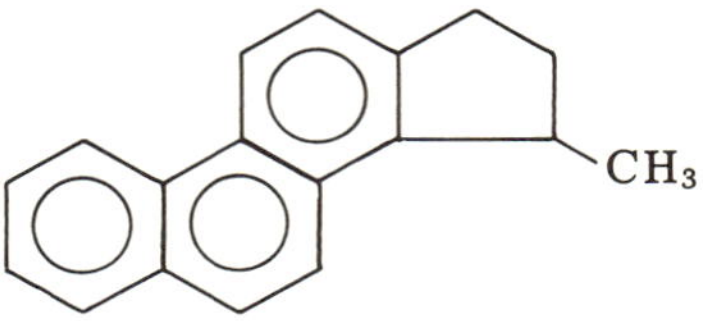 **(77)**

terpenoids, while the alkaloid thebenol yields pyrene. Pine-wood tar contains retene (1-methyl-7-isopropylphenanthrene) and a number of other alkyl phenanthrenes (262). Ir and uv spectral studies can certainly limit the number of structures for these products by suggesting the type of ring system, and possible substitution patterns, while indirect methods of identification, such as gas-chromatographic retention data (262,263), have also been proposed. The unique power of pmr for determining the nature and orientation of alkyl substituents, as well as the nature of the polynuclear skeleton, should make it a more useful adjunct to the other methods than is apparent from its exploitation to date. One example is its use by Bharucha, Weiss, and Reichstein (264) to help identify 3′-methyl-3′-ethylcyclopentenophenanthrene (**79**) as a product of the selenium dehydrogenation of the pregnendiol obtained by reduction of 3β-acetoxy-Δ^5-pregnen-20-one (**78**). A more direct application of pmr in the steroidal

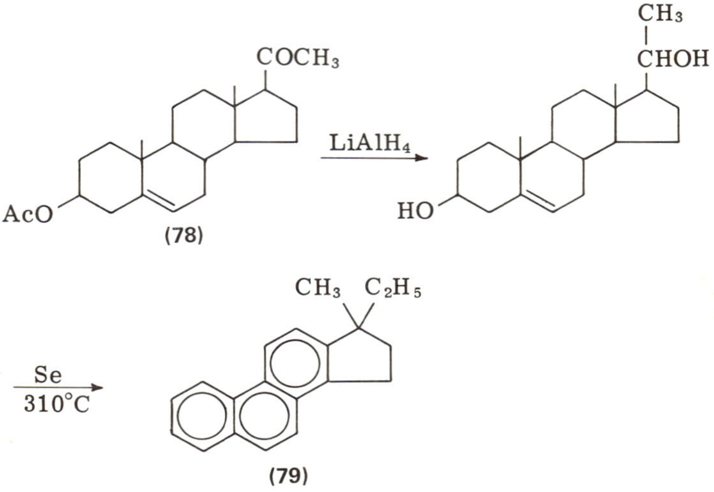

$COCH_3$ CH_3 | CHOH

AcO (**78**) $\xrightarrow{LiAlH_4}$ HO

CH_3 C_2H_5

$\xrightarrow[310°C]{Se}$

(79)

polynuclear field involves ring-current effects in phenanthrenes and dihydrophenanthrenes. These enabled Cross *et al.* (265) to establish incontrovertibly that oestrone methyl ether **(80)** is dehydrogenated to the dihydrophenanthrene **(81)**, which can then be readily oxidized to the corresponding phenanthrene. Similarly, Kaufman (266) proved that 17α-methyl-$\Delta^{1,6,4}$ androstatrien-17-ol-3-one yields a tetrahydrophenanthrene when treated with *p*-toluenesulfonic acid anhydride, while Brown and Turner (267) also used pmr to investigate the aromatization of 17α-methyl-$\Delta^{9(11)}$-testosterone.

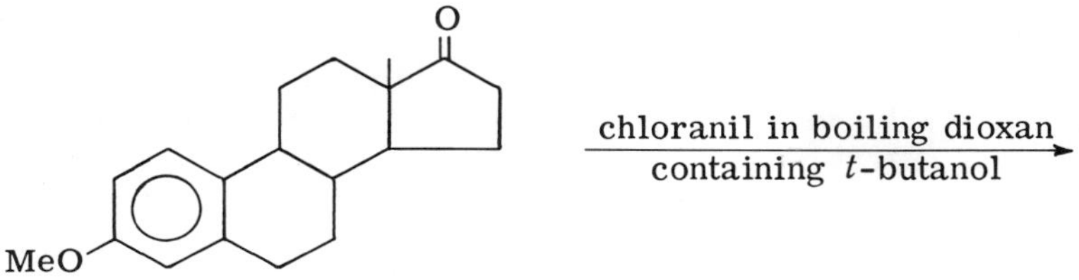

MeO

(80)

chloranil in boiling dioxan containing *t*-butanol

Me COO-*t*-Bu

MeO

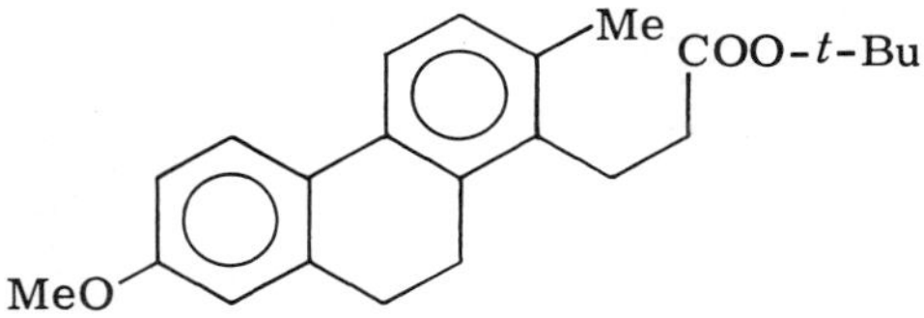

(81)

Good examples of the contribution pmr can make to the elucidation of the structures of polynuclear natural products are the investigations by Gaudemer, Polonsky, and Alais (268) of piloquinone, a substituted phenanthrene-9,10-quinone isolated from *Streptomyces pilosus* (269), and of novel aristolochic acid derivatives from *Aristolochia indica* (270). The 60 MHz spectrum of piloquinone (268) shows the presence of the isocaproyl

$$\underset{}{—CO—}\underset{\alpha}{CH_2}—\underset{\beta}{CH_2}—CH(CH_3)—CH_3$$

group; a 6-proton doublet at $\delta = 0.97$, $J = 5.5$ Hz, arises from methyls of an isopropyl group, the C*H* signal of which is hidden beneath the β-CH_2 signals near $\delta = 1.67$. The α-CH_2 triplet ($J = 7.5$ Hz), deshielded by the adjacent carbonyl, is located at $\delta = 2.87$ ppm. Reduction of the isocaproyl carbonyl to CH_2 to give desoxydihydropiloquinone shifts the methyl singlet at 2.23 (a value typical of CH_3 bonded to an aromatic ring) to $\delta = 2.13$ ppm. This shielding effect suggests that methyl and isocaproyl are *ortho* to each other in piloquinone.

The aromatic region of the spectrum of piloquinone (411-463 Hz downfield from TMS at 60 MHz) shows the presence of four aromatic protons; their disposition was proved from the spectrum of the derived tetraacetate. Here, a low-field singlet ($\delta = 8.72$) is suggestive of H-4 in a 1,2,3 trisubstituted ring and, *via* model compounds, that an acetoxy group is *para* to this proton. The remaining aromatic signals in the tetraacetate are recognizable as an AMX system from three contiguous aromatic protons, one of which resonates at $\delta = 8.48$ ppm, and hence is assigned to H-5 with an acetoxy group at H-8. The extra deshielding of H-4 compared with H-5 in the tetraacetate, but not in desoxydihydropiloquinone where CH_2 replaces CO, shows that the isocaproyl group is in the 3 position in piloquinone, and thus has an *ortho* effect on H-4.

In addition, the resonance frequencies of the phenolic protons in piloquinone ($\delta = 12.28$ and 12.75), which are almost unaltered either on dilution of the solution or in desoxydihydropiloquinone, confirm that the hydroxyl groups are intramolecularly chelated to CO groups other than that of the isocaproyl group. Piloquinone is, therefore, 1,8-dihydroxy-2-methyl-3-isocaproylphenanthrene-9,10-quinone (**82**).

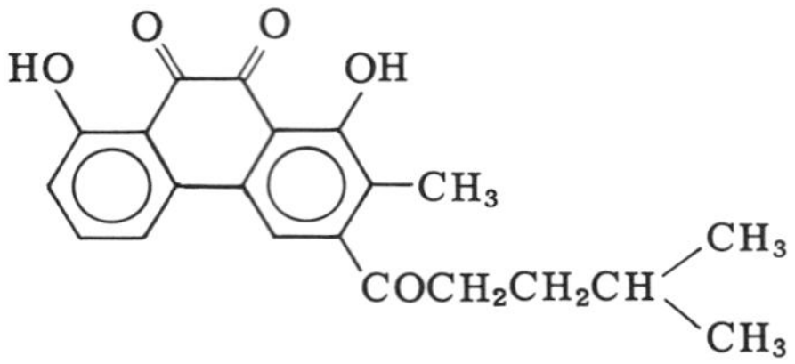

(82)

The pmr spectrum (270) supported the formula $C_{17}H_{11}NO_8$ for aristolochic acid-D (**83**), and showed methoxy, methylenedioxy, aromatic, and two D_2O-exchangeable protons. Methylation with diazomethane yielded a dimethyl derivative (**84**) with uv, ir and pmr spectra similar to those of aristolochic acid (**85**), and suggestive of a 3,4-methylenedioxy-dimethoxy-10-nitro-1-phenanthroic acid methyl ester structure for **84**; comparison with other known aristolochic acid derivatives suggested location of methoxy groups at C-6 and C-8 and this was confirmed by total synthesis. The remaining structural problem was a choice between C-6 and C-8 for the position of the single methoxy group in **83**. Now in this series C-8 methoxy protons resonate 0.1 ppm to low field of C-6 methoxy. The OCH_3 shift in **83** ($\delta = 4.08$) leads to the 8-methoxy structure.

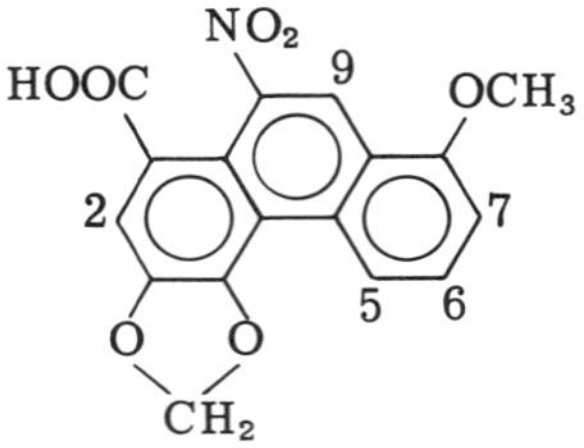

(85) 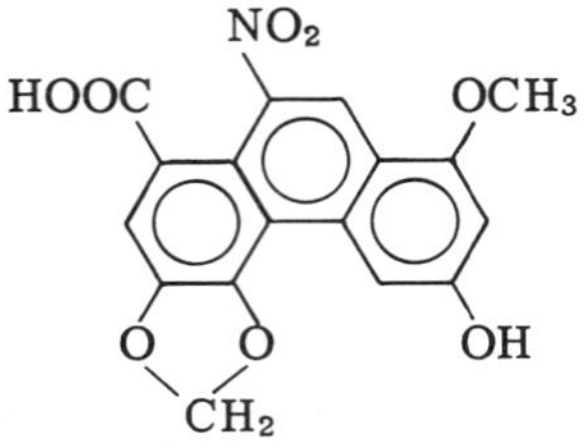 **(83)**

The isolation from the rhizomes of *Tamus communis* L. of the nitrogen-free phenanthrenes **86-88**, which were identified principally from their pmr spectra (271), is relevant to theories of the biosynthesis of aristolochic acid (272,273).

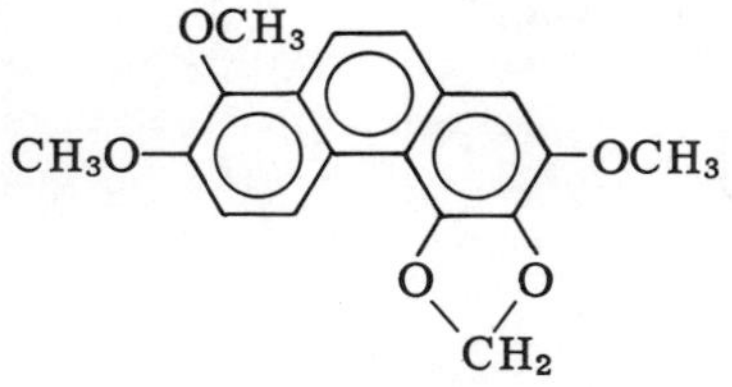

(86)

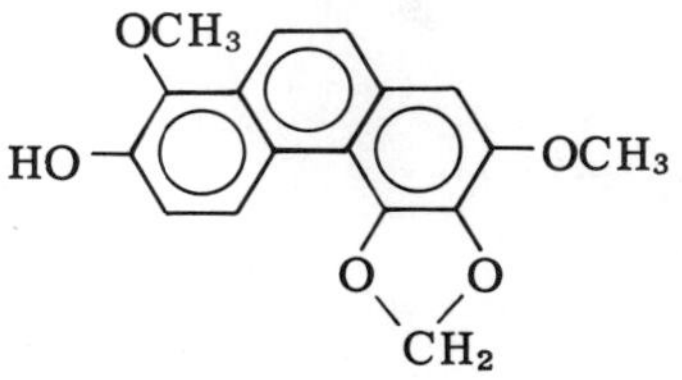

(87)

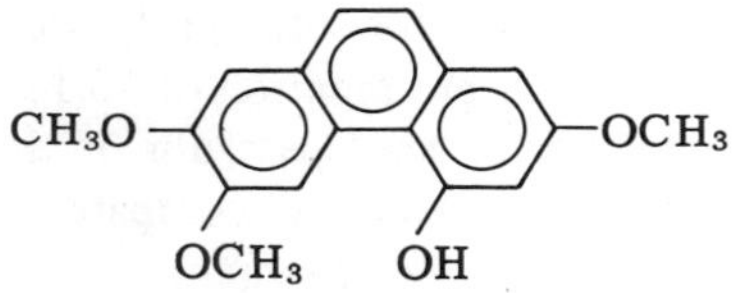

(88)

In addition to isolating aristolochic acid (274) from the roots of *Aristolochia argentina,* Priestap *et al.* extracted a dimethylaminoethylphenanthrene base which pmr showed to have the structure **89a** or **b**. A singlet at $\delta = 2.40$ ppm, intensity six protons, was assigned to the *N*-methyls, while a 4-proton multiplet between $\delta = 2.5$ and $\delta = 3.5$ was attributed to the aliphatic side chain. In view of the elimination by exchange of the broad 1-proton singlet at $\delta = 6.80$, the presence of an OH group was inferred; an OCH_3 singlet was found at 3.82 ppm. In the aromatic region, the H-5 multiplet shifted to $\delta = 9.52$ ppm, and the five other protons resonated at between 7.40 and 8.00 ppm. The structure of the base was established (275) as **89a** by synthesis.

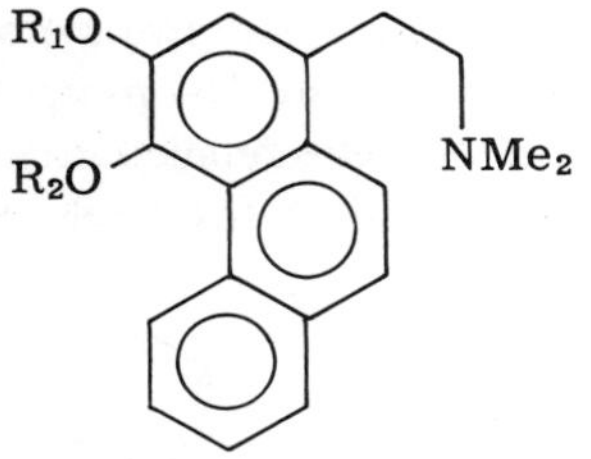

(89a) $R_1 = H$, $R_2 = Me$;
(89b) $R_1 = CH_3$, $R_2 = H$.

A very similar pmr spectrum, with OH replaced by another OCH_3 signal, was observed by Bick and Douglas (276) for atherospermine (**90**), isolated from the bark of *Atherosperma moschatum* Labill., and identical with an alkaloid found (277) in *Cryptocarya angulata.* The pmr spectrum of **90** also enabled the structure to be assigned completely.

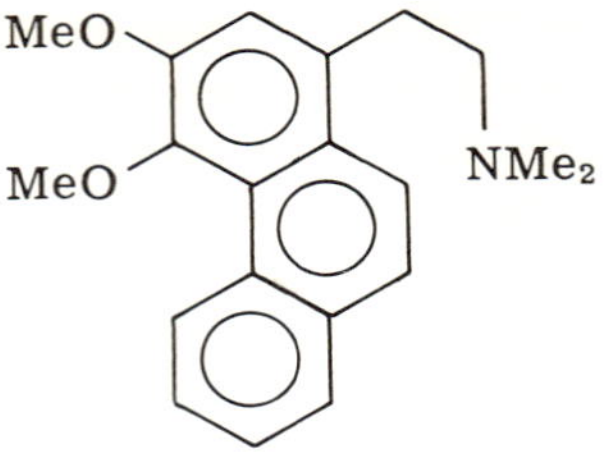

MeO MeO NMe₂

(90)

In spite of limitations imposed by poor solubility, pmr spectroscopy may provide structural information in polynuclear systems containing many fused rings. Violanthrene (**91**) derivatives are sufficiently soluble in tetramethylurea for Nair *et al.* (278) to show that bromination of 16,17-dimethoxyviolanthrene gives the 3,12-dibromo derivative. The spectrum of the product of reductive methylation of the dibromo derivative was compared with the fully assigned spectrum of 5,10,16,17-tetramethoxyviolanthrene.

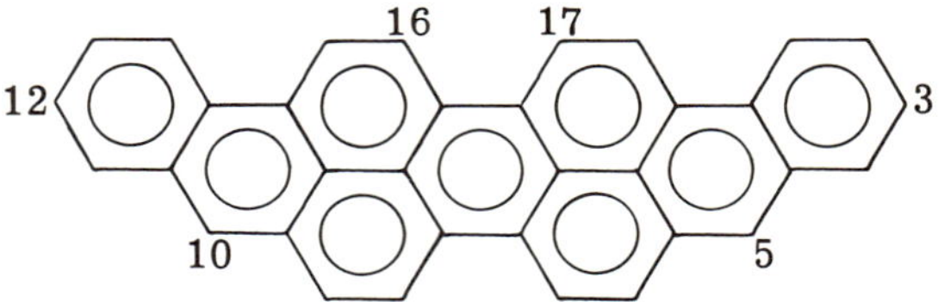

16 17 12 3 10 5

(91)

Quite simple considerations of the pmr spectrum (279) are sufficient to eliminate **92** (280) and **93** (281) as possible structures for isolapachenole, the dimer of lapachenole or 2,2-dimethyl-6-methoxy-7,8-benzchromen. Instead of the ten aromatic protons required for **92** and **93**, the spectrum shows only four α-aromatic protons around 8.1 ppm and five β-aromatic protons around 7.4 ppm; moreover there are four distinct methyl groups (between 1.2 and 1.6 ppm) instead of three predicted for **93**. More detailed examination reveals, in addition to the two methoxy groups, a total of five single protons: one near 1.5 ppm,

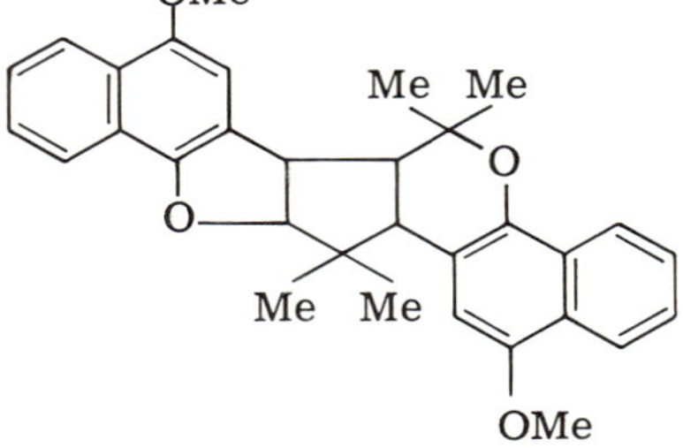

OMe Me Me O O Me Me OMe

(92)

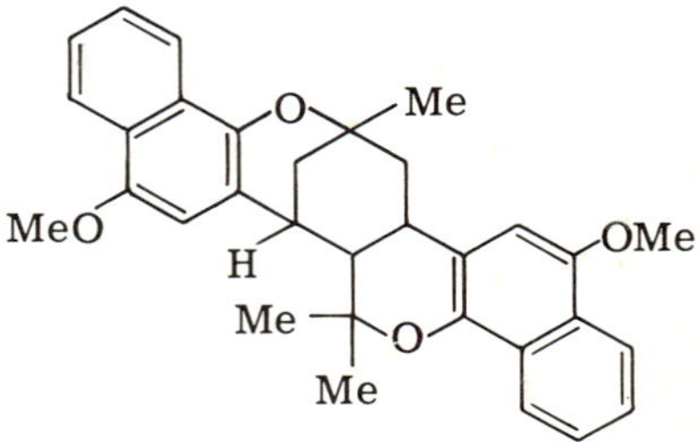

O Me MeO OMe H Me O Me

(93)

masked by the methyls; two in a 2.2-2.4 ppm multiplet; one in a 3.0-3.4 ppm multiplet; and one in a 7.4 Hz doublet at 4.7 ppm. That this last doublet arises from a singlet chemical shift was proved by comparison of 60 and 100 MHz spectra, a procedure which showed that the 6-proton doublet at 4.0 ppm arises from OMe groups with different shifts, and further that the four methyl groups between 1.2 and 1.6 ppm are distinct, not coupled to each other, and consistent with two *gem*-dimethyl groups.

A series of 100-MHz frequency-sweep spin-decoupling experiments confirmed the presence of a single proton at 1.5 ppm under the four methyl groups, and enabled the following coupling relationship to be established between the five single nonaromatic protons (shifts in ppm in parentheses; *J* values are as for first-order spectra):

H–1 (4.7) —7.4 Hz— H–2 (2.3) —10.0 Hz— H–3 (3.2) —11.8 Hz— H–4 (1.5)

H–3 (3.2) —5.5 Hz— H–5 (2.2) —13.0 Hz— H–4 (1.5)

This suggests a terminal methylene group (with shifts 1.5 and 2.2 ppm appropriate to being adjacent to a *gem*-dimethyl) which is linked to three methine groups. The coupling between hydrogens 3, 4, and 5 shows that these hydrogens must be on the same dihydronaphthopyran (**94**). Since H-3 is strongly coupled to H-2 in the other fragment (**95**), the two dihydronaphthopyrans must be joined to give the isolapachenole structure (**96**); finer details of its stereochemistry were inferred from further consideration of the shifts and *J* values (279).

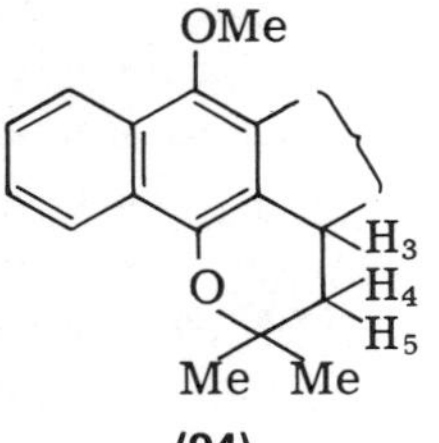

(94)

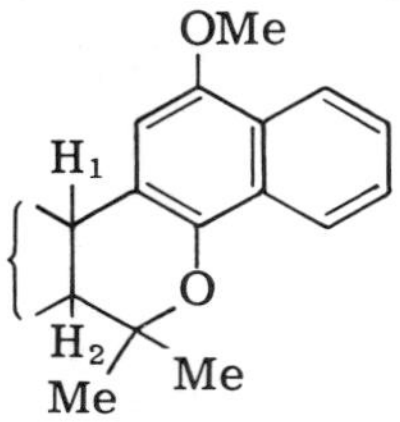

(95)

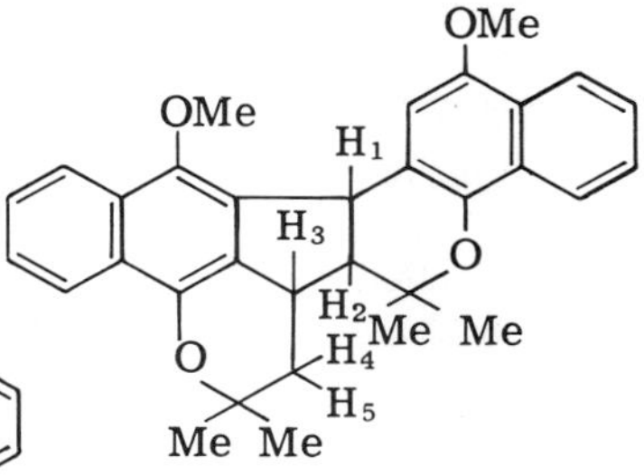

(96)

Recently, pmr and mass spectroscopic data have shown (282) that gardenin, the flavanoid pigment of the resinous exudate of *Gardenia lucida* possesses the structure **97** instead of **98** assigned earlier (283). There is one methyl group at

4.1 ppm and five at 4.0 ppm; downfield, the 12.8 ppm singlet corresponding to 5-OH is consistent with either **97** or **98**, as is the 2-proton singlet at 7.2 ppm assigned to the 2′,6′ protons of the B ring. The 1-proton singlet at 6.6 ppm is much more likely to be the H-3 in **97** than the H-7 in **98**, which should be at rather lower field. Conclusive evidence for **97** was provided by pmr identification of one of the products of treating gardenin methyl ether with ethanolic NaOH. In addition to trimethylgallic acid, the other product, 2-hydroxy-3,4,5,6-tetramethoxyacetophenone gave a spectrum with peaks assigned to one chelated hydroxyl (13.0 ppm), four methoxyl groups (3.7, 3.7, 3.9, and 3.9 ppm), and one acetyl group (2.6 ppm).

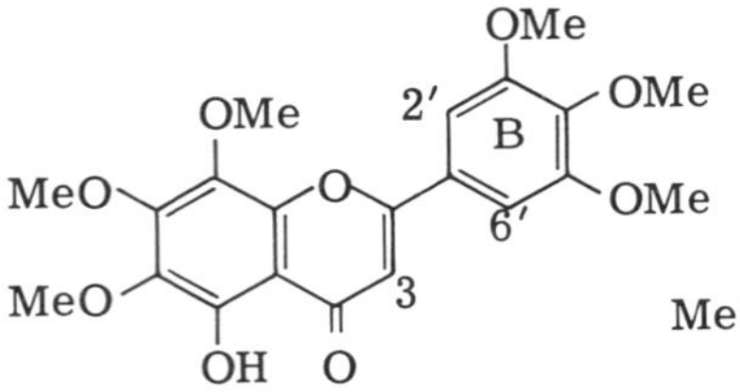

OMe OMe 2′ B OMe MeO OMe 6′ MeO 3 OH O

(97)

OMe OMe 2′ OMe 7 OMe 6′ MeO OMe OH O

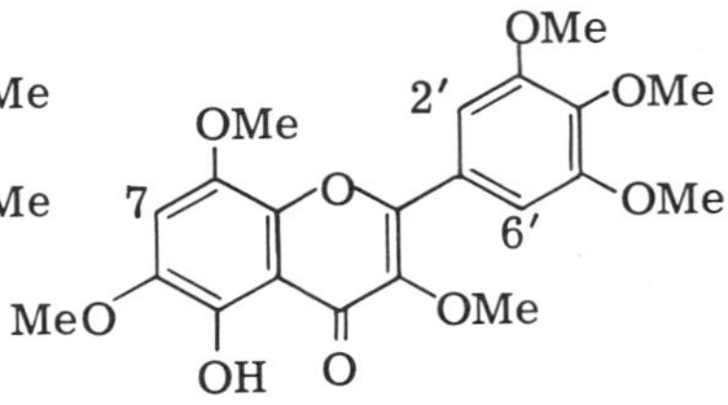

(98)

B. PROTON MAGNETIC RESONANCE IN COAL TAR AND PETROLEUM CHEMISTRY

Coal tar is a highly aromatic multicomponent system containing up to 10^4 separate species of organic molecules, neutral, basic, and acidic (284). The aromatic fractions of petroleum are also complex mixtures of derivatives and heterocyclic analogs of polynuclear aromatic hydrocarbons. Studies of the composition of both types of material, which have been reviewed [coal tar, (285-287); petroleum, (288)] tend to fall into one of two classes. In one approach, the individual components have been separated, purified, and listed [coal tar, (285-287, 289-295); petroleum, (288,296-298)]. The second approach involves the application of a physical method to the tar, pitch, or petroleum, either for estimation of the concentration of known compounds or for quantitative analysis for different structural types or functional groups. Pmr spectroscopy has been of great value in providing methods for determining the structure of the "average molecule" of various fractions separated from the tar or petroleum.

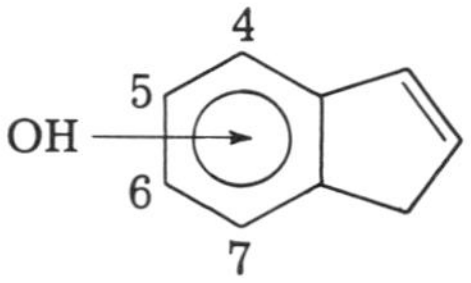

4 5 OH 6 7

(99)

A direct application of pmr to coal-tar compound identification is represented (299,300) by the confirmation of the presence of indenols **(99)**; high-resolution pmr was used to identify synthetic 4- and 7-indenols (300,301), by a comparison with the spectra of 4- and 5-indanols, and of indene and its methyl derivatives (302). Mair (148,296,303) has made extensive use of pmr to identify many polynuclear derivatives in the di- and trinuclear aromatic portions of petroleum. By combining other spectrometric data, particularly from mass spectrometry, Mair was able to identify and estimate the purity of the alkylnaphthalenes, -biphenyls, -acenaphthenes, -fluorenes, -phenanthrenes, and -pyrenes which he had isolated. For example, 1,2- and 2,3-dimethylnaphthalenes were identified, along with 2-ethylnaphthalene, and the absence of 1,8-dimethylnaphthalenes was inferred (296,303) from alkyl proton shifts (150). The spectrum of 3′-methyl-1,2-cyclopentenophenanthrene (**77**) was recognized by comparison with that of the 1′-methyl isomer (**100**), and the identity of 9,10-cyclopentenophenanthrene was deduced by examination of both aromatic aliphatic regions of the pmr spectrum. In later work, the same group used both methyl chemical shifts (150) and spin-coupling patterns in the aromatic region to identify (151) numerous alkylnaphthalenes and biphenyls from a petroleum fraction.

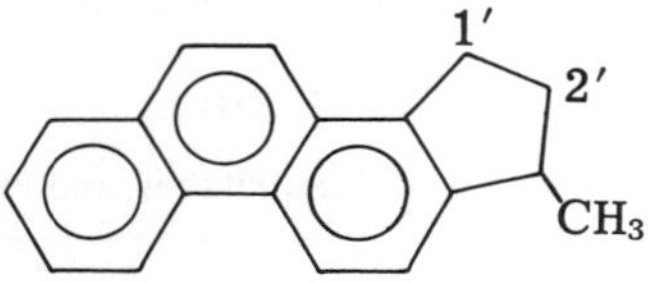

1′
2′
CH_3

(100)

The neutral fraction of tobacco-smoke condensate also contains many polynuclear aromatic hydrocarbons (304-397) but the total tumorigenic activity of the smoke condensate is incompletely accounted for by the unsubstituted carcinogenic polycyclics which have been isolated (308). While synergistic mechanisms may operate, other carcinogens, such as alkyl polynuclears, may also be present. Unfortunately, uv spectrometric methods, which give little information about alkyl substitution (309), have generally been used to infer the presence of the latter compounds. Consequently, claims of having identified the potent carcinogens 7,12-dimethylbenz[*a*]anthracene (310,311) and 3-methylcholanthrene (312) in tobacco-smoke condensates have been subject to some doubt (305,313). There would appear to be considerable scope here for the application of pmr methods to help settle this controversy. One of the very few uses of pmr in the analysis of tobacco-smoke polycyclics so far reported is the identification of 3-, 4-, and 5-methyl and 1,3-dimethylacenaphthylenes by Entwistle and Johnson (314). Pmr also played a key role in identifying 2,3,6-trimethyl-1,4-naphthoquinone (**101**), the first report of the occurrence of quinones in tobacco smoke (315). The compound showed two sets of methyl protons (δ = 2.1, 6H, CH_3–C = C–; and δ = 2.45, 3H, Ar–CH_3), as well as three aromatic protons between δ = 7.4 and 8.0 ppm.

The ability of high-resolution pmr to distinguish hydrogen atoms in different chemical environments, and to estimate their amount directly from the signal intensity, should enable it to be a considerable aid in the investigation of

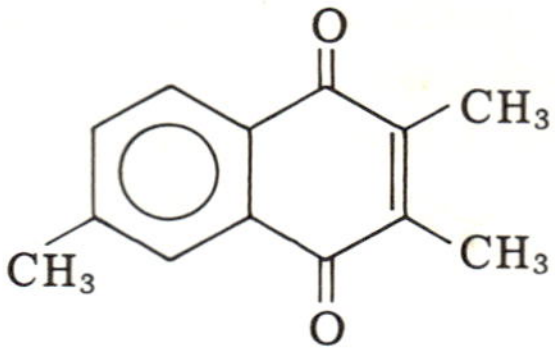

(101)

multicomponent materials such as coal tar and petroleum. Its absence of restrictions as to volatility and molecular weight offers advantages over gas chromatography and other spectroscopic methods. Furthermore, data will in general refer to *all* the material which can be dissolved in a nonproton-containing solvent, rather than to such specific compounds as happen to be separable by contemporary chemical methods or chromatography.

Hydrogen types recognized in coal tars (316-322), shock-heated (322-324) and solvent-extracted (325) coals, petroleum (326-330), and asphalt (331-337), are listed in Table XIII. By resolution of new bands (338), spectroscopy of

TABLE XIII

Hydrogen Types Present in Coal Tars, Pitches, and Petroleum Fractions

δ range, ppm[a]	Assignment in 60 MHz spectra	Symbol	Subdivision possible at 220 MHz
0.5-1.0	Hydrogen of paraffinic methyl groups and methyl groups γ or further from ring	H_γ	–
1.0-1.7	Methylene groups β or further from ring; paraffinic methylene; methyl groups β to ring; naphthenic hydrogen	H_β	Naphthenic hydrogen: 1.4-1.7
1.7-3.3	Hydrogen on carbon atoms α to ring	H_α	Hydrogen of tetralins and indanes: 1.7-2.0 Methylenes of acenaphthenes and indenes: 3.0-3.3
3.3-4.2	Methylene groups α to two rings (e.g., of fluorene)	H_F	
4.7-5.4	Olefinic		
6.2-9.9	Aromatic and phenolic	H_A	Sterically hindered aromatic hydrogen: >8.1

[a]20% solution in $CDCl_3$ or CS_2.

coal-carbonization products at 220 MHz allows further subdivision of some of these hydrogen types. In general, petroleum- and coal-derived products do not give sharply defined spectra but, in the low (< 250) molecular-weight portions of coal tars, fine structure is visible, particularly at 220 MHz, which can be attributed to individual compounds (317,320); high-field pmr offers an alternative to the more usual separation/spectroscopic methods of identifying individual components. Line broadening in the spectra of both crude oils (339) and high-molecular-weight coal-tar fractions (321) has been shown to originate from the presence of free radicals. Strong electron spin resonance signals could be detected in the coal-tar samples. Once the distributions of hydrogens between the chemical types of Table XIII have been determined, they are combined with number-average molecular weights, and with elemental, functional-group, and other analyses, in structural analysis schemes so as to arrive at the properties of the average molecule. A method derived by Brown and Ladner (324) for the analysis of coal-carbonization products has been widely applied, both in original and modified form, to coal tars and pitches (317-319,322), asphalts, and other petroleum fractions (331,333,334), despite criticism (286) of its arbitrary assumptions concerning heterocyclic structures and the nature of alkyl side-chains. From pmr and elemental analyses, Brown and Ladner (324) calculated the following average structural parameters: f_{ar}, the ratio of aromatic carbon (C_{ar}) to total carbon (C); σ, the fraction of the total available outer edge positions which is occupied by substituents; and H_{aru}/C_{ar}, the atomic H/C ratio that the average aromatic skeleton would have if each substituent were replaced by a hydrogen atom. These are derived from equations:

$$f_a = \left(\frac{C}{H} - \frac{H_\alpha^*}{x} - \frac{H_\beta^*}{y}\right)\left(\frac{C}{H}\right)^{-1} \tag{8}$$

$$\sigma = \left(\frac{H_\alpha^*}{x} + \frac{O}{H}\right)\left(\frac{H_\alpha^*}{x} + \frac{O}{H} + H_{ar}^*\right)^{-1} \tag{9}$$

$$\frac{H_{aru}}{C_{ar}} = \left(\frac{H_\alpha^*}{x} + \frac{O}{H} + H_{ar}^*\right)\left(\frac{C}{H} - \frac{H_\alpha^*}{x} - \frac{H_\beta}{y}\right)^{-1} \tag{10}$$

where C/H and O/H are atomic ratios; $H_\alpha^* = H\alpha/H$, ratio of "alpha-type" to total hydrogen, determined by pmr; $H_\beta^* = H_\beta/H$, ratio of "beta-type" hydrogen to total hydrogen, again from pmr; x and y are the atomic ratios of hydrogen to carbon in, respectively, α and β structures, both of which are assumed; and $H_{ar}^* = H_{ar}/H$ is the ratio of aromatic to total hydrogen determined by pmr, with the assumption that 60% of the total oxygen is phenolic.

An earlier scheme for the structural analysis of petroleum fractions, due to Williams (326) and Chamberlain (329), and discussed by Chamberlain (328), has been applied by Gardner *et al.* (332) to petroleums and asphalt and, in combination with the Brown and Ladner method, to other asphalt fractions by Peterson (334,340). During studies of the composition of refined tars, and the chemical changes which occur when these tars are exposed in road surfaces

(341), Bartle and Smith proposed a third method (320,321). Possible ranges of numbers of each alkyl group per average molecule are deduced, together with other rather "direct" structural parameters; corrections are made for paraffinic hydrocarbons, phenols, nitrogen bases, and nitrogen, oxygen, and sulfur heterocycles, all based on analysis.

While all these schemes lead to much useful information concerning the number and nature of aliphatic structures and about the *formula* of the aromatic skeleton, no scheme based on pmr can be specific about the *structure* of the skeleton. In particular, ring-joining carbon atoms of the types in biphenyl (**102**) and naphthalene (**103**) cannot be distinguished from the proton chemical shifts.

(102) (103)

^{13}C nmr spectroscopy may be complementary here, provided that integration problems associated with long relaxation times, magnetization transfer, and weak signals can be overcome (342). The ^{13}C spectra of coal and coal-tar materials have been measured and, for example, the aromaticity of the small fraction of bituminous coal which is soluble in CS_2 has been evaluated directly as 0.6 ± 0.1, even without the use of a CAT (computer of average transients) (331,343,344). Knight (345) has applied a structural analysis scheme based on ^{13}C nmr to petroleum fractions. The newly developed technique of noise decoupling, in which ^{13}C spectra are measured while all proton interactions are removed by heteronuclear spin decoupling, may allow further development in this field. A large increase in sensitivity occurs because of the collapse of multiplets into singlets and also as a result of the large positive Overhauser effect, although the variability of the latter may prevent precise integration.

C. CAVITY SUBSTITUENTS IN POLYCYCLIC MOLECULES

Provided that sufficient space is available, Hückel molecular-orbital theory predicts the feasibility of sufficiently large aromatic rings carrying substituents located within the empty cavity of the π-electron cloud. Pmr has proved of immense diagnostic value in studies by Boekelheide and co-workers of the synthesis and chemistry of such molecules (346-349). Thus laborious synthesis (350) yielded **104**, which was treated with zinc dust and acetic anhydride. Examination of the pmr spectrum of the product (**105**) showed immediately that the aim of locating substituents between the π doughnuts had been achieved. The resonance of the methyl groups at $\delta = -4.03$ ppm was 5 ppm higher than in **104**, indicative of a strong aromatic ring around the $(4n + 2)$ π-electron periphery of the molecule. Further indications of the system are provided by (*1*) measurement of the large perpendicular magnetic-susceptibility anisotropy of *trans*-15,16-dimethyldihydropyrene (351); and (*2*) diffraction measurements (352) showing ring planarity and carbon-carbon bond lengths of 1.386-1.401 Å in 2, 7-diacetoxy-*trans*-15,16-dimethyldihydropyrene, compared with 1.397 Å in benzene (353).

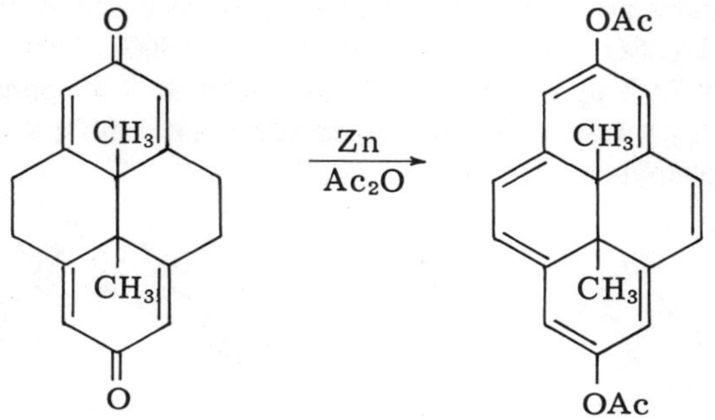

(104) $\xrightarrow[Ac_2O]{Zn}$ (105)

The structure of the parent compound, *trans*-15,16-dimethyldihydropyrene (**106**), was also confirmed by its pmr spectrum (350): a 6-proton methyl singlet at $\delta = -4.25$ ppm, the H-4,5 singlet at 8.67 ppm, and the $\mathbf{AB_2}$ system of H-1,3 between $\delta = 8.67$ and 7.98 ppm ($J = 7.5$ Hz). Changes in the aromatic-proton pattern allowed identification of the products of electrophilic substitution as the 2- and 2,7 derivatives (354); ethyl group shifts were conclusive for *trans*-15,16-diethyldihydropyrene (355), in which $\delta_{CH_3} = -1.85$, and $\delta_{CH_2} = -3.96$ ppm. Chemical shifts show that thermal treatment of **106** gives the rearrangement product (**107**); in its spectrum, methyl groups resonate at $\delta = 1.37$ and 1.13 ppm and there are three aromatic and seven olefinic protons (356). The shift of CH_3-2 may be compared with the 1.10 shift of CH_3 in the model compound **108**.

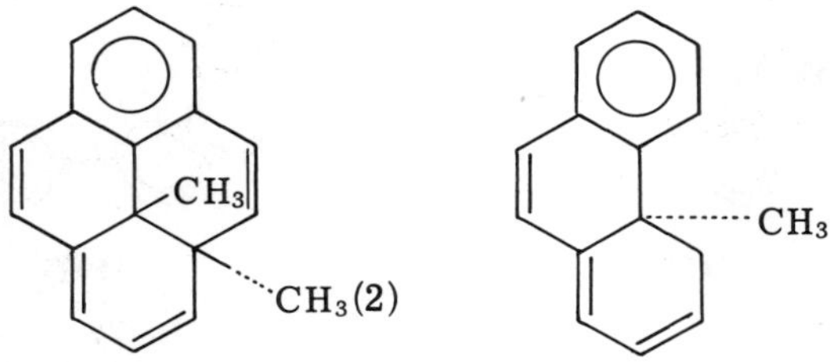

(107) (108)

trans-15,16-Dihydropyrenes undergo valence tautomerism (348,357) about which pmr provides key information (358).

It is convenient to mention here the use by Jenny (359-361) of pmr, usually in conjunction with ir, uv, and mass spectroscopy, to establish the probable conformation of some other novel macrocyclic systems, the [2.2. . .] *meta*cyclophanes. These were synthesized *via* a modified Wurtz reaction, and are typified by [2.2.2.2.2.2.] *meta*cyclophane (**109**), for which the *a, b,* and *c* protons have shifts of 7.2, 7.0, and 6.9 ppm. In [2.2.2.2.2.2.] *meta*cyclophane–all *trans*-hexaene (**110**)–stereospecifically prepared from [2.2.2.2.2.2.] *meta*cyclophane, the AB_2C aromatic protons in the pmr spectrum, recorded at 100 MHz in dilute solution in $(CD_3)_2SO$ with the help of a CAT (360), have shifts $\delta_a = 7.39$, $\delta_b = 7.54$, $\delta_c = 7.92$ ppm (with the vinyl singlet at 7.40 ppm) and couplings $J_{ab} = 7.7$ Hz, $J_{bc} = 1.2$ Hz. The sequence of aromatic shifts is analogous to that in the hexa-*m*-phenylenes (362).

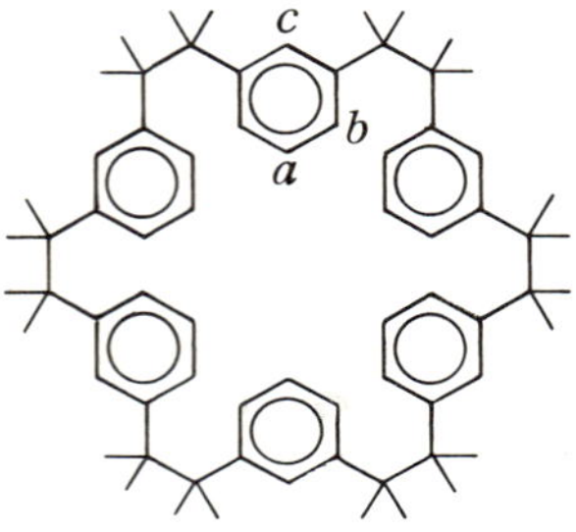

(109)

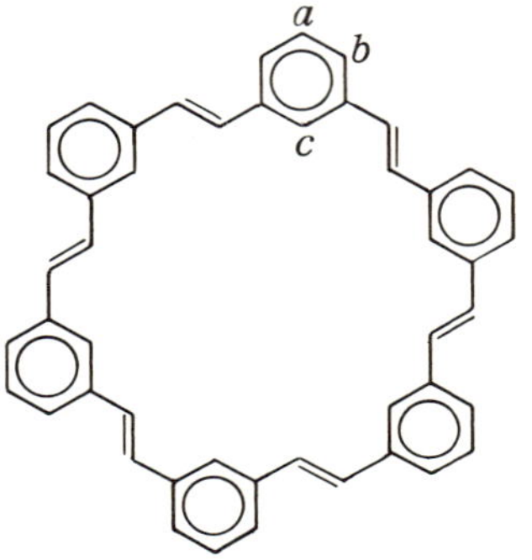

(110)

Other novel cyclic systems to which pmr has been applied include di(benzo[*c*]phenanthrene-3,6-dimethylene) (**111**) (363,364); 3,6′,3′,6″,3″,6-triphenanthrylene (**112**) (365), and [7](2.6)dithia*meta*cyclophane, which in-

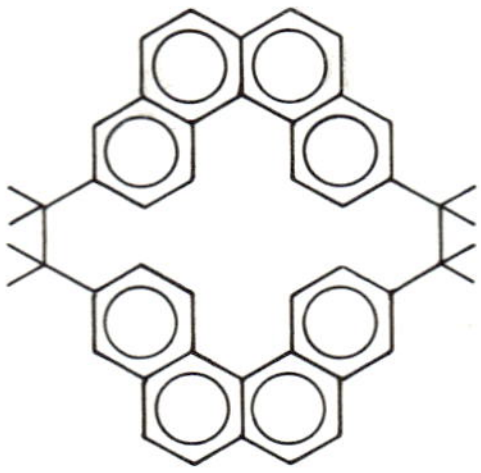

(111)

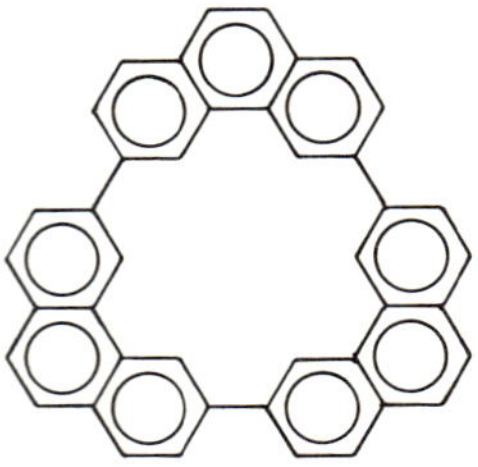

(112)

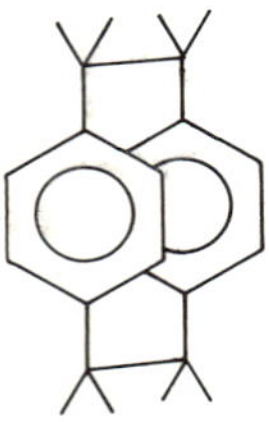

(113)

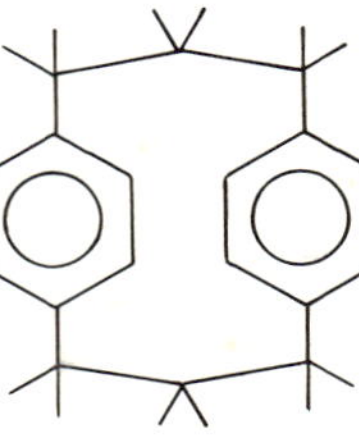

(114)

volves a 10-membered ring with a substituent in the π cloud (366). Pmr has also made a substantial contribution (367,368) to structure determination in the chemistry of [2.2]*para*cyclophane (**113**) (367,369), and [3.3]*para*cyclophane (**114**) (368,370). Tables of SSV for substituent groups in **113** (367) and **114** (368) have appeared and have been compared with values in benzene (368).

IV. Determination of Conformation in Polynuclear Hydrocarbons and Their Derivatives

A. *BIPHENYL AND ITS ANALOGS*

Beharul *et al.* (371) have used pmr both to help establish the structures of the chlorination products of *o*- and *p*-terphenyls (**115,116**) as 2- and 4-monochloro and 2,4″- and 4,4″-dichloro derivatives, and also to determine the extent of catalytic reduction of terphenyl (372). Similarly, by combining pmr with uv and mass spectrometry, Puskas and Fields (373) showed that nitration of mixtures of *o*-xylene or hemimellitene (1,2,3-trimethylbenzene) with more basic methylbenzenes gave nitropolymethylbiphenyls formed to a considerable extent by cross-coupling. However, as well as helping to establish the structure of products, pmr can provide useful information about the detailed conformation of biphenyl and similar compounds.

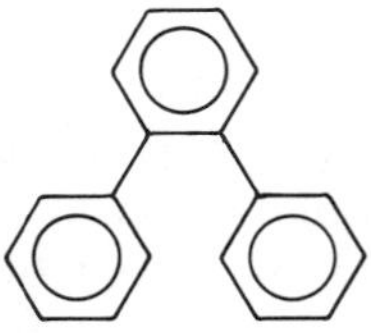

(115)

(116)

The structure of biphenyl has been the subject of numerous experimental and theoretical investigations. In the solid state, the coplanarity of the phenyl rings has long been established (374-379). Robertson (380) confirmed this in a three-dimensional X-ray analysis and attributed a number of ring deformations to intermolecular forces. For the vapor, electron-diffraction measurements (381-383) suggest a different, nonplanar structure, with 42° as the most recent measurement (383) of the dihedral angle. In solution, however, the conformation of biphenyl is not well established (384). From the Kerr constant, Chau *et al.* (385) argued that the rings were coplanar; it is noteworthy that very recent Kerr constant measurements (386) by the same group on the analogous 2-phenylnaphthalene (**117**) in CCl_4 solution lead to an angle of about 20° between the planes of the phenyl and naphthyl systems. The electronic spectrum of this compound is consistent with a dihedral angle of 30° (387).

Ir spectra (388,389) of biphenyl have been interpreted in terms of a nonplanar structure in solution. Electronic spectra (390) suggest some degree of interring conjugation but, on the basis of this and the spectra of a number of

(117)

derivatives, Suzuki (391) inferred that the rings were inclined at an angle of 20°. Although Hoffman *et al.* (392) were first to point out that the anisotropy of magnetic susceptibility of the phenyl rings should lead to a "ring-current" chemical shift between the *ortho* and *meta* protons in biphenyl, of magnitude related to the dihedral angle between the ring planes, they were unable to extract this shift from the complex five-spin spectrum by the spectral-moment (393) method. Kurland and Wise (394) managed to determine (δ_2–δ_3) from the deuterium-decoupled AA′BB′ spectra of biphenyl-4,4′-d_2 in a number of solvents, and found these values were in excellent agreement with the ring-current shift calculated from the Johnson and Bovey (35) tables for a dihedral angle of 45° (381). Similar agreement was obtained for 4,4′-difluoro, dichloro, dibromo, diiodo, and dinitro derivatives.

Bovey *et al.* (205) reported an analysis of the pmr spectrum of biphenyl based on use of the 60 MHz spectrum in the refinement of parameters obtained at 220 MHz. The approach of Mayo and Goldstein (132) was to obtain initial chemical-shift values from four symmetrically deuterated biphenyls, and then to analyze iteratively the 60 MHz spectrum of the parent; inter-ring coupling was excluded since the spectrum of 2,3,4,5,6-pentadeuterobiphenyl was almost identical with that of the unsubstituted compound. Mayo and Goldstein also calculated the anisotropy effect, $\Delta\sigma$ Hz at 60 MHz, with respect to benzene, of the adjacent phenyl group on the protons of biphenyl. This was evaluated, at 10° intervals over the dihedral-angle range 0-90° (Figure 3), from the magnetic potential (395) due to two current loops (396) located 0.7 Å above and below the ring, respectively (397):

$$\Delta\sigma = \frac{-1848}{R^3}\left[(1 - 3\cos^2\theta) + \frac{0.735}{R^2}(35\cos^4\theta - 30\cos^2\theta + 3)\right] \quad (11)$$

where R is the distance from the center of the loop to the proton, and θ is the angle between R and the vector normal to the plane of the current loop. Agreement of the average values for *ortho* and *meta* protons (−19.1 and −7.1 Hz) with observed shifts with respect to benzene (−14.2 and −3.6 Hz) was sufficient to justify the conclusion that there is free rotation in biphenyl; almost the same results were obtained for a dihedral angle of 42° (383). Agreement is even better (−11.5 and −4.3 Hz) if the calculated values are multiplied by a factor of 0.6 (41); this is equivalent to taking the ring-current contribution to the magnetic anisotropy of benzene calculated by MO methods (41) as

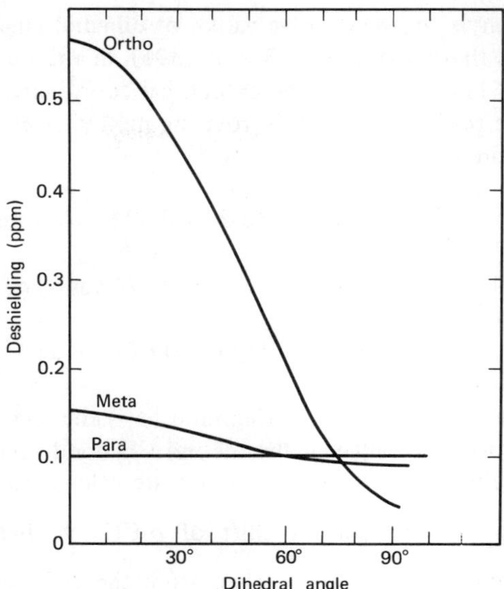

Figure 3 Diamagnetic anisotropic deshielding *versus* dihedral angle for protons of biphenyl (after 132).

-30×10^{-12} m^3 mol^{-1}. This is discussed more fully elsewhere (398). Similar conclusions were reached for *p*-terphenyl (132). Several 2- and 4-halogenobiphenyls have been studied by pmr (399); some resonances in the spectra of the 2-halogeno derivatives are shifted to high-field of those of biphenyl itself, and it seems that steric interaction between the halogen atom and *o*-hydrogens leads to an increase in the dihedral angle between the ring planes and shielding of some protons.

Similar approaches have been made to explain the methyl shifts of methyl biphenyls, For the conformationally mobile (*vide infra*) bridged biphenyls (**118**: X absent; X = O; N–CH_3; CH_2; CO) and 2,2′,6,6′-tetramethylbiphenyl, Mislow

X
CH_2 CH_2
CH_3 CH_3

(118)

et al. (400) plotted δ_{CH_3} against calculated torsional angle; their semi-empirical points agreed rather well with a curve calculated fairly crudely from the Johnson and Bovey (35) tables. Application of this curve to 2- and 4-methyl-, and

2,2′-dimethylbiphenyls, however, gave values of dihedral angles which diverged widely (400) from these deduced by Suzuki (391). In a more refined treatment, Koptyug *et al.* (161) showed that the cylindrical coordinates (Figure 4) $z(MK)$ and $\rho(OK)$ for the position of a freely rotating methyl proton, with respect to the opposite ring, are given by:

$$z = 0.747 \sin \alpha \cos \phi + (2.23 + 0.374 \cos \alpha) \sin \phi \tag{12}$$

$$\rho = \{[0.747 \sin \alpha \sin \phi - (2.03 + 0.374 \cos \alpha) \cos \phi]^2 + [1.90 + 0.647 \cos \alpha]\}^{1/2} \tag{13}$$

where all distances are expressed in ring radii, ϕ is the angle between the ring planes, and α the rotation angle of the methyl CH bond (angle *NRM* in Figure 4). For a given value of ϕ, values of z and ρ are calculated for 36 values of α from 0 to 350°. The ring-current shift of o-CH_3 is then $1/36 \sum_0^{350} \delta_{z,\rho_i}$, for each ϕ, where δ_{z,ρ_i} is the shift taken from the Johnson and Bovey (35) tables. While agreement was good (Table XIV) between observed shifts and those calculated from ϕ values deduced by Suzuki (391), its significance is unclear in view of the known (38,41) overestimation of chemical shifts in aromatic compounds when the Johnson and Bovey (35) tables are used. Yew and Mair (151) used the shielding of methyl protons in 2-methylbiphenyls to help identify alkylbiphenyls from petroleum.

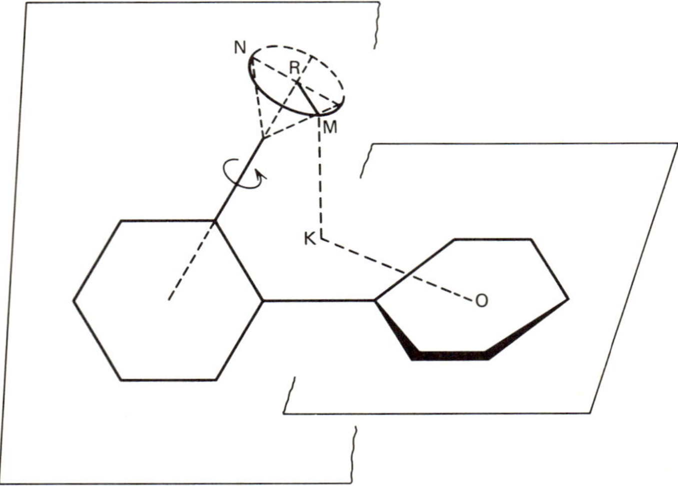

Figure 4 Model of 2-methylbiphenyl molecule for calculation of ring-current effect (after 161).

TABLE XIV

Chemical Shifts (δ ppm) in Methylbiphenyls (161)

Location of CH_3 groups	Chemical shifts m- and p-CH_3	o-CH_3	ϕ[based on Suzuki (391)]	Ring–current shift of o-CH_3 Observed	Calculated
2		2.18	58°	0.12[a]	0.19
2,2′		2.00	70°	0.30[a]	0.31
2,2′,3,3′,4,4′,5,5′	2.23	1.92	70°	0.31[b]	0.31
2,2′,4,4′,6,6′	2.31	1.87	90°	0.44[b]	0.40
2,3,4,5,6,2′,3′,4′,5′,6′	2.23	1.78	90°	0.45[b]	0.40

[a]Relative to CH_3 of 4-methylbiphenyl.

[b]Relative to m-CH_3 and p-CH_3 in same molecule.

The conformations of a number of polyphenyls and related compounds other than biphenyl have been investigated by pmr. Hoffman *et al.* (392) measured chemical shifts directly from the surprisingly simple 40 MHz spectrum of *o*-terphenyl. From calculations based both on the dipole approximation (401) and the corrected (35) model of Waugh and Fessenden (396), they concluded that the δ values were consistent with a configuration in which the lateral phenyl groups are perpendicular to the central ring; deviations of more than 10-15° from 90° were thought unlikely.

In 1,8-diphenylnaphthalene (402) the phenyl protons resonate at δ = 6.85 ppm, and so are all similarly shifted to high field of 1-phenyl- (δ = 7.38 ppm) and 1,7-diphenylnaphthalenes (7.49 ppm); the most likely explanation is mutual shielding of the phenyl protons because of a parallel orientation of the phenyl rings similar to that suggested from uv evidence for 5,6,11,12-tetraphenylnaphthacene (403). For janusene (**119**), in which two of the benzene rings are parallel and about 2.5 Å apart, aromatic signals are found at fields as high as 6.35 ppm (404).

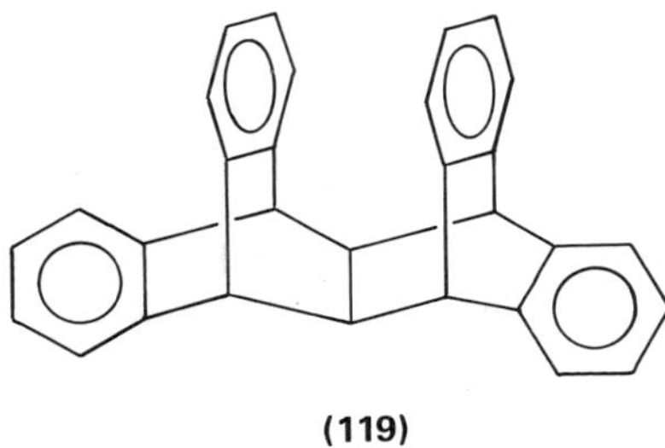

(119)

In the pmr spectrum (405) of 1,4,5,8-tetraphenylnaphthalene (**120**), there are signals at δ = 7.20 (singlet, intensity four protons), assigned to H-2,3,6,7 of the naphthalene system; δ = 7.00 ppm, (triplet, 4H), from the *para* protons of

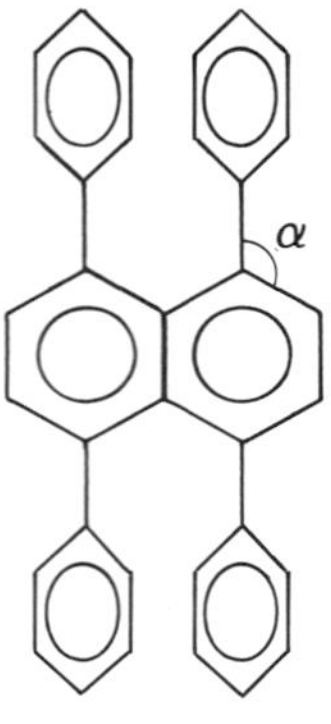

(120)

phenyl rings; and $\delta = 6.73$ ppm (doublet, 16H), from the *ortho* and *meta* protons of phenyls, shifted upfield by interaction between parallel π systems. Rabinovitz *et al.* (405) explained this nonequivalance of the phenyl protons by postulating a strained arrangement (α less than 90°) for the rings. Similar observations (406) have been made for 1,9-diphenylanthracene, 1,9,10-triphenylanthracene, and 1,4,9,10-tetraphenylanthracene (407) which show, respectively, the following signals for phenyl: $\delta = 6.92$ and 6.90 ppm; 7.19 and 6.91 ppm; and 6.98 and 6.90 ppm.

Regan and Miller (408) examined the pmr spectra of a number of 1,4-dialkyl-2,3,4,6-tetraphenylbenzenes (**121**) and 1,4-dialkyl-2,3,9-triphenylanthracenes (**122**), in which through-bond effects were considered negligible, so as to determine the ring-current shielding of alkyl protons held near the axis perpendicular to the plane of the phenyl group. The results showed that shieldings estimated from calculations of the Johnson and Bovey (35) type alone generally at least account for, and, in some cases, greatly exceed, the experimental values.

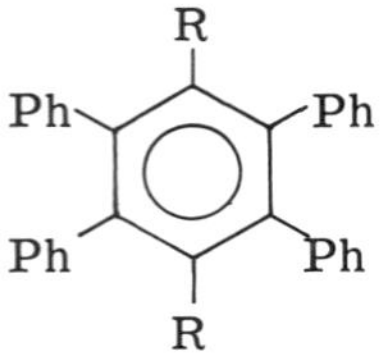

(121)

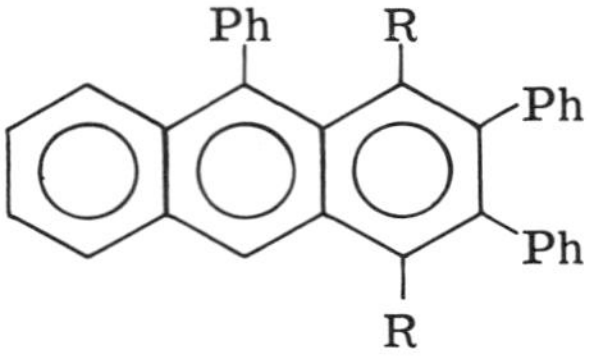

(122)

From a comparison with δ_{CH_3} in 9,10-dimethylanthracene, Dickerman and Haase (409) deduced a value of 1.31 ppm for the ring-current shielding of CH_3 in 1,4,5,8-tetraphenyl-9,10-dimethylanthracene. This also is much less than the shift (4.7 ppm) calculated from the Johnson and Bovey tables. Other cases in which the shielding is less than that calculated are 1-phenyl-9-methyl-, 1,4-diphenyl-9-methyl-, 1-phenyl-9,10-dimethyl-, 9-α-naphthyl-1,4-dimethyl-, and 9-β-naphthyl-1,4-dimethylanthracene (410).

A number of calculations have been made of the rotation barrier in biphenyls; for example, potential energy curves giving shallow minima [1.7 kJ mol^{-1} (411); 12.5 kJ mol^{-1} (412)] near the 40° dihedral angle deduced from electron-diffraction results, have been calculated. Howlett (413) adopted the Westheimer-Mayer model (414,415) for the calculation of the minimum-energy configuration of an overcrowded molecule, and calculated the barrier to rotation in biphenyl as 16 kJ mol^{-1}. Golebriewski and Parczewski (416) developed a procedure for calculating the most stable conformation of biphenyl by minimizing the combined delocalization and steric-repulsion energies.

Pmr methods have been invoked in attempts to confirm these theoretical approaches. From studies of the variations with temperature of $(\delta_2 - \delta_3)$ in biphenyl-4,4′-d_2 in a number of solvents, Kurland and Wise (394) deduced values of between 1.7 and 4.6 kJ mol^{-1} for the barrier to rotation in biphenyl. Mislow observed (400,417,418) that *o*-methylene protons in several bridged biphenyls are nonequivalent since the methylene groups are part of a ring and not free to rotate. For substituted biphenyls such as **123**, Meyer and Meyer (419) pointed out that the *ortho* substituent on one ring is located above the plane of the second; if the second ring holds two different *ortho* groups, then the substituent

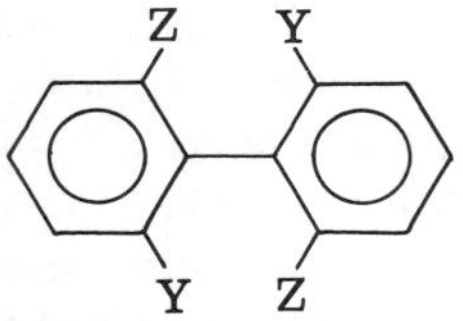

Z Y

Y Z

(123a) Y = CH_2OH; Z = C_2H_5
(123b) Y = CH_2OH; Z = CH_3
(123c) Y = CH_2OCOCH_3; Z = H

on the first is in a chiral environment. Consequently, if a methylene group forms part of a substituent, its two protons will not necessarily be magnetically equivalent; in **123a** the CH_2OH signals appear as an AB quartet with $\delta_A = 4.2$, $\delta_B = 4.05$ ppm, $J_{AB} = 12$ Hz (419) while in **123b** $\delta_A = 4.15$, $\delta_B = 3.93$ ppm, $J_{AB} = 11.6$ Hz (420). Even though the barriers to conformational change in such compounds are much lower than the 84 kJ mol^{-1} necessary (415) for the study (421) of restricted rotation in *ortho* substituted biphenyls at ordinary temperatures by following the racemization of optically active compounds, pmr techniques may still be applicable (422). Meyer and Meyer (419) used the resonances of the methylene groups to determine an Arrhenius activation energy, E_a, for rotation in **123c** of about 54 kJ mol^{-1}. At 400°K, the AB pattern ($\delta_A - \delta_B = 3.5$ Hz, $J_{AB} = 12.6$ Hz at room temperature) collapses to a single line.

At intermediate temperatures, the outer AB lines are broadened, and allow (423) determination of E_a and the average lifetime of the molecule between rotations. Colebrook and Jahnke (424) determined E_a and other kinetic data for the rotation of the dimethyl (**124a**) and diacetyl (**124b**) derivatives of podototarin,

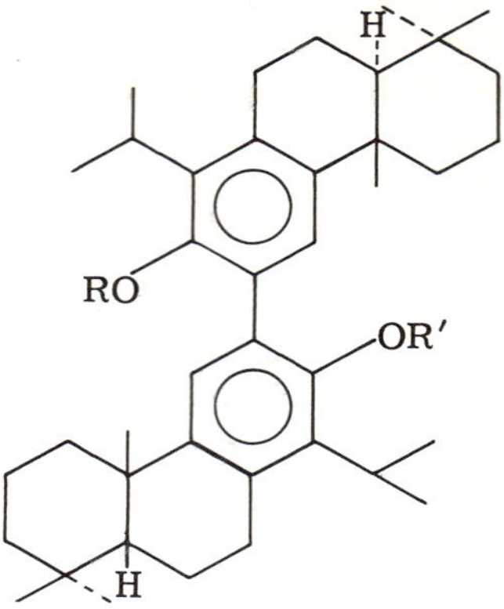

(124a) $R = R' = CH_3$
(124b) $R = R' = Ac$

a naturally occurring biphenyl obtained from the species *Podocarpus totara* G. Benn (425,426). It had been estimated previously, from uv spectra and pmr chemical shifts at room temperature (425,426), that podototarin has a preferred dihedral angle of about 70°. Below 268°K, **124a** shows two superimposed spectra with two peaks of unequal intensity arising from OCH_3 signals centered near $\delta = 3.28$ ppm. Structure **124b** has a similar spectrum (427) in which the two acetyl methyl groups of the two conformers give rise to two peaks at normal temperature. Application to the methyl signals of a least-squares analysis version (428) of the general line-shape methods, developed by Gutowsky and Holm (429) for the exchange between two unequally populated sites, yielded $E_a = 28 \pm 2$ kJ mol^{-1} for **124a** and 45 ± 5 kJ mol^{-1} for **124b**.

At temperatures below 293°K, House and Baske (430) found that the dimethyl ester (**125**) gives separate methyl signals. Evidently, equilibration between different conformation is slowed down, although other explanations are possible

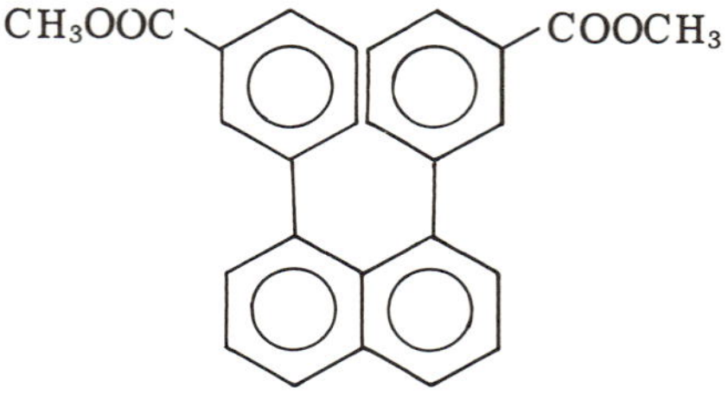

CH_3OOC $COOCH_3$

(125)

than those involving rotation of the phenylnaphthalene bonds. An example of magnetic nonequivalence as a result of hindered rotation in a biphenyl-type hydrocarbon is provided by the distortion of the α-proton triplet in 1,4-di-*n*-

propyl-2,3,5,6-tetraphenylbenzene (**121**; R = $CH_2CH_2CH_3$). Regan and Miller (408) regarded this as analogous to the nonequivalence of the similar protons in hexa-*n*-propylbenzene (431).

The pmr spectra of 3H indole derivatives with CX_2Y groups on C-3, for example, **126**, are consistent with magnetic nonequivalence of *gem*-X-groups. (432). For ethyl and benzyl derivatives, an equilibrium between two energetically similar rotamers is assumed, but for isopropyl one rotamer is strongly favored at room temperature.

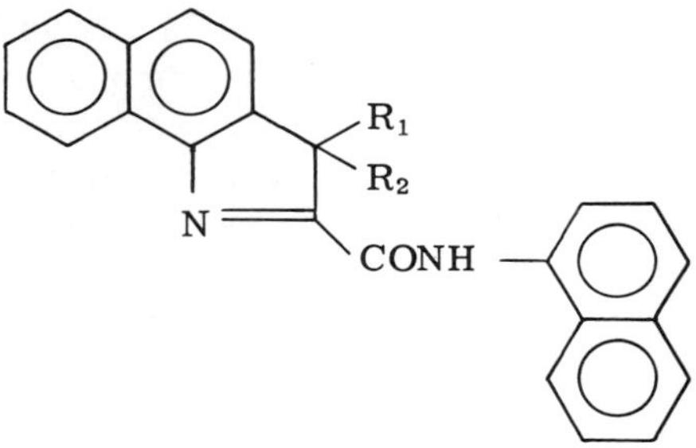

R_1
R_2
N
CONH

(126)

B. MAGNETIC NONEQUIVALENCE IN FUSED POLYNUCLEAR COMPOUNDS

Examples under this heading are rather less numerous than in the biphenyl series. Reisse *et al.* (433) reported that the chemical shifts of $(CH_3)_2CH$ differed by as much as 1.45 ppm in the 9-isopropylbenz[*a*]fluorenes (**127**). As discussed more fully below, the isopropyl group of the diester (**128**) of phenanthrene-4,5-dicarboxylic acid forms an AX_3Y_3 system at 230°K, which changes to AX_6 at 293°K (434).

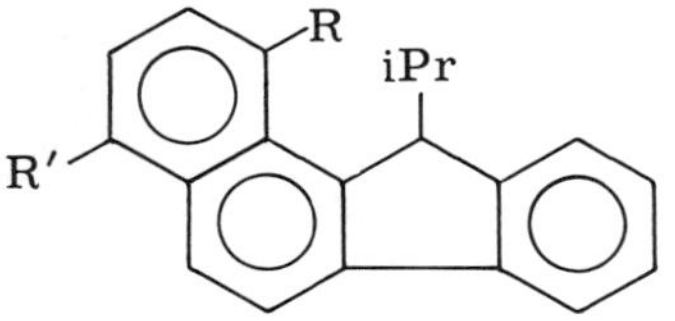

R
iPr
R′

(127) R = H or CH_3
R′= H or CH_3

While a recent refinement of the crystal structure of phenanthrene suggests only very small deviations from planarity (435), the more hindered 4,5 disubstituted derivatives are expected (436) to be substantially nonplanar and to be optically active, as are the helicenes (166-168,437), and derivatives of benzo[*c*]phenanthrene (438,439) and dibenzo[*c,g*]phenanthrene (440,441). Newman and Hussey (442) resolved (optically active) 4,5,8-trimethylphenanthrene-l-acetic acid (**129**), which racemized rapidly in solution at room temperature. Munday and Sutherland (434) have determined activation parameters for

$COOCH(CH_3)_2$
$COOCH_3$

(128)

CH_3 CH_2COOH
CH_3 CH_3

(129)

conformational changes in the 4,5-disubstituted phenanthrenes **(130a-e)** from the temperature dependence (Table XV) of their pmr spectra. In some of the compounds, exchange rates were determined at a number of temperatures, either from the spectrum line shape [by comparison with computed shapes (422,443)] or from the broadening of lines from nonequivalent methylene protons (444). The results (Table XV) were interpreted in terms of inversion of a nonplanar conformation. Relative free-energy barriers to conformational inversion were comparable with values obtained by polarimetry on similar compounds.

TABLE XV

Pmr Data with Coalescence and Activation Parameters for Conformational Changes in 4,5-Disubstituted Phenanthrenes (434)

Compound	Solvent	Temp, °K	Spin system	δ, ppm		J, Hz	Coalescence temperature, °K	ΔG‡, kJ mol^{-1}	E_a, kJ mol^{-1}
130a	Pyridine	253	ABX_3	4.68[a]	4.48[b]	10.4[d]	290 ± 2	64 ± 1 (290)	
				1.07[c]		7.0[e,f]			
		309	A_2X_3	4.47[a]	1.07[c]	7.0			
130b	Pyridine	244	AB	5.83[a]	5.62[b]	12.1	289 ± 2	62 (273)	65 ± 5
		358	A_2	5.59[a]					
130c	Pyridine	256	AX_3Y_3	5.59[a]	1.27[c]	6.0[e,g]	269 ± 2	62 ± 1 (269)	
		327	AX_6	5.47[a]	1.20[c]	6.0			
130d	$CDCl_3$	230	AX_3Y_3	5.22[a]	1.29[c]	6.0[e,g]	275 ± 3	61 (273)	74 ± 6
		293	AX_6	5.15[a]	1.19[c]	6.0			
130e	$C_6D_5NO_2$	308	AB	5.64[a]	5.31[b]	12.4	352 ± 2	76 (273)	77 ± 6
		393	A_2	5.48[a]					

[a] δ_A.
[b] δ_B.
[c] δ_X.
[d] J_{AB}.
[e] J_{AX}.
[f] J_{BX}.
[g] J_{AY}.

R_1 R_2

	R_1	R_2
(130a)	R_1 = COOH;	R_2 = $COOCH_2CH_3$
(130b)	R_1 = COOH;	R_2 = $COOCH_2Ph$
(130c)	R_1 = COOH;	R_2 = $COOCH(CH_3)_2$
(130d)	R_1 = $COOCH_3$;	R_2 = $COOCH(CH_3)_2$
(130e)	R_1 = CH_2OAc;	R_2 = CH_2OAc

C. CONFORMATION OF HYDROAROMATIC POLYNUCLEAR COMPOUNDS

Hydro derivatives of polynuclear aromatic hydrocarbons provide examples of conformational interchange. For example, the allylic and benzylic protons in 1,2-dihydronaphthalenes **(131)** may assume "pseudo-axial" and "pseudo-equatorial" conformations **(132)**. Inversion between conformers **132a** and **132b** must be rapid on the pmr time scale since cooling of **131** to 218°K did not freeze out the conformers (445). Similar observations have been made on 9,10-dihydrophenanthrene (116) for which the ^{13}C satellite pmr spectrum has also been studied (446). Cook *et al.* analyzed the spectra of a number of substituted 1,2-dihydronaphthalenes and related model compounds by an iterative computer method and the use of decoupling (445). From coupling constants obtained in this way it was shown that substituents in the 1 position exist preferentially "pseudo-axial."

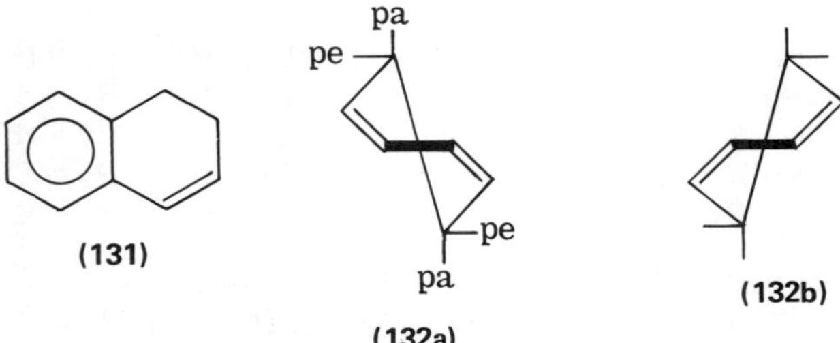

(131) **(132a)** **(132b)**

4,5-Dimethyl-9,10-dihydrophenanthrene **(133)** has been resolved (447), and pmr suggests (400) that the methyl groups are held out of the planes of the rings; the methylene protons give an A_2B_2 pattern in $CDCl_3$ solution at 100 MHz. The activation energy of racemization (447) and some conformational kinetic isotope effects have been measured by polarimetry (448). On the other hand, derivatives of 9,10-dihydroanthracene undergo conformational interchange to which pmr has been applied. Curtin *et al.* (449) found that the degree of

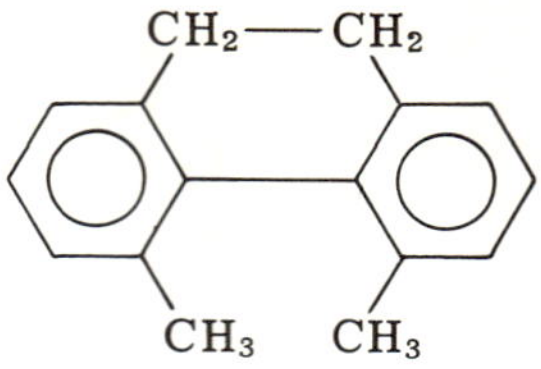

CH_2—CH_2

CH_3 CH_3

(133)

equivalence of the methyl protons in the 10,10-dimethyl-9-methylene-9,10-dihydroanthracenes **(134a-c)** varied with the substituents at the methylene carbon. Presumably there is a hindered boat-boat interconversion of the central ring, in which the methyls occupy axial and equatorial positions; investigation by the Gutowsky and Holm (429) method led to the conclusion that there was a similar inhibition to inversion in these compounds to that found (450) in the hindered rotation of biphenyls. Holubec and Jonas (451) reinvestigated the kinetics of ring inversions in **134b**, **d**, and **e** by applying the more rigorous complete line-shape analysis method (452,453) to the methyl group protons. $\Delta G\ddagger$ was found to be 80 kJ mol^{-1} in **134b**, and, in **134d** and **134e**, 66 and 73 kJ mol^{-1}, respectively. These differences were interpreted in terms of nonbonded interactions between the methylene substituents and the *peri* substituents, R_1 and R_4.

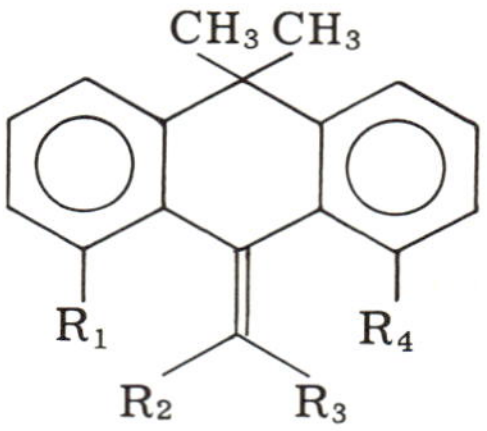

CH_3 CH_3

R_1 R_4

R_2 R_3

(134a)	R_1 = H;	R_2 = COOMe;	R_3 = Br;	R_4 = H
(134b)	R_1 = H;	R_2 = Br;	R_3 = Br;	R_4 = H
(134c)	R_1 = H;	R_2 = Ph;	R_3 = Br;	R_4 = H
(134d)	R_1 = Cl;	R_2 = H;	R_3 = H;	R_4 = Cl
(134e)	R_1 = H;	R_2 = Cl;	R_3 = Cl;	R_4 = H

The pmr spectrum (454) of 9-*t*-butyl-9,10-dihydroanthracene **(135)** shows that the molecule has fixed geometry at 306°K; H-10 and H-10′ form an AB system, with J_{AB} = 19 Hz and δ_A = 4.23 and δ_B = 3.79 ppm, and H-9′ is a singlet at δ = 3.72 ppm. The "normal" (0.89) value for the *t*-butyl protons implies that the substituent has the "pseudo-equatorial" (e') conformation, since models suggest a significant low-field ring-current shift for the "pseudo-axial" arrangement (a'). This assignment is consistent with resonance positions of δ = 3.72 and 3.79 for the benzylic a' protons, and a downfield shift to δ = 4.23 for the e' proton, in line with the general finding (455) in cyclohexane derivatives. Two possibilities are suggested by the pmr spectrum of 9,10-di-*t*-

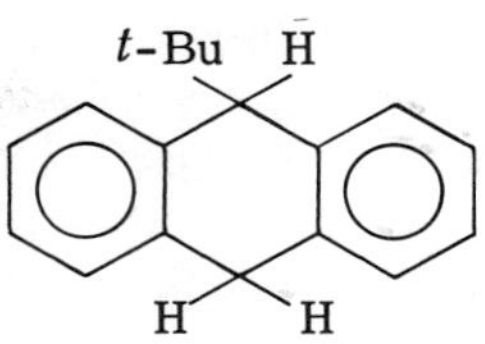

t-Bu H

H H

(135)

butyl-9,10-dihydroanthracene (454): either the central ring is stabilized in a planar conformation, as has been shown (456) by X-ray crystallography for solid 9,9′,10,10′-tetrachloro-9,10-dihydroanthracene; or, even at 223°K, there is a rapid interconversion of the two equivalent *trans* forms, $a' - e' \leftrightharpoons e' - a'$. As in *trans*-9,10-dimethyl-9,10-dihydroanthracene (457), the benzylic resonance is at $\delta = 3.94$ ppm; this is almost exactly the mean of the a' and e' shifts in the mono-*t*-butyl derivative (**135**).

Pmr shows, however, that the dimethylene compound **136** has a stable "boat" geometry, since the CH_2 resonance is constant in width and shift (458) between 253 and 473°K. The parent of **134-136**, 9,10-dihydroanthracene, shows (176,459) a sharp methylene singlet in CS_2 solution down to 213°K. Since the "boat" structure of the central ring suggests nonequivalence of the CH_2 protons, "boat-boat" interconversion must be rapid on the pmr time scale, even at this temperature. Substitution of an ethyl group (459) either quenches this motion or greatly favors one stereoisomer, even at 313°K; the 10-protons form an AB system ($J = 18$ Hz) in 9-ethyl-9,10-dihydroanthracene.

Ph CH_2 Ph

Ph CH_2 Ph

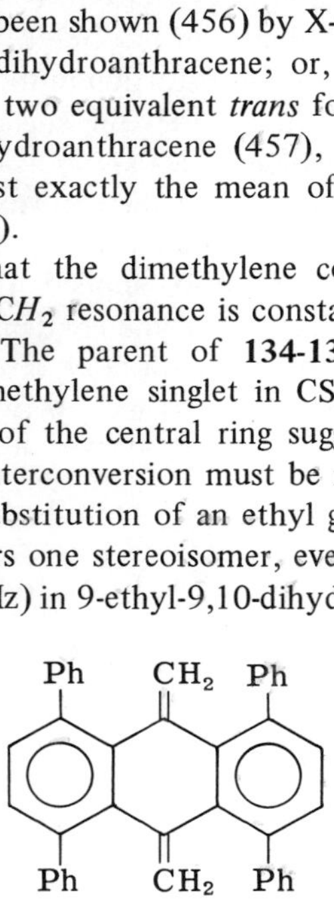

(136)

By acid-induced dimerization of the methoxypropenylbenzene isoelemicin (**137**), MacMillan and Walker (460) derived diisoelemicin, 1,2,3,5,6,7-hexamethoxy-9,10-diethyl-9,10-dihydroanthracene (**138**), to which the 9,10-*trans*

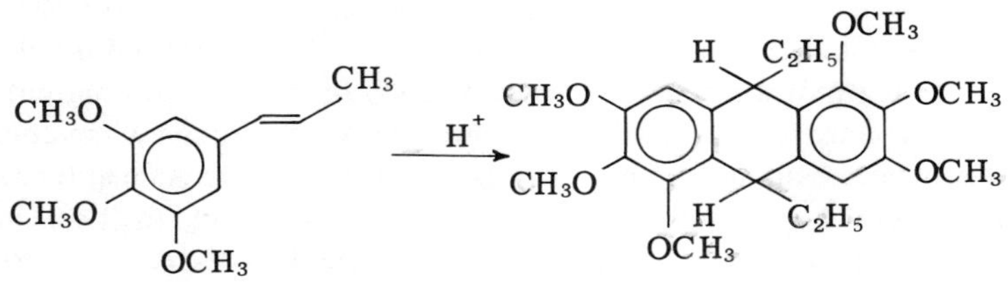

CH_3 CH_3O CH_3O OCH_3 $\xrightarrow{H^+}$ OCH_3 H C_2H_5 CH_3O OCH_3 OCH_3 CH_3O H C_2H_5 OCH_3

(137) (138)

stereochemistry is allocated from the high-field (0.2 ppm) triplet methyl resonance. In related work (461), the same authors have established the conformations of *cis*- and *trans*-9,10-diethyl-9,10-dihydro-9,10-dihydroxyanthracenes (462).

The 1-deutero-2-bromoacenaphthenes provide another example of the use of orthodox coupling-constant and chemical-shift arguments to assess conformation in the hydroaromatic series. Dewar and Fahey (463) found $J_{1,2}$ to be 7.4 and 1.9 Hz, respectively, for the *cis* and *trans* forms, from which Hayward and Csizmadia (464) calculated the dihedral angles (ϕ) to be 18 or 24° (*cis*) and 119 or 117° (*trans*), according to the value (465) selected for J_0 in the Karplus (466) equation. If allowance is made for distortion in the dihedral angles, owing to the bulk of the bromines, one would expect the angles in the parent to be about 0° for *cis* and 120° for *trans*. These values are as assumed (467,468) for the fully eclipsed bisectional bonds expected in the essentially planar rigid conformation which X-ray structural analysis (469) indicates.

The Karplus equation was also applied (470) to 1(10-anthron-9-yl)-2(9-methoxy-10-anthryl)ethane (**139**), the condensation product of 9-methoxy-10-methylanthracene. The structure of **139** is established by its pmr spectrum: 16 aromatic protons with signals between δ = 8.2 and 6.8 ppm;

2 [9-methoxy-10-methylanthracene: OCH_3, CH_3] $\xrightarrow[CuBr]{CuCl}$ [O, H(a), CH_2, CH_2, OCH_3]

(139)

one 6-Hz CH_2 triplet (2 protons) at 3.45 ppm, and another near 2.3 ppm; and a 4-proton singlet at δ = 4.2 ppm, from OCH_3 and H_a. The absence of detectable coupling between H_a and CH_2 suggests that the H_2-C-C-H dihedral angle is 60°. Chemical shifts and coupling constants in the 9,9′,10,10′-tetrahydro-9,9′-biphenanthryls (**140**) indicate that they belong (446) to the 9,9′ racemic series; the 9,9′ coupling constants for **140b** and **140c** are typical *trans* ethane couplings. For the same compounds, the two aromatic protons that resonate at higher field than the other 14 are taken to be H-8 and H-8′ shielded by the benzene rings in the racemic modification of **140** which has the *trans* axial-axial conformation. Values of J and δ also suggest that the 9,9′ link is formed from "pseudo-axial" bonds, and that the 10 and 10′ substituents occupy "pseudo-axial" positions.

5,6-Dihydro-5,6-dimethylchrysene-5,6,-diol (**141**) can be prepared by the action of methyl magnesium iodide on 5,6-chrysenequinone (471). Cohen, Millar, and Richards (472) were able to show from its pmr spectrum that the diol assumes the *trans* racemic form in which both methyls are "pseudo-axial." There is an H-4 multiplet at $\delta = 9.32$ ppm and singlet methyls at $\delta = 1.49$ and 1.67 ppm. Similarity of methyl shifts implies similar environments. The downfield shift of H-4 from its customary 8.80 ppm is attributed to an intramolecular hydrogen bond (also indicated by ir) between H of the OH on C-5 and O of OH on C-6; this causes the lone pair of the O on C-5 to approach and deshield H-4.

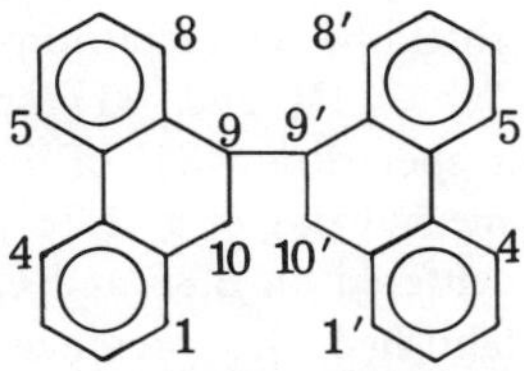

(**140a**) unsubstituted
(**140b**) 10–NO_2, 10′–OH
(**140c**) 10–NO_2, 10′–ONO_2

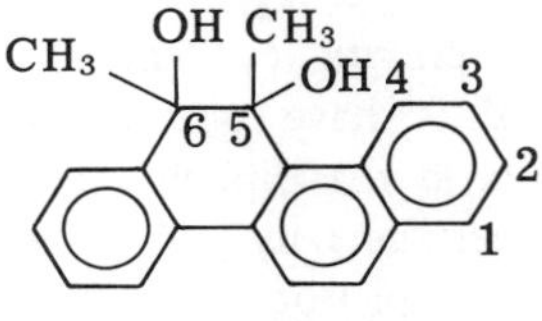

(**141**)

A high-energy barrier to rotation about a carbon-carbon sp^3-sp^2 bond has been reported in di-*t*-butyl-*p*-methoxyphenylcarbinol (473). In the hydroaromatic series, strong hindrance to rotation about an sp^3-sp^2 bond has been deduced (474) from the variable-temperature pmr spectrum of 9-mesitylfluorene-9-ol (**142**); in *m*-dibromobenzene solution, Chandross and Sheley report (474) three sharp methyl peaks up to 423°K but, in 1,2,4-trichlorobenzene solution, Rieker and Kessler (475) find that the methyls have broadened at 373°K and coalesced at 418°K. 9-Mesitylfluorene itself appears (474) to have its rotation very restricted up to at least 475°K. In 9-chloro-9-mesitylfluorene, for which the pmr spectrum in many solvents shows three sharp methyl peaks below 273°K and a narrow singlet above 313°K, the mechanism of conformational interchange (i.e., bond rotation or planar transition state) is not yet clear (474,475).

In both $(CD_3)_2SO$ and $CCl_2{=}CCl_2$ (477) solutions, 9-(α-naphthyl)-fluorene shows two signals for H-9, at 303°K and below, which broaden at higher

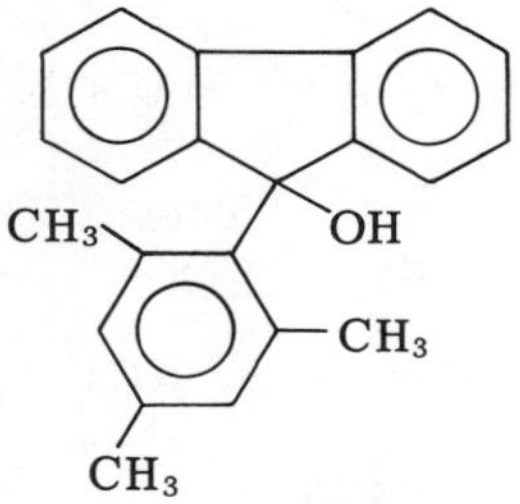

(**142**)

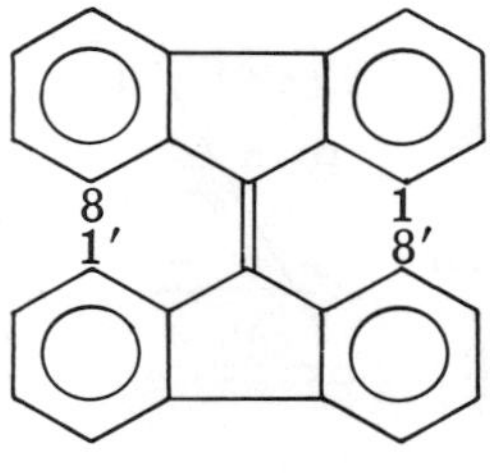

(**143**)

temperatures and coalesce at 363°K to a single peak (476,477); $\Delta G^{\ddagger}$ is estimated at 75 kJ mol^{-1} for rotation about the sp^3-sp^2 bond. Internal motions have been investigated for a number of other 9- and 9,9-substituted fluorenes (478,479). In 9-*t*-butyl-9,9′-bifluorenyl (474), there is pmr evidence of hindered rotation about two bonds; the width of the *t*-butyl resonance is broad at low temperatures, narrows, and then passes through a maximum at about 293°K before narrowing again at high temperatures. In the thermochromic dibiphenyleneethene **(143)**, the 1,1′,8,8′ protons resonate at such a low field (8.23 ppm in $CDCl_3$ solution) that an interaction between them must be assumed (480).

Brief reference will be made here to biphenylenes, although the contribution of pmr has not perhaps been commensurate with the growing importance of this branch of polynuclear chemistry. Empirical shielding contributions to the chemical shift have already been listed in Tables III and VI for typical substituents in the biphenylene series. The pmr spectrum (481) of diaminodimethylbiphenylene in TFA showed a 6-proton methyl singlet at 2.28 ppm and four aromatic protons forming an AB system centered on 6.85 and 6.65 ppm ($J = 4.5$ Hz). While the symmetry of this established the presence of two equivalent pairs of *meta* protons, it failed to distinguish between **144a** and **144b**.

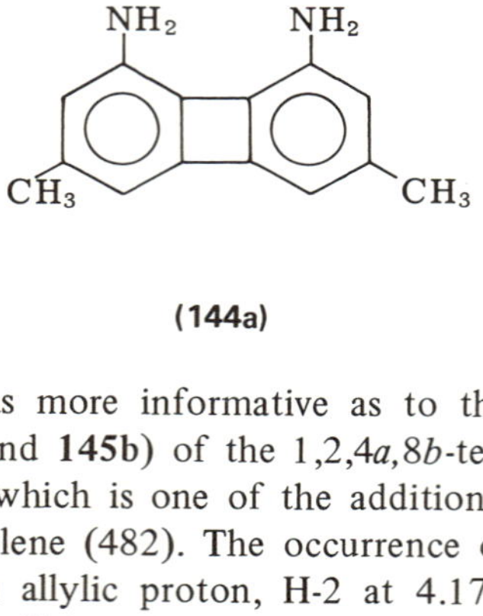

NH2 NH2 CH3 CH3

(144a)

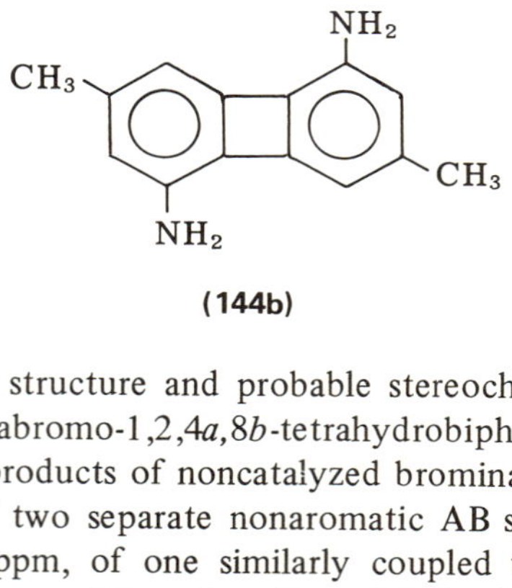

NH2 CH3 CH3 NH2

(144b)

Pmr was more informative as to the structure and probable stereochemistry (**145a** and **145b**) of the 1,2,4*a*,8*b*-tetrabromo-1,2,4*a*,8*b*-tetrahydrobiphenylene isomer which is one of the addition products of noncatalyzed bromination of biphenylene (482). The occurrence of two separate nonaromatic AB systems, with an allylic proton, H-2 at 4.17 ppm, of one similarly coupled to both protons, H-3 at 6.39 and H-4 at 5.77 ppm, of the other ($J_{2,3} = 2.3$ Hz, $J_{2,4} = 1.9$

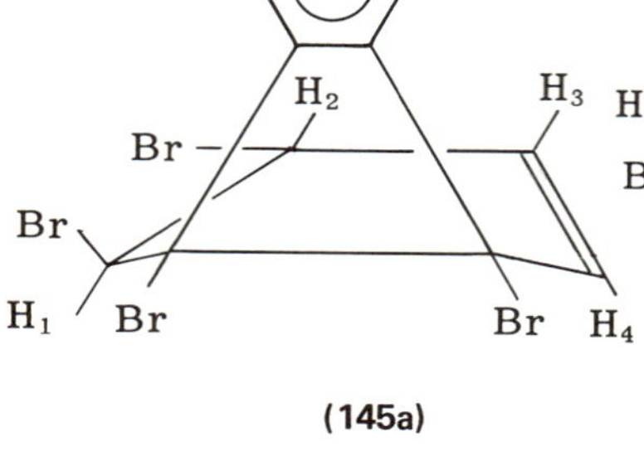

H2 H3 Br Br H1 Br Br H4

(145a)

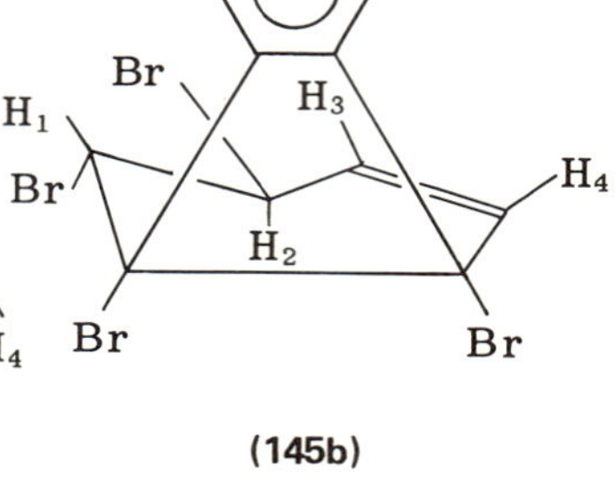

Br H1 H3 Br H4 H2 Br Br

(145b)

Hz), establishes addition without skeletal isomerization. The magnitude of the coupling, $J_{1,2} = 9.7$ Hz, in the AB system at higher field from the saturated protons, H-1 at 4.60 and H-2 at 4.17 ppm, corresponds (466) to a diaxial arrangement and points to one of the conformation **145a** or **145b**. Of these, **145a** is favored on the grounds that shielding by the benzene ring will help to explain the resonance of the allylic proton H-2 at higher field than H-1.

On the basis chiefly of the observed doublet benzhydryl proton resonance, Hertzler, Eisenbraun, and Flanagan (483) were able to show that sulfuric acid-catalyzed dimerization of 1,2-dihydronaphthalene, and also the action of phosphoric oxide on a mixture of 1,2-dihydronaphthalene and 1,2,3,4-tetrahydronaphthalene, both yield stereoisomers of 4,5,6,6*a*,6*b*,7,8,12*b*-octahydrobenzo[*j*] fluoranthene **(146a)** rather than structure **146b** (484,485). Of the two **146a** isomers, the *cis,syn* one melting at 426°K **(146d)** has the benzhydryl resonance shifted down to 4.17 ppm, while the *cis,anti* isomer **(146c)** melting at 366°K has this resonance at 3.96 ppm. In **146d** the broadening of the saturated proton multiplets is interpreted as a consequence of the conformational rigidity of the saturated rings brought about by steric interaction of aromatic protons H-1 and H-12 across the large bay. Furthermore, the adjacent aromatic ring plane appears to cause one H-8 saturated hydrogen to resonate at about 3 ppm.

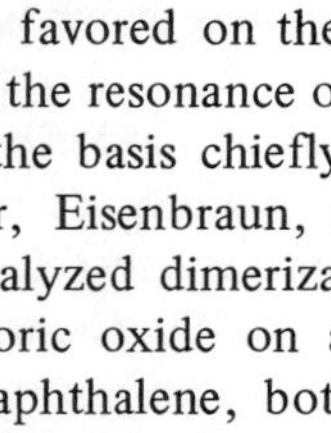

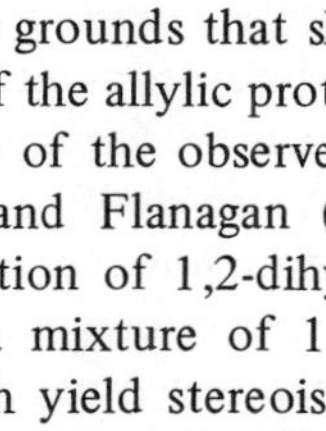

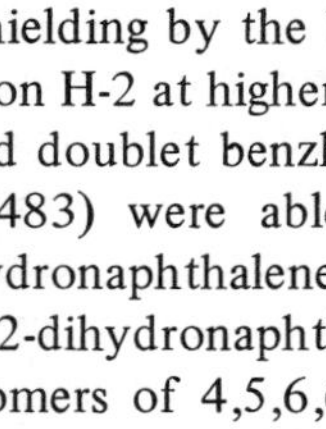

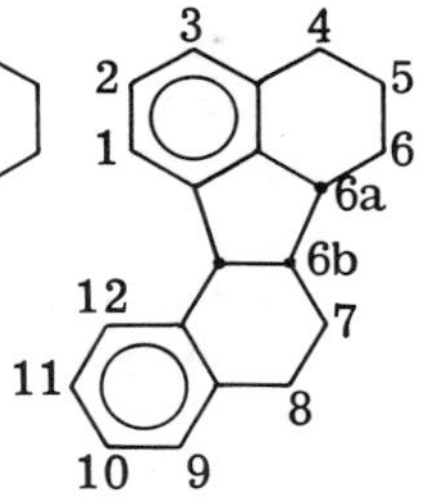

3 4 2 5 1 6 6a 6b 12 7 11 8 10 9

(146a) **(146b)** **(146c)** **(146d)**

D. *CONFORMATION OF SUBSTITUENT GROUPS*

Pmr spectra of polycyclic derivatives containing substituent groups (other than phenyl, discussed in Section IV.A) which are not axially symmetric can provide information about the conformation of such groups with respect to the mean plane of the aromatic rings.

Generally, protons *peri* to a substituent are deshielded (406). Dudek (84) interpreted the large deshieldings found for H-8 of 1-hydroxy,1-methoxy, and 1-amino substituted naphthalenes in terms of a type of hydrogen bond between the *peri* hydrogen and the adjacent oxygen or nitrogen atom lone pair; a *trans* arrangement of the substituent is implied. Zweig *et al.* (83) extended Dudek's measurements to the anthracene series. They found that the *peri* shift in 9,10-dimethoxyanthracene is about the same as in 1,4-dimethoxynaphthalene, but noted a larger *peri* effect for 9,10-*bis*(methylthio)anthracene than might be anticipated from a comparison with 1,4-*bis*(methylthio)naphthalene. Since methyl groups are prevented from being in the plane of the ring in the anthracene derivatives, but not in the naphthalene compounds, the conclusion was that the *peri* effect of $S{-}CH_3$ is anisotropic in origin and depends on "the sulphur bond-angle relative to the plane of the aromatic system."

The *peri* effect of the −COR group has been widely studied. In 9−COR phenanthrenes (73,89) the shift to low field of H-8 is fairly constant for R = CH_3, OH, and OCH_3 or OC_2H_5; if R = H, however, a much larger deshielding of H-8 is observed. The sequency of *peri* effects in 9−COR phenanthrenes ($CHO > COOCH_3 > COOH > COCH_3$), on the other hand, differs from that sensed by H-10 in 1-substituted pyrenes (75): $COOCH_3 > CHO > COCH_3$. Bartle and Smith (73) found that the *peri* effect in 9−COR phenanthrenes was greatly reduced in presence of a 10 substituent. For example, H-8 of 9,10-dimethoxycarbonylphenanthrene is deshielded by only 0.30 ppm, compared with 1.19_5 in the 9-methoxycarbonyl derivative; the *peri* effect of 9−COOH is only −0.07 ppm, when *ortho* to phenyl, but −1.12 ppm in 3-ethyl-9-carbonylphenanthrene. It was suggested that the basic cause of these smaller deshieldings was alteration of the preferred conformation of −COOR with respect to the aromatic plane by steric interaction between the *ortho* groups; this would change the effects of the anisotropy of the magnetic susceptibility and possibly any electric-field shifts. By a similar process, H-5 is shielded in the highly sterically hindered 4-methoxycarbonylphenanthrene (73) as is H-12 in the analogous 1-methoxycarbonyltriphenylene (81).

Dudek (84) also discussed the conformation of 5-acetylacenaphthene, and concluded that the pmr evidence favors the *trans* methyl arrangement **(147)**. Correspondingly, the *peri* shift of −1.25 ppm induced by the carbonyl group in 1-formylnaphthalene **(148)** is substantially reduced (to −0.55 ppm) in 2-hydroxy-1-formylnaphthalene **(149)** in which hydrogen bonding (87) locks the carbonyl group *trans* to the *peri* H-8. With hydrogen bonding absent in 2-methoxy-1-formylnaphthalene **(150)**, the strong *peri* effect is restored to −1.57 ppm.

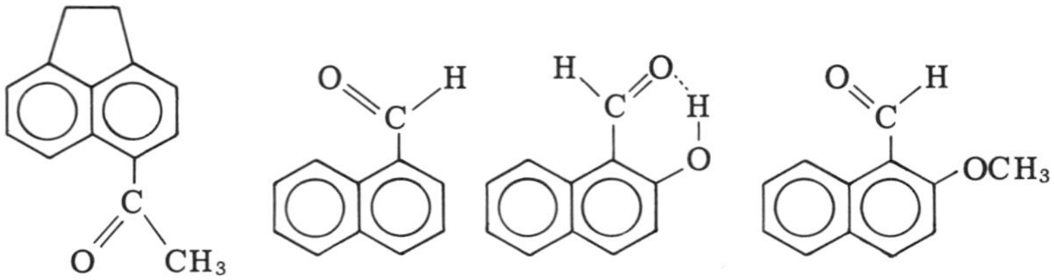

(147) **(148)** **(149)** **(150)**

Cameron, Kingston, and Schutz (171) argued that the pmr spectra of 9-formylanthracene and 9-acetylanthracene show how the CHO group can approach more nearly to the *peri* H-1 and H-8 than the more bulky $COCH_3$

group. In the 9-formyl compound, the H-1,8 resonances are deshielded by 0.99 ppm compared with H-4,5; in 9-acetylanthracene, H-1,8 and H-4,5 have rather similar shifts. The same authors accepted the nonplanarity of 1,5-diacetylanthracene from a consideration of its uv spectrum and those of related compounds (486,487) and used the solvent effect of C_6D_6 to determine which of the conformational isomers (**151a** and **151b**) is the more nearly correct representation. They noted values of the solvent shift induced by C_6D_6 relative to $CDCl_3$, and found that for the related compound (**25**), in which ring constraint leads to a conformation analogous to **151b**, all shifts were large and to high field except for those *peri* to carbonyl (H-2 and H-6). These observations are wholly consistent with those made for other carbonyl compounds (488); H-2 and H-6 of 1,5-diacetylanthracene, however, show large high-field solvent shifts, while H-9 and H-10 exhibit equally large shifts in the opposite direction; **151a** is therefore accepted as a representation for this compound. A similar conclusion was reached for 1-acetylnaphthalene (406,489) from dipole-moment measurements.

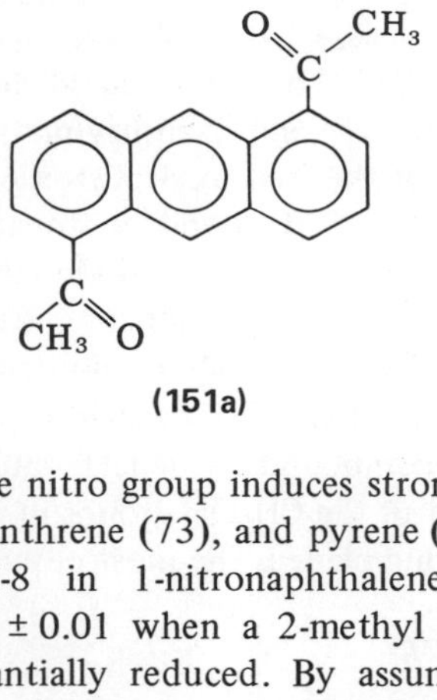

(151a) 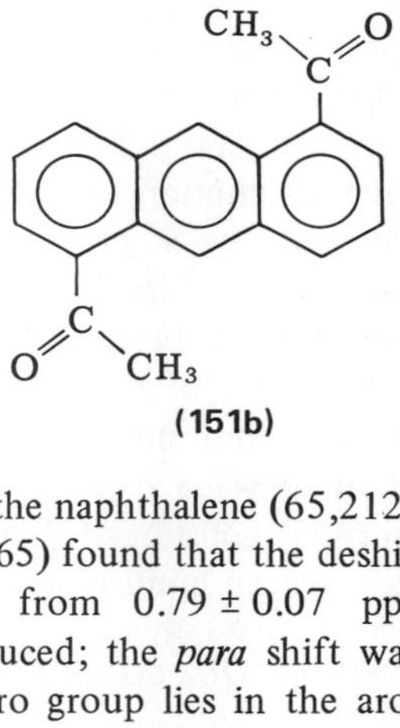 **(151b)**

The nitro group induces strong *peri* effects in the naphthalene (65,212,490), phenanthrene (73), and pyrene (75) series. Wells (65) found that the deshielding of H-8 in 1-nitronaphthalenes was reduced from 0.79 ± 0.07 ppm to 0.025 ± 0.01 when a 2-methyl group was introduced; the *para* shift was also substantially reduced. By assuming that the nitro group lies in the aromatic plane in the unhindered compounds and is perpendicular to it in the hindered compounds, he was able to calculate a value, -13.3×10^{-12} m^3 mol^{-1} for $\chi_p - \chi_T$, the difference between the principal magnetic susceptibilities of the nitro group respectively perpendicular to the $C{-}NO_2$ plane and transverse to the carbon-nitrogen bond. Yamaguchi (491) estimated the principal magnetic susceptibilities of the nitro group from the CH_3 shifts in polymethyl nitrobenzenes; he found $\chi_p - \chi_T = -11 \times 10^{-12}$ to -17×10^{-12} m^3 mol^{-1}, depending on the location of the point dipole (492). The presence of a bromine atom *ortho* to nitro in 9-bromo-10-nitrophenanthrene diminishes the *peri* effect of the nitro groups in a similar fashion.

Wells (65) also noted different chemical-shift influences on the methyl protons of methylnitronaphthalenes depending on the degree of rotation of the nitro group. The CH_3 of 3-methyl-2-nitronaphthalene is deshielded by 0.21 ppm, but in 2-methyl-1-nitronaphthalenes the nitro group is rotated out of the ring plane, and a small (0 to 0.04 ppm) shielding is observed.

The low-field *peri* proton absorption (8.20 ppm) in the spectrum of 1-cyclopropylnaphthalene may be consistent with preferred conformations in which the cyclopropane ring does not achieve bisected geometry (493). On the other hand, the signals of $(CH_3)_3$ in *peri*-di-*t*-butylnaphthalenes appear as sharp singlets, so that locking and/or restricted rotation is avoided (494). Since the *t*-butyl chemical shift is smaller (1.22-1.30 ppm) than in "normal" *t*-butylnaphthalenes, a bending of the substituent groups away from each other above and below the ring plane to relieve crowding is suggested (494).

Ouellette and Van Leuwen (495) have applied ingenious chemical-shift arguments in a discussion of the conformation of the CH_2OH group in a number of polycyclic aryl methanols. δ_{OH} and δ_{CH_2} were determined in $(CD_3)_2SO$ solution and correlations investigated with the function $\Sigma I_n R_n^{-3}$, where R_n is the distance from the proton in question to the center of ring *n*, and I_n is the ring current for ring *n* in the unsubstituted hydrocarbon.* For δ_{CH_2}, the correlation coefficient was high, as might be expected from the similar result obtained by the same authors (496) for δ_{CH_3} *versus* ΣIR^{-3} plots in methylarenes. On the other hand, δ_{OH} correlated with ΣIR^{-3} only for benzyl alcohol, 2-naphthylmethanol, 2- and 3-phenanthrylmethanol, and 2-anthrylmethanol, where the assumption that the average position of the hydroxyl proton can be approximated by a projection on the extended C—C bond is apparently legitimate. Substantial deviations were found for δ_{OH} of 1-naphthylmethanol, 9-phenanthrylmethanol, and 1-anthrylmethanol; the populations of sites in which O*H* is out of the ring plane are larger. The greater steric hindrance in 9-anthrylmethanol and 2-methyl-1-naphthylmethanol is reflected in still poorer correlation between δ_{OH} and ΣIR^{-3} for these compounds. Ouellette and Van Leuwen (495) considered that the conformation of the CH_2OH groups in which hydrogen-oxygen bond is perpendicular to the ring plane is the most populated.

E. MISCELLANEOUS CONFORMATIONAL PROBLEMS

It is convenient to include in this section a brief mention of the magnetic nonequivalence of methyl groups in tri-*o*-carvacrotide (**152**), which embraces an unusual 12-membered ring (498); the benzene rings can take up helical- or propeller-shaped conformations. At 283°K, there are four aromatic methyl peaks: three from nonequivalent groups in the helical form and one from three equivalent groups in the more symmetrical propeller form. Increase of temperature reduces the methyl peaks to two at 328°K and one at 373°K. For interchange between propeller and helix, E_a = 95 kJ mol^{-1} and

---

*While effects of substituent groups on ring currents (63,64) were ignored, this approach allowed 'experimental' ring currents to be derived for benz[*a*]anthracene (496) in rather good agreement with those computed by an MO method (497).

$\Delta G\ddagger$ = 86 kJ mol^{-1}, and for helix-helix interconversion, E_a = 89 kJ mol^{-1} and $\Delta G\ddagger$ = 74 kJ mol^{-1}.

Systems such as 6-methyl-14*H*-dinaphtho [1,8-*b,c* : 1′,8′-*f,g*] thiocin **(153)** and its derivatives incorporate novel medium rings between *peri*-naphthalenes. Instead of showing proton groups in intensity ratio 11:2:3, the pmr spectrum (499) of the sulfide **(154)** shows a complex multiplet at δ = 7.5, a doublet at 4.1, and a singlet at 2.8 ppm in intensity ratio 12:1:3. Double-resonance and spin-tickling experiments confirmed the suspicion that the *syn*-hydrogen of the methylene group was under the aromatic resonance and that it was chemically shifted by 3.88 ppm from the *anti*-hydrogen of the methylene bridge, a remarkable difference for geminal hydrogens.

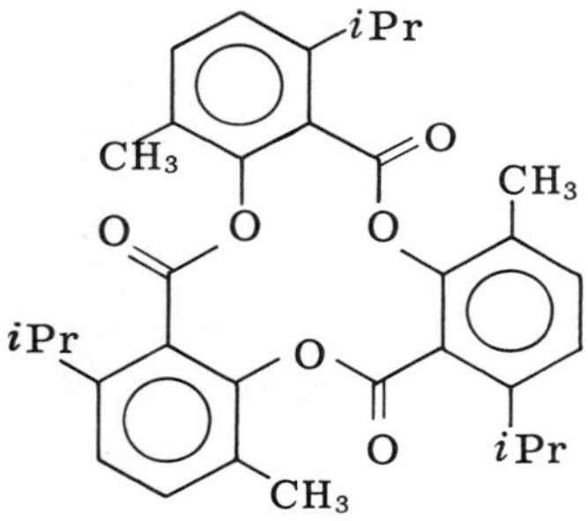

iPr CH₃ O CH₃ O O O iPr O O iPr CH₃

(152)

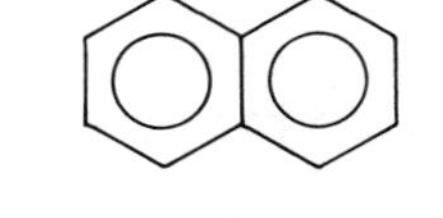

CH₃ S C=O

(153)

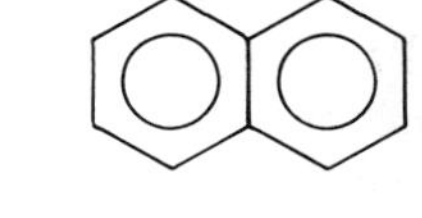

CH₃ S CH₂

(154)

Up to 433°K there is no sign of incipient coalescence of the AB spectrum, so that the barrier to conformational inversion of the 8-membered ring must be high. When **154** is oxidized with *m*-chloroperbenzoic acid to the *syn*-sulfoxide **(155)**, the rapid isomerization to **156** at room temperature can be followed from the intensity ratios of the methyl resonances. These lead to $k_1 = 1.06 \pm 0.09 \times 10^{-3}\ sec^{-1}$ for the first-order rate constant of the isomerization and $\Delta G\ddagger$ = 106 kJ mol^{-1}. Johnson and Vegh (499) suggest that steric overcrowding between the sulfoxide oxygen and the *syn*-hydrogen causes isomerization to proceed *via* pyramidal inversion at sulfur rather than ring inversion.

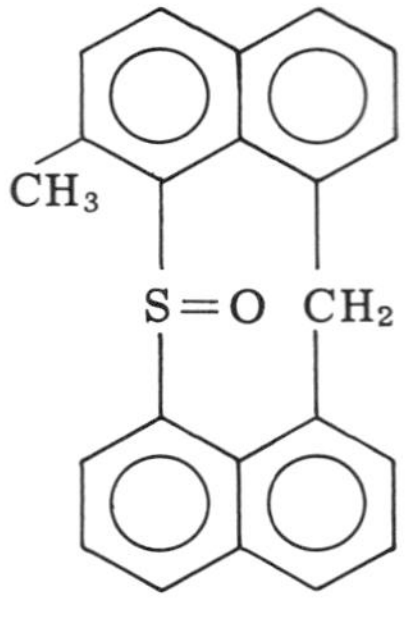

(155)

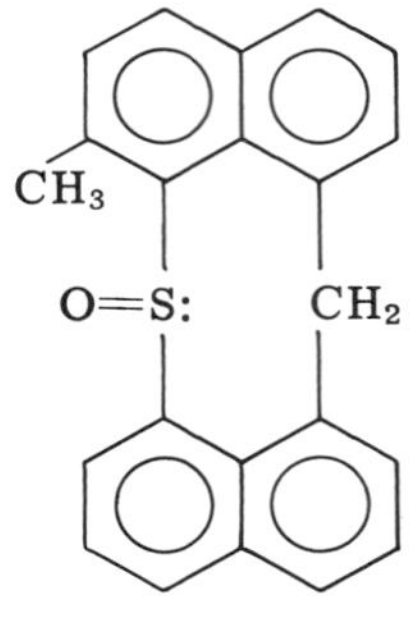

(156)

Magnitudes of coupling constants are the crux of the argument (500) favoring **157a** as against **157b** for the structure of the adduct formed by reaction of equimolecular proportions of 3-(2-hydroxyphenyl)-1,3-diphenylpropan-1-one with 3-(2-hydroxyphenyl)-1,1-diphenylprop-2-en-1-ol. Compared with the strength of the unresolved aromatic band, the resonances from the other hydrogens, H-1 to H-4, are weak, but they can be seen to involve the following 1-proton multiplets: (*1*) a sharp 10-Hz doublet at 5.80 ppm, (*2*) a less sharp 10.5-Hz doublet at 4.21 ppm, (*3*) a quadruplet (4, 6, and 4 Hz) at 2.72 ppm, and (*4*) a rather similar quartet at 3.22 ppm. Now structure **157b** would require at most very weak coupling between H-1 and H-3 but strong H-1-H-2 and H-3-H-4 coupling. Change of solvents helps to establish that (*3*) and (*4*) correspond to H-2 and H-3, respectively. A model compound indicates that the (*1*) doublet comes from H-1, and structure **157b** is rejected. Stereochemical models show that the relatively large J_{34} of 10.5 Hz is plausible in terms of structure **157a**.

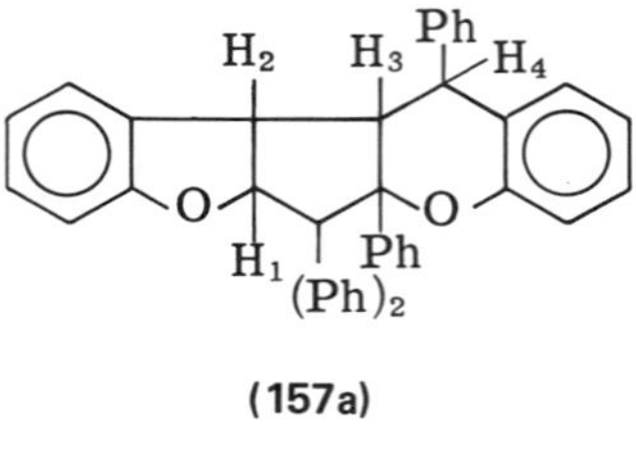

(157a)

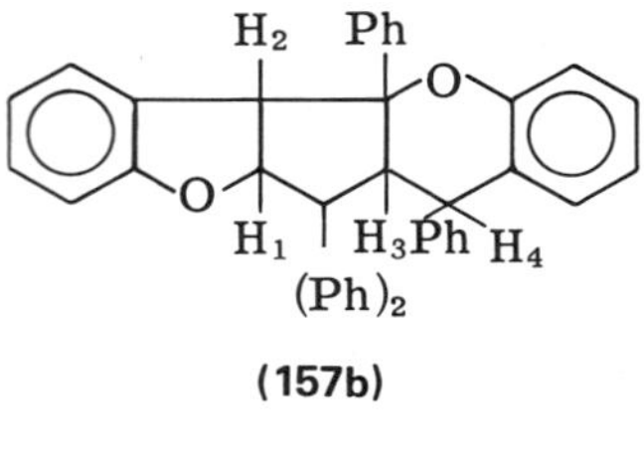

(157b)

V. Conclusion

This chapter has been concerned only with steady-state high-resolution pmr spectroscopy in which the observable parameters have been the positions (frequencies) and sizes (areas under peaks) of the resonances appropriate to the chemical shifts and coupling constants of the nuclear spin system. So far, little use has been made of transient experiments in the examination of polycyclic hydrocarbons. In these, the times for decay (the spin-lattice [T_1] and spin-spin [T_2] relaxation times) are measured after the molecular system has been excited by an intense radio-frequency field. Relaxation-rate studies have been made to locate the binding sites by which small molecules are attached to polymer molecules (501), so that the technique should have relevance to the interaction between carcinogenic polycyclic hydrocarbons and molecules of biological interest (502,503).

An effort has been made in this review to cover some of the more interesting and illustrative applications of pmr to polycyclics published up to mid-1969. It could not attempt to be comprehensive. Despite increasing recourse to high-field resonance spectra, multiple-resonance experiments, variation of solvent shifts and concentrations, and the use of moderately sophisticated theoretical calculations, the full spectral analysis, necessary for the derivation of chemical shifts and coupling constants in a complex polycyclic system, can still present formidable problems. However, as the fund of solved spectra grows, so does the facility with which pmr can be applied with confidence to structural problems in polycyclic chemistry and in branches of natural-product chemistry involving polynuclears. In mixtures as complicated as those encountered in coal-tar and petroleum chemistry, the discernment of certain individual compounds by pmr is remarkable. Among single compounds, there has already been notable and telling use of pmr in developing fields such as the cavity aromatics and in hydroaromatics. Increasingly, conformational problems are assuming the attention of many organic chemists and, as the quantitative physical-organic chemistry of pmr becomes more predictable, it is perhaps in this field that some of the most distinctive contributions of pmr may be made to polycyclic hydrocarbons.

ACKNOWLEDGMENTS

K. D. Bartle acknowledges the award of fellowships by Imperial Chemical Industries, Ltd., at the University of Bradford, England, and by the Swedish and Scandinavian Tobacco Companies at the Institute of Analytical Chemistry, University of Stockholm, Sweden.

REFERENCES

1. J. W. Emsley, J. Feeney, and L. H. Sutcliffe, *High Resolution Nuclear Magnetic Resonance Spectroscopy,* Pergamon, Oxford, 1965.
2. N. Sheppard, J. A. Elvidge, E. O. Bishop, C. N. Banwell, R. J. Abraham, and J. Feeney in D. W. Mathieson, ed., *Nuclear Magnetic Resonance for Organic Chemists,* Academic Press, London, 1967.
3. J. B. Stothers, in *Techniques of Organic Chemistry,* Part I, Vol. II (A. Weissberger, ed.), Wiley-Interscience, New York, 1963, Chap. 4.
4. H. Conroy, in *Advances in Organic Chemistry,* Vol. 2 (R. A. Raphael, E. C. Taylor, and H. Wynberg, eds.), Wiley-Interscience, New York, 1960, p. 265.

5. J. K. Becconsall and M. C. McIvor, *Chem. Brit.*, **5**, 147 (1969).
6. C. W. Haigh and R. B. Mallion, *J. Mol. Spec.*, **29**, 478 (1969).
7. K. D. Bartle, D. W. Jones, and R. S. Matthews, *Spectrochim. Acta,* **25A,** 1603 (1969).
8. J. A. Pople, W. G. Schneider, and H. J. Bernstein, *High-Resolution Nuclear Magnetic Resonance,* McGraw-Hill, New York, 1959.
9. G. van D. Tiers, *J. Phys. Chem.*, **62**, 1151 (1958).
10. P. Laszlo, *Prog. NMR Spectrosc.*, **3**, 231 (1967).
11. J. Homer, *Tetrahedron,* **23**, 4065 (1967).
12. P. Laszlo, A. Speert, R. O. Hinger, and J. Reisse, *J. Chem. Phys.*, **48,** 1732 (1968).
13. W. T. Raynes and M. A. Raza, *Mol. Phys.*, **17,** 157 (1969).
14. R. H. Martin, N. Defay, F. Geerts-Evrard, and D. Bogaert-Verhoogen, *Tetrahedron, Suppl.* 8 (1), 181 (1966).
15. R. H. Martin, N. Defay, F. Geerts-Evrard, and S. Delavarene, *Tetrahedron,* **20,** 1073 (1964).
16. K. D. Bartle, D. W. Jones, and R. S. Matthews, *J. Chem. Soc., A,* **1969,** 876.
17. D. H. Williams, J. Ronayne, and R. G. Wilson, *Chem. Commun.*, **1967,** 1089.
18. K. D. Bartle, D. W. Jones, and S. Maričić, *Croat. Chem. Acta,* **40**, 227 (1968).
19. N. F. Ramsey, *Phys. Rev.*, **78,** 699 (1951).
20. W. N. Lipscomb, *Advan. Magnetic Resonance,* **2,** 138 (1966).
21. A. Saika and C. P. Slichter, *J. Chem. Phys.*, **22,** 26 (1954).
22. J. A. Pople, *Discussions Faraday Soc.*, **34,** 7 (1962).
23. K. D. Bartle and D. W. Jones, *Trans. Faraday Soc.*, **63,** 2868 (1967).
24. K. D. Bartle and J. A. S. Smith, *Spectrochim. Acta,* **23A,** 1689 (1967).
25. G. G. Hall, A. Hardisson, and L. M. Jackman, *Tetrahedron, Suppl.* **2**(19), 101 (1963); *Discussions Faraday Soc.*, **34,** 15 (1962).
26. P. J. Black, R. D. Brown, and M. L. Heffernan, *Aust. J. Chem.*, **20,** 1305 (1967).
27. J. I. Musher, *J. Chem. Phys.*, **43,** 4081 (1965); **46,** 1219 (1967); *Advan. Magnetic Resonance,* **2,** 177 (1966).
28. J. M. Gaidis and R. West, *J. Chem. Phys.*, **46,** 1218 (1967).
29. J. D. Memory, *Quantum Theory of Magnetic Resonance Parameters,* McGraw-Hill, New York, 1968, Chap. 6.
30. L. Pauling, *J. Chem. Phys.*, **4,** 673 (1936).
31. F. London, *J. Phys. Radium,* **8,** 397 (1937).
32. J. A. Pople, *Mol. Phys.*, **1,** 175 (1958); *J. Chem. Phys.*, **37,** 53 (1962).
33. L. Salem, *The Molecular Orbital Theory of Conjugated Systems*, Benjamin, New York, 1966, Chap. 6.
34. R. McWeeney, *Mol. Phys.*, **1,** 311 (1958).
35. C. E. Johnson and F. H. Bovey, *J. Chem. Phys.*, **29,** 1012 (1958).
36. D. G. Farnum and C. F. Wilcox, *J. Amer. Chem. Soc.*, **89,** 5379 (1967).
37. C. W. Haigh, *Mol. Phys.*, in press.
38. N. Jonathan, S. Gordon, and B. P. Dailey, *J. Chem. Phys.*, **36**, 2443 (1962).
39. J. Hoarau, *Ann. Chim.* (Paris), **1,** 560 (1956).
40. J. A. Pople, *J. Chem. Phys.*, **41,** 2559 (1964).
41. B. P. Dailey, *J. Chem. Phys.*, **41,** 2304 (1964).
42. R. H. Martin, *Tetrahedron,* **20,** 897 (1964).
43. J. D. Memory, G. W. Parker, and J. C. Halsey, *J. Chem. Phys.*, **45,** 3567 (1966).
44. K. D. Bartle and D. W. Jones, *J. Chem. Soc., A,* **1969,** 437.
45. R. C. Fahey and G. C. Graham, *J. Phys. Chem.*, **69,** 4417 (1965).
46. T. B. Cobb and J. D. Memory, *J. Chem. Phys.*, **47,** 2020 (1967).
47. T. J. Batterham, L. Tsai, and H. Ziffer, *Aust. J. Chem.*, **18,** 1959 (1965).
48. K. D. Bartle, D. W. Jones, and J. E. Pearson, *J. Mol. Spectrosc.*, **24**, 330 (1967).
49. K. D. Bartle and D. W. Jones, *J. Mol. Struct.*, **1,** 131 (1967).
50. M. J. S. Dewar and A. P. Marchand, *J. Amer. Chem. Soc.*, **88,** 3318 (1966).
51. J. R. Ballantine and C. T. Pillinger, *Tetrahedron,* **23,** 1691 (1967).

52. U. E. Malter, C. Pascual, E. Pretsch, A. Pross, W. Simon, and S. Sternhell, *Tetrahedron,* **25**, 691 (1969).
53. C. Pascual, J. Meier, and W. Simon, *Helv. Chim. Acta,* 49, 164 (1966).
54. B. Richardson and T. Schaefer, *Can. J. Chem.,* **46**, 2195 (1968).
55. T. K. Wu and B. P. Dailey, *J. Chem. Phys.,* **41**, 2706 (1964).
56. J. C. Schug, *J. Chem. Phys.,* **46**, 2447 (1967).
57. J. C. Schug and B. P. Smith, *J. Chem. Phys.,* **47**, 849 (1967).
58. A. A. Bothner-By and J. A. Pople, *Ann. Rev. Phys. Chem.,* **16**, 43 (1965).
59. T. Nomura and Y. Takeuchi, *Tetrahedron Lett.,* **1968**, 5665.
60. T. Schaefer, W. F. Reynolds, and T. Yonemoto, *Can. J. Chem.,* **41**, 2969 (1963).
61. V. M. S. Gil and W. A. Gibbons, *Mol. Phys.,* **8**, 199 (1964); **9**, 163 (1965).
62. R. F. Zürcher, in *N.M.R. in Chemistry* (B. Pesce, ed.), Academic Press, New York, 1965.
63. R. J. Abraham, *Discussions Faraday Soc.,* **34**, 70 (1962).
64. R. P. Figeys and R. Flammang, *Mol. Phys.,* **12**, 581 (1967).
65. P. R. Wells, *Aust. J. Chem.,* **17**, 967 (1964).
66. W. B. Smith and J. L. Roark, *J. Amer. Chem. Soc.,* **89**, 5018 (1967).
67. F. Hruska, H. M. Hutton, and J. Schaefer, *Can. J. Chem.,* **43**, 2392 (1965).
68. R. H. Martin, J. P. van Trappen, N. Defay, and J. F. W. McOmie, *Tetrahedron,* **20**, 2373 (1964).
69. P. Durand, J. Parello, N. P. Buu-Hoï, and L. Alais, *Bull. Soc. Chim. Fr.,* **1963**, 2438.
70. J. L. Garnett, *Tetrahedron Lett.,* **1961**, 516.
71. H. Spiesecke and W. G. Schneider, *J. Chem. Phys.,* **35**, 731 (1961).
72. P. Diehl, *Helv. Chim. Acta,* **44**, 829 (1961).
73. K. D. Bartle and J. A. S. Smith, *Spectrochim. Acta,* **23A**, 1715 (1967).
74. K. Hayamizo and O. Yamamoto, *J. Mol. Spectrosc.,* **28**, 89 (1968).
75. R. N. Martin, R. Flammang, and M. Arbaoui, *Bull Soc. Chim. Belges,* **74**, 418 (1965).
76. E. Mühle, *Ann. Phys.* (Leipzig), **18**, 130 (1966).
77. E. V. Blackburn, C. E. Loader and C. J. Timmons, *J. Chem. Soc.,C,* **1968**, 1576.
78. J. Paviot and J. Hoarau, *J. Chim. Phys.,* **64**, 1415 (1967).
79. S. Castellano and R. Kostelnik, *Tetrahedron Lett.,* **1967**, 5205.
80. W. B. Smith and S. Chiranjeevi, *J. Phys. Chem.,* **70**, 3505 (1966).
81. R. H. Martin, N. Defay, and F. Geerts-Evrard, *Tetrahedron,* **20**, 1091 (1964).
82. A. J. Fry, E. Katzmann, and P. A. Leermakers, *Spectrochim. Acta,* **24A**, 951 (1968).
83. A. Zweig, J. E. Lancaster, and M. T. Neglia, *Tetrahedron,* **23**, 2577 (1967).
84. G. O. Dudek, *Spectrochim. Acta,* **19**, 691 (1963).
85. K. Takase, T. Asao, Y. Takagi, and T. Nozoe, *Chem. Commun.,* **1968**, 368.
86. R. H. Martin, N. Defay, and F. Geerts-Evrard, *Tetrahedron,* **21**, 2421 (1965).
87. R. H. Martin, N. Defay, and F. Geerts-Evrard, *Tetrahedron,* **20**, 1505 (1964).
88. Reference 1, p. 1140.
89. R. H. Martin, N. Defay, F. Geerts-Evrard, and H. P. Figeys, *Bull. Soc. Chim. Belges,* **73**, 199 (1964).
90. A. J. Fry, B. W. Bowen, and P. A. Leermakers, *J. Org. Chem.,* **32**, 1970 (1967).
91. K. D. Bartle and D. W. Jones, unpublished measurements.
92. M. P. Williamson, R. J. Kostelnik, and S. M. Castellano, *J. Chem. Phys.,* **49**, 2218 (1968).
93. G. S. Reddy, *Z. Naturforsch.,* **219**, 609 (1966).
94. C. MacLean and E. L. Mackor, *Mol. Phys.,* **3**, 223 (1960).
95. K. D. Bartle and D. W. Jones, *J. Mol. Spectrosc.,* **32**, 353 (1969).
96. K. D. Bartle, H. Heaney, D. W. Jones, and P. Lees, *Spectrochim. Acta,* **22**, 941 (1966).
97. D. Meuche, B. Molloy, D. H. Reid, and E. Heilbronner, *Helv. Chim. Acta,* **46**, 2483 (1963).
98. D. Cagniant, *Bull. Soc. Chim. Fr.,* **1966**, 2325.
99. K. Hayamizo and O. Yamamoto, *J. Mol. Spectrosc.,* **29**, 183 (1969).

100. A. Cornu, J. J. Ulrich, and K. Persaud, *Chim. Anal.*, **47**, 357 (1965).
101. H. E. Zieger and E. M. Laski, *Tetrahedron Lett.*, **1966**, 3801.
102. R. H. Martin and J. C. Nouls, *Tetrahedron Lett.*, **1966**, 2727.
103. W. Brügel, *Nuclear Magnetic Resonance Spectra and Chemical Structure*, Academic Press, New York, 1967.
104. T. Schaefer, G. Kotowycz, H. M. Hutton, and J. W. S. Lee, *Can. J. Chem.*, **46**, 2531 (1968).
105. M. J. S. Dewar and R. C. Fahey, *J. Amer. Chem. Soc.*, **85**, 2704 (1963).
106. M. L. Heffernan, A. J. Jones, and P. J. Black, *Aust. J. Chem.*, **20**, 589 (1967).
107. C. N. Banwell and N. Sheppard, *Discussions Faraday Soc.*, **34**, 115 (1962).
108. A. D. Buckingham and K. A. McLauchlan, *Proc. Chem. Soc.*, **1963**, 144.
109. K. D. Bartle, D. W. Jones, and R. S. Matthews, *J. Mol. Struct.*, **4**, 445 (1969).
110. W. B. Smith, W. H. Watson, and S. Chiranjeevi, *J. Amer. Chem. Soc.*, **89**, 1438 (1967).
111. W. B. Smith, private communication, 1968.
112. J. P. Maher and D. F. Evans, *J. Chem. Soc.*, **1965**, 637.
113. A. R. Katritzky and R. E. Reavill, *Rec. Trav. Chim. Pays-Bas*, **83**, 1230 (1964).
114. R. Flammang, H. P. Figeys, and R. H. Martin, *Tetrahedron*, **24**, 1171 (1968).
115. S. Sternhell, *Rev. Pure Appl. Chem.*, **14**, 15 (1964); *Quart. Rev.*, **23**, 236 (1969).
116. K. D. Bartle, D. W. Jones, and R. S. Matthews, *Tetrahedron*, **25**, 2701 (1969).
117. K. D. Bartle, D. W. Jones, and R. S. Matthews, *Rev. Pure Appl. Chem.*, **19**, 191 (1969).
118. D. J. Blears, S. S. Danyluk, and T. Schaefer, *Can. J. Chem.*, **46**, 654 (1968).
119. H. J. Bernstein, J. A. Pople, and W. G. Schneider, *Can. J. Chem.*, **35**, 65 (1957).
120. M. van Gorkom and G. E. Hall, *Quart. Rev.*, **22**, 14 (1968).
121. P. L. Richards, *Pure Appl. Chem.*, **11**, 261 (1965).
122. Y. K. Pan and M. T. Rogers, *Rev. Pure Appl. Chem.*, **18**, 17 (1968).
123. E. O. Bishop, *Ann. Rev. NMR Spectrosc.*, **1**, 91 (1968).
124. B. Dischler and C. Englert, *Z. Naturforsch.*, **16a**, 1180 (1961).
125. R. J. Abraham, *J. Chem. Soc.*, **1965**, 256.
126. H. P. Figeys, *Chem. Commun.*, **1967**, 495.
127. T. J. Batterham, L. Tsai, and H. Ziffer, *Aust. J. Chem.*, **17**, 163 (1964).
128. C. A. Reilly and J. D. Swalen, *J. Chem. Phys.*, **34**, 980 (1961).
129. P. Faller, *Bull. Soc. Chim. Fr.*, **1967**, 387.
130. J. Lee and L. H. Sutcliffe, *Trans. Faraday. Soc.*, **54**, 308 (1958).
131. K. D. Bartle, Ph.D. Thesis, University of Leeds, England, 1965.
132. R. E. Mayo and J. H. Goldstein, *Mol. Phys.*, **10**, 301 (1966).
133. W. G. Schneider, H. J. Bernstein, and J. A. Pople, *J. Amer. Chem. Soc.*, **80**, 3497 (1958).
134. H. Spiesecke and W. G. Schneider, *Tetrahedron Lett.*, **1961**, 468.
135. E. Clar, Ü. Sanigök, and H. Zander, *Tetrahedron*, **24**, 2817 (1968).
136. P. M. G. Bavin, K. D. Bartle, and J. A. S. Smith, *Tetrahedron*, **21**, 1087 (1965).
137. J. D. Roberts, *An Introduction to the Analysis of Spin-Spin Splitting in High Resolution Nuclear Magnetic Resonance Spectra*, Benjamin, New York, 1962, p. 71.
138. P. L. Corio, *Structure of High-Resolution NMR Spectra*, Academic Press, New York, 1966, p. 299.
139. J. B. Kim and J. M. Anderson, *Spectrochim. Acta*, **24A**, 792 (1968).
140. P. R. Houlton and W. Kemp, *J. Chem. Soc., C*, **1967**, 591.
141. K. D. Bartle, H. Heaney, D. W. Jones, and P. Lees, *Tetrahedron*, **21**, 3289 (1965).
142. F. A. Vingiello and A. K. Youssef, *Chem. Commun.*, **1968**, 68.
143. D. Lavit-Lamy and N. P. Buu-Hoï, *Chem. Commun.*, **1966**, 92.
144. F. A. Vingiello and A. K. Youssef, *Chem. Commun.*, **1967**, 863.
145. A. Rahman and O. L. Tombesi, *Chem. Ind.* (London), 1968, 1840.
146. F. H. Marquardt, *J. Chem. Soc.*, **1965**, 1517.

147. K. D. Bartle, H. Heaney, P. Lees, and D. W. Jones, unpublished measurements, 1966.
148. B. J. Mair and J. L. Martinez-Pico, *Proc. Amer. Petrol Inst.*, **42**, 173 (1962).
149. W. Carruthers, H. N. M. Stewart, P. G. Hansell, and K. M. Kelly, *J. Chem. Soc.,C,* **1967**, 2607.
150. F. F. Yew, R. J. Kurland, and B. J. Mair, *Anal. Chem.*, **36**, 843 (1964).
151. F. F. Yew and B. J. Mair, *Anal. Chem.*, **38**, 231 (1966).
152. P. Cannone and A. Regnault, *Can. J. Chem.*, **45**, 1267 (1967).
153. A. Regnault, *Tetrahedron Lett.*, **1968**, 355.
154. B. A. Nagasampagi, R. C. Pandey, V. S. Pansare, J. R. Prahlad, and S. Dev., *Tetrahedron Lett.*, **1964**, 411.
155. R. Freyman, M. Dvolaitsky, and J. Jacques, *C. R. Acad. Sci., Paris,* **253**, 1436 (1961).
156. M. Dvolaitsky, and J. Jacques, *C. R. Acad. Sci., Paris,* **252**, 1336 (1961).
157. P. H. Gore, C. K. Thadani, and S. Thorburn, *J. Chem. Soc.,C,* **1968**, 2503.
158. K. Elbs, *J. Prakt. Chem.*, **41**, 121 (1890).
159. C. MacLean and E. L. Mackor, *Mol. Phys.*, **4**, 241 (1961).
160. K. D. Bartle and D. W. Jones, *J. Phys. Chem.*, **73**, 293 (1969).
161. V. A. Koptyug, A. I. Rezvukhia, A. P. Krysin, and I. S. Isaev, *Z. Strukt. Khim.*, **8**, 622 (1967).
162. V. A. Koptyug, A. P. Krysin, and M. I. Garfinkel, *Izv. So. AN SSSR Ser. Khim.*, **5**, 12 (1967).
163. R. H. Martin, N. Defay, F. Geerts-Evrard, and S. Delavarenne, *Bull Soc. Chim. Belges,* **73**, 189 (1964).
164. S. Delavarenne, private communication in Ref. 163.
165. C. S. Wood and F. B. Mallory, *J. Org. Chem.*, **29**, 3373 (1964).
166. M. Flammang-Barbieux, J. Nasielski, and R. H. Martin, *Tetrahedron Lett.*, **1967**, 743.
167. R. H. Martin, M. Flammang-Barbieux, J. P. Cosyn, and M. Gelbcke, *Tetrahedron Lett.*, **1968**, 3507.
168. R. H. Martin, G. Morren, and J. J. Schurter, *Tetrahedron Lett.*, **1969**, 3683.
169. B. M. Trost, S. F. Nelsen, and D. R. Britelli, *Tetrahedron Lett.*, **1967**, 3959; *J. Amer. Chem. Soc.*, **89**, 4244 (1967).
170. J. J. R. Reed, *Anal. Chem.*, **39**, 1589 (1967).
171. D. W. Cameron, D. G. I. Kingston, and P. E. Schutz, *J. Chem. Soc.,C,* **1967**, 2113.
172. H. Dannenberg and H.-J. Kessler, *Justus Liebigs Ann. Chem.*, **606**, 184 (1957).
173. N. Campbell and A. H. Scott, *J. Chem. Soc.,C,* **1966**, 1050.
174. I. Ya. Postovskii and N. P. Bednyagina, *J. Gen. Chem. (USSR)*, **7**, 2919 (1937).
175. D. W. Cameron and P. E. Schutz, *J. Chem. Soc.,C,* **1967**, 2121.
176. W. B. Smith and B. A. Shoulders, *J. Phys. Chem.*, **69**, 2022 (1965).
177. M. Schach v. Wittenau and R. K. Blackwood, *J. Org. Chem.*, **31**, 613 (1966).
178. J. S. Meek, F. M. Dewey, and M. W. Hanna, *J. Org. Chem.*, **32**, 69 (1967).
179. G. S. Reddy and J. H. Goldstein, *J. Chem. Phys.*, **39**, 3509 (1963).
180. E. B. Whipple, Ph.D. dissertation, Emory University, Atlanta, Georgia, 1959.
181. R. H. Martin, N. Defay, F. Geerts-Evrard, P. H. Given, J. R. Jones, and R. W. Wedel, *Tetrahedron,* **21**, 1833 (1965).
182. E. G. Davis and S. Smiles, *J. Chem. Soc.*, **1910**, 1290.
183. W. Knapp, *Monatsch.*, **71**, 440 (1938).
184. L. F. Fieser and D. H. Sachs, *J. Amer. Chem. Soc.*, **90**, 4129 (1968).
185. H. Wichelhaus, *Chem. Ber.*, **30**, 2199 (1897).
186. S. C. Hooker and L. F. Fieser, *J. Amer. Chem. Soc.*, **58**, 1216 (1936).
187. P. G. Copeland, R. E. Dean, and D. McNeil, *J. Chem. Soc.*, **1960**, 1689.
188. F. Weygand and J. Frank, *Chem. Ber.*, **84**, 591 (1951).
189. H. G. Heller, D. Auld, and K. Salisbury, *J. Chem. Soc., C,* **1967**, 682.
190. H. G. Heller, D. Auld, and K. Salisbury, *J. Chem. Soc., C,* **1967**, 1552.
191. H. G. Heller, D. Auld, and K. Salisbury, *J. Chem. Soc., C,* **1967**, 2457.

192. M. F. Ansell and R. A. Murray, *Chem. Commun.*, **1968**, 1583.
193. L. F. Fieser and J. T. Dunn, *J. Amer. Chem. Soc.*, **59**, 1016, 1021 (1937).
194. J. Grimshaw and J. S. Ramsay, *J. Chem. Soc., C,* **1966**, 654.
195. J. Grimshaw and E. J. F. Rea, *J. Chem. Soc., C,* **1967**, 2628.
196. P. M. G. Bavin, *Can. J. Chem.*, **38**, 911 (1960).
197. P. H. Gore, *Chem. Rev.*, **55**, 229 (1955).
198. E. B. Girdler, P. M. Gore, and J. A. Hoskins, *J. Chem. Soc., C,* **1966**, 518.
199. M. Bil and A. Bernstein, *Chem. Ind.* (London), **1967**, 1221.
200. D. W. Cameron, R. I. T. Cromartie, D. G. I. Kingston, and G. B. U. Subramanian, *J. Chem. Soc.*, **1964**, 4565.
201. D. W. Cameron and P. E. Schutz, *J. Chem. Soc., C,* **1967**, 95.
202. D. W. Cameron, D. G. I. Kingston, and P. E. Schutz, *J. Chem. Soc., C,* **1967**, 99.
203. D. W. Cameron and P. E. Schutz, *J. Chem. Soc., C,* **1967**, 2118.
204. R. E. Erickson and B. Schnipp, *Chem. Ind.* (London), **1966**, 161.
205. F. A. Bovey, F. P. Hood, E. Pier, and H. E. Weaver, *J. Amer. Chem. Soc.*, **87**, 2060 (1965).
206. R. R. Fraser, *Can. J. Chem.*, **38**, 2226 (1960).
207. J. Martin and B. P. Dailey, *J. Chem. Phys.*, **37**, 2594 (1962).
208. T. Schaefer and W. G. Schneider, *J. Chem. Phys.*, **32**, 1218 (1960).
209. G. W. Smith, *J. Mol. Spectrosc.*, **12**, 146 (1964).
210. Y. Nomura, Y. Takeuchi, and N. Nakagawa, *Tetrahedron Lett.*, **1969**, 639.
211. E. Berliner, J. B. Kim, and M. Link, *J. Org. Chem.*, **33**, 1160 (1968).
212. P. R. Wells and P. G. E. Alcorn, *Aust. J. Chem.*, **16**, 1108 (1963).
213. J. C. Richer and C. Lamarre, *Can. J. Chem.*, **43**, 715 (1965).
214. J. C. Richer and Y. Pepin, *Can. J. Chem.*, **43**, 3443 (1965).
215. J. C. Richer and N. Baskevitch, *Can. J. Chem.*, **46**, 2975 (1968).
216. J. C. Richer, *Can. J. Chem.*, **46**, 3363 (1968).
217. L. Erichomovitch, M. Menard, F. L. Chubb, Y. Pepin, and R. C. Richer, *Can. J. Chem.*, **44**, 2305 (1966).
218. C. W. J. Chang, R. E. Moore, and P. J. Scherer, *J. Chem. Soc., C,* **1967**, 840.
219. F. Bell and K. R. Buck, *J. Chem. Soc., C,* **1966**, 905.
220. E. Vernon and R. D. Wilson, *Tetrahedron,* **21**, 2719 (1965).
221. B. M. Trost and D. R. Britelli, *J. Org. Chem.*, **32**, 2620 (1967).
222. T. S. Cantrell and H. Shechter, *J. Org. Chem.*, **33**, 114 (1968).
223. P. R. Constantine, L. W. Deady, and R. D. Topsom, *J. Org. Chem.*, **34**, 1113 (1969).
224. A. Bosch and R. K. Brown, *Can. J. Chem.*, **46**, 715 (1968).
225. J. A. Dixon and D. H. Fishman, *J. Amer. Chem. Soc.*, **85**, 1356 (1963).
226. J. A. Dixon, D. H. Fishman, and R. S. Dudinyak, *Tetrehedron Lett.*, **1964**, 613.
227. H. E. Zieger and J. E. Rosenkranz, *J. Org. Chem.*, **29**, 2469 (1964).
228. H. E. Zieger, *J. Org. Chem.*, **31**, 2977 (1966).
229. H. Heaney, *Chem. Rev.*, **62**, 81 (1962).
230. G. Wittig, *Angew. Chem. Int. Ed.*, **4**, 73 (1965).
231. W. L. Dilling, *Tetrahedron Lett.*, **1966**, 939.
232. H. Bouas-Laurent, R. Lapouyade, and F. Moulines, *Tetrahedron Lett.*, **1967**, 5281.
233. B. H. Klanderman and W. V. Perkins, *J. Org. Chem.*, **34**, 630 (1969).
234. C. J. Paget and A. Burger, *J. Org. Chem.*, **30**, 1329 (1965).
235. K. G. Kidd, G. Kotowycz, and T. Schaefer, *Can. J. Chem.*, **45**, 2155 (1967).
236. H. Heaney and P. Lees, *Tetrahedron Lett.*, **1964**, 3049.
237. H. Heaney and P. Lees, *Tetrahedron,* **24**, 3717 (1968).
238. S. H. Graham, D. M. Pugh, and A. J. S. Williams, *J. Chem. Soc., C,* **1969**, 68.
239. G. M. Badger, *Quart. Rev.*, **5**, 147 (1951).
240. G. M. Badger, J. W. Cook, and A. R. M. Gibb, *J. Chem. Soc.*, **1951**, 3456.
241. N. L. Drake and T. R. Sweeney, *J. Org. Chem.*, **11**, 67 (1946).
242. J. W. Cook, G. Dickson, and J. D. Loudon, *J. Chem. Soc.*, **1947**, 746.

243. D. H. Reid, W. H. Stafford, and J. P. Ward, *J. Chem. Soc.*, **1955**, 1193.
244. M. J. S. Dewar and C. R. Ganellin, *J. Chem. Soc.*, **1959**, 3139.
245. C. M. M. Da Silva Corrêa and W. A. Waters, *J. Chem. Soc., C.* **1968**, 1874.
246. C. M. M. Da Silva Corrêa and W. A. Waters, *J. Chem. Soc., C,* **1968**, 1880.
247. P. T. Lansbury, J. F. Bieran, and A. J. Lacher, *J. Amer. Chem. Soc.*, **88**, 1482 (1966).
248. P. T. Lansbury, A. J. Lacher and F. D. Saeva, *J. Amer. Chem. Soc.*, **89**, 4361 (1967).
249. H. Rottendorf and S. Sternhell, *Tetrahedron Lett.*, **1963**, 1289.
250. H. Rottendorf and S. Sternhell, *Aust. J. Chem.*, **17**, 1315 (1964).
251. J. G. Colson, P. T. Lansbury, and F. D. Saeva, *J. Amer. Chem. Soc.*, **89**, 4987 (1967).
252. G. Moreau, *Bull. Soc. Chim. Fr.*, **1969**, 1770.
253. J. Lugtenburg and E. Havinga, *Tetrahedron Lett.*, **1969**, 1505.
254. J. Lugtenburg, private communication, 1969.
255. J. G. Westra, W. G. B. Huysmans, W. J. Mijs, H. A. Gaur, J. Vriend, and J. Smidt, *Rec. Trav. Chim. Pays-Bas,* **87**, 1121 (1968).
256. J. J. McCullough, C. Calvo, and C. W. Huang, *Chem. Commun.*, **1968**, 1176.
257. J. J. McCullough and C. W. Huang, *Can. J. Chem.*, **47**, 757 (1969).
258. T. W. Mattingley, Jr., and A. Zweig, *Tetrahedron Lett.*, **1969**, 621.
259. A. Berg, H. J. Jakobsen, and S. R. Johansen, *Acta Chem. Scand.*, **23**, 567 (1969).
260. N. P. Buu-Hoï and P. Cagniant, *Chem. Ber.*, **77**, 121 (1944).
261. P. A. Plattner and E. C. Armstrong, *Newer Methods of Preparative Organic Chemistry*, Interscience, New York, 1948, p. 21.
262. R. Jönsson, *Talanta,* **15**, 425 (1968).
263. A. J. Solo and S. W. Pelletier, *Anal. Chem.*, **35**, 1584 (1963).
264. M. S. Bharucha, E. Weiss, and T. Reichstein, *Helv. Chim. Acta,* **45**, 103 (1962).
265. A. D. Cross, H. Carpio, and P. Crabbé, *J. Chem. Soc.*, **1963**, 5539.
266. S. Kaufman, *J. Org. Chem.*, **28**, 1390 (1963).
267. W. Brown and B. Turner, *Chem. Commun.*, **1968**, 501.
268. A. Gaudemer, J. Polonsky, and L. Alais, *Bull. Soc. Chim. Fr.*, **1963**, 1918.
269. J. Polonsky, B. C. Johnson, P. Cohen, and E. Lederer, *Bull. Soc. Chim. Fr.*, **1963**, 1909.
270. S. M. Kupchan and J. J. Merianos, *J. Org. Chem.*, **33**, 3735 (1968).
271. J. Reisch, M. Barthory, K. Szendrei, E. Minkere, and I. Novak, *Tetrahedron Lett.*, **1967**, 67.
272. F. Comer, H. P. Tiwari, and I. D. Spenser, *Can. J. Chem.*, **47**, 481 (1969).
273. I. D. Spenser and H. P. Tiwari, *Chem. Commun.*, **1966**, 55.
274. E. A. Ruveda, H. A. Priestap, and V. Deulofeu, *An. Assoc. Quim. Argentina,* **54**, 137 (1966).
275. H. A. Priestap, E. A. Ruveda, S. M. Albonico, and V. Deulofeu, *Chem. Commun.*, **1967**, 754.
276. I. R. C. Bick and G. K. Douglas, *Aust. J. Chem.*, **19**, 1997 (1965).
277. R. G. Cooke and H. F. Haynes, *Aust. J. Chem.*, **7**, 99 (1954).
278. P. M. Nair, T. G. Manjrekar, A. V. R. Rae, and K. Venkataraman, *Chem. Ind.* (London), **1967**, 1524.
279. W. D. Cotterill, R. Livingstone, K. D. Bartle, and D. W. Jones, *Tetrahedron,* **24**, 1981 (1968).
280. R. Livingstone and M. C. Whiting, *J. Chem. Soc.*, **1950**, 3631.
281. C. S. Barnes, J. L. Occolowitz, and M. I. Strong, *Tetrahedron,* **19**, 839 (1963).
282. A. V. Rama Rao and K. Venkataraman, *Indian J. Chem.*, **6**, 677 (1968).
283. P. K. Bose, *J. Indian Chem. Soc.*, **22**, 233 (1945).
284. H. G. Franck, *Brennst.-Chem.*, **36**, 12 (1955).
285. D. McNeil, *Coal Carbonisation Products,* Pergamon, Oxford, 1966.
286. D. McNeil, in *Bituminous Materials,* Vol. 3 (A. J. Hoiberg, ed.), Wiley-Interscience, New York, 1966, p. 139.
287. E. J. Greenhow and C. Sugowdz, *Coal Research in CSIRO,* **1961** (15), 10.

288. F. D. Rossini, *J. Chem. Educ.*, **37**, 554 (1960).
289. K. F. Lang and I. Eigen, *Fortsch. Chem. Forsch.*, **8**, 91 (1967).
290. *Coal Tar Data Book,* 2nd ed., The Coal Tar Research Association, Gomersal, Leeds, England, 1965.
291. H. C. Anderson and W. R. K. Wu, *Properties of Compounds in Coal Carbonization Products,* U.S. Gov. Printing Office, Washington, D.C., 1963.
292. K. C. Linepensel, *The Chemicals Identified in Coal Tar,* Koppers Co., Pittsburgh, 1956.
293. H. E. Schaëfer, *Freiberg Forsch.*, **51A,** 35 (1956).
294. J. F. Weiler, in *Chemistry of Coal Utilization* (H. H. Lowry, ed.), Wiley, New York, 1963, p. 587.
295. K. F. Lang, *Fortsch. Chem. Forsch.*, **7**, 172 (1966).
296. B. J. Mair, *Proc. Amer. Petrol. Inst.*, **44**, 378 (1964).
297. B. J. Mair, *Oil Gas J.*, **62,** 130 (1964).
298. R. H. Hunt and M. J. O'Neil, Jr., *Adv. Petrol. Chem.*, **10,** 2 (1965).
299. S. Friedman, M. L. Kaufman, B. D. Blaustein, I. Wender, and R. E. Dean, Amer. Chem. Soc. 143rd Meeting, January, 1963, *Div. Fuel Chem. Preprints,* 7 (1), 198 (1963).
300. S. Friedman, M. L. Kaufman, B. D. Blaustein, I. Wender, and R. E. Dean, *Tetrahedron,* **21,** 485 (1965).
301. H. L. Retcofsky, J. A. Queiser, and R. A. Friedel, *Chem. Ind.* (London), **1966**, 340.
302. J. A. Elvidge and R. G. Foster, *J. Chem. Soc.*, **1963,** 590.
303. B. J. Mair and T. J. Mayer, *Anal. Chem.*, **36,** 351 (1964).
304. E. L. Wynder and D. Hoffman, *Advan. Cancer Res.*, **8,** 249 (1964).
305. E. L. Wynder and D. Hoffman, *Tobacco and Tobacco Smoke,* Academic Press, New York, 1967, Chap. 12.
306. H. Elmenhorst and G. Reckzeh, *Beitr. Tabakforsch.*, **2,** 180 (1964).
307. R. L. Stedman, *Chem. Rev.*, **68,** 153 (1968).
308. H. Druckrey, *Acta. Med. Scand. Suppl.*, **69B,** 24 (1961).
309. D. Peters, *J. Amer. Chem. Soc.*, **79,** 646, 4182 (1957).
310. A. Pietzsch, *Naturwiss.*, **45,** 445 (1958).
311. A. Pietzsch, *Pharmazie,* **14,** 466 (1959).
312. E. Kröller, *Deut. Lebensm. Rundschau,* **60,** 214 (1964).
313. J. W. Cook, *Royal Institute of Chemistry Lecture Series, No.* **5**, 1961.
314. I. D. Entwistle and R. A. W. Johnstone, *J. Chem. Soc., C,* **1968,** 1818.
315. W. J. Chamberlain and R. L. Stedman, *Phytochemistry,* **7**, 1201 (1968).
316. H. S. Rao, G. S. Murti and A. Lahiri, *Fuel,* **39,** 263 (1960).
317. C. W. Dewalt, Jr. and M. S. Morgan, *Amer. Chem. Soc. Symp. on Tars, Pitches, and Asphalts,* Atlantic City, N.J., Sept. 1962, p. 33.
318. J. D. Brooks and J. R. Stevens, *Fuel,* **43,** 87 (1964).
319. J. D. Brooks and J. W. Smith, *Fuel,* **43,** 125 (1964).
320. K. D. Bartle and J. A. S. Smith, *Fuel,* **44,** 109 (1965).
321. K. D. Bartle and J. A. S. Smith, *Fuel,* **46,** 29 (1967).
322. H. N. Oelert, *Brennst. Chem.*, **48,** 362 (1967).
323. J. K. Brown, W. R. Ladner, and N. Sheppard, *Fuel,* **39,** 79 (1960).
324. J. K. Brown and W. R. Ladner, *Fuel,* **39,** 87 (1960).
325. R. A. Durie, Y. Schewchyk, and S. Sternhell, *Fuel,* **45,** 99 (1966) and refs. therein.
326. R. B. Williams "Characterization of Hydrocarbons in Petroleum by N.M.R. Spectroscopy," *Symp. Composition of Petroleum Oils, ASTM, STP,* 1957, p. 224.
327. R. B. Williams, *Spectrochim. Acta,* **74,** 24 (1959).
328. N. F. Chamberlain, in *Treatise on Analytical Chemistry,* Part I, Vol. 4 (I. M. Kolthoff and P. J. Elving, eds.), Wiley-Interscience, 1963, p. 1885.
329. R. B. Williams and N. F. Chamberlain, *Sixth World Petroleum Congress, Frankfurt,* **5,** 217 (1963).

330. R. Louis, *Erdöl Kohle,* **19,** 281 (1966).
331. R. A. Friedel and H. L. Retcofsky, *Proc. Fifth Conference on Carbon,* Vol. 2, Pergamon, London, 1963, p. 149.
332. R. A. Gardner, H. F. Hardman, A. L. Jones, and R. B. Williams, *J. Chem. Eng. Data,* **4,** 155 (1959).
333. T. F. Yen and J. G. Erdman, *Div. Petroleum Chem., Amer. Chem. Soc., Preprints,* **7**(3), 99 (1962).
334. J. W. Ramsay, F. R. McDonald, and J. C. Peterson, *Ind. Eng. Chem., Prod. Res. Develop.,* **6,** 231 (1967).
335. L. W. Corbett and R. E. Swarbrick, *Proc. Assoc. Asphalt Paving Technologists,* **27,** 107 (1958).
336. R. S. Winniford and M. Bersohn, *Div. of Fuel Chem., Amer. Chem. Soc., Preprints,* **1,** 21 (1962).
337. D. E. Wetmore, C. K. Hancock, and R. N. Traxler, *Anal. Chem.,* **38,** 225 (1966).
338. K. D. Bartle and D. W. Jones, *Fuel,* **48,** 21 (1969).
339. J. J. Delman and G. J. Béné, *J. Chem. Phys.,* **30,** 1634 (1959).
340. R. V. Helm and J. C. Petersen, *Anal. Chem.,* **40,** 110 (1968).
341. K. D. Bartle, J. A. S. Smith, and W. G. Wilman, *J. Appl. Chem.,* **19,** 283 (1969) and briefly reported in *Chem. Ind.* (London), **1966,** 762.
342. J. C. Hammel and J. A. S. Smith, private communication, 1967.
343. R. A. Friedel and H. L. Retcofsky, *Chem. Ind.* (London), **1966,** 455.
344. H. L. Retcofsky and R. A. Friedel, *Fuel,* **47,** 488 (1968).
345. S. A. Knight, *Chem. Ind.* (London), **1967,** 1920.
346. V. Boekelheide and J. B. Phillips, *J. Amer. Chem. Soc.,* **85,** 1545 (1963).
347. V. Boekelheide and J. B. Phillips, *Proc. Nat. Acad. Sci. U.S.,* **51,** 550 (1964).
348. H. R. Blattmann, D. Neuche, E. Heilbronner, R. J. Molyneux, and V. Boekelheide, *J. Amer. Chem. Soc.,* **87,** 130 (1965).
349. H. B. Renfroe, L. A. R. Hall, and J. Gurney, *Abstracts 153rd National Meeting of the Amer. Chem. Soc.,* Miami Beach, Florida, April 1967, Paper O-184.
350. V. Boekelheide and J. B. Phillips, *J. Amer. Chem. Soc.,* **89,** 1695 (1967).
351. H. J. Dauber and J. L. Laity, *Abstracts 150th National Meeting, Amer. Chem. Soc.,* Atlantic City, New Jersey, 1965, p. 325.
352. A. W. Hanson, *Acta Cryst.,* **18,** 599 (1965).
353. B. P. Stoicheff, *Can. J. Phys.,* **32,** 339 (1954).
354. J. B. Phillips, R. J. Molyneux, E. Sturm, and V. Boekelheide, *J. Amer. Chem. Soc.,* **89,** 1704 (1967).
355. V. Boekelheide and T. Miyasaka, *J. Amer. Chem. Soc.,* **89,** 1709 (1967).
356. V. Boekelheide and E. Sturm, *J. Amer. Chem. Soc.,* **91,** 902 (1969).
357. B. A. Hess, Jr., A. S. Bailey, and V. Boekelheide, *J. Amer. Chem. Soc.,* **89,** 2746 (1967).
358. H. B. Renfroe, *J. Amer. Chem. Soc.,* **90,** 2193 (1968).
359. K. Burri and W. Jenny, *Helv. Chim. Acta,* **50,** 1978 (1967).
360. K. Burri and W. Jenny, *Helv. Chim. Acta,* **50,** 3542 (1967).
361. W. Jenny and K. Burri, *Chimia,* **21,** 186 (1967).
362. H. A. Staab and F. Binning, *Chem. Ber.,* **100,** 293 (1967).
363. R. Peter and W. Jenny, *Chimia,* **19, 45** (1965).
364. R. Peter and W. Jenny, *Helv. Chim. Acta,* **49,** 2123 (1966).
365. H. A. Staab and H. Braunling, *Tetrahedron Lett.,* **1965,** 45.
366. R. H. Mitchell and V. Boekelheide, *Tetrahedron Lett.,* **1969,** 2013.
367. H. J. Reich and D. J. Cram, *J. Amer. Chem. Soc.,* **91,** 3534 (1969).
368. M. Sheehan and D. J. Cram, *J. Amer. Chem. Soc.,* **91,** 3544 (1969).
369. H. J. Reich and D. J. Cram, *J. Amer. Chem. Soc.,* **91,** 3505, 3517, 3527 (1969).
370. M. Sheehan and D. J. Cram, *J. Amer. Chem. Soc.,* **91,** 3553 (1969).

371. Y. Beharul, J. Billion, and G. Descotes, *Bull. Soc. Chim. Fr.*, **1968**, 4255.
372. Y. Beharul, G. Descotes, and J. Sabadie, *Bull. Soc. Chim. Fr.*, **1968**, 4259.
373. I. Puskas and E. K. Fields, *J. Org. Chem.*, **32**, 589 (1967).
374. J. Dhar, *Indian J. Phys.*, **7, 43** (1932).
375. A. Kitaigorodsky, *Acta Physicochim. URSS*, **21**, 575 (1946).
376. I. C. Karle and L. O. Brockway, *J. Amer. Chem. Soc.*, **66**, 1974 (1944).
377. J. Trotter, *Acta Cryst.*, **14**, 1135 (1961).
378. S. H. Rizvi and J. Trotter, *Proc. Chem. Soc.*, **1961**, 122.
379. A. Hargreaves and S. H. Rizvi, *Acta Cryst.*, **15**, 365 (1962).
380. G. B. Robertson, *Nature*, **191**, 593 (1961); **192**, 1026 (1961).
381. O. Bastiansen, *Acta Chem. Scand.*, **3**, 408 (1949).
382. A. Almenningen and O. Bastiansen, *Det. Kegl. Norske Videnskabers Selskabs Skrofter*, **1958** (4).
383. O. Bastiansen and M. Traetteberg, *Tetrahedron*, **17**, 147 (1962).
384. G. W. Wheland, in *Resonance in Organic Chemistry* (M. S. Newman, ed.), Wiley, New York, 1955, p. 163.
385. J. Y. Chau, C. G. LeFèvre and R. J. W. Le Fèvre, *J. Chem. Soc.*, **1959**, 2066.
386. E. D. Bergman, M. Rabinovitz, M. J. Aroney, R. J. W. LeFèvre, L. Radom, and G. L. D. Ritchie, *J. Chem. Soc., B*, **1968**, 1551.
387. H. E. Holloway, R. V. Nauman, and J. H. Wharton, *J. Phys. Chem.*, **72**, 4474 (1968).
388. J. Dale, *Acta Chem. Scand.*, **11**, 640 (1957).
389. D. Steele and E. R. Lippincott, *J. Mol. Spectrosc.*, **6**, 238 (1961).
390. M. J. O'Shaughnessy and W. H. Rodebush, *J. Amer. Chem. Soc.*, **62**, 2906 (1940).
391. H. Suzuki, *Bull. Chem. Soc. Japan*, **32**, 1340, 1350, 1357 (1959).
392. R. A. Hoffman, P.-O. Kinnell, and G. Bergström, *Arkiv Kemi*, **15**, 533 (1960).
393. W. Anderson and H. M. McConnell, *J. Chem. Phys.*, **26**, 1496 (1957).
394. R. J. Kurland and W. B. Wise, *J. Amer. Chem. Soc.*, **86**, 1877 (1964).
395. C. A. Coulson, *Electricity*, Wiley-Interscience, New York, 1961.
396. J. S. Waugh and R. W. Fessenden, *J. Amer. Chem. Soc.*, **79**, 846 (1957).
397. J. C. Slater, *Phys. Rev.*, **36**, 57 (1930).
398. K. D. Bartle, D. W. Jones, and R. S. Matthews, *J. Med. Chem.*, **12**, 1062 (1969).
399. S. Brownstein, *J. Amer. Chem. Soc.*, **80**, 2300 (1958).
400. K. Mislow, M. A. W. Glass, H. B. Hopps, E. Simon, and G. H. Wahl, Jr., *J. Amer. Chem. Soc.*, **86**, 1710 (1964).
401. H. J. Bernstein, W. G. Schneider, and J. A. Pople, *Proc. Roy. Soc., Ser. A*, **236**, 515 (1956).
402. H. O. House, R. W. Magin, and R. W. Thompson, *J. Org. Chem.*, **28**, 2403 (1963).
403. H. H. Jaffé and O. Chalvet, *J. Amer. Chem. Soc.*, **85**, 1561 (1963).
404. S. J. Cristol and D. C. Lewis, *J. Amer. Chem. Soc.*, **89**, 1476 (1967).
405. N. Rabinovitz, I. Agranat, and E. D. Bergmann, *Tetrahedron Lett.*, **1965**, 4133.
406. V. Balasubramaniyan, *Chem. Rev.*, **66**, 567 (1966).
407. S. C. Dickerman, D. DeSouza, and P. H. Wolf, *J. Org. Chem.*, **30**, 1981 (1965).
408. T. H. Regan and J. B. Miller, *J. Org. Chem.*, **32**, 502 (1967).
409. S. C. Dickerman and J. R. Haase, *J. Amer. Chem. Soc.*, **89**, 5438 (1967).
410. J. R. Haase, unpublished measurements quoted in Ref. 409.
411. F. Adrian, *J. Chem. Phys.*, **28**, 608 (1958).
412. I. Fischer-Hjalmars, *Tetrahedron*, **19**, 1805 (1963).
413. K. E. Howlett, *J. Chem. Soc.*, **1960**, 1055.
414. F. H. Westheimer and J. E. Mayer, *J. Chem. Phys.*, **14**, 733 (1946).
415. F. H. Westheimer in *Steric Effects in Organic Chemistry* (M. S. Newman, ed.), Wiley, New York, 1956, Chap. 12.
416. A. Golebriewski and A. Parczewski, *Theor. Chim. Acta*, **7**, 171 (1967).

417. K. Mislow and M. A. W. Glass, *J. Amer. Chem. Soc.*, **83,** 2780 (1961).
418. K. Mislow, E. Simon, and H. B. Hopps, *Tetrahedron Lett.*, **1962,** 1011.
419. W. L. Meyer and R. B. Meyer, *J. Amer. Chem. Soc.*, **85,** 2170 (1963).
420. K. Mislow, private communication quoted in Ref. 419.
421. W. Theilacker and H. Bohm, *Angew. Chem. Int. Ed.*, **6,** 251 (1967).
422. C. S. Johnson, *Advan. Magnetic Resonance,* **1,** 33 (1965).
423. J. Kaplan, *J. Chem. Phys.*, **28,** 278 (1958).
424. L. D. Colebrook and J. A. Jahnke, *J. Amer. Chem. Soc.*, **90,** 4687 (1968).
425. R. C. Cambie, W. R. J. Simpson, and L. D. Colebrook, *Chem. Ind.* (London), **1962,** 1757.
426. R. C. Cambie, W. R. J. Simpson, and L. D. Colebrook, *Tetrahedron,* **19,** 209 (1963).
427. C. R. Bennett and R. C. Cambie, *Phytochem.*, **6,** 883 (1967).
428. A. Allerhand, H. S. Gutowsky, J. Jonas, and R. A. Meinzer, *J. Amer. Chem. Soc.*, **88,** 3185 (1966).
429. H. S. Gutowsky and C. H. Holm, *J. Chem. Phys.*, **25,** 1228 (1956).
430. H. O. House and R. W. Bashe, *J. Org. Chem.*, **32,** 784 (1967).
431. H. Hopff and A. Gati, *Helv. Chim. Acta,* **48,** 509 (1965).
432. H. Kessler and B. Zeeh, *Tetrahedron,* **24,** 6825 (1968).
433. A. Reisse, D. Cagniant, and P. Cagniant, *Bull. Chim. Soc. Fr.*, **1969,** 2110.
434. R. Munday and I. O. Sutherland, *J. Chem. Soc., B,* **1968,** 80.
435. D. W. Jones and J. Yerkess, *J. Cryst. Mol. Struc.*, **1,** 17 (1971).
436. A. J. Kitaigorodsky and V. G. Dashevsky, *Tetrahedron,* **24,** 5917 (1968).
437. M. S. Newman and D. J. Lednicer, *J. Amer. Chem. Soc.*, **78,** 4765 (1956).
438. M. S. Newman and W. B. Wheatley, *J. Amer. Chem. Soc.*, **70,** 1913 (1948).
439. M. S. Newman and R. M. Wise, *J. Amer. Chem. Soc.*, **78,** 450 (1956).
440. F. Bell and D. H. Waring, *J. Chem. Soc.*, **1949,** 2689.
441. M. Crawford, R. A. M. Mackinnon, and V. R. Supanekar, *J. Chem. Soc.*, **1959,** 2807.
442. M. S. Newman and A. S. Hussey, *J. Amer. Chem. Soc.*, **69,** 3023 (1947).
443. J. Heidberg, J. A. Weil, G. A. Janusonis, and J. K. Anderson, *J. Chem. Phys.*, **41,** 1033 (1964).
444. F. B. Mallory, S. L. Manatt, and C. S. Wood, *J. Amer. Chem. Soc.*, **87,** 5483 (1965).
445. M. J. Cook, A. R. Katritzky, F. C. Pennington, and B. M. Semple, *J. Chem. Soc., B,* **1969,** 523; M. J. Cook, *Tetrahedron Lett.*, **1969,** 2893.
446. D. Cohen, I. T. Millar, H. Heaney, P. R. Constantine, A. R. Katritzky, B. M. Semple, and M. J. Sewell, *J. Chem. Soc., B,* **1967,** 1248.
447. K. Mislow and H. B. Hopps, *J. Amer. Chem. Soc.*, **84,** 3018 (1962).
448. K. Mislow, R. Graeve, A. J. Gordon, and G. H. Wahl, *J. Amer. Chem. Soc.*, **86,** 1733 (1964).
449. D. Y. Curtin, C. G. Carlson, and C. G. McGarty, *Can. J. Chem.*, **42,** 565 (1964).
450. E. L. Eliel, *Stereochemistry of Carbon Compounds,* McGraw-Hill, New York, 1962, p. 160.
451. Z. M. Holubec and J. Jonas, *J. Amer. Chem. Soc.*, **90,** 5986 (1968).
452. J. Jonas, A. Allerhand, and H. S. Gutowsky, *J. Chem. Phys.*, **42,** 3396 (1965).
453. H. S. Gutowsky, J. Jonas, and T. H. Siddall, III, *J. Amer. Chem. Soc.*, 89, 4300 (1967).
454. W. Carruthers and G. E. Hall, *J. Chem. Soc., B,* **1966,** 861.
455. N. S. Bhacca and D. H. Williams, *Applications of N.M.R. Spectroscopy in Organic Chemistry,* Holden-Day, San Francisco, 1964, p. 47.
456. Y. Yannoni, A. P. Krukonis, and J. Silverman, *Science,* **148,** 231 (1965).
457. L. M. Jackman and J. W. Lown, *J. Chem. Soc.*, **1962,** 3776.
458. S. C. Dickerman, J. H. Berg, J. R. Haase, and R. Varma, *J. Amer. Chem. Soc.*, **89,** 5457 (1967).

459. D. Nicholls and M. Szwarc, *J. Amer. Chem. Soc.,* **88,** 5757 (1966).
460. J. MacMillan and E. R. H. Walker, paper at Chemical Society Nuclear Magnetic Resonance Discussion Group, April 1969.
461. J. MacMillan and E. R. H. Walker, *Chem. Commun.,* **1969,** 1031, 1032.
462. D. Cohen, L. Hewitt, and I. T. Millar, *Chem. Ind.* (London), **1966,** 1532.
463. M. J. S. Dewar and R. C. Fahey, *J. Amer. Chem. Soc.,* **84,** 2012 (1962).
464. L. D. Hayward and I. C. Csizmadia, *Tetrahedron,* **19,** 2111 (1963).
465. R. J. Abraham, L. D. Hall, L. Hough, and K. A. McLauchlan, *J. Chem. Soc.,* **1962,** 3699.
466. M. Karplus, *J. Chem. Phys.,* **30,** 11 (1959).
467. F. V. Brutcher, Jr., and N. Pearson, *Chem. Ind.* (London), **1957,** 1295.
468. F. V. Brutcher, Jr., T. Roberts, S. J. Barr, and N. Pearson, *J. Amer. Chem. Soc.,* **81,** 4915 (1959).
469. H. W. W. Ehrlich, *Acta Cryst.,* **10,** 699 (1957).
470. D. C. Nonhebel and J. A. Russell, *Chem. Ind.* (London), **1968,** 1841; *Tetrahedron,* **25,** 3493 (1969).
471. S. Hauptman, K. Dietrich, and H. Bier, *J. Prakt. Chem.,* **29,** 206 (1965).
472. D. Cohen, I. T. Millar, and K. E. Richards, *J. Chem. Soc., C,* **1967,** 1499.
473. G. P. Newsoroff and S. Sternhell, *Tetrahedron Lett.,* **1967,** 2539.
474. E. A. Chandross and C. F. Sheley, *J. Amer. Chem. Soc.,* **90,** 4345 (1968).
475. A. Rieker and H. Kessler, *Tetrahedron Lett.,* **1969,** 1227.
476. P. M. G. Bavin, K. D. Bartle, D. W. Jones, and R. L'Amie, *Tetrahedron,* **26,** 911 (1970).
477. T. H. Siddall, III, and W. E. Stewart, *J. Org. Chem.,* **34,** 233 (1969).
478. T. H. Siddall, III, and W. E. Stewart, *Tetrahedron Lett.,* **1968,** 5011.
479. K. D. Bartle, P. M. G. Bavin, and D. W. Jones, *J. Chem. Soc., B,* **1971,** 388.
480. M. Rabinovitz, I. Agranat, and E. D. Bergmann, *Tetrahedron Lett.,* **1965,** 1265.
481. C. R. Harrison, J. F. W. McOmie, J. B. Searle, J. P. Smith, J. O. Jenkins, and J. W. Smith, *J. Chem. Soc., C,* **1966,** 1769.
482. J. W. Barton and K. E. Whittaker, *J. Chem. Soc., C,* **1968,** 28.
483. D. V. Hertzler, E. J. Eisenbraun, and P. W. K. Flanagan, *Chem. Ind.* (London), **1969,** 877.
484. C. D. Nenitzescu and M. Avram, *J. Amer. Chem. Soc.,* **72,** 3486 (1950).
485. M. Crawford and V. R. Supanekar, *J. Chem. Soc., C,* **1968,** 2329.
486. H. F. Bassilios, M. Shawley, and A. Y. Salem, *Rec. Trav. Chim. Pays-Bas,* **81,** 679 (1962); **82,** 298 (1963).
487. M. Martynoff, *C. R. Acad. Sci., Paris,* **244,** 205 (1957).
488. J. H. Bouve, D. W. Cameron, P. E. Schutz, D. H. Williams, and N. S. Bhacca, *Tetrahedron,* **22,** 1771 (1966).
489. V. Balasubramaniyan. Ph.D. Thesis, Annamali University, India, 1964.
490. P. R. Wells, *J. Chem. Soc.,* **1963,** 1967.
491. J. Yamaguchi, *Mol. Phys.,* **6,** 1065 (1963).
492. H. M. McConnell, *J. Chem. Phys.,* **27,** 225 (1957).
493. R. C. Hahn, P. H. Howard, S. M. Kong, G. A. Lorenzo, and N. L. Miller, *J. Amer. Chem. Soc.,* **91,** 3558 (1969).
494. R. W. Franck and E. G. Leser, *J. Amer. Chem. Soc.,* **91,** 1577 (1969).
495. R. J. Ouellette and B. G. Van Leuwen, *Tetrahedron,* **25,** 819 (1969).
496. R. J. Ouellette and B. G. Van Leuwen, *J. Org. Chem.,* **34,** 62 (1969).
497. J. D. Memory, *Biochim. Biophys. Acta,* **66,** 168 (1963).
498. A. P. Downing, W. D. Ollis, and I. O. Sutherland, *Chem. Commun.,* **1967,** 171.
499. C. R. Johnson and D. C. Vegh, *Chem. Commun.,* **1969,** 557.
500. J. Cottam, R. Livingstone, M. Walshaw, K. D. Bartle, and D. W. Jones, *J. Chem. Soc.,* **1965,** 5261.

501. O. Jardetzky, *Advances in Chemical Physics,* Vol. 7, Wiley, London, 1964, p. 499.
502. P. Brookes and P. D. Lawley, *Nature,* **202,** 781 (1964).
503. R. S. Umans, S. A. Lesko, Jr., and P. O. P. Ts'o, *Nature,* **221,** 763 (1969).

Additional References

This review was compiled from a comprehensive literature survey up to the middle of 1969. Titles and references of other significant papers in the field available between then and mid-1970 are listed below under the same paragraph headings as the main text.

I.A NUCLEAR MAGNETIC RESONANCE SPECTROSCOPY

1. "The Chemical Shift: A Review," F. H. A. Rummens, *Org. Mag. Res.,* **2,** 209 (1970).

I.B CHEMICAL SHIFTS

1. "Substituent Effects on Proton Resonances and Vicinal H,H Coupling Constants in Benzene Derivatives," W. Bremser and H. Günther, *Org. Mag. Res.,* **1,** 435 (1969).
2. "Some Examples of the Application of N.M.R. Spectroscopy to the Determination of the Structure of Polysubstituted Benzenes," D. Cagniant, A. Reisse, and P. Cagniant, *Bull. Chim. Soc. Fr.,* **1969,** 2129.
3. "The Proton Magnetic Resonance Spectral Parameters of Some 2- and 2,3-Substituted Naphthoquinones and Their Relation to Electronic Structure," R. W. Crecely, K. M. Crecely, and J. H. Goldstein, *J. Mol. Spectrosc.,* **32,** 407 (1969).
4. "Uncoupled Hartree-Fock Calculations of the Ring-Current Contribution to the Chemical Shifts in Conjugated Hydrocarbons," H. G. ff. Roberts, *Theoret. Chim. Acta,* **15,** 63 (1969).
5. "Substituent Effects in N.M.R.: Tetramethylbiphenyls," Y. Nomura and Y. Takeuchi, *J. Chem. Soc., B,* **1970,** 956.
6. "Substituent Effects on Aromatic Proton Chemical Shifts," W. B. Smith, A. M. Ihrig, and J. L. Roark, *J. Phys. Chem.,* **74,** 812 (1970).
7. "High-Resolution Proton Magnetic Resonance Spectra of Fluorenes: 9-Substituted Fluorenes," K. D. Bartle, P. M. G. Bavin, and D. W. Jones, *J. Chem. Soc., B,* **1971,** 388.
8. "An Empirical Re-assessment of the Semi-Classical Theory of π-Electron Ring Currents in Planar-Condensed Benzenoid Hydrocarbons," R. B. Mallion, read before Chemical Society Autumn Meeting, London, 1970, *J. Chem. Soc., B,* **681** (1971).

I.C SPIN-SPIN COUPLING

1. "Long-Range Spin-Spin Coupling," M. Barfield and B. Chakrabarti, *Chem. Rev.,* **69,** 757 (1969).
2. "Exact Determination of the H,H Coupling Constants in Naphthalene and Anthracene," J. B. Pawliczek and H. Günther, *Z. Naturforsch.,* **24b,** 1068 (1969).
3. "The Location of Double Bonds in Fluoranthene, Perylene, and 1,1′-Dinaphthyl by N.M.R.," E. Clar, A. Mullen, and Ü. Sanigök, *Tetrahedron,* **25,** 5639 (1969).
4. "H-H Coupling Constants and π-Electron Structure in Unsaturated Six-Membered Rings," J. B. Pawliczek and H. Günther, *Tetrahedron,* **26,** 1755 (1970).
5. "The Location of Double Bonds in Alkyl Pyrenes," E. Clar, B. A. McAndrew, and Ü. Sanigök, *Tetrahedron,* **26,** 2099 (1970).
6. "The Location of Double Bonds in Methyl Acenes," E. Clar and C. C. Mackay, *Tetrahedron Lett.,* **1970,** 871.
7. "Analysis of the Proton Magnetic Resonance Spectra of Some Strained Benzocycloalkenes and Benzocycloalkenediones. Effect of Strain on Proton-Proton, Spin-Spin

Coupling Constants," M. A. Cooper and S. L. Manatt, *J. Amer. Chem. Soc.,* **92,** 1605 (1970).
8. "The Effect of Steric Compression on Proton-Proton Spin-Spin Coupling Constants. Further Evidence and Mechanistic Considerations," M. A. Cooper and S. L. Manatt, *J. Amer. Chem. Soc.,* **92,** 4646 (1970).

I.D ANALYSIS OF P.M.R. SPECTRA OF POLYNUCLEARS (see also 2, 7 and 8 of I.C)

1. "Proton Magnetic Resonance Spectra of Planar-Condensed Benzenoid Hydrocarbons," C. W. Haigh and R. B. Mallion, *Mol. Phys.,* **18,** 737, 751, 767 (1970).
2. "Notation for Spin Systems in Nuclear Magnetic Resonance Spectroscopy," C. W. Haigh, *J. Chem. Soc., A,* **1970,** 1682.

II.A SYMMETRY CONSIDERATIONS

1. "Formation of Polycyclic Quinones in Oxidations of 3,3′,4,4′-Tetramethoxybiphenyl and Veratrole," O. C. Musgrave and C. J. Webster, *Chem. Commun.,* **1969,** 712.

II.B CHEMICAL SHIFTS

1. "Synthesis of 4-Azahexahelicene," R. H. Martin and M. Deblecker, *Tetrahedron Lett.,* **1969,** 3597.
2. "The Photocyclisation of 1,2-Diarylethylenes; Hexahelicene-7-*d*," R. H. Martin and J. J. Schurter, *Tetrahedron Lett.,* **1969,** 3679.
3. "[13] Helicene and [13] Helicene-10,21-d_2," R. H. Martin, G. Morren, and J. J. Schurter, *Tetrahedron Lett.,* **1969,** 3683.
4. "Preparation of Polysubstituted Anthracenes and Phenanthrenes," A. Regnault and P. Canonne, *Tetrahedron,* **25,** 2349 (1969).
5. "Substituted Naphthalenes as Dienes in Diels-Alder Reactions at High Pressure," H. Plieninger, D. Wild, and J. Westphal, *Tetrahedron,* **25,** 5561 (1969).
6. "New Methods of Synthesis of 6-Methylaceanthrene and 1-Methylpyrene," J.-P. Hoeffinger, P. Jacquignon, and N. P. Buu-Hoï, *Bull. Soc. Chim. Fr.,* **1969,** 3099.
7. "The Photochemistry of Bifluorenylidene," G. P. de Gunst, *Rec. Trav. Chim. Pays-Bas,* **88,** 801 (1969).
8. "Potentially Carcinogenic Cyclopenta[*a*]phenanthrenes," M. M. Coombs, S. B. Jaitly, and E. H. Crawley, *J. Chem. Soc., C,* **1970,** 1266.
9. "The Nuclear Halogenation of Fluorene, Fluorenone, and Acenaphthenequinone by *N*-Bromosuccinimide and *N*-Chlorosuccinimide," F. Dewhurst and P. K. J. Shah, *J. Chem. Soc., C,* **1970,** 1266.
10. "Diels-Alder Reactions of 5-Substituted Naphthacenes," J. S. Meek and F. M. Dewey, *J. Org. Chem.,* **35,** 1315 (1970).

II.C COMPLETE INTERPRETATION OF PROTON MAGNETIC RESONANCE SPECTRUM

1. "The Isomerisation of Methylenediquinones by Bases," F. M. Dean and L. E. Houghton, *Tetrahedron Lett.,* **1969,** 3579.
2. "The Photo-addition of Diphenylene to Three Phenanthrenes and the Photo-dimerization of 4*H*-Cyclopenta[*d,e,f*]phenanthrene," G. Sugowdz, P. J. Collin, and W. H. F. Sasse, *Tetrahedron Lett.,* **1969,** 3843.
3. "Reaction of Indenes and Fluorene with the System Fluorodichloromethane/Ethylene Oxide/Tetraethylammonium Bromide," F. Nerdel, G. Blume, and P. Weyerstahl, *Tetrahedron Lett.,* **1969,** 3867.

4. "Synthesis of Compounds Related to Cyclohepta[*d,e,f*]fluorene," R. Munday and I. O. Sutherland, *J. Chem. Soc., C,* **1969,** 1427.
5. "Reactions of Cupric Halides with Organic Compounds," A. D. Mosnaim and D. C. Nonhebel, *Tetrahedron,* **25,** 3493 (1969).
6. "Photocyclo-addition of 2-Acetylnaphthalene to Methyl Cinnamate," D. R. Arnold, L. B. Gillis, and E. B. Whipple, *Chem. Commun.,* **1969,** 918.
7. "The Structures of the Photodimers of 9-Nitroanthracene, 9-Aminoanthracene, and 9-Anthrylisocyanate," O. C. Chapman and K. Lee, *J. Org. Chem.,* **34,** 4166 (1969).
8. "Total Synthesis of 5,6-Benz*peri*naphthane," A. Rahman and J. C. Podesta, *Chem. Ind.* (London), **1970,** 1173.
9. "Total Synthesis of 5,6-Benzindane," A. Rahman and J. C. Podesta, *Chem. Ind.* (London), **1970,** 1204.
10. "Arylaliphatic Hydrocarbons," A. Reisse, D. Cagniant, and P. Cagniant, *Bull. Soc. Chim. Fr.,* **1970,** 742.
11. "Novel Dimeric Products from 10-Methyleneanthrone," W. H. Starnes, Jr., *J. Org. Chem.,* **35,** 1974 (1970).
12. "A Trimeric $C_{48}H_{30}$ Hydrocarbon of Unusual Structural Interest Derived from 9,10-Dihydro-9,10-ethenoanthracene," C. F. Huebner, R. T. Rickett, M. Brzechffa, and S. L. Schwartz, *Tetrahedron Lett.,* **1970,** 359.
13. "Benzo[*c*]phenanthrenes and Chrysenes from Styryl Naphthalenes," E. V. Blackburn, C. E. Loader, and C. J. Timmons, *J. Chem. Soc., C,* **1970,** 163.
14. "Stilbene Analogues which Fail to Undergo Cyclisation to Phenanthrenes," E. V. Blackburn and C. J. Timmons, *J. Chem. Soc., C,* **1970,** 172.
15. "Rigid Polycyclic Systems. The Reaction Between Cyclopentadiene and Acenaphthylene and the Spectral Properties of the Adducts and Derived Alcohols and Ketones," R. Baker and T. J. Mason, *J. Chem. Soc., C,* **1970,** 596.
16. "Three Isomeric Grisan Derivatives from the Oxidative Cyclisation of Benzylidene Bis-2-naphthol," D. J. Bennett, F. M. Dean, and A. W. Price, *J. Chem. Soc., C,* **1970,** 1557.
17. "1,3- and 3,3-Bis(methoxyphenyl)naphthol[2,1-*b*]pyrans," W. D. Cotterill, R. Livingstone, and M. V. Walshawe, *J. Chem. Soc., C,* **1970,** 1758.
18. "Aromatic Steroids," R. C. Cambie and V. F. Carlisle, *J. Chem. Soc., C,* **1969,** 1234,1240; **1970,** 1706.
19. "Formation of Xanthen and Fluorene Derivatives by Competitive 1,4- and 1,3-Cyclo Additions of Quinonyl Methyl Anions to Quinones," F. M. Dean and L. E. Houghton, *J. Chem. Soc., C,* **1970,** 722.
20. "Partially Hydrogenated Derivatives of 7,12-Dimethylbenz[a]anthracene," E. Bisagni, J.-J. Pepin, and J. Andre-Louisfert, *Bull. Soc. Chim. Fr.,* **1970,** 3023, 3031.

II.D NUCLEAR MULTIPLE RESONANCE METHODS

1. "Identification of 1,5- and 8-Nitro-2-methoxy-6-methylnaphthalenes from Nitration of 2-Methoxy-6-methylnaphthalenes from N.M.R. and I.N.D.O.R. experiments," D. T. Clark and D. J. Fairweather, *Tetrahedron,* **25,** 5525 (1969).
2. "Assignments of Configuration by Nuclear Overhauser Effects in Some Dibenzthiepin Derivatives," R. R. Fraser and F. J. Schuber, *Can. J. Chem.,* **48,** 633 (1970).

III.A STRUCTURAL STUDIES OF POLYNUCLEAR NATURAL PRODUCTS AND RELATED COMPOUNDS

1. "Isolation and Structure of 4-Hydroxypiloquinone, A Second *Ortho*phenanthrenequinone from *Streptomyces pilosus*," M. Lounasmaa and J. Zylber, *Bull. Soc. Chim. Fr.,* **1969,** 3100.

2. "On the Biosynthesis of Piloquinone, an *Ortho*phenanthrene Quinone Produced by *Streptomyces pilosus*," J. Zylber, E. Zissman, J. Polonsky, E. Lederer, and M. A. Merrien, *Eur. J. Biochem.*, **10,** 278 (1969).
3. "Lignans and Related Phenols. Observations on the Spectroscopic Characterisation and Conformational Analysis of Apo-compounds of the Arylnaphthalene Class," D. C. Ayres, *Can. J. Chem.*, **47,** 2075 (1969).
4. "Phenolic Constituents of Elm Wood. 2-Naphthoic Acid Derivatives from *Ulmus thomasii*," C.-L. Chen and F. D. Hostettler, *Tetrahedron,* **25,** 3223 (1969).
5. "The Structure of Arenicochromine, a Trihydroxymethoxy Benzpyrenequinone," I. Morimoto, M. I. N. Shaikh, R. H. Thomson, and D. G. Williamson, *J. Chem. Soc., D.* **1970,** 550.
6. "Methoxynaphthaldehydes as Constituents of the Heartwood of *Diospyros quiloensis* and Their Synthesis by the Stobbe Condensation," S. H. Cooper, A. D. Kemp, and J. Tannock, *J. Chem. Soc., C,* **1970,** 626.

III.B PROTON MAGNETIC RESONANCE IN COAL TAR AND PETROLEUM CHEMISTRY

1. "Properties, Composition, and Chemical Structure of High-Boiling Petroleum Hydrocarbons," C. Kajdös, *Erdol Kohle,* **22,** 459 (1969).
2. "A Structural Investigation of the Constituents of Athabasca Bitumen by P.M.R. Spectroscopy," J. G. Speight, *Fuel,* **49,** 76 (1970).
3. "Thermal Cracking of Athabasca Bitumen, Athabasca Asphaltenes, and Athabasca De-asphalted Heavy Oil," J. G. Speight, *Fuel,* **49,** 134 (1970).
4. "New Results of the Investigations of Coal Structure by Spectroscopic Methods," H. H. Oelert and E. A. Hemmer, *Erdöl Kohle,* **23,** 163 (1970).
5. "Identification of Polycyclic Naphthenic Mono- and Di-aromatic Crude Oil Carboxylic Acids," W. K. Seifert and R. M. Tester, *Anal. Chem.,* **42,** 180 (1970).

III.C CAVITY SUBSTITUENTS IN POLYCYCLIC MOLECULES

1. "[2.2] (1,4)Naphthaleno(2,5)furanophane," H. H. Wasserman and P. M. Keehn, *Tetrahedron Lett.*, **1969,** 3227.
2. "Synthesis and Ring Inversion of Substituted Dithiametacyclophanes," F. Vögtle, *Tetrahedron,* **25,** 3231 (1969).
3. "Preparations and Reactions of Some [2,2] Paracyclophane Derivatives," Y. L. Yeh and W. F. Gorham, *J. Org. Chem.*, **34,** 2366 (1969).
4. "*Syn*-[2.2] paracyclonaphthane," H. H. Wasserman and P. M. Keehn, *J. Amer. Chem. Soc.,* **91,** 2374 (1969).
5. "Naphthalenophanes," F. Vögtle, *Justus Liebigs Ann. Chem.,* **734,** 102 (1970).
6. "1,3-Bridged Aromatic Systems," W. E. Parham, D. R. Johnson, C. T. Hughes, M. K. Meilahn, and J. K. Rinehart, *J. Org. Chem.,* **35,** 1048 (1970); W. E. Parham, R. W. Davenport, and J. K. Rinehart, *J. Org. Chem.,* **35,** 2662 (1970).
7. "Syntheses, Spectra, and Identification of Isomeric, Fused Ring Paracyclophane Derivatives," M. J. Nugent and T. L. Vigo, *J. Org. Chem.,* **35,** 2203 (1970).
8. "A Multilayered [2.2] Paracyclophane," D. L. Longune and H. S. Chow, *J. Amer. Chem. Soc.,* **92,** 994 (1970).
9. "Barriers to Ring Rotation in [2.2] Metaparacyclophane Derivatives." D. T. Hefelfinger and B. J. Cram, *J. Amer. Chem. Soc.,* **92,** 1072 (1970).
10. "Syntheses of [2.2] Metacyclophane-1,9-diene and *trans*-15,16-Dihydropyrene," R. H. Mitchell and V. Boekelheide, *J. Amer. Chem. Soc.,* **92,** 3511 (1970).
11. "A Convenient Synthesis of 15,16-Dihydropyrene," R. H. Mitchell and V. Boekelheide, *Tetrahedron Lett.,* **1970,** 1197.

IV.A BIPHENYL AND ITS ANALOGUES

1. "High-Resolution PMR of Phenanthrenes. Phenyl Phenanthrenes," K. D. Bartle, P. M. G. Bavin, and D. W. Jones, *Org. Mag. Res.*, **2,** 259 (1970).
2. "The Rotational Barrier in 1,8-Diaryl naphthalenes," H. O. House, W. J. Campbell, and M. Gall, *J. Org. Chem.*, **35,** 1815 (1970).

IV.C CONFORMATION OF HYDROAROMATIC POLYNUCLEAR COMPOUNDS

1. "A Graphical Approach to Conformational Studies," M. J. Cook, *Tetrahedron Lett.*, **1969,** 2893.
2. "Stereochemistry and Conformation of 9,10-Diisopropyl-9,10-dihydroanthracenes," E. Zieger, D. J. Schaeffer, and R. M. Padronaggio, *Tetrahedron Lett.*, **1969,** 5027.
3. "Stereochemistry of *meso* substituted 9,10-Dihydroanthracenes," A. L. Ternay, Jr., A. W. Brinkman, S. Evans, and J. Herrmann, *Chem. Commun.*, **1969,** 654.
4. "Ring Inversion in 9,10-Dihydrophenanthrene Derivatives," P. W. Rabideau, R. G. Harvey, and J. B. Stothers, *Chem. Commun.*, **1969,** 1005; *J. Org. Chem.*, **35,** 25 (1970).
5. "Preferred Axial Conformation of the Pendant Substituent in 1-Aryldihydronaphthalenes and Aryltetralones," D. C. Ayres and J. A. Harris, *Chem. Commun.*, **1969,** 1135.
6. "Isolable Stereoisomeric Methylene Dihydroanthracenes," D. Y. Curtin and Z. M. Holubec, *J. Org. Chem.*, **34,** 3093 (1969).
7. Metal-Ammonia Reduction: Single-Stage Reduction of Polycyclic Aromatic Hydrocarbons," R. G. Harvey, L. Arzadon, J. Grant, and K. Urberg, *J. Amer. Chem. Soc.*, **91,** 4545 (1969).
8. "Study of 9,10-Dihydroanthracenes by NMR," R. Lapouyade and P. Labandibar, *Tetrahedron Lett.*, **1970,** 1589.
9. "Stereochemistry of 9-Alkyl-9,10-dihydroanthracenes. Confirmation by Nuclear Overhauser Enhancements and Long-Range Coupling Constants," A. W. Brinkman, M. Gordon, R. G. Harvey, P. W. Rabideau, J. B. Stothers, and A. L. Ternay, Jr., *J. Amer. Chem. Soc.*, **92,** 592 (1970).
10. "Rotational Barriers and Conformation Studies in 9-Arylxanthyl Derivatives," S. V. McKinley, P. A. Grieco, A. E. Young, and H. H. Freedman, *J. Amer. Chem. Soc.*, **92,** 7145 (1970).

IV.D CONFORMATION OF SUBSTITUENT GROUPS

1. "Experimental and Theoretical Estimate of the Steric Hindrance Effect of the *t*-Butyl Group on Aromatic Proton Chemical Shifts in Substituted 1-Nitronaphthalenes," S. Odiot, J. C. Richer, and F. Tonnard, *Can. J. Chem.*, **47,** 2805 (1969).
2. "Proximity Effects in *peri*-Substituted Naphthalenes," D. C. Kleinfelter and P. H. Chen, *J. Org. Chem.*, **34,** 1741 (1969).
3. "Proton N.M.R. Spectra of Arylmethyl Systems," M. D. Bentley and M. J. S. Dewar, *J. Org. Chem.*, **35,** 2707 (1970).
4. "Conformational Analysis. Conformations of Methyl-, Ethyl-, and Isopropylarenes," R. J. Ouellette, B. K. Sinha, J. Stolfo, C. Levin, and S. Williams, *J. Amer. Chem. Soc.*, **92,** 7145 (1970).

IV.E MISCELLANEOUS CONFORMATIONAL PROBLEMS

1. "Stereochemistry and Stereospecific Photochemical Rearrangement Reaction of 1,2-Bisdiphenylmethylene-3-methylindane," H. G. Heller and K. Salisbury, *J. Chem. Soc., C,* **1970,** 873.
2. "Steric Effects in the P.M.R. Spectra of Fluorenes and Bifluorenyls," K. D. Bartle, P. M. G. Bavin, D. W. Jones, and R. L'Amie, *Tetrahedron,* **27,** 911 (1970).

Author Index

Subject Index